MATERIALS FOR CIVIL AND HIGHWAY ENGINEERS

FOURTH EDITION

BY

KENNETH N. DERUCHER

*Dean, College of Engineering, Computer
Science and Technology*
California State University, Chico

GEORGE P. KORFIATIS
A. SAMER EZELDIN

*Department of Civil, Environmental, and
Coastal Engineering*
Stevens Institute of Technology

PRENTICE HALL, *Upper Saddle River, New Jersey 07458*

Library of Congress Cataloging-in-Publication Data

Derucher, Kenneth N.
  Materials for civil and highway engineers / Kenneth N. Derucher,
  George P. Korfiatis, A. Samer Ezeldin
    p. cm.
  Includes bibliographical references and index.
  ISBN: 0-13-905043-4
CIP data available

Acquisitions editor: **WILLIAM STENQUIST**
Production editor: **EDWARD DEFELIPPIS**
Editor-in-chief: **MARCIA HORTON**
Managing editor: **BAYANI MENDOZA DE LEON**
Assistant Vice President of Production and Manufacturing: **DAVID W. RICCARDI**
Manufacturing buyer: **PAT BROWN**
Manufacturing Manager: **TRUDY PISCOTTI**
Creative Director: **PAULA MAYLAHN**
Art Director: **JAYNE CONTE**
Cover designer: **BRUCE KENSELAAR**
Copy Editor: **MICHAEL CONDOURIS**

©1981, 1988, 1994, 1998 by Prentice-Hall, Inc.
Simon & Schuster / A Viacom Company
Upper Saddle River, New Jersey 07458

Printed in the United States of America

10 9 8 7 6 5 4

ISBN: 0-13-905043-4

Prentice-Hall International (UK) Limited, *London*
Prentice-Hall of Australia Pty. Limited, *Sydney*
Prentice-Hall Canada Inc., *Toronto*
Prentice-Hall Hispanoamericana, S.A., *Mexico*
Prentice-Hall of India Private Limited, *New Delhi*
Prentice-Hall of Japan, Inc., *Tokyo*
Simon & Schuster Asia Pte. Ltd., *Singapore*
Editora Prentice-Hall do Brasil, Ltda., *Rio de Janeiro*

PREFACE

Many books have been written on engineering materials for material scientists, but none has been written for civil engineers. Unfortunately, civil engineers have had to rely on the available books for lack of better or more appropriate texts.

Materials for Civil and Highway Engineers was designed specifically for civil engineers. Unlike other texts, *Materials for Civil and Highway Engineers* covers all materials used in civil engineering. The text discusses the engineering performance of soil, aggregate, cements, concrete, timber, asphalt, metals, and plastics, as well as environmental considerations.

Materials for Civil and Highway Engineers is intended for use by second- or third-year students taking introductory courses in civil engineering materials. The text is meant for use at a four-year institute, but it would be equally acceptable at a two-year school. *Materials for Civil and Highway Engineers* complements the laboratories that are a major part of civil engineering materials courses.

The text's wide variety of subjects enables the instructor to cover several topics within one semester. The text provides some theory but is centered around the practical approach. The instructor may use the text to cover the basics and supplement this with personal notes emphasizing relevance where needed. The essay-type homework problems were left as they were in the previous edition. Many instructors create their own examples and problems for the classroom based on specific points they wish to address and emphasize.

Chapter 1 is devoted to soils as used in civil and highway engineering. Chapter 2 concentrates on mineral aggregates as the foundation for portland cement concrete and bituminous concrete. Chapter 3 deals with, in addition to portland cement, many types of cement. Chapters 4 to 7 are concerned with concrete, its general strength, its general behavior, the design of concrete mixes, its failures, and latest applications. Chapter 8 is devoted to timber classifications, physical characteristics, mechanical properties, decay, durability, and preservation. Chapter 9 covers asphalt and asphaltic cements. Chapters 10 and 11 cover metals, the metallic state, and ferrous metals. Chapter 12 is concerned with plastics and new advances dealing with FRP. Chapter 13 briefly discusses glass, concrete block, brick, mortar, and shotcrete to round out those materials utilized by Civil and Highway Engineers. Chapters 14 and 15 cover environmental considerations in construction and design and stimulate the thought process about the need to be concerned about the environment.

To assist the instructor, a number of the more important American Society for Testing and Materials (ASTM) standards are presented in the appendices. The student in a first materials course needs to become familiar with such standards, as they will become part of his or her professional career.

Kenneth N. Derucher
George P. Korfiatis
A. Samer Ezeldin

CONTENTS

Preface   *iii*

1 Soil   **1**

Origin and Geologic Properties   1

Soil Composition and Properties   3

Soil Classification   17

Soil-Water Interaction   20

Soil Compressibility   32

Soil Strength   35

Soil Improvement   41

Problems   49

References   50

2 Mineral Aggregates   **51**

Aggregate Types and Processing   51

Aggregate as a Base-Course Material   55

Aggregates for Portland Cement Concrete   56

Aggregates for Bituminous Mixtures   61

Aggregate Beneficiation   62

ASTM Test Specifications for Aggregates   63

Problems   75

References   75

3 Cements   **77**

Lime   77

Natural Cements   80

Portland Cements   81

Portland-Pozzolan Cements   90

Portland Blast-Furnace-Slag Cement   92

Alumina Cements   92

Expansive Cements   93

Special Portland Cements   93

Problems   94

References   94

4 Concrete: Strength and Behavior   **95**

Introduction   95

Uniaxial Behavior   95

Biaxial Behavior   105

Creep   106

Shrinkage   107

Fatigue   110

Bond Strength   111

Shearing and Torsional Strengths   114

Durability   114

Problems   124

References   125

5 Design Procedure in Making Concrete   **127**

Concrete Materials   127

Principal Requirements for Concrete   128

Influences of Ingredients on Properties of Concrete   129

Proportioning Concrete Mixes   131

Proportioning Materials   141

Mixing and Depositing   141

Compressive, Tensile, and Flexural Strength   143

Problems   143

References   144

6 Proportioning Structural Concrete Mixtures with Fly Ash and Other Pozzolans   **146**

Background   146

Procedure   148

Applicability to Lightweight Concrete   157

Sample Computations   159

Conclusions   161

References   161

7 Advances in Concrete Technology   **162**

Admixtures   162

Mineral Admixtures   165

High-Strength Concrete   166

Fiber-Reinforced Cementitious Composites   170

Waste and By-Products Use in Concrete   178

Problems   180

References   181

8 Timber   **182**

Classes of Trees   182

Wood Structure   183

Physical Characteristics of Wood   185

Mechanical Properties of Timber   189

Decay. Durability, and Preservation of Timber   194

Problems   199

References   200

9 Asphalt Cements   **202**

Asphalt Cements   202

Methods of Testing   205

Modifications of Asphalt Cements   208

Road Tars   210

Proportioning Asphaltic Mixes   212

Problems   217

References   219

10 Metallic State   **220**

Crystalline Nature of Metals   220

Mechanical Behavior of Metals   227

Testing and Evaluation of Metals   231

Problems   238

References   238

11 Ferrous Metals   **239**

General   239

Ferrous Metals   241

Manufacture of Steel   242

Structure of Iron and Steel   254

Heat Treatments   256

Physical Properties of Steels   263

Alloy Steels   266

Nonferrous Metals   268

Problems   268

References   269

12   Plastics   **270**

Classification   270

Types of Plastics   271

Manufacture of Organic Plastics   278

Properties of Organic Plastics   279

Advanced FRP Composites in Construction   284

Problems   286

References   268

13   Miscellaneous Materials   **287**

Glass   287

Concrete Block   288

Brick   289

Mortar   291

Shot-crete   291

Problems   294

References   294

14   Design for the Environment   **295**

Introduction   295

The Principle of Sustainable Development   296

Life Cycle Assessment   297

Material Selection   301

Material Minimization   304

Environmental Quality Standards   307

Problems   308

References   308

15 Environmental Considerations in Construction   **309**

Regulatory Issues   309

Materials for Construction of Waste Impoundments   312

Contaminated Soils and Remedial Options   314

References   317

Appendices   **318**

Soil   319

Aggregate and Concrete   365

Wood   398

Asphalt   417

Metal   423

Index   **465**

1 SOIL

Soil encountered close to the earth's surface is one of the most important and widely used engineering materials. Its frequency of use and the importance of its functions rank it among the top materials in the fields of civil, highway, and architectural engineering.

Soil can be used in engineering works as imported fill or as in-situ material. As imported fill, soil is brought to the site from a different location and is placed in a controlled manner to fill excavations, build earthen structures, or to support various other types of engineering structures. As in-situ material, the soil is used in its original location to perform similar functions.

This chapter is intended to give an overview of soil composition, properties, and use as an engineering material.

ORIGIN AND GEOLOGIC PROPERTIES

Minerals—Rocks—Soils

The outer crust of the earth, which extends 10 to 15 km in depth, is composed of eight predominant chemical elements. The relative abundance of these elements in the earth's crust is shown in Table 1-1.

These elements are found in the soil and rock of the crust, assembled in solid homogeneous chemical compounds of regular architecture called *minerals*. Minerals form rocks that in turn are exposed to *weathering* to produce soil. The five most predominant minerals that are found near the earth's surface are shown in Table 1-2.

Geologists classify rocks according to their origin into three groups:

1. *Igneous rocks.* These are rocks that are formed by the solidification of molten magma ejected to the surface of the earth by volcanic eruption. The magma cools as

1

TABLE 1-1 RELATIVE ABUNDANCE
OF ELEMENTS IN THE
EARTH'S CRUST
(BY WEIGHT)

Element	Percentage
Oxygen	46.0
Silicon	28.0
Aluminum	8.0
Iron	6.0
Magnesium	4.0
Calcium	2.4
Potassium	2.3
Sodium	2.1

it moves through areas of lower temperature and solidifies when it comes into contact with the atmosphere. The rate of magma cooling determines the type of rock that will be formed. Some of the most commonly encountered igneous rocks are granite, basalt, gabbro, rhyolite, obsidian, pumice, and diorite.

2. *Sedimentary rocks.* These rocks are formed by compaction and cementation of soil deposits formed by weathering, or by various chemical processes. Pressure is exerted by the weight of overlying deposits (*overburden pressure*) and cementation is caused by cementing agents that are found dissolved in groundwater. Some of the most commonly encountered sedimentary rocks are shale, sandstone, mudstone, limestone, gypsum, and dolomite.

3. *Metamorphic rocks.* These rocks are formed by a process called *metamorphism* (change of form). This entails the transformation of all types of rocks—igneous, sedimentary, and metamorphic—by heat and pressure. Some of the most commonly encountered metamorphic rocks are gneiss from metamorphism of granite, marble from limestone, and dolomite slate from shale.

All three types of rocks can be subjected to weathering, a process by which rock breaks down to smaller grains to form soil. Weathering can take place due to changes in mineral composition (*chemical weathering*), changes in temperature (*mechanical weathering*), or a combination of the two.

Soils are also classified with respect to the mode by which they are transported.

1. Residual soils: Formed by weathering and remaining at their place of origin.
2. Transported soils or sediments:
   A. Soils transported by water
      Alluvial: transported by streams and rivers.
      Deltaic: transported by rivers and deposited in delta areas.
      Marine: formed by deposition and transport in the sea.
      Lacustrine: deposited in lake bottoms.
   B. Soils transported by ice (glacial deposits): transported and deposited by glacial movement.
   C. Soils transported by wind (aeolian deposits): such as loess and dune sand.

TABLE 1-2　PREDOMINANT ROCK MINERALS

Mineral	Percentage	Chemical Composition
1. Feldspars	30	
Plagioclase (calcium or sodium feldspar)		$Ca(Al_2Si_2O_3)$, $Na(AlSi_3O_8)$
Orthoclase (potassium feldspar)		$K(AlSi_3O_8)$
2. Quartz (silicon dioxide)	28	SiO_2
3. Micas	18	
Biotite		$K(Mg,Fe)_3AlSi_3O_{10}(OH)_2$
Muscovite		$KAl_3Si_3O_{10}(OH)_2$
4. Calcium Carbonates		
Calcite	9	$CaCo_3$
Dolomite	9	$CaMg(CO_3)_2$
5. Iron oxides	4	

    D. Gravity-transported soils (colluvial deposits): transported by gravity during slope failures and landslides.

The various geologic processes that take place with time bring certain changes to the character of residual soils. These changes are responsible for the development of a natural stratification of surface soils, which is called the *soil profile,* or weathering profile. In general, there are three distinct zones or horizons in the natural soil profile.

    The *A horizon* is the top surficial soil layer that includes humus (topsoil).

    The *B horizon* is the subsurface layer composed mostly of fine colloidal soil particles transported from the A horizon.

    The *C horizon* contains the less-weathered soils that extend into the bedrock.

    Each horizon has a particular importance in highway, airfield, and building design and construction.

SOIL COMPOSITION AND PROPERTIES

Particle Size

Soils are classified into four major categories in relation to their particle (grain) size: *gravel, sand, silt,* and *clay.*

    Several agencies have developed grain size limits to identify these soil categories. Table 1-3 shows the limits developed by four agencies and that are most frequently encountered in literature concerning grain size limits.

    The Unified Soil Classification system is the most widely used in the United States and abroad.

Particle Size Distribution

Soil in its natural state consists of a conglomeration of solid grains of various sizes. The distribution of the particle size range is a very important characteristic of a given material. This

TABLE 1-3   GRAIN SIZE LIMITS FOR SOIL IDENTIFICATION

Soil	Grain Size—mm and (in.)			
	USC[a]	MIT[b]	AASHTO[c]	USDA[d]
Gravel[e]	76.2 (3) to 4.75 (0.2)	>2 (0.08)	76.2 (3) to 2 (0.08)	>2 (0.08)
Sand	4.75 (0.2) to 0.075 (0.003)	2 (0.08) to 0.06 (0.002)	2 (0.08) to 0.075 (0.003)	2 (0.08) to 0.05 (0.002)
Silt	} Fines	0.06 (0.002) to 0.002 (8×10^{-5})	0.075 (0.003) to 0.002 (8×10^{-5})	0.05(0.002) to 0.002 (8×10^{-5})
Clay	<0.075 (0.003)	<0.002 (8×10^{-5})	<0.002 (8×10^{-5})	<0.002 (8×10^{-5})

[a] Unified Soil Classification System (U.S. Army Corp of Engineers).
[b] Massachusetts Institute of Technology.
[c] American Association of State Highway and Transportation Officials.
[d] U.S. Department of Agriculture.
[e] Rock fragments larger than gravel size are called cobbles or boulders.

distribution is generally depicted graphically on a curve called the *gradation* or *grain size distribution curve.*The grain size distribution curve is a semilogarithmic plot of the grain diameter (log scale) versus the weight expressed as a percent of the dry weight. Typical grain size distribution curves are shown in Figure 1-1.

There are two laboratory methods used to determine the grain size distribution of soils: *mechanical* or *sieve analysis* and *hydrometer analysis.*

Mechanical (Sieve) Analysis.   This test is performed on soils with particle sizes larger than 0.075 mm in diameter. It consists of shaking the dried soil through a series of sieves of progressively smaller diameter openings. The weight of the soil retained in each sieve is measured, and the percentage of soil passing through each sieve is computed based on the total dry weight of the sample. The grain size distribution curve is then plotted. The U.S. standard sieve numbers and size of the openings are given in Table 1-4.

The grain size distribution curve is very important for soil identification and classification of granular soils. Three parameters that are used to describe the grain size distribution curve are the following:

1. The effective size, D_{10}.
2. The coefficient of gradation,

$$C_c = \frac{(D_{30})^2}{D_{60}\,D_{10}}$$

3. The uniformity coefficient,

$$C_u = \frac{D_{60}}{D_{10}}$$

where D_{10}, D_{30}, and D_{60} = the diameters corresponding to 10, 30, and 60 percent finer, respectively, in the grain size distribution curve.

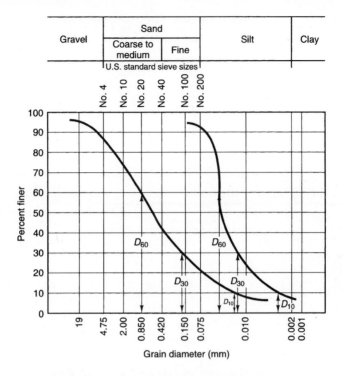

Figure 1-1 Typical Grain Size Distribution Curves

These parameters are used for classification of granular soils.

Hydrometer Analysis. This method is valid for soil with particle diameters of less than 0.075 mm. It is based on Stokes's law, the principle of the rate of settling of suspended solids in a liquid solution. Stokes's law of settling is the basis for the computation of grain size distribution. According to Stokes's law, the settling velocity of suspended particles is proportional to the square of their diameter.

The hydrometer test entails the measurement of the specific gravity of a dispersed soil suspension, contained in a sedimentation cylinder, by a hydrometer bulb immersed in the suspension. As larger soil particles settle, the specific gravity of the suspension decreases, approaching the specific gravity of water. The hydrometer readings are then correlated to the percentage finer (equivalent to the percentage passing in the sieve analysis) and the grain size distribution curve is plotted similarly to the method used in sieve analysis.

The standard test methods for both sieve and hydrometer analyses are described in detail in ASTM Designation D 422.

The grain size distribution of soils is a very important parameter used to classify soil and to estimate several of its properties, such as compaction and drainage characteristics.

Weight-Volume Relationships

Soil as found in nature is a three-phase medium, which makes it a unique engineering material. It is composed of solid mineral grains, water (moisture), and air. The water and

TABLE 1-4   U.S. STANDARD SIEVES AND SIZE
OF OPENINGS

Sieve Size or Number	
Size (in.)	Opening Size (mm)
4	101.60
3	76.10
2	50.80
$1\frac{1}{2}$	38.10
$1\frac{1}{4}$	32.00
1	25.40
$\frac{3}{4}$	19.00
$\frac{1}{2}$	16.00
$\frac{3}{8}$	9.51
Number	
4	4.750
6	3.350
8	2.360
10	2.000
16	1.180
20	0.850
30	0.600
40	0.425
50	0.300
60	0.250
80	0.180
100	0.150
140	0.106
170	0.088
200	0.075
270	0.053

air (or vapor) occupy the openings (voids or pores) that surround the solid grains.

To visualize and better understand the basic definitions that follow, consider the elementary unit volume shown in Figure 1-2, which shows the three phases as if segregated.

It is clear that the total volume (V) of this element is

$$V = V_s + V_w + V_a = V_s + V_v \tag{1.1}$$

where V = total element volume
V_s = volume of solids
V_w = volume of water
V_a = volume of air
V_v = volume of voids

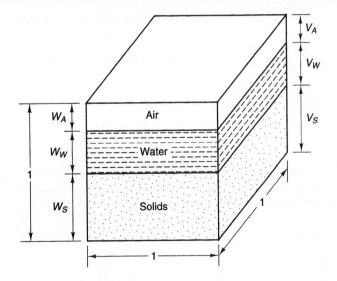

Figure 1–2   Elementary Unit Volume

Similarly, by considering the weight of the air to be negligible, the total weight can be expressed as

$$W = W_s + W_w + W_a \text{ or } W = W_s + W_w \qquad (1.2)$$

where W_s =   weight of solids
W_w =   weight of water

The following definitions can be made based on these elementary relationships:

1. *Void ratio, e,* is defined as the volume of voids divided by the volume of solids.

$$e = \frac{V_v}{V_s} \qquad (1.3)$$

2. *Porosity, n,* is defined as the volume of voids divided by the total volume.

$$n = \frac{V_v}{V} \qquad (1.4)$$

3. *Water or moisture content, w,* is defined as

$$\text{percentage } w = \frac{W_w}{W_s} \times 100 \qquad (1.5)$$

4. *Degree of saturation, S,* is defined as

$$\text{percentage } S = \frac{V_w}{V_v} \times 100 \tag{1.6}$$

5. *Unit weight,* γ, is defined as

$$\gamma = \frac{W}{V} \tag{1.7}$$

The unit weight, γ, is sometimes called moist, wet, total or in-situ unit weight since it contains the weight of the moisture present in the voids. In two special cases the unit weight is of particular importance.

A. When the soil is free of any moisture the *dry unit weight,* γ_d, is defined as

$$\gamma_d = \frac{W_s}{V} \tag{1.8}$$

B. When the soil is fully saturated (all voids are filled with water) Equation 1.7 yields the *saturated unit weight,* γ_{sat}.

In analogy with the unit weight the density of the soil ρ is defined as

$$\rho = \frac{M}{V} \tag{1.9}$$

and

$$\rho_d = \frac{M_s}{V} \tag{1.10}$$

where M =   total mass (water and solids)
           M_s =   mass of solids

6. *Specific gravity* of the soil solids (G_s) is defined as

$$G_s = \frac{\gamma_s}{\gamma_w} = \frac{W_s}{V_s \, \gamma_w} \tag{1.11}$$

where γ_s =   unit weight of soil solids
           γ_w =   unit weight of water (62.4 lb/ft³)

Combinations of these definitions (Equations 1.1 to 1.11) yield numerous expressions for the computation of the various parameters. Table 1-5 is a compilation of these expressions for saturated soils. Table 1-6 shows the specific gravity of selected minerals. The specific gravity of the most commonly occurring soil types ranges from 2.6 to 2.7.

TABLE 1-5 EXPRESSIONS FOR WEIGHT-VOLUME RELATIONS FOR SATURATED SOILS (AFTER JUMIKIS, 1962)

Given Parameters	Equations For					
	Dry Unit Weight γ_d	Saturated Moisture Content, w %	Void Ratio e	Specific Gravity G_s	Porosity n	Saturated Unit Weight γ_{sat}
$\gamma_w,\ \gamma_{sat},\ w$	$\dfrac{\gamma_{sat}}{1+w}$		$\dfrac{w\gamma_{sat}}{\gamma_w - w(\gamma_{sat}-\gamma_w)}$	$\dfrac{\gamma_{sat}}{\gamma_w - w(\gamma_{sat}-\gamma_w)}$	$\dfrac{w\gamma_{sat}}{(1+w)\gamma_w}$	
$\gamma_w,\ \gamma_{sat},\ e$	$\gamma_{sat} - \dfrac{e}{1+e}\gamma_w$	$\dfrac{e\gamma_w}{\gamma_{sat}+e(\gamma_{sat}-\gamma_w)}$		$(1+e)\dfrac{\gamma_{sat}}{\gamma_w} - e$	$\dfrac{e}{1+e}$	
$\gamma_w,\ \gamma_{sat},\ n$	$\gamma_{sat} - n\gamma_w$	$\dfrac{n\gamma_w}{\gamma_{sat}-n\gamma_w}$	$\dfrac{n}{1-n}$	$\dfrac{\gamma_{sat}-n\gamma_w}{(1-n)\gamma_w}$		
$\gamma_w,\ G_s,\ \gamma_d$		$\left(\dfrac{1}{\gamma_d}-\dfrac{1}{G_s\gamma_w}\right)\gamma_w$	$\dfrac{G_s\gamma_w}{\gamma_d}-1$		$1-\dfrac{\gamma_d}{G_s\gamma_w}$	$\left(1-\dfrac{1}{G_s}\right)\gamma_d+\gamma_w$
$\gamma_w,\ G_s,\ w$	$\dfrac{G_s}{1+wG_s}\gamma_w$		wG_s		$\dfrac{wG_s}{1+wG_s}$	$\left(\dfrac{1+w}{1+wG_s}\,G_s\right)\gamma_w$
$\gamma_w,\ G_s,\ e$	$\dfrac{G_s}{1+e}\gamma_w$	$\dfrac{e}{G_s}$			$\dfrac{e}{1+e}$	$\dfrac{G_s+e}{1+e}\gamma_w$
$\gamma_w,\ G_s,\ \gamma_{sat}$	$\dfrac{\gamma_{sat}-\gamma_w}{G_s-1}G_s$	$\dfrac{G_s\gamma_w-\gamma_{sat}}{(\gamma_{sat}-\gamma_w)G_s}$	$\dfrac{G_s\gamma_w-\gamma_{sat}}{\gamma_{sat}-\gamma_w}$		$\dfrac{G_s\gamma_w-\gamma_{sat}}{(G_s-1)\gamma_w}$	
$\gamma_w,\ G_s,\ n$	$G_s(1-n)\gamma_w$	$\dfrac{n}{G_s(1-n)}$	$\dfrac{n}{1-n}$			$[G_s-n(G_s-1)]\gamma_w$
$\gamma_w,\ w,\ n$	$\dfrac{n}{w}\gamma_w$		$\dfrac{n}{1-n}$	$\dfrac{n}{(1-n)w}$		$n\left(\dfrac{1+w}{w}\right)\gamma_w$
$\gamma_w,\ w,\ e$	$\dfrac{e}{(1+e)w}\gamma_w$			$\dfrac{e}{w}$	$\dfrac{e}{1+e}$	$\dfrac{e}{w}\left(\dfrac{1+w}{1+e}\right)\gamma_w$
$\gamma_w,\ \gamma_d,\ w$			$\dfrac{w\gamma_d}{\gamma_w-w\gamma_d}$	$\dfrac{\gamma_d}{\gamma_w-w\gamma_d}$	$\dfrac{w\gamma_d}{\gamma_w}$	$(1+w)\gamma_d$
$\gamma_w,\ \gamma_d,\ e$		$\left(\dfrac{e}{1+e}\right)\left(\dfrac{\gamma_w}{\gamma_d}\right)$		$(1+e)\dfrac{\gamma_d}{\gamma_w}$	$\dfrac{e}{1+e}$	$\dfrac{e\gamma_w}{1+e}+\gamma_d$
$\gamma_w,\ \gamma_d,\ \gamma_{sat}$		$\dfrac{\gamma_{sat}}{\gamma_d}-1$	$\dfrac{\gamma_{sat}-\gamma_d}{\gamma_w+\gamma_d-\gamma_{sat}}$	$\dfrac{\gamma_d}{\gamma_w+\gamma_d-\gamma_{sat}}$	$\dfrac{\gamma_{sat}-\gamma_d}{\gamma_w}$	
$\gamma_w,\ \gamma_d,\ n$		$\dfrac{n\gamma_w}{\gamma_d}$	$\dfrac{n}{1-n}$	$\dfrac{\gamma_d}{(1-n)\gamma_w}$		$\gamma_d+n\gamma_w$

TABLE 1–6 TYPICAL SPECIFIC GRAVITY
VALUES FOR SELECTED
SOIL MINERALS

Mineral	Specific Gravity
Bentonite	2.15
Calcite	2.90
Chlorite	2.80
Biotite (mica)	3.00
Muscovite (mica)	2.80
Hornblende	3.30
Kaolinite	2.62
Illite	2.60
Quartz	2.60
Gibbsite	2.40
Feldspar	2.50
Montmorillonite	2.40
Anhydrite	3.00
Dolomite	2.90

Cohesive and Cohesionless Soils

Cohesion is an inherent physical property of a soil that causes individual grains to stick to each other when subjected to wetting or drying. As a result a certain force is required to break the soil apart when it is dry. Soils that exhibit this property are called *cohesive* soils. Soils in which the individual grains are not held together by any forces when dry are called *cohesionless*.

The presence of clay minerals gives soil deposits their cohesive properties. The amount of cohesion depends on the grain size distribution of the soil and the amount of clay minerals present. In general, most soils with the majority of grains passing through the No. 200 sieve will exhibit cohesion.

The forces responsible for keeping fine soil grains together are ionic bonds, van der Waals bonds, hydrogen bonds, and gravitational attraction.

Clays and Clay Minerals

As shown earlier in Table 1-3, clays are defined as soils having particle diameters of less than 0.002 mm. This is not, however, the only requirement. For a soil to be classified as clay it must also be able to exhibit plasticity, that is, to behave as a plastic medium for a certain range of moisture content.

Clay minerals are mainly aluminum, iron, or magnesium silicates arranged in crystalline structures composed of two basic units, silica tetrahedron and aluminum octahedron. The silica tetrahedron unit [Figure 1-3(a)] consists of a silicon atom surrounded by four oxygen atoms. The octahedron unit is composed of six hydroxyls surrounding either a magnesium atom or an aluminum atom as shown in Figure 1-3(b). A combination of silica tetrahedron units composes a *silica sheet.* Sheets composed of octahedron units are called *gibbsite* sheets when the metallic atom is aluminum or *brucite* sheets when the metallic atom is magnesium.

A physical characteristic of clay minerals is the *specific surface,* which is defined as the surface area of the mineral per unit mass.

There are three predominant clay minerals.

1. *Kaolinite.* Kaolinite consists of silica and gibbsite sheets stacked in alternating layers as shown in Figure 1-4(a). Each layer has a thickness of approximately 7 Å (1 Å = 1×10^{-10} m). The specific surface of kaolinite is approximately 15 m²/g.

2. *Illite.* Illite consists of a gibbsite sheet embedded between two silica sheets with potassium ions located between each layer as shown in Figure 1-4(b). Illite is derived primarily from micaceous minerals and is sometimes called *clay mica.* The thickness of each illite layer is 10 Å and the specific surface is about 80 m²/g. Another clay mineral in the illite family is *vermiculite.* It contains a brucite sheet in place of the gibbsite and calcium and magnesium ions along with potassium at the layer interfaces.

3. *Montmorillonite.* This mineral is composed of an octahedron sheet sandwiched between two tetrahedra sheets [Figure 1-4(c)]. Several different metallic atoms such as zinc, iron, lithium, or magnesium can be found in place of the aluminum atom in

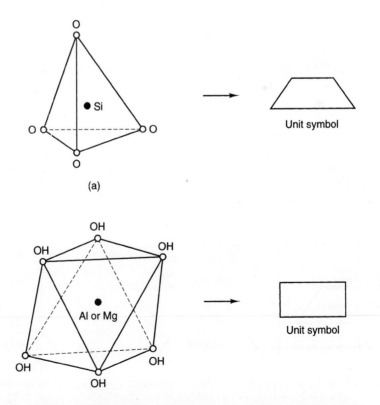

Figure 1–3 Clay Mineral Basic Units: (a) Tetrahedron Unit; (b) Octahedron Unit

the octahedra sheets and aluminum can be found in place of silica in the tetrahedra sheets. The thickness of the montmorillonite unit is approximately 10 Å and its surface area is 800 m^2/g. Bentonite is a member of the montmorillonite mineral group that has found wide applications in soil drilling explorations because of its expansive properties when exposed to water.

Clay particles in their natural state are always surrounded by water. The binding of water in the solid surface of clay minerals is called *adsorption*. This phenomenon greatly influences the physical properties of the clay. The primary cause of adsorption is the unbalanced electrostatic charge that exists in the mineral surfaces. The clay mineral surfaces generally have a net negative charge that may vary in intensity.

The water molecule is a neutral but also dipole molecule, with one pole having a positive charge and the other a negative charge. The water dipole is aligned when it is within the influential zone of the mineral's electrostatic field. The orientation is such that the positively charged pole of the water molecule is attracted by the negatively charged mineral surface. The attraction is strongest closest to the mineral surface and it diminishes farther away until it becomes so weak that the orientation of the dipoles ceases to exist and water molecules are randomly distributed. The zone that contains oriented water molecules is called the *diffuse double layer*. The larger the specific surface of the mineral, the larger is the negative charge at its surface, and therefore the thicker is the diffuse double layer.

Besides being bonded to the adsorbed water, clay minerals are bonded to each other by hydrogen bonding and van der Waals forces. If an external force squeezes some of the adsorbed water out and brings the particles closer together, the attraction becomes larger and the interparticle bond becomes stronger. This is the property of plasticity that is a characteristic of clay minerals.

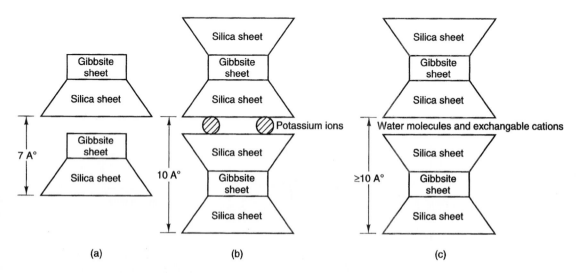

Figure 1–4 Predominant Clay Minerals: (a) Kaolinite; (b) Illite; (c) Montmorillonite

Soil Consistency and Atterberg Limits

As was described in the previous section, the water retained in the voids of fine-grained soils can greatly affect their engineering properties. Depending on the amount of water present in the voids the soil consistency can change from that of a brittle solid to a liquid state. To delineate the different states of soil consistency the Swedish soil scientist A. Atterberg proposed five consistency limits called *Atterberg Limits.* The consistency limits are based on the water content of the soil and are as follows:

1. The *liquid limit (LL)* is defined as the water content above which the soil behaves as a viscous liquid (its shearing strength is negligible).

    In soils engineering the liquid limit is measured in the laboratory using a liquid limit machine. It is arbitrarily defined as the water content at which a standard trapezoidal groove, cut in moist soil contained in a special cup, closes after 25 blows in the liquid limit machine.
2. The *plastic limit (PL)* is defined as the water content below which the soil ceases to behave as a plastic medium, and begins to exhibit the properties of a semisolid medium. The plastic limit is determined in the laboratory as the water content at which the soil cracks and breaks apart when rolled by hand into a thread having a diameter of $\frac{1}{8}$ inch.
3. The *shrinkage limit (SL)* is defined as the water content at which the soil reaches its minimum theoretical volume and no more volume reduction takes place with continued drying.

The plastic, liquid, and shrinkage limits have found wide applications in soil mechanics. Two more limits introduced by Atterberg, the *cohesion limit* and *sticky limit,* are utilized mostly in agricultural applications. Figure 1-5 presents the various states of soil consistency and their corresponding consistency limits.

The *plasticity index (PI)* was defined by Atterberg to describe the range of water content over which the soil behaves as a plastic medium. The plasticity index is defined as

$$PI = LL - PL \qquad\qquad (1.12)$$

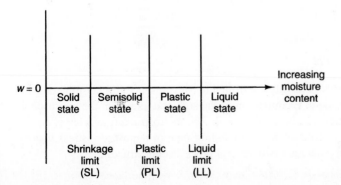

Figure 1–5 Atterberg Limits and the Corresponding States of Soil Consistency

Another index used to relate the water content of the soil to the liquid and plastic limit is the *liquidity index* (*LI*). The liquidity index can be used to describe the engineering behavior of soil when subjected to a shearing force and is defined as

$$LI = \frac{w - PL}{PI} \qquad (1.13)$$

where w =   the natural water content of the solid.

The following material behavior patterns can be correlated to the liquidity index.

Liquidity index	Engineering behavior
$LI < 0$	Brittle solid
$0 < LI \leq 1$	Plastic
$LI > 1$	Viscous liquid

The Atterberg limits and indices are very valuable for the classification of soils and they have been empirically correlated to several soil properties with wide applications in soil mechanics. Another parameter often used to express the plasticity of clay minerals is the *activity (A)*, which is defined as

$$A = \frac{PI}{\text{Percentage of clay size particles } (\% < 0.002 \text{ mm})} \qquad (1.14)$$

The activity is also a measure of the water retention capacity of the clay minerals and can be used to identify them.

Some typical values for the predominant clay minerals are as follows:

Mineral	Activity
Kaolinite	< 1
Illite	0.5-1.5
Montmorillonite	1-6

Standard laboratory test procedures for the liquid and plastic limits are described in the ASTM Designation D 4318.

Structure of Cohesive Soils

The structure of cohesive soils is largely controlled by the presence of clay minerals and the various forces acting on them. Examination of the soil structure from the microscopic point of view reveals the microstructure of the soil. The microstructure of clay is dependent on the geologic history of the material, including type of deposition, stress history, and deposition environment. There are two general types of clay structure.

Dispersed structures [Figure 1-6(a)] are formed when the repulsive forces between grains cause them to be deposited in an oriented fashion with maximum possible distances separating each of them.

Flocculated structures [Figure 1-6(b)] are formed when soil deposition occurs in a chemical environment that aids the flocculation of initially dispersed grains. Flocculation is the process by which individual particles lose their repulsive forces when they are in suspension and attract each other to form clusters called *flocs*. The loss of the repulsive forces is caused by the presence of an electrolyte, such as salt water. As flocs become larger they settle under gravity to form flocculent structures. Clay deposits formed in marine environments are usually highly flocculent.

Examination of the soil from the macroscopic point of view reveals the soil's macrostructure. The macrostructure is used to determine patterns of soil deposition, discontinuities in the soil structure caused by cracking, and the effects of external forces on the soil's structural behavior.

Structure of Granular Soils

Cohesionless soils are generally encountered as single-grain structures. The most important characteristic of a single-grain structure is the arrangement of the individual grains, or *packing*. A soil's state of denseness or looseness depends on three factors: the grain shape, the grain size distribution, and the grain relative positions (packing).

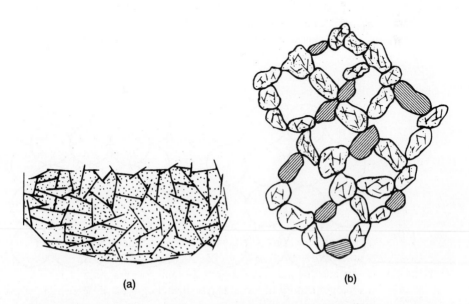

(a)   (b)

Figure 1-6 Structure of Clay Soil: (a) Dispersed Structure; (b) Flocculated Structure

Grain shape can range from angular to very round. This affects the packing of the grains and therefore the void ratio. A wide range of void ratios can be obtained by changing the mode of grain packing. This can be illustrated by packing perfectly spherical particles as shown in Figure 1-7. In Figure 1-7(a), the spheres are packed in a cubical arrangement with a void ratio of $e = 0.91$. In Figure 1-7(b), the same spheres are packed in a pyramidal arrangement with a void ratio of $e = 0.35$.

The grain size distribution is also an important factor affecting the packing of granular soils. A soil is well graded if there is an assortment of small particles to fill the voids formed by larger particles, thus allowing for denser packing [Figure 1-8(a)]. On the other hand, the soil is poorly graded if the voids are not occupied by grains yielding a looser sample with higher void ratio [Figure 1-8(b)]. The *relative density* is used as a measure of the in-situ void ratio of a soil. It is defined as

$$D_r = \frac{e_{max} - e}{e_{max} - e_{min}} \tag{1.15}$$

where e =   in-situ void ratio

          $e_{max, min}$ =   maximum and minimum void ratios as determined in the laboratory.

The relative density can also be expressed in terms of the dry unit weight of the soil as

$$D_r = \frac{\gamma_d - \gamma_{dmin}}{\gamma_{dmax} - \gamma_{dmin}} \frac{\gamma_{dmax}}{\gamma d} \tag{1.16}$$

where γ_d =   in-situ dry unit weight

          $\gamma_{dmax, min}$ =   maximum and minimum dry unit weights as determined in the laboratory

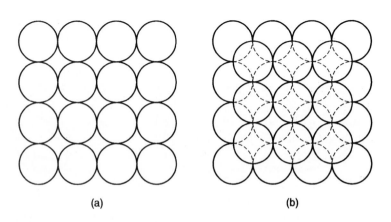

(a)   (b)

Figure 1-7   (a) Cubical Grain Packing with a Void Ratio of $e = 0.91$; (b) Pyramidal Grain Packing with a Void Ratio of $e = 0.35$

Soil Classification

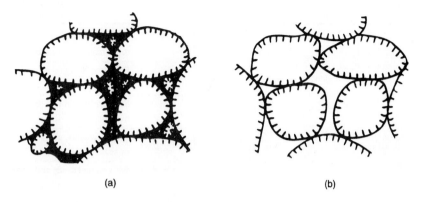

(a)   (b)

Figure 1-8   Soil Gradation: (a) Well-Graded Soil; (b) Poorly Graded Soil

SOIL CLASSIFICATION

Soil classification is the grouping of different types of soils with similar properties. There are several soil classification systems. These systems provide a common language used to briefly and accurately communicate the type of a particular soil as related to its engineering use.

The most widely used classification systems in engineering are the Unified Soil Classification system (USC) and the American Association of State Highway and Transportation Officials (AASHTO) soil classification system. The basis for soil classification in both systems are the index properties of the soil—the grain size distribution and plasticity.

1. *Unified Soils Classification system.* This classification system was first proposed by Casagrande in 1942 in airfield construction works performed by the Army Corps of Engineers. It has since been slightly modified and today is used widely in geotechnical engineering. It is based on grouping and subgrouping of similar soils based on their basic index properties. The USC system is shown in Table1-7. The soil grouping of the USC system can be described as follows:

   A. *Coarse-grained soils.* Composed mainly of gravels and sands with less than 50% passing through the No. 200 sieve. The letter symbol prefixes used are: G to describe gravels, (50 percent or more grains are retained on No. 4 sieve) and S to describe sand (more than 50 percent passes through No. 4 sieve). The suffixes W, P, M, and C are used to describe well graded, poorly graded, silt, and clay content, respectively, of the gravels or sand mixtures.

   B. *Fine-grained soils.* Consisting of silts and clays with 50 percent or more passing through the No. 200 sieve. The prefixes M, C, and O are used to describe inorganic silts, inorganic clays, and organic silts and clays, respectively. The suffix L is used to describe fines of low plasticity (liquid limit of 50 percent or less) and the suffix H is used to describe fines of high plasticity (liquid limit greater than 50 percent). The plasticity chart shown in Table 1-7 is used for the classification of fine-grained soils.

TABLE 1-7 UNIFIED SOILS CLASSIFICATION SYSEM (AFTER ASTM)

Major Divisions			Group Symbols	Typical Names
Coarse-Grained Soils: More than 50% retained on No. 200 sieve	Gravels 50% or m of coarse fraction retained on No. 4 sieve	Clean Gravels	GW	Well-graded gravels and gravel-sand mixtures, little or no fines
			GP	Poorly graded gravels and gravel-sand mixtures, little or no fines
		Gravels with Fines	GM	Silty gravels, gravel-sand-silt mixtures
			GC	Clayey gravels, gravel-sand-clay mixtures
	Sands More than 50% of coarse fraction passes on No. 4 sieve	Clean sands	SW	Well-graded sands and gravelly sands, little or no fines
			SP	Poorly graded sands and gravelly sands, little or no fines
		Sands with Fines	SM	Silty sands, sand-silt mixtures
			SC	Clayey sands, sand-clay mixtures
Fine-Grained Soils: 50% or more passes No. 200 sieve	Silt and Clays Liquid limit 50% or less		ML	Inorganic silts, very fine sands, rock flour, silty or clayey fine sands
			CL	Inorganic clays of low to medium plasticity, gravelly clays, sandy clays, silty clays, lean clays
			OL	Organic silts and organic silty clays of low plasticity
	Silts and Clays Liquid limit greater than 50%		MH	Inorganic silts, micaceous or diatomaceous fine sands or silts, elastic silts
			CH	Inorganic clays of high plasticity, fat clays
			OH	Organic clays of medium to high plasticity
Highly Organic Soils			PT	Peat, muck, and other highly organic soils

TABLE 1-7 (*Continued*)

			Classification Criteria	
Classification on Basis of Percentage of fines	*Less than 5% pass No. 200 sieve*	*GW, GP, SW, SP*	$C_M = D_{60}/D_{10}$ Greater than 4 $C_C = \dfrac{(D_{30})^2}{D_{10} \times D_{60}}$ Between 2 and 3	
			Not meeting both criteria for GW	
	More than 12% pass No. 200 sieve	*GW, GC, SM, SC*	Atterberg limits plot below "A" line and plasticity index less than 4	Atterberg limits plotting in hatched area are borderline classifications requiring use of dual symbols
			Atterberg limits plot above "A" line and plasticity index greater than 7	
	5% to 12% pass No. 200 sieve	*Borderline classification requiring use of dual sumbols*	$C_M = D_{60}/D_{10}$ Greater than 6 $C_C = \dfrac{(D_{30})^2}{D_{10} \times D_{60}}$ Between 1 and 3	
			Not meeting both hcriteria for GW	
			Atterberg limits plot below "A" line and plasticity index less than 4	Atterberg limits plotting in hatched area are borderline classifications requiring use of dual symbols
			Atterberg limits plot above "A" line and plasticity index greater than 7	

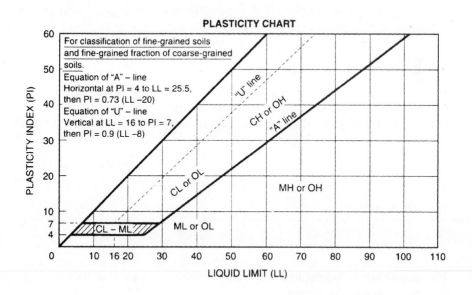

PLASTICITY CHART

For classification of fine-grained soils and fine-grained fraction of coarse-grained soils.

Equation of "A" – line
Horizontal at PI = 4 to LL = 25.5,
then PI = 0.73 (LL −20)
Equation of "U" – line
Vertical at LL = 16 to PI = 7,
then PI = 0.9 (LL −8)

C. *Highly organic soils.* The symbol PT for peat is used to describe soils of high organic content. For more details on the USC system, the student should refer to ASTM D 2487.

2. *AASHTO Classification system.* The AASHTO Classification system was originally developed by the Bureau of Public Roads. It has been revised several times and was brought to its present form by the Highway Research Board. It is described in ASTM Designation D 3282 and AASHTO method M-145 and is shown in Table 1-8. The soils under this system are classified into seven major categories, A-1 to A-7. The soils in groups A-1, A-2, and A-3 are granular soils with 35 percent or less passing through the No. 200 sieve and the soils in groups A-4, A-5, A-6, and A-7 are fine-grained soils with more than 35 percent passing through the No. 200 sieve. The AASHTO system is more frequently used to classify soils used as subgrade materials in highway construction.

SOIL-WATER INTERACTION

As was pointed out in previous sections, soil in its natural state contains water. The degree of saturation in soils can range from a few percent in the upper soil layers, which are subject to evaporation, to 100 percent within the groundwater table. When the voids in the soil are partially filled with water the soil is said to be partially saturated. When the voids are completely filled, the soil is said to be fully saturated.

The water or moisture present in the soil can be either static (*stationary*) or dynamic (*flowing*). Flow can take place both in partially saturated and fully saturated soils. The interaction of both stationary and flowing water with soil will be examined in this section.

The Concept of Effective Stress

To analyze several soil engineering problems we need to know the stress distribution along a soil profile. Consider the soil profile shown in Figure 1-9.

Let us first find what is the vertical stress at point A. This point is subjected to the total weight of the soil column having height H_1. The *vertical stress* at this point is therefore

$$\sigma_A = \gamma_d H_1 \tag{1.17}$$

similarly, at point B,

$$\sigma_B = \gamma_d (H_1 + H_2) = \gamma_d H \tag{1.18}$$

The vertical stress resulting from the weight of the overlying soil is often called *geostatic pressure* or *overburden pressure*.

Let's now consider point C. The total stress at point C is

$$\sigma_c = \gamma_d H + \gamma_{sat} H_3 \tag{1.19}$$

TABLE 1-8 AASHTO SOIL CLASSIFICATION SYSTEM

General Classification	Granular Materials (35% or less passing No. 200)							Silt-Clay Materials (More than 35% passing No. 200)			
	A-1		A-3	A-2				A-4	A-5	A-6	A-7
Group Classification	A-1-a	A-1-b	A-3	A-2-4	A-2-5	A-2-6	A-2-7	A-4	A-5	A-6	A-7-5, A-7-6
Sieve analysis, percentage passing											
No. 10 (2.00 mm)	50 max	—	—								
No. 40 (425 μm)	30 max	50 max	51 min								
No. 200 (75 μm)	15 max	25 max	10 max	35 max	35 max	35 max	35 max	36 min	36 min	36 min	36 min
Characteristics of fraction passing No. 40 (425 μm)											
Liquid limit		—		40 max	41 min	40 max	41 min	40 max	41 min	40 max	41 min
Plasticity index		6 max	N.P.	10 max	10 max	11 min	11 min	10 max	10 max	11 min	11 min[a]
Usual types of significant constituent materials	Stone fragments, gravel and sand		Fine sand	Silty or Clayey Gravel and Sand				Silty Soils		Clayey Soils	
General rating as subgrade	Excellent to good							Fair to poor			

[a] Plasticity index of A-7-5 subgroup is equal to or less than LL minus 30. Plasticity index of A-7-6 subgroup is greater than LL minus 30.

Reprinted with permission of American Association of State Highway and Transportation Officials.

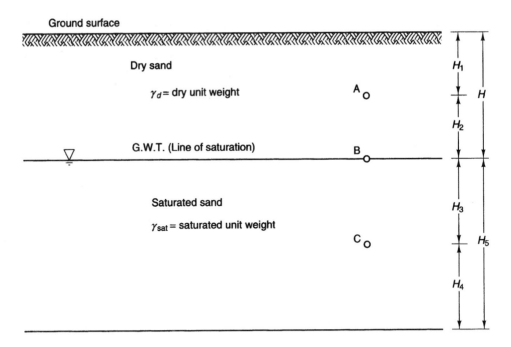

Figure 1-9 Soil Profile

The vertical stress at point C consists of two parts. One part is carried by the water, which at this location fills all the interconnected voids and is acting as a continuum. This portion is called the *neutral stress* or *pore water pressure* and is equal to the *hydrostatic stress,* or

$$u = \gamma_w H_3 \tag{1.20}$$

where γ_w = the unit weight of water.

The other part is carried by the solid structure and is transmitted from grain to grain. This is the *intergranular* or *effective stress* ($\bar{\sigma}$).

The effective stress at point C therefore is

$$\bar{\sigma} = \sigma_c - u = (H\gamma_d + H_3 \gamma_{sat}) - H_3\gamma_w$$

or

$$\bar{\sigma}_c = H\gamma_d + H_3 (\gamma_{sat} - \gamma_w)$$

or

$$\sigma_c = H\gamma_d + H_3 \gamma_{sub} \tag{1.21}$$

where γ_{sub} has been defined as the submerged unit weight $= \gamma_{sat} - \gamma_w$.

The microscopic view of the soil arrangement at the vicinity of point C shown in Figure 1-10 will aid in the visualization of the concept of the effective stress. Here we see that the hydrostatic forces are acting in all directions with the same intensity and therefore only the forces transmitted from grain to grain at the points of contact need to be considered. In the case of dry soil the neutral stress is zero, so the effective stress is equal to the total stress.

In this case we considered a situation where the groundwater was stationary. If the water is moving through the pores, seepage forces must also be considered. In almost all engineering analyses involving overburden stresses the effective stresses are used.

Flow of Water in Soil: Darcy's Law

In the context of soil engineering, the property of soil that controls the passage of water through its pores is called *permeability*. It follows from this definition that soils with large interconnected pores will have higher permeability than soils with smaller pores. If we therefore assume that all other parameters in a soil are kept constant, we expect that permeability will increase with increasing void ratio.

The factors that influence the flow of fluids through soils are:

1. The size, shape, and arrangement of the pore spaces.
2. The relative density of the soil.
3. The density and viscosity of the fluid.
4. The mineralogical and electrochemical properties of the soil particles.

Flow in soils takes place due to the existence of a pressure (*hydraulic head*) difference between two points in the flow field. The direction of flow is from regions of high pressure to low pressure, and the velocity of flow depends on the relative magnitude of the hydraulic head difference and the permeability of the soil.

In the mid-eighteenth century, the French scientist H. Darcy performed experiments to investigate the flow of water through sand filters. The arrangement of Darcy's experiment is shown in Figure 1-11.

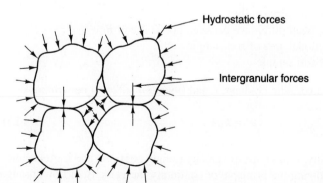

Hydrostatic forces

Intergranular forces

Figure 1-10 Microscopic View of the Soil Arrangement and Forces at the Area Surrounding Point C in Figure 1-9.

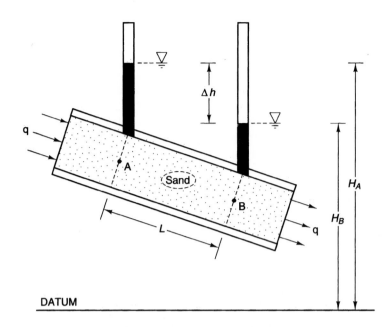

Figure 1-11   Set up of Darcy's Experiment

Flow is passed through a cylindrical column of inside diameter, D, containing a soil sample having length, L, at a rate, q. Darcy observed that the flow rate per unit time, q, is proportional to the change in hydraulic head drop, Δh, and the cross-sectional area of the soil sample and inversely proportional to the length, L, over which the head drop, Δh, takes place. Referring to Figure 1-10:

$$q = \frac{V}{t} = k\frac{\Delta h}{L}A \qquad (1.22)$$

where q = volumetric flow rate per unit time
        V = volume
        t = time
        k = coefficient of permeability
        Δh = hydraulic head difference between points A and B
        A = cross-sectional area of soil sample
        L = length of soil sample

The ratio $\Delta h/L$ is called the hydraulic gradient, i, and Equation 1.22 can be written as

$$q = kiA \qquad (1.23)$$

The coefficient of permeability, k, has units of velocity (length per time) and is often called *hydraulic conductivity*. Following the principle of continuity the velocity of flow through the soil is

$$v = \frac{q}{A} = ki \tag{1.24}$$

The velocity is often called the *Darcy* or *discharge velocity*. It is an average velocity since it is obtained by dividing the flow rate by the gross cross-sectional area of the sample, which includes both soil grains and pores. Because flow takes place through the pores only, the actual or interstitial velocity is

$$v_{act} = \frac{v}{n}$$

where $n =$ the soil porosity.

Permeability is one of the most important soil properties and is used to estimate the flow through soils in several civil and highway engineering applications, such as seepage through earthen dams, site dewatering, and groundwater pollution.

It must be noted here that permeability is a tensor parameter and its value often depends on the direction along which it is measured. Soils can be classified as follows with respect to permeability:

Homogeneous:	if the coefficient of permeability is independent of position within a soil formation.
Heterogeneous:	if the coefficient of permeability is dependent on position within a soil formation.
Isotropic:	if the coefficient of permeability is independent of the direction of measurement at a point in a soil formation.
Anisotropic:	if the coefficient of permeability varies with the direction of measurement at a point in a soil formation.

Some typical values of the coefficient of permeability for various types of soils are given in Table 1-9.

Permeability testing. Permeability is one of the most widely varying properties of engineering materials. As shown in Table 1-9, the coefficient of permeability can take values varying over several orders of magnitude. It is therefore necessary to be able to

TABLE 1-9 PERMEABILITY RANGES FOR VARIOUS TYPES OF SOILS (AFTER TERZAGHI AND PECK, 1967)

Soil Type	Coefficient of Permeability (cm/sec)	Relative Permeability
Gravel	Greater than 10^{-1}	High
Sandy gravel	10^{-1}—10^{-3}	Medium
Sand	10^{-3}—10^{-5}	Low
Silty clay	10^{-5}—10^{-7}	Very low
Clay	Less than 10^{-7}	Practically impermeable

measure the coefficient of permeability with relative accuracy in order to reliably predict the rate of flow through soils. Permeability testing can either be performed in the laboratory or in the field.

Laboratory testing can be performed quickly and is relatively inexpensive, especially for granular soils. Laboratory tests, however, often do not produce reliable results, mainly because of the dependence of the permeability on the in-situ structure of the soil. Test specimens recovered from the field are often disturbed, resulting in an alteration of the in-situ soil structure stratification, grain orientation, and relative density. Such disturbances and remolding of the soil often result in laboratory permeability values different from the in-situ values.

In such situations field tests produce more reliable results but are generally more time consuming and more expensive.

Laboratory Tests. In the laboratory, permeability is measured with devices called *permeameters* by maintaining a flow through a relatively small soil sample. The flow rate and head loss are measured to determine the coefficient of permeability.

Two types of tests are usually employed, the constant head and falling head tests. The constant head permeameter shown in Figure 1-12 is used to measure the permeability of granular soils under low head conditions. Water flows upward through the soil sample under constant head and is collected in a volumetric cylinder at the effluent port. The coefficient of permeability can be computed by

$$k = \frac{VL}{tAh} \tag{1.25}$$

where V = volume of fluid passed through the sample during time t
            L = length of soil specimen
            A = cross-sectional area of soil specimen
            h = hydraulic head difference across soil sample

The constant head test is described in ASTM Designation D 2434.

The falling head permeameter is shown in Figure 1-13. In this apparatus, water supplied by a long tube flows upward through the soil and is collected at the effluent port in a volumetric cylinder. The head drop per time in the tube is recorded and the coefficient of permeability is computed as

$$k = \frac{R_t^2 L}{R_s^2 t} \ell n \frac{h_1}{h_2} \tag{1.26}$$

where R_t = radius of the tube
            R_s = radius of the soil sample
            L = length of the soil sample
            h_1 = initial head in the tube
            h_2 = final head in the tube
            t = time required for the water level to drop from h_1 to h_2

Soil-Water Interaction

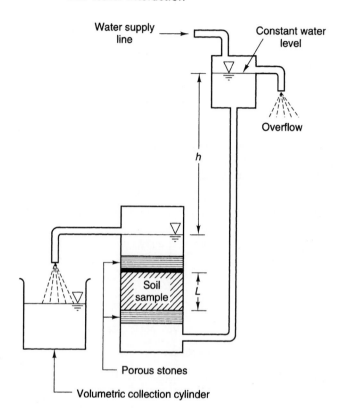

Figure 1-12 Constant Head Permeameter

Falling head permeability tests are performed on soils having relatively low permeability, such as silts.

Field Tests. Various field methods are available for in-situ measurement of soil permeability. These include pumping tests, tracer tests, and borehole tests. The advantage of field tests is that the soil is tested under undisturbed in-situ conditions. Various parameters that could influence the results of the test, such as the location of the groundwater table, the overburden stresses, and the soil stratification are not altered as in the case of laboratory tests.

Some cased borehole testing methods are included in this book. These tests are performed in a cased boring drilled in the ground. The water level variations with time, as water is pumped into or out of the boring, are used to compute the soil permeability. The schematic of four types of borehole tests and the respective equations used for determination of the coefficient of permeability are shown in Table 1-10.

Capillary Rise in Soils

Capillarity is a phenomenon that is attributed to a fluid property that is called *surface tension.* In soils, surface tension occurs between the interfaces of water, soil grains, and air,

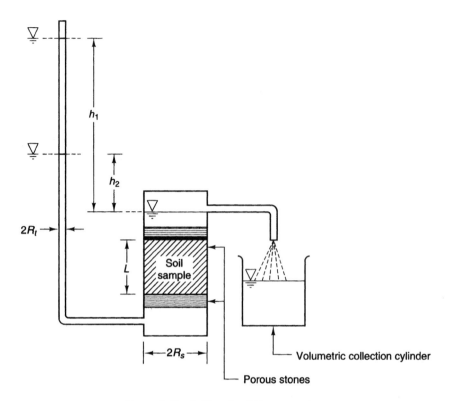

Figure 1-13   Falling Head Permeameter

and results from the differences in the attractive forces between molecules at the surfaces of these materials. As a result of surface tension, the surface of the water bends, forming a curved surface called the *meniscus* as shown in Figure 1-14. Because of the pore water surface tension, water will rise to a certain height above the surface of saturation. This is illustrated in Figure 1-15, which depicts a typical soil profile with a groundwater table. The height of capillary rise is often called the *capillary fringe*. In analogy with capillary rise in thin tubes the following expression can be used to estimate the height of capillary rise in soils.

$$h_c = \frac{0.31}{0.2\,D_{10}}\ (cm) \tag{1.27}$$

where h_c =   height of capillary rise in cm
$\quad\quad\ D_{10}$ =   10 percent particle size of the soil from the grain size distribution
curve in cm

It is observed from this expression that the height of capillary rise is inversely proportional to the effective grain size, D_{10}, which is in turn directly related to the pore size. It fol-

TABLE 1-10   METHODS FOR PERFORMING CASED BOREHOLE PERMEABILILY TESTS

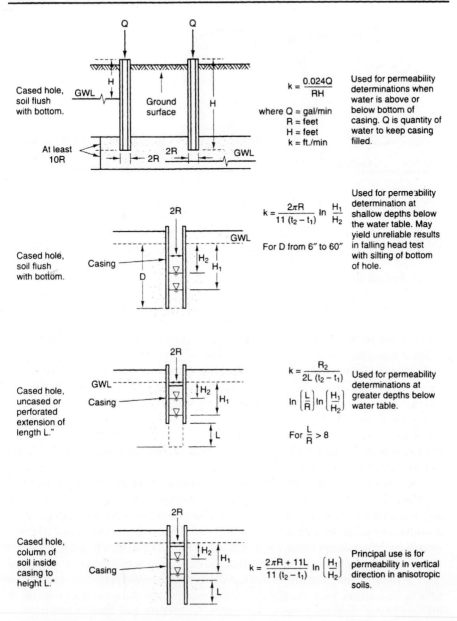

$$k = \frac{0.024Q}{RH}$$

where Q = gal/min
R = feet
H = feet
k = ft./min

Used for permeability determinations when water is above or below bottom of casing. Q is quantity of water to keep casing filled.

$$k = \frac{2\pi R}{11\,(t_2 - t_1)}\, \ln \frac{H_1}{H_2}$$

For D from 6″ to 60″

Used for permeability determination at shallow depths below the water table. May yield unreliable results in falling head test with silting of bottom of hole.

$$k = \frac{R_2}{2L\,(t_2 - t_1)}$$

$$\ln \left(\frac{L}{R} \right) \ln \left(\frac{H_1}{H_2} \right)$$

For $\frac{L}{R} > 8$

Used for permeability determinations at greater depths below water table.

$$k = \frac{2\pi R + 11L}{11\,(t_2 - t_1)}\, \ln \left(\frac{H_1}{H_2} \right)$$

Principal use is for permeability in vertical direction in anisotropic soils.

(Reprinted from McCarthy by permission)

29

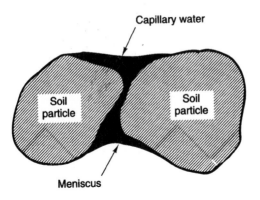

Capillary water

Soil
particle

Soil
particle

Meniscus

Figure 1-14 Capillary Water Held
Between Two Soil Particles Because
of Surface Tension

lows that smaller effective grain sizes will correspond to smaller pore openings and larger capillary rises. Typical heights of capillary rise for various types of soils are shown in Table 1-11.

Frost Action in Soils

When the ambient air temperature falls below the freezing point soil will begin to freeze. The temperature profile in the soil will be similar to that depicted in Figure 1-16. The temperature will be lowest at the soil surface and will increase with depth until it reaches the residual subsurface above-freezing temperatures.

Freezing temperatures in the subsurface environment cause the formation of *ice lenses* in the pore spaces. Water in the soil pores expands as it freezes, creating a volume increase in the soil mass called *frost heave.* One would expect that the total volume change in a soil mass would be equal to the volume increase due to transformation of water from the liquid to the solid state, which is approximately 10 percent. But field experiments in the 1920s showed that frost heave is significantly larger than that attributed to the volumetric expansion of freezing.

Research in the past 50 years has concentrated on the mechanisms of ice lense formation to explain and quantify frost heave. Observations and tests have indicated that as ice lenses are formed they attract capillary water from the surrounding soil and from the underlying water table, and they increase in size. The mechanisms of lense formation and growth are rather complicated, but in general the effects are detrimental to the overall soil structure. Fine-grained soils are more susceptible to frost action than coarse-grained soils.

Two conditions that can be detrimental to structures supported by soils are associated with frost action. First, the frost heave can induce large differential movements that can lift foundations supporting various structures. Second, the moisture content increases dramatically after thawing, thus reducing soil strength.

Frost penetration can vary from zero inches in the southern portions of the United States to up to 110 in. in the northern states. Structural foundations must always be placed below the maximum frost penetration depth to prevent frost heave. Building codes give the minimum depth of soil cover for various regions of the country.

TABLE 1-11   TYPICAL VALUES OF CAPILLARY
RISE FOR VARIOUS SOILS

Soil	Capillary Rise (cm)
Coarse gravel	0.5
Fine gravel	7
Coarse sand	15
Medium sand	25
Fine sand	45
Silt	110
Clay	> 200

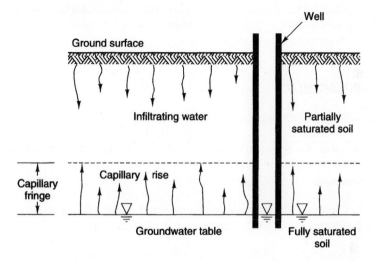

Figure 1-15   Typical Soil Profile Illustrating Capillary Rise

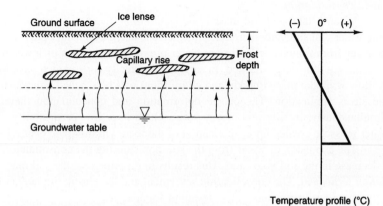

Figure 1-16   Typical Soil Profile Illustrating Capillary Rise Under Freezing
Conditions and Variation of Soil Temperature with Depth

TABLE 1-12 FROST SUSCEPTIBILITY OF VARIOUS TYPES OF SOILS

Soil	USC Classification	Frost Susceptibility
Gravels and sands	GW, GP, SW, SP	None to low
Silty and clayey gravels	GM, GC	Low to medium
Silty and clayey sands	SM, SC	Low to high
Inorganic and organic clays of high plasticity	CH, OH	Medium
Inorganic and organic silts of high plasticity and inorganic clays	OL, MH, CL	Medium to high
Inorganic silts of low plasticity	ML	Medium to very high

The effect of frost heave and soil strength reduction is evident in roadway pavements during spring thaw, when potholes appear. Table 1-12 gives the degree of frost susceptibility of various types of soils.

SOIL COMPRESSIBILITY

As with any other engineering material, soil will undergo a certain deformation (strain) when subjected to a load. *Elastic* materials deform immediately upon load application. In certain soils the stress-strain relationship is time dependent. Materials that exhibit such behavior are called *visco-elastic*. There are two major mechanisms responsible for soil deformation.

1. Change in the grain shape (*distortion* or elastic deformation).
2. Change in the soil volume (*compressibility*).

Distortion (elastic deformation) takes place immediately on the load application and is the main deformation mechanism in coarse-grained soils. Compressibility is the volume change resulting from the expulsion of water from the pores of the soil when subjected to a stress. This time-dependent behavior, which is characteristic of fine-grained soils, is called *consolidation.*

When a saturated soil mass is subjected to an increase in stress the pressure in the pore water will suddenly increase. This increase in the pore water pressure will cause flow to occur from regions of high pore water pressures to regions of lower pressures until a pressure equilibrium is reached. In sandy and gravelly soils, which have high permeability, flow will take place quickly and equilibrium will be reached almost immediately after the stress application. The elastic deformation and consolidation therefore take place simultaneously. In silty and clayey soils, which have low permeability, the excess pore water pressure created upon application of the load will dissipate over a long period.

The expulsion of water from the soil pores during the consolidation process causes a decrease in the soil void ratio. This results in settlement of the soil mass called *consolidation settlement.* The consolidation settlement and the elastic (immediate) settlement are very important parameters in the design of foundations.

In order to compute the consolidation settlement the compressibility characteristics of the soil must be known. The consolidation process is simulated in the laboratory in devices called *consolidometers* or oedometers. A typical consolidometer is shown in

Figure 1-17. A relatively undisturbed compressible soil specimen is placed on the brass confining ring and is trimmed carefully. Two porous stones having diameter slightly smaller than the ring diameter are placed on the top and bottom of the sample. The entire assembly is then placed in the holding container. Increments of load are then applied to the specimen via a loading plate and the vertical deformation is measured with a deflection dial.

The data collected from a consolidation test are presented in two types of plots. The first is a plot of dial reading versus the logarithm of time (*time-deformation curve*). One such curve is constructed for each loading increment. A typical time-deformation curve is shown in Figure 1-18. The time-deformation curves are used to compute the coefficient of consolidation, a parameter that is used to determine the time rate of consolidation.

The second curve is a plot of void ratio or strain versus the logarithm of the stress applied to the specimen (*stress-deformation curve*). A typical stress-deformation curve is shown in Figure 1-19. The stress-deformation curve is used to determine the compression index C_c which in turn is used to compute the ultimate consolidation settlement.

The compression index is defined as

$$C_c = \frac{\Delta e}{\Delta(\log \bar{\sigma})} \tag{1.28}$$

where Δe = the change of void ratio
$\Delta(\log \bar{\sigma})$ = the corresponding change in consolidation stress

It is observed from this equation that the compression index is the slope of the virgin compression curve as shown in Figure 1-19.

It must be noted that consolidation is not a reversible process. On release of the load the sample will rebound slightly but it will never return to its original void ratio, as is shown by the unloading curve in Figure 1-19. This indicates that the past history of the soil sample will have an effect on its compressibility characteristics at present. Based on this observation compressible soils can be classified into two categories.

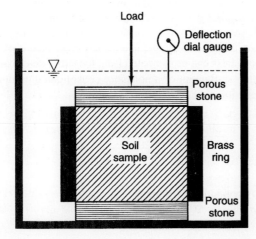

Figure 1-17 Typical Consolidometer Apparatus

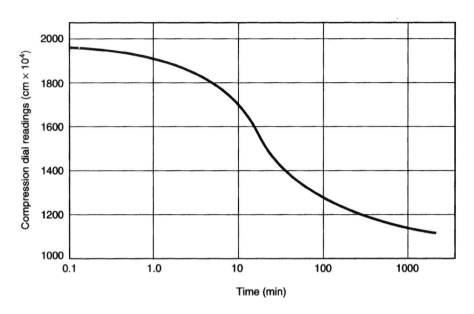

Figure 1-18 Typical Time-Deformation Curve

1. *Normally consolidated.* Soils that have never experienced larger stresses than the existing overburden stresses. Theoretically, such soils should not undergo any consolidation until the consolidation stress exceeds the existing overburden stress. When the consolidation stress becomes larger than the overburden stress the soil mass will be consolidated for the first time, known as virgin compression (Figure 1-19). The difference between the theoretical expected stress-deformation curve and the actual laboratory behavior depicted in Figure 1-19 is due to the soil disturbance during sampling and handling.
2. *Overconsolidated.* Soils that have been subjected in the past to larger stresses than the existing overburden stresses. This excess stress is called *preconsolidation stress.* The most common cause of overconsolidation of natural clay deposits is the extremely high pressure applied at the earth's surface by glaciers.

The past stress history of a compressible soil sample is reflected in the results of a consolidation test and the preconsolidation stress can be estimated from the stress-deformation curve.

The ultimate consolidation settlement of a normally consolidated compressible soil layer that is subjected to a stress increase of magnitude $\Delta\sigma$ is

$$S = \frac{H}{1 + e_o} C_c \log\left(\frac{\overline{\sigma}_o + \Delta\sigma}{\overline{\sigma}_o}\right) \tag{1.29}$$

where   S   =   ultimate consolidation settlement
        e_o =   initial void ratio
        H   =   thickness of the compressible layer

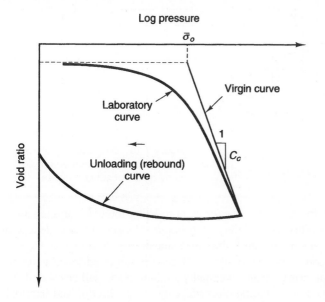

Figure 1-19 Typical
Stress-Deformation Curve

C_c = compression index
$\bar{\sigma}_o$ = effective overburden stress
$\Delta\sigma$ = consolidation stress resulting from application of a load

To compute the ultimate consolidation settlement for overconsolidated soils the preconsolidation stress must be considered.

The standard procedure for the consolidation test is described in the ASTM
Designation D 2435.

SOIL STRENGTH

The ability of soil to resist imposed loads is derived from its shear strength. Loads can be
imposed on the soil by foundations of various structures or by the soil's own weight. The
soil shearing strength is a very important parameter that is used in the design of shallow
and deep foundations, earth retaining structures, slope stability, and numerous other
aspects of civil and highway engineering.

Several tests have been developed for the determination of soil shearing strength
both in the laboratory and in the field. Laboratory tests are conducted on soil samples
recovered from the construction site, in either an undisturbed or a reconstructed state. Field
tests are performed in-situ on soils in their natural undisturbed state. In general, shear
strength depends on the type of soil, degree of saturation, void ratio, and stress history.

Shearing Strength of Cohesionless Soils

It is known from basic mechanics that a rigid body of weight, N, resting on a horizontal
surface and subjected to a horizontal force, H, will start moving when the resisting frictional
force, F, developed at the interface, is exceeded (Figure 1-20).

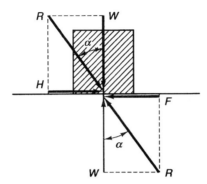

Figure 1-20 Development of Friction and Impending Motion of a Rigid Body on a Horizontal Surface

At the point of impending movement the resultant resisting force, R, forms an angle, α, with the vertical direction. This is called the *friction angle*. The tangent of angle α is the coefficient of friction. In granular soils the failure mechanism is quite similar to the principle of sliding friction. When a dry cohesionless soil mass is subjected to an increasing load, it will undergo shear deformation and eventually a failure plane will be created. The shearing strength is developed along the failure surface due to the friction and interlocking of grains to resist the movement.

The shear stress-shear strain behavior of the soil will depend on the initial density of the sample. Figure 1-21 shows typical shear stress-shear strain curves for two samples of the same soil having different initial densities (void ratios). It is observed that the two samples have the same ultimate strength but the denser sample exhibits a peak strength that is higher than the ultimate. This difference is due to the higher degree of particle interlocking in the denser sample.

The relationship of the normal stress to the shear stress is

$$\tau = \sigma \tan \phi \tag{1.30}$$

where τ =   shear stress
        σ =   normal stress
        π =   angle of internal friction

A plot of Equation 1.30 is shown in Figure 1-22. The equation plots a line that defines a failure envelope called the *Mohr-Coulomb failure envelope*. In terms of Mohr's circle analysis, a stress condition for which the Mohr's circle touches the failure envelope indicates an impending failure condition, whereas a Mohr's circle that is below the failure envelope line indicates that the full shearing strength has not been mobilized (safe condition). The angle of internal friction, ϕ, depends on the type of soil. Some representative values of ϕ are given in Table 1-13.

Moist cohesionless soils exhibit a strength relationship of the form

$$\tau = \sigma \tan \phi + c_{ap} \tag{1.31}$$

where c_{ap} =   apparent cohesion

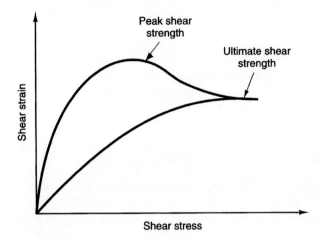

Figure 1-21 Typical Shear Stress-Shear Strain Curves for Soils with Different Void Ratios

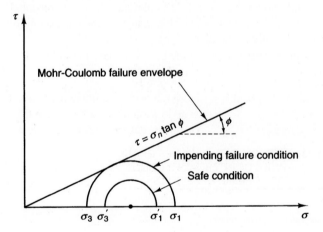

Figure 1-22 Mohr-Coulomb Envelope for Dry Cohesionless Soils

The apparent cohesion results from the capillary stresses in the soil, and it disappears when the soil dries out.

Shearing Strength of Cohesive Soils

The shearing strength of cohesive soil is mainly derived from its cohesion. The two major factors that influence the strength-deformation characteristics of cohesive soils are the stress history of the soil and the drainage characteristics during loading. Because most cohesive (clay) soils are saturated in their natural state, the effective stress is considered in the computation of the shearing strength. The Mohr-Coulomb envelope for saturated soils is defined as

$$\tau = \overline{\sigma} \tan \phi + c \qquad (1.32)$$

where $\bar{\sigma}$ =   effective vertical stress (total stress minus pore water pressure)
          c =   cohesion

A plot of Equation 1.32 is shown in Figure 1-23. The cohesion of inorganic silts and sands is zero. The cohesion of normally consolidated clays is approximately zero and for overconsolidated clays it is greater than zero. Some representative values of cohesion for various types of clays are given in Table 1-14.

Laboratory testing.   The three most frequently used laboratory tests for shear strength are direct shear, unconfined compression, and triaxial.

Direct Shear Test.   ASTM D 3080. In this test a soil sample is placed on a shear box arranged as shown in Figure 1-24. The shear box is split in two halves. The sample is compressed by two porous stones and a constant normal force is applied to the top of the sample via a loading plate. The bottom half of the box is kept stationary while the top half is subjected to a shear force until the sample fails. A shear stress-shear displacement graph similar to the one shown in Figure 1-21 is plotted. The test is repeated for different normal forces and the normal stresses are plotted against the corresponding shear stresses at failure to obtain the Mohr-Coulomb strength envelope.

One disadvantage of the direct shear test is that the soil is forced to fail along a predetermined plane, which may not necessarily be the weakest plane.

The direct shear test can be used to test both granular and cohesive soils.

TABLE 1-13   REPRESENTATIVE VALUES OF THE
ANGLE OF INTERNAL FRICTION
FOR COHESIONLESS SOILS

Soil Type	ϕ (degrees)
Gravel and sand-gravel mixtures	34-38
Sands (well graded)	32-35
Sands (poorly graded)	29-33
Sand-silt mixtures	26-32
Nonplastic silt	25-30

TABLE 1-14   REPRESENTATIVE VALUES
OF COHESION c FOR CLAYS

Clay Condition	c (psf)
Soft	250-500
Medium	500-1000
Stiff	1000-2000
Very stiff	2000-4000
Hard	> 4000

Soil Strength

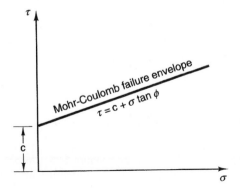

Figure 1-23 Mohr-Coulomb Envelope for Saturated Soils

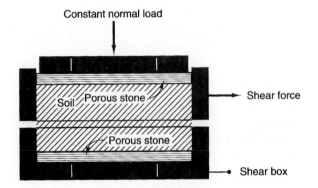

Figure 1-24 Direct Shear Test Arrangement

Unconfined Compression Test. (ASTM D 2166) This test is used to determine the unconfined compressive strength of cohesive soils. Soils can be treated in their original undisturbed state, remolded, or compacted. A cylindrical soil specimen is placed on a compression machine and is subjected to a strain-controlled axial load, Q, as shown in Figure 1-25. The load is applied at a relatively fast rate so that the pore water does not have adequate time to drain and therefore consolidate the sample. This condition is called *undrained unconsolidated*. The deformation of the sample is recorded and the loading is continued until the sample fails in shear or experiences excessive deformation. A stress-deformation curve is plotted and the ultimate strength (q_u) is recorded. Because the sample is not confined laterally, the total minor principal stress at failure is zero, and the total major principal stress is σ_1. The Mohr circle for this condition is plotted as shown in Figure 1-26. It follows that the undrained shearing strength is half of the ultimate strength:

$$c_u = \frac{q_u}{2} \tag{1.33}$$

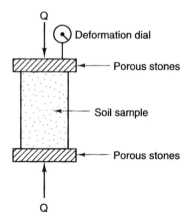

Figure 1-25   Unconfined
Compression Test Arrangement

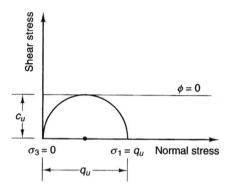

Figure 1-26   Mohr's Circle for
Unconfined Compression Test

The unconfined compression test has the advantage of providing quickly and inexpensive-
ly an approximate compressive strength for soils that possess enough cohesion. A disadvantage of this test is that the way the soil specimen is tested is not representative of the
in-situ conditions because of the absence of a confining pressure that represents the overburden stress.

   Triaxial Shear Test.   The triaxial shear test is the most reliable method to
determine the shear strength of soils. In this test a cylindrical soil sample is encased in a
thin rubber membrane and is placed in a chamber containing water. The water in the
chamber is pressurized to apply a confining pressure to the sample. Shear failure is
caused by an axial stress applied at the top of the sample. The axial stress is applied by
incremental application of axial loads or by application of a constant rate deformation
(strain-controlled test). The triaxial test apparatus is equipped with drainage lines and
devices to measure the pore pressure variations in the sample and sample deformation.
The stress condition that a sample is subjected to during triaxial testing is shown in
Figure 1-27. The axial stress applied to the sample is called the *deviator stress*. The following three standard triaxial tests are usually performed.

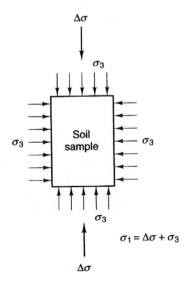

Figure 1-27 State of Stress in a Soil Sample Subjected to Triaxial Testing

1. Unconsolidated—Undrained Triaxial Compression Test (UU test). (ASTM D 2850). During this test, drainage of the pore water from the sample is not permitted. The axial load is applied quickly and the undrained shearing strength is obtained. The undrained shearing strength is used for design in cases where soil is subjected to large loads very suddenly.

2. Consolidated—Undrained Triaxial Compression Test (CU test). During this test, the sample is first consolidated by applying a confining stress and allowing drainage. After the sample has been consolidated the drainage lines are closed, and the axial stress is applied until the sample fails.

3. Consolidated—Drained Triaxial Compression Test (CD test). In this test the sample is first consolidated by applying a confining stress. After consolidation is complete, and with the drainage lines remaining open, the deviator stress is applied at a very slow rate so that complete dissipation of the pore water pressure is obtained. This test results in determination of the drained shearing strength of the soil and is representative of field situations where in-situ soils are subjected to slow rates of loading.

SOIL IMPROVEMENT

In many areas soil deposits are not suitable to provide support for various engineering structures. Such deposits may be composed of soft clays, organic soils and loose sands, or silts. In many cases the option of relocating the structure may not be feasible. The choices in such situations are limited to two. The unsuitable soils can be excavated and replaced by good quality soils or they can be improved in-situ.

In the first case soil is used as construction material to fill the excavation. When placed properly and compacted, a soil mass can provide adequate support for engineering structures. As a construction material soil is not limited to filling excavations, but also can be used to build various earthen structures such as highway embankments, earth dams and foundations for buildings, roadways, and parking areas. When soil is used as a construction material it is called *compacted fill* or *structural fill*.

In the second case the soil is stabilized in-situ. Stabilization is generally defined as the improvement of the engineering properties of soils in-situ by mechanical or chemical means or both. There are numerous soil stabilization methods and techniques currently in practice. The type of technique to be chosen for a particular site depends on a variety of factors, including the type of soil to be stabilized, the extent of required stabilization, the type of structure to be built, the availability of materials, and the environmental effects.

In this section the most commonly used methods of soil improvement are summarized.

Soil Compaction

Compaction is the densification of loose soils by mechanical means. Densification results in a decrease of void ratio and therefore in an increase of shearing strength and volume stability. The soil improvements realized are reduction or elimination of settlements, increase in the soil bearing capacity, and reduction or elimination of volume changes that can occur due to frost heave or swelling.

Soil compaction is divided into two categories on the basis of the extent of treatment: shallow compaction and deep compaction. Shallow compaction is performed on soft in-situ soils or on earth fills. Deep compaction is performed on in-situ soils at greater depths.

R. R. Proctor developed the fundamentals of soil compaction in the early 1930s. He established that the degree of compaction of a soil depends on the soil dry unit weight, moisture content, and grain size distribution and on the magnitude of the mechanical energy imparted on the soil. When water is added to a soil during compaction it acts as a lubricating agent, causing soil grains to slip on each other and attain more efficient packing (higher dry unit weight). As the water content is increased, a point is reached beyond which any addition of water to the soil mass will result in a decrease of the dry unit weight (dry density). This is because water will occupy void spaces otherwise occupied by solid grains. The moisture content at which the maximum dry unit weight is obtained for a given compactive energy is called the *optimum moisture content*. A representative dry unit weight-moisture content relationship is shown in Figure 1-28.

In the laboratory the dry unit weight-moisture content relationship can be established by a test called the *Proctor Compaction Test*. In this test the soil is compacted on a compaction mold having a volume of $1/30$ ft^3 by a hammer falling from a certain height.

Two types of tests are commonly used—the Standard Proctor Test and the Modified Proctor Test. In the Standard Proctor Test (ASTM D 698) the soil is compacted in three layers of equal height. Each layer is subjected to 25 blows by a hammer that weighs 5.5 lb (2.49 kg) and free falls 12 in. (305 mm). The test is repeated with the soil at different moisture contents until enough points are obtained to plot the dry unit weight-moisture content curve. With known moisture content and the weight of the compacted soil in the mold, the dry unit weight is computed by

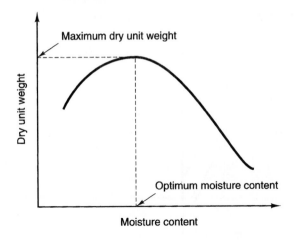

Figure 1-28   Variation of Dry Unit Weight of Soil with Moisture Content in a Compaction Test

$$\gamma_d = \frac{W}{\left(1 + \dfrac{w}{100}\right) V_m}$$ (1.34)

where W =   weight of compacted soil in the mold
        w =   moisture content in percentage
        V_m =   volume of compaction mold

The Modified Proctor Test (ASTM D 1557) is similar in execution to the standard test. In the modified test the soil is subjected to a higher impact energy. The soil is compacted in five equal layers by a hammer having a weight of 10 lb (4.54 kg) falling 18 in. (457 mm). For a given moisture content the maximum dry unit weight is obtained when there is no air contained in the pores of the soil. The maximum dry unit weight for the zero air-void condition is computed by

$$\gamma_{zav} = \frac{\gamma_w}{\dfrac{w}{100} + \dfrac{1}{G_s}}$$ (1.35)

where γ_w =   unit weight of water
        G_s =   specific gravity of soil solids
        w =   moisture content in percentage

The zero air-void curve is usually plotted on the same plot with the compaction curve and it serves as a guideline. The compaction curve should never cross the zero air-void curve. A compaction curve and zero air-void curve for a silty soil is shown in Figure 1-29.

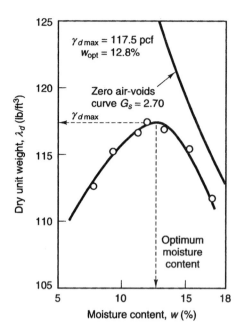

Figure 1-29 Standard Proctor Compaction Test Results for a Silty Soil

In addition to the moisture content, compaction is affected by the soil type and the compactive energy. The soil characteristics that most influence compaction are the grain size distribution, content of clay minerals, and specific gravity. Well-graded soils compact more efficiently than poorly graded soils. Figure 1-30 shows typical compaction curves of various soil types.

The effect of the compactive energy is shown in Figure 1-31, where the compaction curves obtained using the standard and modified Proctor tests are plotted for a silty sand. Since in the standard test less energy is imparted on the sample, the resulting maximum dry unit weight is less than the one obtained by the modified Proctor test. Besides impact compaction, other methods of compaction in the laboratory include kneading, vibration, and static compaction. For more information on these types of compaction the reader should refer to a soil mechanics book.

In the field, shallow compaction is performed by heavy compacting equipment called *rollers*. The type of roller used for a particular job depends on the type of soils to be compacted. The most commonly used rollers are

1. *Pneumatic rubber tire rollers.* These rollers are effective in compacting both cohesive and cohesionless soils. They are heavy equipment supported on rows of closely spaced rubber tires. They compact primarily by kneading, and the contact pressure under the tires can range from 80 to 100 lb/m² (585 to 690 kg/m²). Pneumatic rollers are usually fitted with ballast containers to vary the weight of the machine and thus the contact pressure.

2. *Sheep's-foot rollers.* These compactors consist of a steel drum manufactured with small projections. They compact soil by a combination of tamping and kneading. In some types of rollers the drum can be filled with water to provide greater weight. As

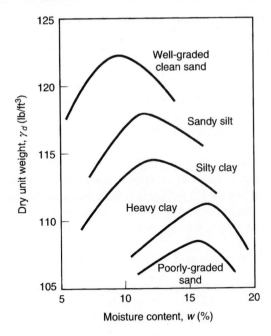

Figure 1-30 Typical Compaction Curves of Various Types of Soil

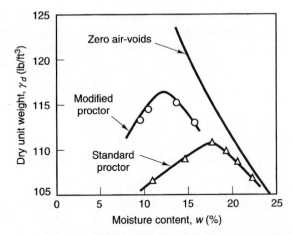

Figure 1-31 Standard and Modified Proctor Test Results for a Silty Sand

rolling takes place, the surfaces of the projections knead the soil. Constant pressures under the drum can range from 100 psi (700 kN/m²) to 600 psi (4200 kN/m²). Sheep's-foot rollers are suitable for compaction of silty and clayey materials.

3. *Vibratory rollers.* These rollers consist of vibratory drums and are manufactured as self-propelled units or tow-back units. Vibration in the drum is generated by a motor that drives a set of eccentric weights located inside the drum. Both smooth surface and sheep's foot drums are available. Vibratory rollers are most efficient in compacting granular soils.

Fill placement and compaction generally requires procedures that include laboratory test-ing and field inspection and testing. The dry unit weight-moisture content relationship of the fill soil is initially determined in the laboratory. Either the standard or modified Proctor test is used according to the job specifications. The fill is usually placed in layers (lifts) having a loose thickness of 12 in. The job specification in most cases requires that the soil be compacted to a dry unit weight higher than 90 to 95 percent of the maximum dry unit weight as determined by the laboratory compaction curves. Each lift is com-pacted by several passes of a compaction roller. Field tests are then performed to deter-mine the in-situ (compacted) soil density. If the in-situ dry unit weight is not within the specified percentage of the maximum dry unit weight, the soil is compacted more. The number of roller passes required to result in acceptable compaction is determined. As long as the fill moisture content and the lift thickness remain the same the compaction is controlled by making sure that the roller makes a sufficient number of passes over each lift. Periodically, in-situ dry unit weight tests are performed to verify that the in-situ fill dry weight is within specifications.

There are three methods commonly used to determine the in-situ soil density.

1. *Sand Cone Method* (ASTM D 1556). This test consists of a glass or plastic jar with a metal cone attached to its top. The jar is filled with dry Ottawa sand of very uni-form grain size with known dry unit weight. The weight of the sand in the jar and the sand that is required to fill the cone is measured. In the field a rectangular metal plate with a circular hole having diameter equal to the diameter of the cone is used as a guide to excavate a small hole in the soil to be tested. All of the loose soil is recovered from the excavated hole and its weight and moisture content are deter-mined. The sand cone assembly is then placed on the metal plate as shown in Figure 1-32. The sand is allowed to "flow" into the hole. The weight of sand required to fill the hole is determined and is divided by the dry unit weight of the sand to give the volume of the hole. The in-situ dry unit weight of the soil is then determined by

$$\gamma_d = \frac{\text{dry weight of soil excavated from hole}}{\text{volume of hole}} \qquad (1.36)$$

The test procedure is described in detail in the ASTM Standard D 1556.

2. *Rubber Balloon Method* (ASTM D 2167). This method is based on the same princi-ple as the sand cone method. In this test the volume of the hole is determined by the volume of water, contained in a rubber balloon, that is required to fill the hole.

3. *Nuclear Density Method*. Nuclear density meters have been developed for the direct determination of the compacted dry unit weight of soils. By detecting emitted gamma-rays passing through the soil mass they measure the density and moisture content of the soil.

Deep Compaction

The most popular method of massive in-situ soil densification is *vibroflotation*. This tech-nique involves the use of a device called vibrofloat. The vibrofloat is a metal cylinder, usu-ally 6 ft (183 cm) long and 16 in. (40.6 cm) in diameter, which is suspended from a crane.

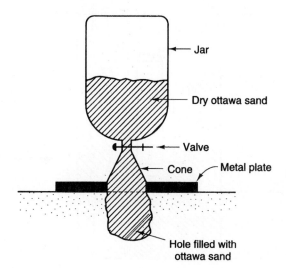

Jar

Dry ottawa sand

Valve

Cone — Metal plate

Hole filled with
ottawa sand

Figure 1-32 Sand Cone Test
Arrangement

An eccentric set of weights inside the cylinder creates the vibration, which is used to propel the vibrofloat into the soil. Jet ports are located at the top and bottom of the vibrofloat from which water under pressure of 60 psi (444 kPa) can be ejected. The vibrofloat is propelled into the loose soil by the action of the vibration and by the water jet action at its lower end. Once the desired depth is reached, the lower water jets are shut off, the upper jets are turned on, and the vibrofloat is slowly retracted. A cavity is formed around the vibrofloat due to vibration as the surrounding soil is densifying. The cavity is filled with sand shoveled from the ground surface. The radial extent of the densification can range from 3 ft (1 m) to 10 ft (3 m). Figure 1-33 shows the sequence of vibroflotation. This method of deep compaction is effective in soils having less than 20 percent fines and no more than 3 percent active clays. Other methods of deep compaction are also available.

Dynamic consolidation is a method by which soils are densified by heavy tamping. This is obtained by dropping heavy weights on the soil. Weights varying from 5 to 40 tons are dropped by a crane from heights ranging from 20 to 100 ft (6 to 30 m) at a predetermined pattern to densify the soil.

Piles are also used to densify soils. Piles are driven into the soil, and densification is obtained by displacement and vibration.

Soil Stabilization

Stabilization is the process by which a stabilizing agent (*admixture*) is added to the natural soil deposit to improve its engineering behavior. One or more soil properties can be improved by admixture stabilization. There are several types of admixtures used for soil stabilization purposes. One of the most common stabilization methods used in highway construction is the blending of coarse-grained materials with fine-grained soils (called binders) to produce a soil mixture that possesses both cohesion and friction. The appropriate blending of soils of different grain size distribution can produce a soil having much better engineering properties.

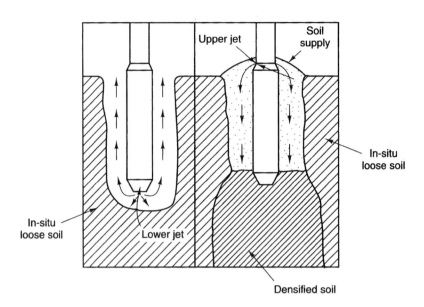

Figure 1-33 Principle of Vibroflotation: (a) Advancement of the Vibrofloat;
(b) Retraction of the Vibrofloat

Portland cement is also commonly used as a stabilizing agent. An increase in soil strength occurs from cementation of the soil particles upon hydration of the cement. The soil is mixed with the cement in-situ and the blend is compacted by rollers. The resulting mixture is called soil cement and usually contains 7 to 15 percent cement by weight. Similar stabilization is obtained by using asphalt cement as the admixture agent. Both portland and bituminous cement stabilization have found wide applications in roadway and parking lot construction.

Lime and calcium chloride are used as stabilizing agents for improvement of fine-grained soils. During lime stabilization the soil is mixed with lime and, after curing for a few days, is remixed and compacted. This results in improvement of the soil strength and reduction of the swelling characteristics of the clay.

Several types of chemicals have been used as cementing agents for soils. The most frequently used agents are silicates. Grouting is the most common method of chemical stabilization. Grouting refers to injection of a liquid stabilizing agent into soil. Pressurized grout is forced to flow through a natural soil deposit and fill the pores. The grout solidifies, bonding the soil grains. An advantage of grouting is that the soil is stabilized in its natural state without being disturbed. A limitation of grouting is that it is effective only in soils with relatively high permeability. Often the purpose of chemical grouting is to minimize the soil permeability and close fractures in rock formations.

A special method of soil stabilization is earth reinforcement. Fabrics composed of synthetic fibers (geotextiles) or metal strips are embedded into soil to improve its strength. Several types of fabrics and reinforcing strips are currently available. They have been used extensively to stabilize slopes and to build embankments, dams, and earth retaining walls.

P R O B L E M S

1.1 The moist unit weight for a sandy soil is 18.9 kN/m³. If the moisture content is
13.5 percent and the specific gravity of soil solids is 2.69, find the following:
  a. Dry unit weight.
  b. Void ratio.
  c. Porosity.
  d. Degree of saturation.

1.2. A laboratory study showed that a soil has the following physical properties:

Sieve No.	Soil, Percentage Passing
4	100
10	100
20	98.1
40	96.5
60	95
100	90.4
200	85.5

moisture content	=	31 percent
moist unit weight	=	20 kN/m³
specific gravity of the solids	=	2.75
plastic limit	=	48.65 percent
liquid limit	=	24.21 percent

  a. Determine the plasticity and liquidity indices and properties.
  b. Classify the soil using the USC system.

1.3. The results of grain size analysis for a soil sample are as follows:

Sieve No.	Soil, Percentage Finer
4	100
10	99.44
20	86.07
40	36.07
60	20.30
100	16.70
200	12.80

  a. Plot the gradation curve.
  b. Calculate C_u, and C_c.
  c. Classify the soil, usiong the USC and AASHTO tables.

1.4. Which phenomenon is responsible for capillary rise in soils?

1.5. Daytona Beach in Florida is often used as a car race track. What phenomenon
makes the sandy beach so compact as to support moving vehicles?

1.6. What property is responsible for clay soils forming chunks after they dry out?

1.7. The following data are collected from a constant head permeability test:

sample diameter	=	6.5 cm
sample length	=	15.4 cm
total saturated weight	=	1000 g

test duration = 5 min
average volume of water collected = 775 cm³
constant head = 46 cm
compute the coefficient of permeability

1.8. Explain the difference between elastic and consolidation settlement in soils.

1.9. Explain why well-graded soils compact more efficiently than poorly graded soils.

1.10. Laboratory results from direct shear tests on 6 samples of silty soil are

No.	Normal Stress σ (psi)	Max Shear Stress τ_{max} (psi)
1	3.27	5.54
2	7.25	7.68
3	9.1	8.81
4	11.51	10.09
5	14.64	11.8
6	16.2	12.8

a. Plot the Mohr-Coulomb envelope.

b. Determine the angle of internal friction ϕ and the cohesion c.

1.11. Explain the soil properties responsible for shearing strength.

REFERENCES

American Society of Civil Engineers, *"Soil Improvement, History Capabilities and Outlook,"* *Report by the Committee on Placement and Improvement of Soils,* Geotechnical Engineering Division, ASCE, 1978.

American Society for Testing and Materials, *Annual Book of Standards,* Volume 04.08, Soil and Rock; Building Stones, 1986.

BOWLES, J. E., *Physical and Geotechnical Properties of Soils,* 2nd ed., McGraw-Hill Book Company, New York, 1984.

DAS, B. M., *Advanced Soil Mechanics,* McGraw-Hill Book Company, New York, 1983.

DAS, B. M., *Principles of Geotechnical Engineering,* Prindle, Webber, and Schmidt, A Division of Waldsworth, Boston, 1985.

FREEZE, R. A. and CHERRY, F. A., *Groundwater,* Prentice Hall, Inc., Englewood Cliffs, N.J., 1979.

HOLTZ, R. D. and KOVACS, W. D., *An Introduction to Geotechnical Engineering,* Prentice Hall, Inc., Englewood Cliffs, N.J., 1981.

JUMIKIS, A. R., *Soil Mechanics,* D. Van Nostrand Co., Inc., Princeton, N.J., 1962.

MCCARTHY, D. F., *Essentials of Soil Mechanics and Foundations,* 2nd ed., Reston Publishing Co., 1982.

SOWERS, G. F., *Soil Mechanics and Foundations: Geotechnical Engineering,* 4th ed., Macmillan Publishing Co., Inc., New York, 1979.

TERZAGHI, K. and PECK, R. B., *Soil Mechanics in Engineering Practice,* 2nd ed., John Wiley & Sons, Inc., New York, 1967.

TODD, D. K., *Groundwater Hydrology,* 2nd ed., John Wiley & Sons, Inc., New York, 1980.

WINTERKORN, H. F., and FANG, H.Y., *Foundation Engineering Handbook,* Van Nostrand Reinhold Co., New York, 1978.

2 MINERAL AGGREGATES

Aggregates play a very important role in the design and construction of highway and airport pavements, as the underlying material on which the pavement rests. They are also important as an ingredient in rigid (concrete) and flexible (asphalt) pavements or structures.

Aggregates are the most important factor in the cost of pavement construction, accounting for more than 30 percent of the total cost. Aggregates make up approximately 65 to 85 percent of a concrete structure and 92 to 96 percent of an asphalt structure.

AGGREGATE TYPES AND PROCESSING

Aggregate is a combination of sand, gravel, crushed stone, slag, or other material of mineral composition, used in combination with a binding medium to form such materials as bituminous and portland cement concrete, macadam, mastic, mortar, and plaster, or alone, as in railroad ballast, filter beds, and various manufacturing processes such as fluxing.

Aggregates may be further classified as natural or manufactured. *Natural aggregates* are taken from natural deposits without change in their nature during production, with the exception of crushing, sizing, grading, or washing. In this group, crushed stone, gravel, and sand are the most common, although pumice, shells, iron ore, and limerock may also be included. *Manufactured aggregates* include blast furnace slag, clay, shale, and lightweight aggregates.

A further classification would be to divide the aggregate into two types: fine and coarse. According to ASTM C125 (Concrete and Concrete Aggregates), *fine aggregate* is defined as aggregate passing a $\frac{3}{8}$-in. (9.5-mm) sieve and almost entirely passing a No. 4 (4.75-mm) sieve and predominantly retained on the No. 200 (75-μm) sieve or that portion of an aggregate passing the No. 4 (4.75-mm) sieve and retained on the No. 200 (75-μm)

51

sieve. *Coarse aggregate* is defined as aggregate predominantly retained on the No. 4 (4.75-mm) sieve or that portion of an aggregate retained on the No. 4 (4.75-mm) sieve. These definitions are for concrete aggregates; for bituminous concrete mixtures the dividing line between fine and coarse aggregate is the No. 8 (9.5-mm) or the No. 10 (11.8-mm) sieve.

Processing

The main fundamental rule of good aggregate processing is to obtain aggregates of the highest quality at the least cost. Each process is completed with these objectives in mind. The processes include, but are not limited to, excavation, transportation, washing, crushing, and sizing. Processing begins with excavation and quarrying of the material and ends upon being stockpiled or delivered to the site.

In the excavation process the overburden is removed (if applicable), as its presence in aggregate in the form of silt or clay cannot be tolerated. The removal of the overburden is carried out through the use of power shovels, draglines, or scrapers. Overburden removal is usually considered only if there is depth of 50 ft. (15.24 m) or more. If the overburden is light, it will wash out in the processing of the aggregate.

After the aggregate is excavated, it is transported by rail, truck, or conveyor belt to the processing plant.

At the processing plant, unacceptable (deleterious) materials are removed. A deleterious material is a material that may prove harmful to the final product for which the aggregate is to be used. One method of removing deleterious materials (clay, mud, leaves, etc.) is to wash the raw material. Sometimes conveyor belts are used to haul the aggregate through flumes that are flushed with water.

The next process is to reduce the size of the stone or gravel. In this process many types of crushers are used. The oldest is the jaw crusher, which consists of a fixed jaw and a reciprocating jaw, which are suited for hard rocks of all types. Newer crushers have a higher capacity than the jaw crusher, but this is the only disadvantage of the jaw crusher. The usual practice is to reduce the size of the rock at a ratio of 1:6 or less.

For sizing, vibratory sieves are used for coarse material and hydraulic classification devices for fine material. The screens vary in design, capacity, and efficiency. In the screening process about 70 percent of the material will pass through the screen, so that the goals of high efficiency and capacity are met. In most cases some removal of oversize particles, called scrapping, will take place.

Particles

The screened aggregate particles may be rounded or angular. Gravel consists of naturally rounded particles resulting from disintegration and abrasion of rock or processing of weakly bound conglomerate. Sand consists of rock particles that have been disintegrated naturally; the grains are generally angular but have been subjected to weathering. Sand is fine aggregate resulting from natural disintegration and abrasion of rock or processing of completely friable sandstone. Crushed stone is a product of the artificial crushing of rocks, boulders, or large cobblestones, substantially all faces of which result from the crushing operation. Stone sand is a finely crushed rock corresponding to sand in size. Gravel and crushed stone are considered to be coarse aggregate.

Aggregate particles vary considerably in texture. Gravel particles have a very smooth texture, whereas crushed stone has a rough surface texture. Aggregate particles have considerable variations in porosity. Usually, crushed stone and gravel have low porosity.

Gradation and Aggregate Building

The gradual gradation in size from coarse to fine is a key property of aggregates. Aggregate gradation affects the workability of portland cement concrete mixes and the stability and durability of bituminous concrete mixes, as well as the stability, drainage, and frost resistance of base courses. Therefore, aggregates should be tailored to their proposed usage: in concrete or bituminous or as base courses. For specific examples, refer to succeeding chapters.

Aggregates may be dense, gap graded, uniform, well graded, or open graded. The terms "dense" and "well graded" are essentially the same, as are "gap," "uniform," and "open graded." Figure 2-1 illustrates the five types of gradations. Other methods of expressing size distribution have been developed. One such method is the Fuller-Thompson method. An empirical formula that can be used to determine the gradation:

$$P = 100\left(\frac{d}{D}\right)^N$$

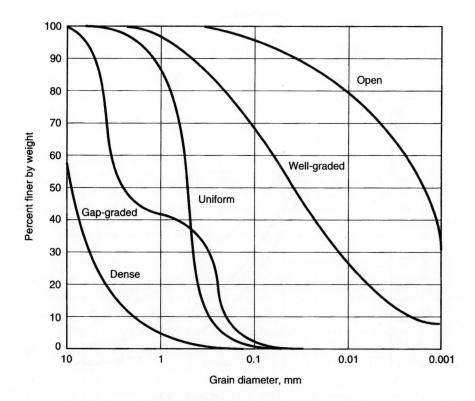

Figure 2-1 Five Types of Gradation

where P = percentage finer than the sieve
          d = sieve size in question
          D = maximum size aggregate to be used (top size)
          N = coefficient of adjustment, which adjusts the curve for fineness or
                  coarseness

   Figure 2-2 shows a typical plot utilizing this formula at values of N equal to 0.3, 0.5, and 0.7. The maximum size of the aggregate is 1 in. (2.54 cm). Notice that a fine gradation is represented by $N = 0.3$ and a coarse-graded material is represented by $N = 0.7$. Therefore, a dense material would be $N = 0.5$, as the Fuller-Thompson experiments indicate. Originally, the work of Fuller and Thompson included the cement in a cement-aggregate mixture, but the relationship (equation) has become known as *Fuller's maximum density curve*, regardless of the constituents. The maximum density curve is only an approximation, because the actual gradation may depend on the nature of the material. However, if employed properly, it can be a valuable tool as a point of beginning in designing aggregate blends for maximum density.
   Other research to determine maximum density of aggregates has been conducted. One theory states that if aggregates were screened into three sizes, coarse, medium, and fine, the combination with the greatest density would be a mixture of approximately two parts coarse aggregate and one part fine aggregate, with no medium fraction. Further research showed that high densities could be obtained by using equal portions of each size.

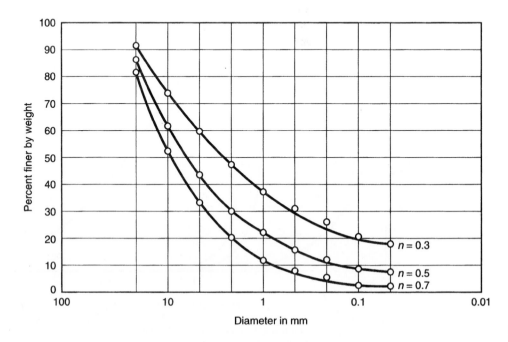

Figure 2-2 Fuller-Thompson Curves

Manufactured Aggregates

Manufactured aggregates are man-made aggregates. One such aggregate, air-cooled blast-furnace slag, has become one of the most common manufactured aggregates to be used in highway construction. ASTM C125 (Concrete and Concrete Aggregates) defines *air-cooled blast-furnace slag* as the material resulting from solidification of molten blast-furnace slag under atmospheric conditions; subsequent cooling may be accelerated by application of water to the solidified surface. *Blast-furnace slag* is defined as the non-metallic product, consisting essentially of silicates and alunminosilicates of calcium and other bases, that is developed in a molten condition simultaneously with iron in a blast furnace. It is not uncommon to find this manufactured aggregate used in industrial areas, especially if steel mills are present, as in the states of Alabama, Indiana, Maryland, New York, Ohio, and Pennsylvania. Slag is lighter in weight than are natural aggregates. Slags have a specific gravity between 2.0 and 2.5, whereas natural aggregates have a specific gravity between 2.3 and 3.2.

Although the definition of manufactured aggregate given here is sufficient for most uses, others further classify manufactured aggregates as those man-made aggregates that have resulted as a direct product rather than as a by-product. In this sense, slag would not be a manufactured aggregate but an artificial aggregate. Thus, an artificial aggregate is a man-made aggregate that results as a by-product from the manufacturing of some other product. Lightweight aggregates are considered artificial aggregates. Lightweight aggregates may be cinders, clay, shale, shells, or slag. These types of aggregates are used to produce lightweight concrete in a structure where dead weight is important.

AGGREGATE AS A BASE-COURSE MATERIAL

The importance of aggregates in a base course lying between the compacted subgrade and the portland cement concrete slab or the bituminous concrete slab cannot be overstated. An improper design of a base course can lead to structural failure of the slab. Base courses may also include several types of undercourses, such as subbases, filter beds, and leveling courses. Base courses serve a variety of purposes depending upon construction practices and the environment. They serve to provide structural capacity to bituminous concrete slabs, drainage for portland cement concrete slabs, and low susceptibility to frost. As previously indicated, gradation is the key factor in the success of aggregates as a base course.

Gradation

The gradation of the aggregate can affect structural capacity, drainage, and frost susceptibility. Thus the control over gradation is a principal concern for most engineers. Related to this control is the hardness of the aggregate as soft, weak, or friable articles undergo aggregate degradation (a process whereby fines are generated by aggregate breakdown during the placement or the use of the material).

According to Krebs and Walker, three general types of aggregate mixtures can be recognized with respect to fines:

   1. Aggregates only, no fines.

2. Fines just filling the voids of aggregate fraction.

3. Fines overfilling the voids of aggregate fraction.

In the first case the aggregate derives its strength from grain-to-grain contact of the aggregate particles. In this situation the base-course material would be unstable unless it was confined, but it does provide excellent drainage and is completely non-frost-susceptible.

In the second case, also, the aggregate derives its strength from grain-to-grain contact of the aggregate particles. However, in this situation the base-course material would be stable even if unconfined, because of the inherent cohesive properties of the fine content that fills the voids between the aggregate particles. Further, the drainage is adequate and can be non-frost-susceptible.

In the final case, the strength is derived from the grain-to-grain contact of the fines rather than of the aggregate particles; thus, a strength reduction occurs. The drainage characteristics of this base course would be poor and it would be very frost-susceptible.

In most highway construction techniques the base course falls between types 1 and 2 for best practice. Thus, a base-course material should have sufficient fines to just fill the voids within the aggregate particles, with a gradation curve approaching that of Fuller's maximum density curve.

Particle Strength, Shape, and Texture

To resist the stress of repeated loads and to avoid aggregate degradation, base-course aggregates must exhibit strength and toughness to function for their intended purpose. Open-graded base courses are more susceptible to degradation than dense-graded material. In addition, the hardness of the aggregate adds strength to the base course. Sandstones and shale generally degrade easily, although the individual particles may be strong.

Table 2-1 gives the average value for the physical properties of the principal types of rocks: igneous, sedimentary, and metamorphic. In each group the specific gravity, hardness, and toughness are given together with other properties. The purpose of the table is to give the reader an indication of which materials would serve best as a base-course material.

Of equal importance to base-course material other than strength is particle shape and toughness. Table 2-1 also shows the toughness of this rock. These two properties are very important to base-course materials. Angular, nearly equidimensional particles rough in texture are extremely preferred for base-course material. The angularity contributes to aggregate interlocking, and a rough surface texture prevents movement of one particle upon another. Rounded particles tend to roll over one another as they do not interlock with one another. Smooth-textured aggregate particles allow slippage when they are in contact with each other. Thus, rounded aggregate with a smooth surface texture is the least preferred for base-course usage. Table 2-2 summarizes the engineering properties of various aggregate types.

AGGREGATES FOR PORTLAND CEMENT CONCRETE

Aggregate properties for portland cement concrete are in many cases different from aggregates used for base courses or for use in bituminous concrete. Aggregate gradation becomes a key factor as it controls the workability of the plastic concrete. Further, the aggregates used in portland cement concrete are a blend of fine and coarse aggregate to achieve an economical mix. For coarse aggregate the same definitions apply as previous-

TABLE 2-1   AVERAGE VALUES FOR PHYSICAL PROPERTIES OF THE PRINCIPAL TYPES OF ROCKS[a]

Type of Rock	Bulk Specific Gravity	Absorption[b] (%)	Loss by Abrasion %		Hardness[e]	Toughness[f]
			Deval[c]	Los Angeles[d]		
Igneous						
Granite	2.65	0.3	4.3	38	18	9
Syenite	2.74	0.4	4.1	24	18	14
Diorite	2.92	0.3	3.1	—	18	15
Gabbro	2.96	0.3	3.0	18	18	14
Peridotite	3.31	0.3	4.1	—	15	9
Felsite	2.66	0.8	3.8	18	18	17
Basalt	2.86	0.5	3.1	14	17	19
Diabase	2.96	0.3	2.6	18	18	20
Sedimentary						
Limestone	2.66	0.9	5.7	26	14	8
Dolomite	2.70	1.1	5.5	25	14	9
Shale	1.85-2.5	—	—	—	—	—
Sandstone	2.54	1.8	7.0	38	15	11
Chert	2.50	1.6	8.4	26	19	12
Conglomerate	2.68	1.2	10.0	—	16	8
Breccia	2.59	1.8	6.4	—	17	11
Metamorphic						
Gneiss	2.74	0.3	5.9	45	18	9
Schist	2.85	0.4	5.5	38	17	12
Amphibolite	3.02	0.4	3.9	35	16	14
Slate	2.74	0.5	4.7	20	15	18
Quartzite	2.69	0.3	3.3	28	19	16
Marble	2.63	0.2	6.3	47	13	6
Serpentinite	2.62	0.9	6.3	19	15	14

[a]From Bureau of Public Roads manual, *The Identification of Rock Types* (1950).
[b]After immersion in water at atmospheric temperature and pressure.
[c]AASHTO T3 or ASTM D289.
[d]AASHTO T96 or ASTM C131.
[e]Dorry hardness test, *U.S. Dept. Agr. Bull. 949.*
[f]AASHTO T5 or ASTM D3a.

ly given. For fine aggregate the same definitions as previously given apply, but we introduce a new term, fineness modulus. The *fineness modulus,* which denotes the relative fineness of the sand, is defined as one-hundredth of the sum of the cumulative percentages held on the standard sieves in a sieve test of sand. Six sieves are used in the determination, Nos. 4, 8, 16, 30, 50, and 100. The smaller the value of the fineness modulus, the finer the sand. The fineness modulus for a good sand should range between 2.25 and 3.25.

In portland cement concrete the strength of the aggregate is not as important as it would be if it was used as a base-course material. The aggregate in portland cement concrete acts as a filler so that not as much sand, cement, and water are needed.

The bond between the aggregate and the cementing materials of portland cement concrete is influenced by surface texture. In most cases the bond strength is improved by an aggregate with a rough surface texture. The degree of texture need not be great, but some texture is desired.

TABLE 2–2   SUMMARY OF ENGINEERING PROPERTIES OF ROCKS[a]

Type of Rock	Mechanical Strength	Durability	Chemical Stability	Surface Characteristics	Presence of Undesirable Material	Crushed Shape
Igneous						
Granite, syenite, diorite	Good	Good	Good	Good	Possible	Good
Felsite	Good	Good	Questionable	Fair	Possible	Fair
Basalt, diabase, gabbro	Good	Good	Good	Good	Seldom	Fair
Peridotite	Good	Fair	Questionable	Good	Possible	Good
Sedimentary						
Limestone, dolomite	Fair	Fair	Good	Good	Possible	Good
Sandstone	Fair	Fair	Good	Good	Seldom	Good
Chert	Good	Poor	Poor	Fair	Likely	Poor
Conglomerate, breccia	Fair	Fair	Good	Good	Seldom	Fair
Shale	Poor	Poor		Good	Possible	Fair
Metamorphic						
Gneiss, schist	Good	Good	Good	Good	Seldom	Good
Quartzite	Good	Good	Good	Good	Seldom	Fair
Marble	Fair	Good	Good	Good	Possible	Good
Serpentinite	Fair	Fair	Good	Fair	Possible	Fair
Amphibolite	Good	Good	Good	Good	Seldom	Fair
Slate	Good	Good	Good	Fair	Seldom	Poor

[a]From Bureau of Public Roads manual, *The Identification of Rock Types* (1950).

In most portland cement concrete mixtures, the effect of surface texture is minimal. If one was to compare two types of aggregates, a smooth-surfaced gravel and a rough-surfaced crushed stone, to be utilized in two separate concrete mixes of equal cement factor and consistency, the smooth-surfaced gravel mix would require less water. The strength achieved from using the smooth-surfaced gravel would be the same as the rough-surfaced crushed stone. Thus, in the smooth-surfaced gravel mix, less water resulted in a lower water-cement ratio increasing its strength. The mix containing rough-surfaced crushed stone required more water, increasing the water-cement ratio, resulting in a weaker concrete. Thus, the two effects balanced out and the resulting strengths were equal.

Particle shape influences the workability of the concrete mix, but the interlocking characteristic needed for base-course material is not important here. Angular aggregates require more mortar to fill voids and separate aggregate particles for workability, which results in a higher water-cement ratio for a given cement factor and consistency.

Freezing and Thawing

Probably the single most important reason why concrete fails is due to the effects of freezing and thawing. Other reasons are deleterious materials and chemical reactions. The freezing and thawing phenomenon is discussed in Chapter 7, but some detail is given here.

The freezing and thawing of concrete is much more important at present than it was several years ago with respect to aggregates. Sources of aggregates of known satisfactory performance are being depleted through one means or another. This fact can only lead to three major possibilities.

1. Production of synthetic aggregate.
2. Beneficiation of unsuitable material.
3. Use of manufactured and waste materials as supplements and replacements for conventional aggregates in construction.

Therefore, the need for an understanding of the influence of aggregates on the resistance of concrete to freezing and thawing is becoming increasingly important. If the aggregate is unsound, three major classifications of distress result.

1. Pitting and popouts.
2. D-line cracking deterioration.
3. Map cracking.

Pitting, the disintegration of weak, friable pieces of aggregates due to frost action, usually occurs by the gradual deterioration of deleterious particles near the surface. *Popouts* are caused by the rapid disruption of harder but saturated pieces of rocks or aggregates. Pitting and popouts themselves do not affect the structure in terms of strength, but they allow water to enter the structure through accesses provided by the pitting and popout and ultimately result in deterioration.

D-line cracking is caused by the change in volume of coarse-aggregate particles breaking the bond with the mortar matrix. It is aided by the entrance of water. Ordinarily, the cracks are filled with a calcareous deposit. D-line cracking appears first along transverse joints, then later along longitudinal joints.

Map cracking is a form of disintegration in which random cracks develop in a well-distributed pattern over an entire surface instead of concentrated along joints or free edges as in D-line cracking. D-line cracking and map cracking can aid structural failure of pavements by causing blowups.

As one views the three major classifications of distress, the question arises as to what materials cause these harmful effects. Several aggregates cause deterioration of concrete, including chert, limestone, and shale. But what is the specific nature of these materials that cause deterioration? Chert is a fine-grained hard rock that will usually pass conventional specifications. However, chert is sufficiently weathered as to be deleterious to concrete exposed to freezing and thawing. Shale and argillaceous limestone cause considerable damage to concrete exposed to severe weathering.

The different responses of aggregates to freezing when saturated depend on the pore characteristics of the aggregate and the cement paste. Saturated aggregates of low porosity may accommodate pore-water freezing by simple elastic expansion. Saturated aggregates of moderate to high porosity may fail because the particle dimension exceeds a certain critical size or may cause failure in the paste immediately adjacent to the aggregate particles because of aggregate pore-water displacement. As for the shale and argillaceous limestones, these materials have small interconnected voids and are capable of attaining a high degree of saturation.

The disruption of concrete by aggregates is a result of hydraulic pressures. The hydraulic pressure is a result of the degree of saturation (proportional to total void space filled with water) and the permeability and size of the aggregate particles. On freezing, water expands 9 percent, and if the degree of saturation of the aggregate is critical, 91.7 percent water will be expelled into the paste surrounding the aggregate particles, and potentially destructive hydraulic pressure may develop there also. So the properties of paste, its permeability, air content, and porosity are also involved in the problem. Three additional factors, composition, texture, and structure, also play important roles in the freezing and thawing of concrete. As shown by Powers, maximum hydraulic pressure is greatest when the aggregate has a low coefficient of permeability. At a given porosity, the smaller the mean size of the capillaries, the lower the coefficient of permeability and the higher the resistance to flow at a given ratio of efflux. Thus, smallness of capillaries is one characteristic of an unfavorable capillary system. In short, low water capacity and high permeability (i.e., low porosity and coarse texture) is a favorable combination, resulting in a relatively large critical size. Critical size varies from rock to rock (because it is dependent on capillary size), so no value is given here. The limit of water capacity for elastic accommodations is 0.3 percent. It should be noted that a wide range of critical sizes exist at a given strength.

The critical saturation point is 91.7 percent, as shown by Powers. One would immediately try to keep the water below the 91.7 percent level, but it must be remembered that water is not uniformly distributed in an aggregate particle at the time of freezing. For an aggregate that is not uniformly distributed, the critical saturation may be something less. Critical diameter is usually considered 4 μm (0.004 mm). The surface texture, which is directly related to the porosity of the material, is considered important in aggregates as well as the permeability. A rough, porous texture will produce high loading strengths and a favorable critical size. Irregular particles need more mortar to fill the void space, and they also have a greater demand for water—because more mortar, more sand; more fines, more water. Another aspect to remember is that some aggregates expand when they come in contact with water, which may result in deterioration.

If the situation is such that the aggregate might become critically saturated, the paste is sure to become saturated, and when it does it can quickly become destroyed by freezing, regardless of the kind of cement of which it is made or the extent of curing.

When a specimen of saturated cement paste becomes frozen, it, too, has a critical thickness of about three thousandths of an inch. Pastes are highly porous and have an extremely fine texture. The structure of the paste is such that 72 percent is solid and 28 percent accessible to evaporated water. The cavities in the cement paste are seldom completely saturated at the time of freezing, and therefore hydraulic pressure does not necessarily develop at the time of freezing. Nevertheless, it is possible for the paste to become dilated.

Concrete freezes from the outside to the inside. When freezing starts, the void spaces contain water. If the water is above the critical saturation point of 91.7 percent, the capillaries will freeze and expel water to the surrounding area—probably into the cement paste because of its small capillaries. Because the paste and aggregate cannot accommodate this expansion of 9 percent, the hydraulic pressure is so great that cracking or popouts result. Once an opening is formed in the concrete, water is allowed to enter and deterioration is underway.

The most acceptable explanation of why concrete may fail during freezing and thawing was given by Powers. Concrete contains many air-filled cavities, consisting of entrained air bubbles, accessible pores in the aggregate particles, and thin fissures under the aggregate particles. All empty cavities of the type mentioned, especially air bubbles, are difficult to fill with water. They cannot be filled by capillary action because a liquid cannot flow from a small capillary to fill a larger one. However, pressures caused by freezing water are more than sufficient to free unfrozen water into such spaces. The resistance to movement of water must be the primary source of pressures, for practically all concrete contains enough air-filled space to accommodate the water and its volume of expansion when frozen.

It is therefore apparent that test procedures must be used or developed to determine the soundness of the aggregates used in the making of concrete.

AGGREGATES FOR BITUMINOUS MIXTURES

The influence of aggregates on the properties and performance of bituminous mixes is great. The ideal aggregate for a bituminous mix would have proper gradation and size, be strong and tough, and be angular in particle shape. Other properties would consist of low porosity, surfaces that are free of dirt, rough texture, and hydrophobic nature. Tables 2-1 and 2-2, which illustrate aggregate porosity, abrasion, hardness, toughness, strength, durability, surface characteristics, shape, and so on, also apply to bituminous mixes. The aggregate gradation and size, strength, toughness, and shape are important considerations for stability of the structure. The porosity and the surface characteristics are important to the aggregate-bitumen interaction. The asphalt cement or its product must adhere to the aggregate and at the same time coat all the aggregate particles. If the aggregate particle has a low porosity and is smooth, the asphalt cement will not adhere to the aggregate. Adhesion becomes an extremely important property during periods when the mix is exposed to water. If the aggregate wets easily, the water will compete with the bitumen for adsorption onto the aggregate surface, and the aggregate will separate from the bitumen, which is known as slippage.

Aggregate Gradation and Size

Depending on the specific use of the bituminous mix, the size and gradation of the aggregate varies tremendously. A high-quality bituminous mix that is to be used as a pavement for heavy traffic will generally utilize a dense-graded aggregate (a well-graded aggregate from coarse to fine). In this particular case one would not utilize Fuller's maximum density curve because it does not leave sufficient room for the asphalt cement. Therefore, the best procedure would be to open the grading somewhat more than the maximum. This opening of the gradation is achieved by the addition of fines (material less than the No. 200 sieve).

Strength, Toughness, Shape, and Porosity

The aggregate in a bituminous mix, unlike that in portland cement concrete, supplies most of the stability and thus should have a certain amount of strength and toughness; otherwise loss of stability will result. Open-graded mixes are subject to greater mechanical break-down than a dense-graded mix. Thus, if the material selected for a bituminous mix is of minimal strength, a denser mix of the same material would be utilized. ASTM C131 (Resistance to Degradation of Small Size Coarse Aggregate by Abrasion and Impact in the Los Angeles Machine) attempts to measure the effective strength and toughness of an aggregate. Table 2-1 shows the abrasive results of various aggregates subjected to the Los Angeles machine. The test shows very little correlation with the field performance of the aggregates. It does, however, have some use in distinguishing aggregates that are unsuit-able in surface treatment.

Particle shape is a property of the aggregate that is more important than gradation and size, strength, and toughness when it comes to bituminous mixes. When rounded aggregates are used in an open-mix gradation, very little stability is achieved. Thus, when an open-mix gradation is used, angular aggregate should be used. If rounded aggregate has to be used in a bituminous mix, it should be crushed, thus resulting in a fractured plane.

Porosity of the aggregate strongly affects the economics of a mix. In each mix the aggregate should have a certain amount of porosity. In general, the higher the porosity, the more asphalt will be absorbed into the aggregate, thus requiring a higher percent of asphalt in the mix design. Thus, an aggregate that is porous gives rise to the possibility of selec-tive absorption. In *selective absorption* the oily portions of the asphalt are selectively absorbed, leaving a hard residue on the surface of the aggregate particle. This process could lead to stripping of the aggregate from the asphalt cement.

AGGREGATE BENEFICIATION

In various parts of the country, sound aggregates are scarce and transportation costs for this good aggregate are expensive. In those areas, the aggregate that is available may have cer-tain deleterious materials that prevent these aggregates from passing specifications, or these aggregates have an adverse field performance record. In a few of these cases the aggregate can be beneficiated so that it becomes a useful economical product.

In the beneficiating of aggregates, several processes have been developed:

1. Washing.
2. Heavy-media separation.
3. Elastic fractionation.
4. Jigging.

Washing

Undesirable aggregates are washed to remove the particle coatings or to change the gra-dation. Fine content can be removed by exposing the aggregate to streams of water while it moves over screens or in special wash tanks. Stokes's law can also be used to remove certain fractions of fine aggregate by differential settlement. For water at 77°F (25°C) and using 2.65 as the specific gravity of sand, Stokes's law is

$$V = 9000D^2 \tag{2.2}$$

where V = velocity of settlement
             D = particle diameter

Thus, if sand is introduced near the top of an elongated water tank, various sized fractions can be removed from pockets spaced away from the sand-water stream, which is entering horizontally.

Heavy-Media Separation

Heavy-media separation utilizes the principle that the specific gravity of much deleterious material is lighter than the specific gravity of sound aggregate. In this method, referred to as the *sink float method,* the suspension is composed of water and magnetite or ferrosilicon, or a combination of the two. The suspension is maintained at a specific gravity that will allow the deleterious material to float to the top and the sound material to sink.

Elastic Fractionation

Elastic fractionation is a procedure whereby heavy but soft particles can be removed. Aggregates fall on an inclined plate, and their quality is measured by the distance they bounce from the surface. The bouncing stones are collected in three different compartments, placed so as to collect particles in categories of their bouncing characteristics. Poor, soft, or friable particles bounce only a short distance, whereas the harder, sound aggregate particles bounce much farther. The elastic fractionation process removes only those particles having elastic properties that cause them to bounce poorly. It does not remove deleterious particles that have a high modulus of elasticity. Thus, heavy-media separation should be used in conjunction with elastic fractionation to remove all the deleterious material.

Jigging

Jigging is a specific-gravity method of removing light particles such as coal, lignite, or sticks. Upward pulsations created by air tend to hinder settlement of lighter particles, which are removed by shimming devices. The advantage of the process is that either fine or coarse aggregate may be used.

ASTM TEST SPECIFICATIONS FOR AGGREGATES

In the following section we will look at various ASTM test specifications, with special emphasis on the following categories:

1. Tests concerning the general quality of aggregates.
2. Tests concerning deleterious materials in aggregates.
3. Tests used in the design of portland cement concrete and bituminous mix design.

In each case we will look at the purpose of the test, followed by a brief description of the test and the authors' opinions and conclusions concerning the procedure, results, and validity of the test.

Tests Concerning the General Quality of Aggregates

ASTM C131 (resistance to degradation of small size coarse aggregate by abrasion and impact in the Los Angeles machine).   The purpose of this specification is to test coarse aggregate smaller than 1.5 in. (3.81 cm) for resistance to abrasion using the Los Angeles testing machine and to evaluate base-course aggregates for possible degradation.

In this procedure, the test sample is placed in the Los Angeles testing machine after it has been prepared for testing in accordance with this specification. The machine is rotated at a speed of 30 to 33 rpm for 500 revolutions. The material is discharged from the machine and a preliminary separation of the sample is made on a sieve coarser than the No. 12. The finer portion is sieved using the No. 12 sieve in a manner conforming to the specification. The material coarser than the No. 12 sieve is washed and oven-dried at 221 to 230°F (105 to 110°C) to constant weight and weighed to the nearest gram. The difference between the original weight and the final weight of the test sample is expressed as a percent of the original weight. The value is reported as a percent of wear.

According to the specifications, blacklash or slip in the driving mechanism is very likely to furnish results that are not duplicated by other laboratories—an apparent disadvantage. In 1937, Woolf compared the Los Angeles abrasion results with the service records of coarse aggregates and concluded that the Los Angeles test gives accurate indications of the quality of the material under test and that its use in specifications controlling the acceptance of coarse aggregates is warranted.

However, lack of sufficient data from any one test procedure makes it impossible to suggest limits or specifications for the abrasion resistance on any type of concrete surface. Also, different surfaces may require different abrasion values. For example, a sidewalk would not have to have a high-wearing resistance as compared to one on which heavy roller cars are constantly passing over. Some materials can pass this test but ultimately are dangerous to the concrete.

Unless someone places limits on wearing-resistance values of material such as limestone, blast-furnace slag, hard quartz, and other materials, the test is only a fair representation of what may happen. It is a good test for base-course aggregates, in that it can give an idea of degradation characteristics.

ASTM C88 (soundness of aggregates by use of sodium sulfate or magnesium sulfate).   The purpose of this specification is to determine the potential resistance of an aggregate to weathering.

In this procedure a specific weight of an aggregate having a known sieve analysis is immersed in a solution of sodium or magnesium sulfate for 16 to 18 hours. Next, it is placed in an oven at 230°F (110°C) and dried to constant weight. The procedure is repeated for the desired period (usually 5 or 10 cycles); the sample is cooled, washed, and dried to constant weight; then sieved, weighed, and recorded as the percent of weight lost.

This method furnishes information helpful in judging the soundness of aggregates subjected to weathering action, particularly when adequate information is not available from service records of the material exposed to actual weathering conditions. Attention is called to the fact that test results by the use of the two salts differ considerably, and care must be exercised in fixing proper limits in any specifications that may include requirements for these tests.

ASTM C666 (resistance of concrete to rapid freezing and thawing). The purpose of this specification is to determine how concrete will react under continuous cycles of freezing and thawing and to rank aggregates.

In this procedure two methods are used. Method A involves rapid freezing and thawing in water, and method B involves rapid freezing in air and thawing in water. Immediately after curing, the specimen is brought to a temperature within $-2°F$ and $+4°F$ ($-1.1°C$ and $2.2°C$) of the target thaw temperature that will be used in the freeze-thaw cycle and tested for fundamental transverse frequency, weighed, and measured in accordance with ASTM C215 (Fundamental Transverse, Longitudinal, and Torsional Frequencies of Concrete Specimens). The specimen is protected against loss of moisture between the time or removal from curing and the start of the freeze-thaw test.

The freezing and thawing test is started by placing the specimens in the thawed water at the beginning of the thawing phase of the cycle. After each interval the specimen is tested for the fundamental transverse frequency, weighed, and returned to the apparatus. For the procedure A, the container is rinsed out, clean water is added, and the specimen is returned to the freezer. This procedure is continued for 300 cycles or until the relative dynamic modulus of elasticity reaches 60 percent of the initial modulus. If the test is interrupted, the specimen is stored in the frozen condition or in a wet condition in a refrigerator. For procedure B it is undesirable to store the specimens in the thawed conditions for more than 2 cycles.

For procedure A, the specimen must have at least $\frac{1}{8}$ in. (0.32 cm) of water all around it. In order to do this, it must be kept in a container. If the container is made of metal, the water will freeze from the top down, and as it expands there is no place for the water to go except into the concrete pores. This situation results in scaling and may cause misleading interpretations. If the container is made of rubber or other material that will expand as the water expands, the problem of scaling in the initial stages will not occur. With scaling there is an eventual weight loss, and in the early stages, especially in a metal container, the results may be misleading. Another problem may be the way the specimen is cured: whether it is cured in water or cured in air and saturated. Another problem arises because of the rate of freezing. The Corps of Engineers freeze concrete at a rate of 12 cycles per day; other agencies freeze at a rate of 4 to 6 cycles per day; and still others freeze at a rate of 1 cycle per day. Thus, the thermal properties of the aggregates begin to play an important role in the process. The test should be used to rank various aggregates rather than to determine their performance characteristics when used in concrete.

ASTM C215 (fundamental transverse, longitudinal, and torsional frequencies of concrete specimens).   The purpose of this test is to determine the relationship between strength loss and cycles of freezing and thawing.

Test specimens are made in accordance with ASTM C192 (Making and Curing Concrete Test Specimens in the Laboratory). The weight and the average length will be determined. The most important test for portland cement concrete is the transverse frequency. The specimen is placed on supports such that it may vibrate without restrictions in a free transverse mode. The specimen is forced to vibrate at various frequencies. Record the frequency of the test specimen that results in maximum indication having a well-defined peak on the indicator and at which observation of nodal points indicates fundamental transverse vibration as the fundamental transverse frequency. *Young's modulus* is then calculated as follows:

$$\text{durability factor} = DF_{300} = \frac{PN}{M} = \frac{(\text{relative } E)(N \text{ cycles})}{\text{duration of test}} \qquad (2.3)$$

In this test it is not necessary to perform the test for 300 cycles of freezing and thawing. It is only necessary to perform the test for 150 cycles and then to calculate the durability factor at 50 percent, and the DF_{300} can be calculated. The only problem with this test is to make sure that the specimen is vibrating at its fundamental transverse mode. The test is a good one in that it is nondestructive and gives a relationship between strength loss and the number of cycles of freezing and thawing.

ASTM C597 (pulse velocity through concrete).   The major purpose of this specification is to check the uniformity in mass concrete, to indicate characteristic changes in concrete, and in the survey of field structures estimate the severity of deterioration, cracking, or both.

In this procedure a sound wave is transmitted through the concrete mass and the length of time it takes to travel from one end to the other is recorded. Knowing the time and the path length, the velocity can be computed.

According to the ASTM, results obtained from this test should not be considered as a means of measuring strength or as adequate for establishing the compliance of the modulus of elasticity of field concrete. The primary use of the test is to check dams for weak spots. There have been complaints about its lack of precision, and so far it has not proven to be completely satisfactory.

ASTM C671 (critical dilation of concrete specimens subjected to freezing).   The purpose of this specification is to determine the test period of frost immunity of concrete specimens measured by the water immersion time required to produce critical dilation when subjected to a prescribed slow-freezing procedure.

In this procedure, the test specimen is molded and cured as prescribed by ASTM C192 (Making and Curing Concrete Test Specimens in the Laboratory). Once the test specimen is prepared and conditioned, the test starts. The test cycle consists of cooling the specimen in water-saturated kerosene from 35 to 15°F (1.67 to -9.44°C) at a rate of 5 $\pm$ 1°F (-2.8 ± 0.5°C) per hour; followed immediately by returning the specimen to the 35°F (1.67°C) water bath, where the specimen will remain until the next cycle. Normally, one test cycle would be carried out every 2 weeks. The length changes are measured during the cooling process. The test is continued until critical dilation is exceeded or until the period of interest is over.

Basically, this is one of the most poorly written procedures in the ASTM specifications. If a good aggregate is utilized, the aggregate may never exceed its critical dilation.

ASTM C682 (evaluation of frost resistance of coarse aggregates in air-entrained concrete by critical dilation procedure). The purpose of this procedure is to evaluate the frost resistance of coarse aggregates in air-entrained concrete.

This procedure is basically the same as that for the preceding specification. The only difference is that the sample is prepared in accordance with ASTM C295 (Petrographic Examination of Aggregates for Concrete). The aggregate is graded in accordance with field use; otherwise, equal portions of the No. 4, $\frac{3}{8}$-in., $\frac{1}{2}$-in., and 1-in. sieves are used. Further, the aggregate should be used in this test as it is used in the field. Portland cement should meet the specifications of ASTM C150 (Portland Cement), and the fine aggregate should meet the specification of ASTM C33 (Concrete Aggregates). The mix proportion should be in accordance with the ACI method of mix design with an air content of 6 percent and a slump of 2.5 ± 0.5 in.

This test is supposed to simulate the field performance. The significance of the results in terms of potential field performance will depend upon the degree to which field conditions can be expected to correlate with those employed in the laboratory. Thus, the field conditions must be assessed. Obviously, one of the main problems is to simulate field conditions. Some problems are: degree of saturation, age of concrete when the first freeze comes, length of freezing season, amount of water available during freezing, curing procedure, and condition of the aggregate when it enters the mixer. All of these conditions must be duplicated and thus there are definite chances for error. Also, the project should be planned well in advance and the conditions known before the test is run.

ASTM C672 (scaling resistance of concrete surfaces exposed to deicing chemicals). The purpose of this specification is to evaluate the effect of mix design, surface treatment, curing, and other variables of concrete subjected to scaling because of freezing and thawing, and to determine the resistance to scaling of a horizontal concrete surface subjected to freezing and thawing in the presence of deicing chemicals.

In this specification the concrete at the age of 28 days, after proper curing is covered with approximately 0.25 in. (0.64 cm) of calcium chloride and water solution having a concentration such that each 100 ml of solution contains 4 g of anhydrous calcium chloride. The specimen is then placed in a freezing chamber for 16 to 18 hours.. The specimen is then removed and placed in air at 75 ± 3°F (23 ± 1.7°C) with a relative humidity of 45 to 55 percent for 6 to 8 hours. Water is added to the chamber between each cycle to maintain the depth of the solution. The procedure is repeated daily, and at the end of 5 cycles the surface of the concrete is flushed thoroughly. A visual inspection of the concrete is made with the ratings given in Table 2-3. These ratings are recorded and the test continues.

Generally, this test is performed up to 50 cycles before final evaluation is made. The only problem with this test is that of the rating system. What represents moderate scaling to one person may not to another.

TABLE 2-3 CONCRETE SCALING
RATINGS

Rating	Condition of Surface
0	No scaling
1	Very slight scaling
2	Slight to moderate scaling
3	Moderate scaling
4	Moderate to severe scaling
5	Severe scaling

ASTM C295 (petrographic examination of aggregates for concrete).
The purpose of this specification is to screen the good from the bad aggregates. If has eight specific purposes:

1. Preliminary determination of quality.
2. Establishing properties and probable performance.
3. Correlating samples with aggregates previously tested and used.
4. Selecting and interpreting other tests.
5. Detecting contamination.
6. Determining effects of processing.
7. Determining physical and chemical properties.
8. Describing and classifying constituents.

In this specification the procedure should be carried out by a geologist utilizing x-ray diffraction, differential thermal analysis, electron microscopy, electron diffraction, electron probe, infrared spectroscopy microscope, and the naked eye.

The main purpose of petrographic examinations is to determine physical and chemical properties of aggregates. The relative abundance of specific types of rocks and minerals is established as well as particle shape, surface texture, pore characteristics, hardness, and potential alkali reactivity. Coatings are identified and described; and the presence of contaminating substances is determined.

If the petrographic examination predicts potential alkali reactivity of the aggregate, it is very helpful, in that the time required for this test is substantially shorter than that of the freeze-thaw test or most other available tests.

This specification also sets down a fundamental principle regarding aggregates: if an unfamiliar source is found, it can be compared to known data.

Petrographic examination is the best method by which deleterious and extraneous substances can be detected and determined quantitatively.

ASTM D1075 (effect of water on cohesion of compacted bituminous mixtures). The purpose of this specification is to measure the loss of cohesion resulting from the action of water on compacted bituminous mixtures containing penetration-grade asphalts. In other words, it evaluates the stripping properties of aggregates.

In this specification a 4-in. (10.16-cm) cylindrical specimen 4 in. (10.16 cm) high is tested in accordance with ASTM D1074 (Compressive Strength of Bituminous Mixtures). Then the bulk specific gravity of each specimen is determined. Each set of six test specimens is sorted into two groups of three specimens each so that the average bulk specific gravity is the same in each group. Group 1 is tested in accordance with procedure A and group 2 in accordance with procedure B. In test procedure A, the test specimens are brought to the test temperature of $77 \pm 1.8°F$ ($25 \pm 1°C$) by storing them in an air bath maintained at the test temperature for not less than 4 hours, their compressive strength determined in accordance with ASTM D1074. In test procedure B, the test specimen is immersed in water for 4 days at $120 \pm 1.8°F$ ($49 \pm 1°C$). The specimen is transferred to a second water bath at $77 \pm 1.8°F$ ($25 \pm 1°C$) and stored for 2 hours. At that time the compressive strength is determined and the numerical index of resistance of bituminous mixtures to the detrimental effect of water as the percentage of the original strength that is retained after the immersion period is calculated as follows:

$$\text{index of retained strength (\%)} = \frac{S_2}{S_1} \times 100 \qquad (2.4)$$

where S_1 =   compressive strength of dry specimens (group 1)
        S_2 =   compressive strength of immersed specimens (group 2)

This test gives an excellent indication of the retained strength but sets no definite values as to what is a good or a bad limit.

Test Concerning Deleterious Materials in Aggregates

ASTM C33 (concrete aggregates).   The purpose of this specification is to ensure that satisfactory materials are used in concrete. The specification covers both fine and coarse aggregates but does not cover lightweight aggregates.

The specification establishes definitions for fine and coarse aggregate and places restrictions on grading, deleterious substances, and soundness. The specification also establishes methods for testing and sampling.

This specification is good in that it lays down the fundamental rules for fine and coarse aggregates used in concrete as well as the sampling and testing methods to be followed.

ASTM C142 (clay lumps and friable particles in aggregates).   The purpose of this specification is to measure only particles that might cause unsightly blemishes in concrete surfaces. It is an approximate method for the determination of clay lumps and friable particles in natural aggregates.

Aggregates for this test consist of the material remaining after the completion of ASTM C117 [Materials Finer Than No. 200 (75-μm) Sieve in Mineral Aggregates by Washing]. The aggregate is dried to a constant weight at a temperature of $230°F$ ($105 \pm 5°C$). Weigh the test sample and spread it into a thin layer on the bottom of the container, cover it with water, and examine it for clay lumps or friable particles. Particles that can be broken down with the fingers into finely divided particles are classified as friable particles, provided that they can be removed by wet sieving. The residue is removed and weighed. The amount of clay lumps and friable particles in fine aggregate or individual sizes of coarse aggregate is computed as follows:

$$P = \frac{W - R}{W} \times 100 \qquad\qquad (2.5)$$

where P =   percent of clay lumps or friable particles
         W =   weight of test sample passing the layer of sieves but coarser than the
                   No. 16 sieve
         R =   weight of particles retained on designated sieve

This test is not widely used because of the physical limitations of sorting through all the particles, and also because these particles are merely a symptom of inadequate processing, which can be remedied by improved washing techniques. Also, the techniques of breaking the particles with the fingers vary from one person to another (i.e., the pressure applied by one person is different from that applied by another). Some clay lumps are hard and cannot be broken apart with the fingers. Thus, inaccurate results will be obtained.

ASTM C117 [materials finer than no. 200 (75-μm) sieve in mineral aggregates by washing].   The purpose of this test is to determine the amount of material finer than a No. 200 (75-μm) sieve in aggregate by washing. Clay particles and other aggregate particles that are dispersed by the wash water as well as water-soluble materials will also be removed from the aggregate during the test.

A sample of the aggregate is washed in a prescribed manner and the decanted wash water containing suspended and dissolved materials is passed through a No. 200 (75-μm) sieve. The loss in weight resulting from the wash treatment is calculated as weight percent of the original sample and is reported as the percentage of material finer than a No. 200 (75-μm) sieve by washing.

This test provides a measure of fines, including clay and silt, in concrete aggregates. The test has been criticized as not furnishing an indication of harmful clays, which may increase mixing-water requirements and volume-change tendencies of concrete. Experiments show that individual operators should have a 95 percent probability of checking their results within 0.3 percent. The amount of material passing a No. 200 (75-μm) sieve, by washing to the nearest 0.1 percent, is calculated as follows:

$$A = \frac{B - C}{B} \times 100 \qquad\qquad (2.6)$$

where A =   percentage of material finer than a No. 200 (75-μm) sieve, by washing
         B =   original dry weight of sample, grams
         C =   dry weight of sample, after washing, grams

Excessive quantities of fines in portland cement concrete detract from the quality of the mix by increasing the mixing requirement. In bituminous mixtures, asphalt demand may increase, although the problem is less serious than that in portland cement concrete. In base-course aggregates, this test can be especially important in determining potential susceptibility to frost action. For this reason, this is an important test.

ASTM C123 (light weight pieces in aggregates).   The purpose of this specification is to determine the approximate percentage of lightweight pieces in aggregates by means of sink-float separation in a heavy liquid of suitable specific gravity.

In this procedure the fine aggregate is allowed to dry and cooled to room temperature after following the procedure prescribed in ASTM D75. The material is sieved using a No. 50 sieve and then brought to saturated-surface dry conditions. It is put into a heavy liquid such as kerosene with 1,1,2,2,-tetrabromethane. The particles will be separated by the float-sink method provided that the specific gravities are different enough to permit separation. The liquid is poured off into a second container and passed through a skimmer. Care is taken that only the floating pieces are poured off with the liquid and that none of the sand is decanted onto the skimmer. The liquid is returned to the first container, agitated again, and the decanting process is repeated until the liquid is free of friable particles. The pieces are dried and the weight determined. For a coarse aggregate the particles are sieved using a No. 4 sieve and the foregoing process is repeated.

The materials or chemicals used in combination to form the heavy liquid for the separation process are very toxic and must be handled with great care. The test eliminates particles that might produce concrete of low durability when exposed to freezing and thawing. Experiments have shown that test procedures using gravity levels up to about 2.50 result in adequate aggregate. Coal and lignite are separated at a specific gravity of 2.0. Potentially harmful chert is separated at a specific gravity of 2.35.

ASTM C40 (organic impurities in sands for concrete).   This specification covers an approximate determination of the presence of injurious organic compounds in natural sands that are to be used in cement mortar or concrete. The principal value of the test is to furnish a warning that further tests of the sands are necessary before they are approved for use.

The procedure for this specification involves a color test. The sand and a 3 percent solution of sodium hydroxide are mixed vigorously in a graduate and allowed to stand for 24 hours. The color of the liquid is then compared to the color of a solution of potassium dichromate in sulfuric acid. If the solution of the sand and sodium hydroxide is darker than the potassium dichromate, organic impurities are present in the sand.

This is a good quick test, and if impurities are indicated, the mortar strength test should be performed to determine if the impurities are deleterious. Certain types of organic matter, principally tannic acid and its compounds derived from the decay of vegetable matter, interfere with the hardening and strength development of cement. This test detects this type of material but unfortunately also reacts to other organics, such as bits of wood, which might not be harmful to strength. A negative test is conclusive evidence of freedom from harmful organic matter, but a positive test may or may not foretell difficulty. This is an excellent test in that it is a warning of possible dangers.

ASTM C227 [potential alkali reactivity of cement-aggregate combinations (mortar-bar method)].   The purpose of this specification is to determine if an aggregate will react with the alkalies in the cement. This test is basically to predict the alkali-silica reaction.

The test method consists of molding bars of mortar 1 in. × 1 in. × 12 in. (2.54 cm × 2.54 cm × 30.5 cm) in which the aggregate in question is combined with a cement that is to be used in the field. The proportions should be 1 part cement to 225 parts of graded aggregate by weight. Use enough water to develop a flow of 105 to 120 in accordance with ASTM C109 (Compressive Strength of Hydraulic Cement Mortar), with the exception that the flow table drops $\frac{1}{2}$ in. (1.27 cm) for 10 trips in 6 seconds. After 24 hours in the molds, the lengths of the bars are measured, and they are stored at a constant temperature of 100°F (37.8°C) in sealed containers containing a small amount of water in the bottom but not in contact with the specimens. Length changes are to be measured after 1, 2, 3, 6, 9, and 12 months and, if necessary, every 6 months after. If expansion is less than 0.04 percent in 6 months, it is considered nonreactive. If the expansion is between 0.04 and 0.07 percent, the aggregate is suspicious. If the expansion is between 0.07 and 0.1 percent, the aggregate is reactive.

The obvious disadvantage to this test is the length of time involved to perform it. The test would seem to be reliable in rejecting poor aggregate. The limits given are judgments based on reports through research and petrographic examination. The limits are somewhat low for the purpose of conservatism.

ASTM C289 [potential reactivity of aggregates (chemical method)]. The purpose of this test procedure is to determine the potential reactivity of an aggregate with alkalies in portland cement concrete in a very short time. This is a test for the alkali-silica reaction and is not intended for the alkali-carbonate reaction.

In this procedure the material is ground to the point when it is finer than the No. 50 sieve but coarser than the No. 100 sieve. Twenty-five grams of the material are mixed with 25 ml of a 1 N solution of NaOH in a steel vessel about 2 in. (5.08 cm) in diameter and 2-$\frac{1}{2}$ in. (6.35 cm) high. The vessel is sealed at a temperature of 176°F (80°C) for 24 hours and then the liquid is filtered and tested for alkalinity and dissolved silica.

The results are presented on semilog paper as dissolved silica vs. reduction in alkalinity. If a considerable amount of silica is dissolved and there is a considerable amount of reduction in alkalinity, we have a potentially reactive material. If there is little dissolved silica and little reduction in alkalinity, the aggregate is of good quality.

This test has the advantage of being quick and accurate, as the results can be run within 24 hours. This test is based upon field experience and should only be used as a screening device. The test will not give satisfactory results with carbonate rocks such as ferrous and magnesium rocks. In these two aggregate types, a reduction in ions results and thus the results are no longer reliable.

Tests Used in the Design of Concrete Mixes
(Portland Cement or Bituminous)

ASTM D75 (sampling aggregates).   The purpose of this test is to sample fine and coarse aggregate for the following purposes:

1. Preliminary investigation of the potential source of supply.
2. Control of the product at the source of supply.

3. Control of the operations at the site of use.

4. Acceptance or rejection of the materials.

In this procedure, sampling plans and acceptance and control tests vary with the type of construction in which the material is used. Samples for preliminary investigation tests are obtained by the party responsible for development of the potential source. The sampler must use every precaution to obtain samples that will show the true nature and condition of the materials they represent. Samples must be inspected and sampling taken from conveyor belts, flowing aggregate stream, or stockpiles. The number of samples taken depends on the variations and the properties measured.

Sampling is as important as testing; the rest results depend on the sampling. If the sampler is inexperienced with the techniques involved, the entire test becomes questionable. Thus, in this specification, a person familiar with sampling should perform all sampling procedures.

ASTM C136 (sieve or screen analysis of fine and coarse aggregates).
The purpose of this specification is to determine the particle size of fine and coarse aggregates to be used in various tests.

In this procedure, a weighed sample of dry aggregate is separated through a series of sieves or screens of progressively smaller openings for determination of particle-size distribution.

In this specification, the results are dependent upon individual technique. The test is placed in two categories: mechanical sieving and hand sieving. This excellent test determines the gradation of the aggregates, which is so important in mix design procedures.

ASTM C127 (specific gravity and absorption of coarse aggregate).
The purpose of this specification is to ultimately determine the solid volume of coarse aggregate and the unit volume of the dry rodded aggregate such that a weight-volume characteristic can be determined so that a concrete design mix can be determined. The bulk specific gravity is used to determine the volume occupied by the aggregate.

In this procedure, approximately 5 kg of the aggregate is selected after quartering. After the aggregate is thoroughly washed to remove dust or other coatings from the surface of the particles, the sample is dried to constant weight at a temperature of 212 to 230°F (100 to 110°C) and cooled in air at room temperature for 1 to 3 hours. After cooling, the sample is immersed in water at room temperature for a period of 24 ± 4 hours. Next, the specimen is removed from the water and rolled in a large absorbent cloth towel until all visible films of water are removed. The large particles are wiped by hand. Care is taken not to allow evaporation of water from aggregate pores during the operation of surface drying. The sample is weighed in the saturated-surface dry condition. Then the sample is weighed in water, making sure that the entrapped air is removed. The sample is dried at 212 to 230°F (100 to 110°C) cooled at room temperature for 1 to 3 hours, and weighed. The bulk and apparent specific gravity and the percent absorption are determined as follows:

$$\text{bulk specific gravity} = \frac{A}{B - C} \tag{2.7}$$

where A =   weight of oven-dry specimen in air, grams
         B =   weight of saturated-surface-dry specimen in air, grams
         C =   weight of saturated specimen in water, grams

$$\text{bulk specific gravity (saturated} - \text{surface} - \text{dry)} = \frac{B}{B - C} \times 100 \tag{2.8}$$

$$\text{apparent specific gravity} = \frac{A}{A - C} \tag{2.9}$$

$$\text{absorption} = \frac{B - A}{A} \times 100 \tag{2.10}$$

The specific gravity of aggregates is important, as it is used to determine the aggregate for use in a concrete mix.

ASTM C128 (specific gravity and absorption of fine aggregate).   The purpose of this specification is to determine the bulk and apparent specific gravity of fine aggregate as well as the absorption.
    In this procedure, 500 g of fine aggregate is immersed in a pycnometer that is filled with water to 90 percent of capacity. The pycnometer is rolled, inverted, and agitated to eliminate air bubbles. The temperature is adjusted to 73.4 ± 3°F (23 ± 1.7°C). The total weight of the pycnometer, sample, and water is determined. The fine aggregate is removed, dried to a constant weight at 212 to 230°F (100 to 110°C), cooled at room temperature for 0.5 to 1.5 hours, and weighed. The weight of the pycnometer is determined and the bulk specific gravity, bulk saturated-surface-dry specific gravity, apparent specific gravity, and the absorption are calculated as follows:

$$\text{bulk specific gravity} = \frac{A}{B + 500 - C} \tag{2.11}$$

where A =   weight of oven-dry specimen in air, grams
         B =   weight of pycnometer filled with water, grams
         C =   weight of pycnometer with specimen and water to calibration mark, grams

$$\text{bulk saturated} - \text{surface} - \text{dry specific gravity} = \frac{500}{B + 500 - C} \tag{2.12}$$

$$\text{apparent specific gravity} = \frac{A}{B + A - C} \tag{2.13}$$

$$\text{absorption} = \frac{500 - A}{A} \times 100 \tag{2.14}$$

This is an important test because the volume of aggregate is determined for the concrete mix. The results of this test are used in all concrete-mix design procedures, whether portland cement or bituminous.

ASTM C29 (unit weight of aggregate). This method covers the determination of the unit weight of fine, coarse, or mixed aggregate.

In this procedure, the sample is dried to constant weight in an oven at 220 to 230°F (105 to 110°C) and thoroughly mixed. A cylindrical metal bucket is calibrated using water (knowing that water weighs 62.4 lb/ft³). The measure is filled one-third full and the surface is leveled with the fingers. The layer of aggregate is rodded 25 times with a tamping rod. The strokes are applied evenly over the sample. This procedure is repeated at two-thirds full and at full. The measure is leveled, weighed, and multiplied by the volume of the bucket. This method applies to aggregates of 1.5 in. (3.81 cm) or less. For aggregate over 1.5 in. (3.81 cm), use the jigging method.

The results of this test should check within 1 percent when duplicated. The results of this procedure are important, as they are used in the mix design procedure for both portland cement and bituminous concrete.

P R O B L E M S

2.1. What is the different between a natural aggregate and a manufactured aggregate?

2.2. Aggregates may be classified as fine aggregate or coarse aggregate; explain the difference.

2.3. Explain how aggregates are processed for use as a portland cement concrete ingredient or as a bituminous concrete ingredient.

2.4. How does particle shape affect the use of aggregate in base-course materials? In portland cement concrete? In bituminous concrete?

2.5. Explain the use of Fuller's maximum density curve.

2.6. Why is gradation important in portland cement concrete?

2.7. Which type of aggregates (igneous, sedimentary, or metamorphic) would you expect to be most suitable as a base-course material? Why?

2.8. Review various references on the subject of freezing and thawing and write a short report on how they eventually lead to concrete failure.

2.9. Review several references and explain why aggregate beneficiation is necessary. Include in your report the methods used for aggregate beneficiating.

2.10. Review the ASTM specifications for tests concerning the general quality of aggregates, deleterious materials in aggregates, and the specifications used in the design of portland cement and bituminous concrete mixes, and write a short report on the purposes, procedures, and reasons for the tests.

R E F E R E N C E S

ABDUN-NUR, E. A., "Concrete and Concrete Making Materials," *ASTM Spec. Tech. Publ. 169-A,* 1966, pp. 7–17.

ALLEN, C. W., "Influence of Mineral Aggregates on the Strength and Durability of Concrete," Symposium on Mineral Aggregate, *ASTM Spec. Tech. Publ. 83,* 1948.

BATEMAN, J. H., *Materials of Construction,* Pitman Publishing Corp., New York, 1950, pp. 23–74.

FULLER, W. B., and THOMPSON S. E., "The Laws of Proportioning Concrete," *Trans. Am. Soc. Civil Engrs., 59,* 1907.

HIGHWAY RESEARCH BOARD, *Bibliography on Mineral Aggregates,* Washington, D.C., 1949 (Bibliography No. 6).

JACKSON, F. H. "The Durability of Concrete in Service," *Proc. Am. Concr. Inst., 43* (1942), p. 165.

KREBS, R. D., and WALKER, R. D., *Highway Materials,* McGraw-Hill Book Company, New York, 1971.

MINOR, C. E., "Degradation of the Mineral Aggregates," *ASTM Spec. Tech. Publ. 277,* 1960, pp. 109–121.

NORDBERG, B., "Canada's Most Modern Gravel Plant," *Rock Prod., 55* (Aug. 1952).

PARSONS, W. H., and INSLEY, H., "Observations on Alkali-Aggregate Reaction," *Proc. Am. Concr. Inst., 44* (1948), p. 625.

Pit and Quarry Handbook, 41st ed., Complete Service Publishing Co., Chicago, 1948.

POWERS, T. C., "Basic Considerations Pertaining to Freezing and Thawing Tests," *ASTM Proc., 55* (1955) p. 1132.

STANTON, T. E. "California Experience with the Expansion of Concrete through Reaction between Cement and Aggregate," *J. Am. Concr. Inst., 39* (Jan. 1942).

SWEET, H.S., "Physical and Chemical Tests of Mineral Aggregates and Their Significance," Symposium on Mineral Aggregates, *ASTM Spec. Tech. Publ. 83,* 1948.

VERBECK, V., "Osmotic Studies and Hypothesis Concerning Alkali-Aggregate Reactions," *ASTM Proc., 55* (1955), pp. 1110–1127.

WOOLF, D.O., "Methods for the Determination of Soft Pieces in Aggregates," *Public Roads, 26,* (Apr. 1951), p. 148.

WOOLF, D.O., "Methods for the Determination of Soft Pieces in Aggregates," *Public Roads, 26,* (Apr. 1937).

3

CEMENTS

Cements are materials that exhibit characteristic properties of setting and hardening when mixed to a paste with water. They are a class of products that can be very complex and of somewhat variable composition and constitution.

Cements are divided into two classifications: hydraulic and nonhydraulic. This division is based upon the way in which the cement sets and hardens. The *hydraulic cements* have the ability to set and harden under water. Hydraulic cements include, but are not limited to, the following: hydraulic limes, pozzolan cements, slag cements, natural cements, portland cements, portland-pozzolan cements, portland blast-furnace-slag cements, alumina cements, expansive cements, and a variety of others (white portland cements, colored cements, oil-well cements, regulated cements, waterproofed cements, hydrophobic cements, antibacterial cements, barium and strontium cements).

Nonhydraulic cements do not have the ability to set and harden under water but require air to harden. The main example of a nonhydraulic cement would be a lime.

LIME

Lime, one the oldest known cementing materials, is readily available and rather inexpensive. Lime is produced by burning limestone (calcium carbonate) with impurities such as magnesia, silica, iron, alkalies, alumina, and sulfur. This burning process takes place in either a vertical or a rotary kiln at a temperature of 1800°F (980°C). Calcium carbonate is decomposed into calcium oxide and carbon dioxide according to the following reaction:

$$CaCO_3 \rightarrow CaO + CO_2 \qquad (3.1)$$

The calcium oxide that is formed is called *quicklime*, which, when in the presence of water, reacts to form calcium hydroxide together with a great evolution of heat:

$$CaO + H_2O \rightarrow Ca(OH)_2 + heat \tag{3.2}$$

This process is called *slaking* and the product calcium hydroxide is called *slaked lime* or *hydrated lime*. The rate of reaction depends mainly on the purity of the lime. The higher the purity of lime, the greater its reactivity with water. Commercial quicklime is classified into three groups: quick, medium, and slow slaking.

Depending upon the amount of water added during the slaking process, lime putty or hydrated lime may be formed. Hydrated lime is produced by adding just enough water (one-third of its weight) to quicklime. Lime putty is formed when an overextended amount of water is added to the quicklime.

Both lime putty and hydrated lime are always mixed with mortar sand in proportions of 1 part lime to 3 parts sand by volume, to prevent excessive shrinkage.

The setting of lime mortar is the result of the loss of water either by absorption of water by the block, brick, or whatever, or by evaporation. The hardening is caused by the reaction of carbon dioxide in the air with the hydrated lime as follows:

$$Ca(OH)_2 + CO_2 \rightarrow CaCO_3 + H_2O \tag{3.3}$$

This results in the formation of calcium carbonate crystals, which bind the heterogeneous mixture into a coherent mass. The hardening process is slow and may take several years to develop its full strength. However, it needs the free circulation of air to provide the necessary carbon dioxide to penetrate the innermost portion of the mortar for hardening to take effect.

Hydraulic Limes

Hydraulic limes are made by burning siliceous or argillaceous limestone whose clinker after calcination (in a continuous kiln) contains a sufficient percentage of lime silicate to give hydraulic properties to the product, but which normally contain so much free lime that the mass of clinker will slake on the addition of water.

As the content of alumina and silica in the lime increases, the rate of slaking and heat evolution decreases to a point where no reaction occurs between water and lime. At high temperature, the alumina and silica combine with calcium oxide, calcium silicates, and aluminates, which do not easily combine with water when it is in lump form. Therefore, quicklime is added in the slaking process and the large lumps are broken up into fine powder due to the expansion of the quicklime. The final product consists of lime silicate and about one-fourth hydrated lime. The material in the fine form can now readily combine with water. Hydraulic limes do exhibit hydraulic properties but are not suited to subaqueous construction because they require free access of air during hardening. The air is necessary to ensure carbonation of free calcium hydroxide present in large amounts in calcium carbonate; otherwise, the full strength of the lime cannot be developed.

Hydraulic limes are used for browning plaster coats or for stucco and similar uses.

Pozzolan Cements

A *pozzolan*, according to ASTM C595 (Blended Hydraulic Cements), is a siliceous or siliceous and aluminous material that, in itself, possesses little or no cementitious value but will, in finely divided form and in the presence of moisture, react chemically with calcium hydroxide at ordinary temperatures to form compounds possessing cementitious properties. Pozzolans are further classified as natural pozzolans or artificial pozzolans. Natural pozzolans are further classified into two groups. The first group is made up of pumicite, obsidian, scoria, tuff, santorin, and trass, which are derived from volcanic rocks. The second group of natural pozzolans contains large quantities of finely dispersed, amorphous silica that react with lime in the presence of water to form hydrated silicates, which accounts for their hydraulic properties.

Artificial pozzolans include fly ash, boiler slag, and by-products from the treatment of bauxite ore.

Pozzolan cements are manufactured by direct grinding of the volcanic rocks or by calcining and grinding clays, shales, and diatomaceous earth. Pozzolan cements include all cementing materials that are made by the incorporation of pozzolans with hydrated lime in which no subsequent calcination is needed.

The requirements for pozzolan use in blended cements are given in ASTM C595 (Blended Hydraulic Cements) and are repeated in part here in Table 3-1. Whenever a pozzolan is used in cement, it must conform to the requirements of ASTM C311 (Sampling and Testing Fly Ash or Natural Pozzolans for Use as a Mineral Admixture in Portland Cement Concrete) and/or ASTM C618 (Specifications for Fly Ash and Raw or Calcinated Natural Pozzolan for Use as a Mineral Admixture in Portland Cement Concrete).

Thus far, very little use has been found for pozzolan cements in the area of structural concrete. Their main use has been where mass is required rather than strength. It has also found limited use when mass concrete is needed with little heat of hydration. However, Chapter 6 will provide a discussion on proportioning such mixes.

Slag Cements

Slag cements are hydraulic cements consisting mostly of an intimate and uniform blend of granulated blast-furnace slag and hydrated lime in which the slag constituent is at least 60 percent of the weight of the slag cement. The mixture is often not calcined.

Basically, two types of slag cement exist. The first is designated Type S slag cement by ASTM C595 (Blended Hydraulic Cements). Type S slag cement may be used in combination with portland cement in making concrete and in combination with hydrated lime

TABLE 3-1 PHYSICAL REQUIREMENTS FOR POZZOLANS

Fineness	
Amount retained when wet-seived on No. 325 (45-μm) sieve, max. (%)	20.0
Pozzolan activity index with portland cement, at 28 days, min. (%)	75.0

in making masonry mortar. The second type of slag cement is designated Type SA slag cement by ASTM C595 (Blended Hydraulic Cements). Type SA slag cement, which is air-entrained slag cement, has the same general uses as Type S.

Blast-furnace slags are a nonmetallic product, consisting essentially of silicates and aluminosilicates of calcium and of other bases, developed in a molten condition simultaneously with iron in a blast furnace. The blast-furnace slags, which are suitable for use in slag cements, are fusible lime silicates derived as waste products from the operation of blast furnaces in smelting iron from its ore.

When slag cements are manufactured, they go through a variety of operations, such as granulation (which not only renders the slag more hydraulic, but at the same time reduces the harmful sulfides), drying of the slag, preparation of the hydrated lime, proportioning the mix, mixing, and final grinding.

Slag cements are of limited importance in structural concrete, buy may find success in projects requiring large masses of concrete masonry where weight and bulk are more important than strength. It may also find use as a masonry cement in that it does not have a staining effect because of its low alkali content.

NATURAL CEMENTS

A natural cement was at one time defined by ASTM C10 (Natural Cements) as a hydraulic cement produced by calcining a naturally occurring argillaceous limestone at a temperature below the sintering point and then grinding to a fine powder. The amounts of silica, alumina, and iron oxide present are sufficient to combine with all the calcium oxide to form the corresponding calcium silicates and aluminates, which account for the hydraulic properties of natural cements.

Two types of natural cement existed according to ASTM C10 (Natural Cements), Type N and Type NA. Type N natural cement is for use with portland cement in general concrete construction. Type NA natural cement is air-entrained cement and has the same uses as Type N.

Natural cements are made by the calcination of a natural clay limestone, which is made up of clay material (13 to 35 percent), silica (10 to 20 percent), and a balance of alumina and iron oxide. The clay material gives the cement its hydraulic properties.

After calcination and possible slaking to remove the free lime, the clinker is ground into a fine powder known as a natural cement, with the following average composition:

SiO_2	CaO	MgO	Fe_2O_3	Al_2O_3
22–29%	31–57%	1.5–22%	1.5–3.2%	5.2–8.8%

Thus, a quick view of the composition would indicate the possibilities of wide variation among mechanical properties. The physical requirements of natural cements are shown in Table 3-2.

Natural cements should not be used in exposed areas but may be used as a substitute for portland cements in mortars and concrete when the stresses encountered will never be high or in situations in which weight and mass are more essential than strength.

TABLE 3-2 PHYSICAL REQUIREMENTS OF NATURAL CEMENTS

Properties	Type N	Type NA
Fineness, specific surface (cm²/g)		
Air permeability apparatus		
Average value, min.	6000	6000
Minimum value, any one sample	5500	5500
Soundness		
Autoclave expansion, of blend of 75% natural cement and 25% portland cement by weight, max. (%)	0.8	0.8
Time of setting, Vicat test		
Time (min)	30	30
Air entrainment		
No agent used, max. (vol. %)	12	
Using air-entrainment agent (vol. %)	—	19±3
Compressive strength, min. [psi MPz]		
Compressive strength of mortar cubes, composed of 1 part natural cement and 1 part standard sand by weight; must be equal to or higher than the values specified for the following ages:		
1 day in moist air, 6 days in water	500 (3.4)	500 (3.4)
1 day in moist air, 27 days in water	1000 (6.9)	1000 (6.9)

PORTLAND CEMENTS

Portland cement is one of the most widely used construction materials and is the most important hydraulic cement. It is used in concrete, mortar, plaster, stucco, and grout. It is used in all types of structural concrete (walls, floors, bridges, tunnels, subways, etc.), whether reinforced or not. It is further used in all types of masonry (foundations, footings, dams, retaining walls, and pavements). When portland cement is mixed with sand and lime, it serves as mortar for laying brick or stone, or as plaster or stucco for interior or exterior walls. When portland cement is mixed with coarse aggregate (aggregate larger than the No. 4 sieve) and fine aggregate (sand) together with enough water to ensure a good consistency, concrete results.

Portland cement is defined, according to ASTM C150, as a hydraulic cement produced by pulverizing clinker consisting essentially of hydraulic calcium silicates, usually containing one or more of the forms of calcium sulfate as an interground addition. The approximate proportions for portland cement are as follows:

Lime (CaO) 60–65%

Silica (SiO_2) 20–25%

Iron oxide and alumina (Fe_2O_3 and Al_2O_3) 7–12%

History of Portland Cement

The name *portland cement* was proposed by Joseph Aspdin in 1824. The name came about because the powdery material, which he patented, set up with water and sand and resembled a natural limestone quarried on the Isle of Portland in England.

The first portland cement manufactured in the United States was produced by David Saylor at Coplay, Pennsylvania, in 1875 by calcining, in vertical kilns at a high temperature, a mixture of argillaceous limestone rock with a pure limestone. With the increase in demand for quantity and quality, rotary kilns came into production in 1899. The production of portland cement in the United States has increased from 10 million barrels per year in 1900 to 400 million barrels per year in 1970. World production is approximately six times that of the United States.

Raw Materials

The raw materials of portland cement may be classified into three groups: calcareous, argillocalcareous, and argillaceous. Table 3-3 illustrates the materials that make up each group. From the table and a little knowledge of geology, it is evident that the essential constituents of portland cement are lime, silica, and alumina. Lime does not occur in nature but is found in a suitable form in a carbonate. Silica and alumina are found free in nature in the form of clay, shale, or slate.

Limestone (calcium carbonate) contains impurities of magnesia, silica, iron, alkalies, and sulfur. Magnesia in the form of carbonate of magnesia often occurs in limestone and, if present in the amount of 5 percent or more, will make the limestone unsuitable. Silica by itself does not combine with lime in the kiln; thus, small quantities of free silica in the limestone makes the limestone unacceptable. However, if silica is combined with alumina in the limestone, it will combine with lime in the kiln and is acceptable.

Iron in limestone can occur either as an oxide (Fe_2O_3) or as a sulfide (FeS_2). If the iron is in the form of an oxide, it acts as a flux in combining the lime and silica in the kiln. As a sulfide it reacts very strangely and can prove to be quite injurious to the production of portland cement. If the iron in the form of sulfide is present in limestone by 4 percent or greater, the limestone should be rejected.

The makeup of limestone is such that it contains alkalies in the form of soda and potash. These alkalies are not harmful and are usually driven off in the kiln.

One of the major roles of alumina in limestone is to combine with silica such that

TABLE 3-3 RAW MATERIALS IN PORTLAND CEMENTS

Calcareous ($CaCO_3 > 75\%$)	Argillocalcareous ($CaCO_3 = 40$ to 75%)	Argillaceous ($CaCO_3 < 40\%$)
Limestone	Clayey limestone	Slate
Chalk	Clayey chalk	Shale
Shells	Clayey marl	Clay

the limestone will combine with the lime in the kiln and thus make the limestone an acceptable product.

Sulfur, the final impurity in limestone, exists in two forms: lime sulfate and iron pyrite. If each is present in amounts of 3 percent or more, the limestone should be rejected.

Chalk is a variety of limestone formed from pelagic, or floating, organisms that are very fine-grained, porous, and friable. It is white or very light colored and consists almost entirely of calcite. The rock is made up of the calcite shells of microorganisms partially cemented by structureless calcite. The best known chalks are those of the Cretaceous, exposed in cliffs on both sides of the English Channel. The Selma (Cretaceous) chalk of Alabama, Mississippi, and Tennessee and the Niobrara chalk of the same age in Nebraska are well-known deposits in the United States.

Marl is an argillaceous, nonindurated calcium carbonate deposit that is commonly gray or blue-gray. It is somewhat friable, and in some respects resembles chalk, with which it is interbedded in some localities. It is formed in some freshwater lakes, partially by the action of aquatic plants. The clay content of marls varies, and all gradations between small amounts of clay (marly limestone) and large amounts (marly clay) are found.

Shales are clays that have been solidified in a laminated structure and have the property of splitting into thin sheets. They have limited application in the production of portland cement.

Slates are clays that have become hardened by pressure. They have been formed from deposits of sedimentary clay. Shales are preferred over soft clays in the production of portland cement because segregation of the shale and limestone is less likely to occur.

Clays are formed from the debris resulting from the decay of rocks. Clays take the form of three groups with respect to methods of transportation: residual, sedimentary, and glacial. Clays left where the rocks decayed are *residual.* Clays that have been transported and deposited by stream action are *sedimentary. Glacial* clays are those deposited by glacial movement. In any clay the silica content should not be less than 55 to 65 percent, and the amount of alumina and iron oxide combined should be between one-third and one-half the amount of silica.

Manufacture of Portland Cement

Portland cement is the most important hydraulic cement used in construction for mortars, plasters, grouts, and concrete. The manufacture of portland cement occurs through a series of steps (quarrying, crushing, grinding, mixing, calcining, addition of retarder, and packing). Portland cement is made by burning an intimate mixture composed mainly of calcareous, argillocalcareous, or argillaceous materials, at a clinker temperature of 2800°F (1550°C). This partially sintered clinker is then ground to a very fine powder with a very small amount of gypsum (2 to 4 percent) as a retarder. The cement is then packaged into 94-lb. (43-kg) bags or into a hypothetical barrel which contains four sacks [376 lb. (170 kg)] and is approximately 4 ft.^{3} (0.1 m^3) loose volume or 1.912 ft.^{3} (0.05 m^3) solid volume with a specific gravity of 3.15; or it may be bulk-stored. The cement-making process shown in Figure 3-1 is the process as it appears today, that is, as most industries are set up.

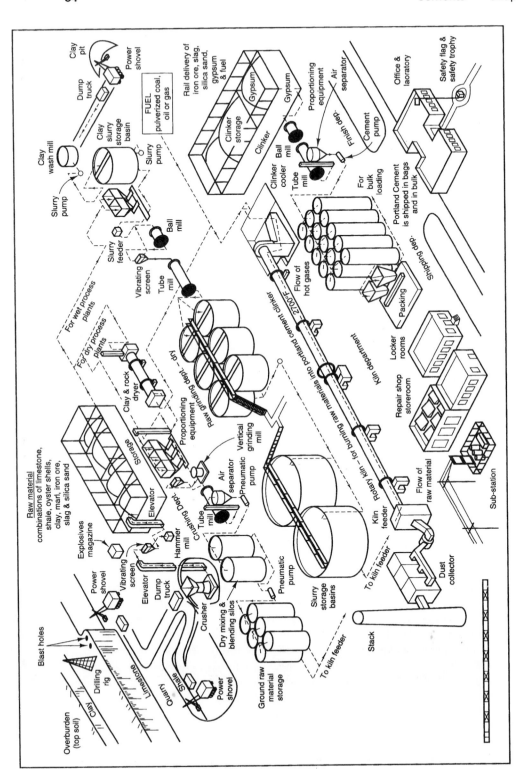

Figure 3-1 Flowchart of Manufacturing Process of Portland Cement (Courtesy of PCA)

Composition of Portland Cement

Eight types of portland cement are recognized by the ASTM under specification ASTM C150:

Type I	For use when the special properties specified for any other type are not required
Type IA	Air-entraining cement for the same uses as Type I, when air entrainment is desired
Type II	For general use, more especially when moderate sulfate resistance or moderate heat of hydration is desired
Type IIA	Air-entraining cement of the same uses as Type II, where air entrainment is desired
Type III	For use when high early strength is desired
Type IIIA	Air-entraining cement for the same use as Type III, where air entrainment is desired
Type IV	For use when a low heat of hydration is desired
Type V	For use when high sulfate resistance is desired

If a chemical analysis were to be performed on any one of the eight portland cements, the composition would be calcium oxide, silica, alumina, iron oxide, magnesium oxide, sulfur trioxide, and others. The chemical analysis would further reveal that these oxides exist in portland cement as calcium silicates and aluminates, such as tricalcium silicate ($3CaO \cdot SiO_2$), dicalcium silicate ($2CaO \cdot SiO_2$), tricalcium aluminate ($3CaO \cdot Al_2O_3$), and tetracalcium aluminoferrite ($4CaO \cdot Al_2O_3 \ Fe_2O_3$). In the nomenclature of the cement industry, these four compounds are written as follows: C_3S, C_2S, C_3A, and C_4AF, respectively. Tables 3-4 and 3-5 show the standard chemical requirements and the optional chemical requirements for portland cement.

TABLE 3-4 STANDARD CHEMICAL REQUIREMENTS

	Cement Type				
	I and IA	II and IIA	III and IIIA	IV	V
Silicon dioxide (SiO_2), min. (%)	—	20	—	—	—
Aluminum oxide (Al_2O_3), max. (%)	—	6	—	—	—
Ferric oxide (Fe_2O_3), max. (%)	—	6	—	6.5	—
Magnesium oxide (MgO), max. (%)	6	6	6	6	6
Sulfur trioxide (SO_3), max. (%)					
When ($3CaO \cdot Al_2O_3$) is 8% or less	3	3	3.5	2.3	2.3
When ($3CaO \cdot Al_2O_3$) is more than 8%	3.5	—	4.5	—	—
Loss on ignition, max. (%)	3.0	3	3	2.5	3
Insoluble residue, max. (%)	0.75	0.75	0.75	0.75	0.75
Tricalcium silicate ($3CaO \cdot SiO_2$), max. (%)	—	—	—	35	—
Dicalcium silicate ($2CaO \cdot SiO_2$), max. (%)	—	—	—	40	—
Tricalcium aluminate ($3CaO \cdot Al_2O_3$), max. (%)	—	8	15	7	5
Tetracalcium aluminoferrite plus twice the tricalcium aluminate	—	—	—	—	20

TABLE 3-5 OPTIONAL CHEMICAL REQUIREMENTS

| | Cement Type | | | | | |
	I and IA	II and IIA	III and IIIA	IV	V	Remarks
Tricalcium aluminate ($3CaO \cdot Al_2O_3$), max. (%)	—	—	8	—	—	Moderate sulfate resistance
Tricalcium aluminate ($3CaO \cdot Al_2O_3$), max. (%)	—	—	5	—	—	High sulfate resistance
Sum of tricalcium silicate and tricalcium aluminate, max. (%)	—	58	—	—	—	Moderate heat of hydration
Alkalies ($Na_2O + 0.658K_2O$), max. (%)	0.6	0.6	0.6	0.6	0.6	Low-alkali cement

Properties of the Main Compounds

As indicated, portland cement is made up of four main compounds: tricalcium silicate, dicalcium silicate, tricalcium aluminate, and tetracalcium aluminoferrite. These compounds are present in the clinker in the form of interlocking crystals. Figure 3-2 shows a phase diagram $CaO-SiO_2-Al_2O_3$ ($C_3S-C_2S-C_3A$) in which the field of portland cement lies between. Also shown in the field are aluminous and blast-furnace-slag cements. Table 3-6 lists the characteristics of the major compounds in portland cement.

The most desirable constituent is that of tricalcium silicate (C_3S), because it hardens rapidly and accounts for the high early strength of the cement. When water is added to tricalcium silicate, a rapid reaction occurs as follows:

$$3CaO \cdot SiO_2 + H_2O = 2CaO \cdot SiO_2 \cdot x \, H_2O + Ca(OH)_2 \qquad (3.4)$$

The results indicate a less basic amorphous hydrated calcium silicate and a crystalline calcium hydroxide. The calcium silicate that is formed is the product to which early strength is attributed.

Dicalcium silicate hardens slowly and contributes largely to strength increase at ages beyond 1 week. In the presence of water, dicalcium silicate ($2CaO \cdot SiO_2$) hydrates slowly and forms a hydrated calcium silicate ($2CaO-SiO_2 \cdot x \, H_2O$).

Tricalcium aluminate liberates a large amount of heat during the first few days of hardening. It also contributes to early-strength development. Tricalcium aluminate ($3CaO \cdot Al_2O_3$) hydrates with water to form a hydrated tricalcium aluminate ($3CaO \cdot Al_2O_3 \cdot 6H_2O$). If gypsum is added it acts as a retarder, and the heat of evolution is less and the setting occurs more slowly. This is due to the fact that gypsum, when present, results in the formation of calcium sulfoaluminate ($3CaO \cdot Al_2O_3 \cdot 3CaSO_4$) rather than hydrated tricalcium aluminate.

Tetracalcium aluminoferrite formation reduces the clinkering temperature, thereby assisting in the manufacture of portland cement. It hydrates rather rapidly but contributes

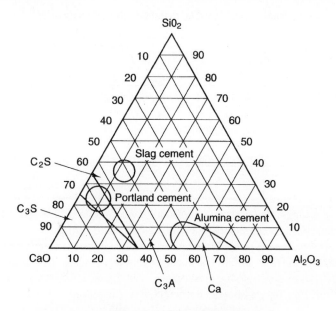

Figure 3-2 Phase Diagram

TABLE 3-6 CHARACTERISTICS OF THE MAJOR COMPOUNDS IN PORTLAND CEMENT

Values	Tricalcium Silicate, $3CaO \cdot SiO_2$ (C_3S)	Dicalcium Silicate, $2CaO \cdot SiO_2$ (C_2S)	Tricalcium Aluminate, $3CaO \cdot Al_2O_3$ (C_3A)	Tetracalcium Aluminoferrite, $4CaO \cdot Al_2O_3 \cdot Fe_2O_3$ (C_4AF)
Cementing value	Good	Good	Poor	Poor
Rate of reaction	Medium	Slow	Fast	Slow
Amount of heat liberated	Medium	Small	Large	Small

very little to strength. Table 3-7 shows typical compound composition for the various portland cements.

Types of Portland Cement

As previously mentioned, eight types of portland cement are recognized by ASTM under Specification ASTM C150. The standard five types of portland cement (this excludes the three that are air-entrained) are also recognized by the Canadian Standards Association (CSA) and given specific names. We next investigate each type separately.

ASTM Type I or CSA normal portland cement is a general-purpose cement. It is used when the special properties specified for any other type are not required. It is used where there would be no severe climate changes or severe exposure to sulfate attack from water or soil. Its uses include reinforced-concrete buildings, pavements, sidewalks,

TABLE 3-7   TYPICAL COMPOUND COMPOSITION
FOR VARIOUS PORTLAND CEMENTS

Types of Portland Cement	Compound Composition %			
	C_3S	C_2S	C_3A	C_4AF
I. Normal	50	24	11	8
II. Moderate	42	33	5	13
III. High early strength	60	13	9	8
IV. Low heat	26	50	5	12
V. Sulfate resisting	40	40	4	9

bridges, railings, tanks, reservoirs, floors, curbs, culverts, and retaining walls. In general, it is used in nearly all situations calling for portland cement.

ASTM Type II or CSA moderate portland cement is a general-purpose cement to be used when moderate sulfate resistance or moderate heat of hydration is desired. It is used in structures of considerable mass, such as abutments and piers and retaining walls. Its use also minimizes temperature rise when concrete is placed in warm weather.

ASTM Type III or CSA high-early-strength portland cement is used when high early strength is desired, usually less than 1 week. It is usually used when a structure must be put into service as quickly as possible. This cement is made by changing the proportions of raw materials, by finer grinding, and by better burning, such that the dicalcium silicate is less and the tricalcium silicate is greater.

ASTM Type IV or CSA low-heat-of-hydration portland cement is used when a low heat of hydration is required. This type of cement develops strength at a slower rate than does the ASTM Type I. However, it is intended for mass structures such as large gravity dams, where the temperature rise on a continuous pour is great. If the temperature were not minimized, large cracks or flaws would appear and the structure might prove to be unsound.

ASTM Type V or CSA sulfate-resisting portland cement is used when high sulfate resistance is desired. It is used when concrete is to be exposed to severe sulfate action by soil or water.

The three types of air-entraining cements, Types IA, IIA, and IIIA, as given by ASTM C150, are used in concrete for improved resistance to freezing and thawing action and to action of salt scaling by chemical attack. Typical air-entraining agents include Vinsol resin, Airolon, and Darex AEA.

Table 3-8 compares the strengths of the various types of cements with Type I at five different moist-curing intervals.

Properties of Portland Cement

Most specifications for portland cement, such as ASTM specifications, place specific chemical composition and physical property requirements on the cement, as shown in Tables 3-9 and 3-10. Next, we will discuss the fineness, soundness, time of setting, com-

TABLE 3-8 APPROXIMATE RELATIVE STRENGTHS OF CONCRETE AS AFFECTED
BY TYPE OF CEMENT

| Types of | Compressive Strength (% of Normal Portland Cement) | | | | |
Portland Cement	1 Day	3 Days	7 Days	28 Days	3 months
I. Normal	100	100	100	100	100
II. Modified	75	80	85	90	100
III. High early strength	190	190	120	110	100
IV. Low heat	55	55	55	75	100
V. Sulfate resisting	65	65	75	85	100

pressive strength, heat of hydration, loss of ignition, and specific gravity of portland cement. Most of these tests are covered by ASTM specifications.

The *fineness* of the cement affects the rate of hydration. The greater the cement fineness, the greater the rate of hydration and hence the greater the strength development during the first 7 days. To measure the fineness of the cement, the Wagner turbidimeter or the Blaine air-permeability apparatus is used.

Soundness of hardened cement paste is a measure of the potential expansion of the several constitiuent parts or the ability to retain its volume after setting. Lack of soundness (unsoundness) is attributed to excessive amounts of hard-burned free lime or magnesia. This free lime takes on water and at some later date develops expansive forces. Most specifications call for the use of an autoclave (high-pressure steam boiler) to indicate the soundness or unsoundness of the cement.

Time of setting is measured by the Gilmore or Vicat apparatus, which is used to determine the rate at which portland cement hardens: in other word, to determine if a cement paste remains plastic long enough to permit normal placing of the concrete. The length of time that concrete remains plastic is dependent upon the chemical composition, fineness, water content, and temperature.

Compressive strength of portland cement is determined by mixing the cement specimen with a uniform silica sand and water in prescribed proportions and molding the mixture into 2 in. × 2 in. × 2 in. (5.08 cm × 5.08 cm ×5.08 cm) cubes. The cubes are cured and then tested in compression to give an indication of the strength-developing characteristics of the portland cement.

Heat of hydration is the heat generated when cement and water react. The amount of heat generated is dependent chiefly on the chemical composition, fineness of the cement, and the temperature of curing time.

Loss of ignition of portland cement is determined by heating a cement sample of known weight to a full red heat of 1652°F (900°C) until a constant weight is obtained. The weight loss of the sample is then determined. Ignition is the indication of prehydration and carbonation, which may be caused by improper or prolonged storage.

The *specific gravity* of portland cement is generally about 3.15.

TABLE 3-9 OPTIONAL CHEMICAL REQUIREMENTS

	Cement Type[a]					
	I and IA	II and IIA	III and IIIA	IV	V	Remarks
Tricalcium aluminate ($3CaO \cdot Al_2O_3$)[b], max. (%)	—	—	8	—	—	For moderate sulfate resistance
Tricalcium aluminate ($3CaO \cdot Al_2O_3$)[b], max. (%)	—	—	5	—	—	For high sulfate resistance
Sum of tricalcium silicate and tricalcium aluminate[b], max (%)	—	58[c]	—	—	—	For moderate heat of hydration
Alkalies (Na_2O + $0.658K_2O$), max. (%)	0.6[d]	0.6[d]	0.6[d]	0.6[d]	0.6[d]	Low-alkali cement

[a]Attention is called to the fact that cements conforming to the requirements for all of these types may be carried in stock in some areas. In advance of specifying the use of other than Type I cement, it should be determined whether the proposed type of cement is or can be made available.

[b]The expressing of chemical limitations by means of calculated assumed compounds does not necessarily mean that the oxides are actually or entirely present as such compounds.

When the alumina–ferrite oxide ratio is less than 0.64, a calcium aluminoferrite solid solution [expressed as $ss(C_4AF + C_2F)$] is formed. Contents of this solid solution and of tricalcium silicate should be calculated by the following formulas:

tricalcium silicate = (4.071 × percent CaO) − C7.600 × percent SiO_2) −
(6.718 × percent Al_2O_3) − (1.430 × percent Fe_2O_3) −
(2.852 × percent SO_3)

dicalcium silicate = (2.867 × percent SiO_2) − (0.7544 × percent C_3S)

tricalcium aluminate = (2.650 × Al_2O_3) − (1.692 × percent Fe_2O_3)

tetracalcium aluminoferrite = 3.043 × percent Fe_2O_3

When the alumina-ferrite oxide ratio is less than 0.64, a calcium aluminoferrite solid solution [expressed as $ss(C_4AF + C_2F)$] is formed. Contents of this solid solution and of tricalcium silicate should be calculated by the following formulas:

$ss(C_4AF + C_2F)$ = (2.100 × percent Al_2O_3) + (1.702 × percent Fe_2O_3)

tricalcium silicate = (4.071 × percent CaO) − (7.600 × percent SiO_2) −
(4.479 × percent Al_2O_3) − (2.859 × percent Fe_2O_3) − (2.852 × percent SO_3)

No tricalcium aluminate will be present in cements of this composition. Dicalcium silicate shall be calculated as previously shown.

In the calculation of $C_3$4, the values of Al_2O_3 and Fe_2O_3 determined to the nearest 0.01 percent should be used. In the calculation of other compounds, the oxides determined to the nearest 0.1 percent shall be used.

All values calculated as described in this note should be reported to the nearest 1 percent.

[c]This limit applies when moderate heat of hydration is required and tests for heat of hydration are not requested.

[d]This limit may be specified when the cement is to be used in concrete with aggregates that may be deleteriously reactive.

PORTLAND-POZZOLAN CEMENTS

Portland-pozzolan cements are hydraulic cements consisting of an intimate and uniform blend of portland cement or portland blast-furnace-slag cement and fine pozzolan produced either by intergrinding portland cement clinker and pozzolan, by blending portland cement or portland blast-furnace-slag cement and finely divided pozzolan, or a combination of intergrinding and blending, in which the pozzolan constituent is between 15 and 40 percent by weight of the portland-pozzolan cement.

Portland-pozzolan cements consists of two types: Type IP, and Type P, each with three optional provisions: Type IP portland-pozzolan cement and option Type IP-A air-

TABLE 3-10 STANDARD PHYSICAL REQUIREMENTS

	Cement Type[a]							
	I	IA	II	IIA	III	IIIA	IV	V
Air content of mortar[b], (vol%)								
Maximum	12	22	12	22	12	22	12	12
Minimum	—	16	—	16	—	16	—	—
Fineness, specific surface (cm²/g) (alternative methods)[c]								
Turbidimeter test, min.	160	160	160	160	—	—	160	160
Air permeability test, min.	280	280	280	280	—	—	280	280
Autoclave expansion, max. (%)	0.8	0.8	0.8	0.8	0.8	0.8	0.8	0.8
Compressive strength [psi (MPa)], not less than the values shown for the following ages:[d]								
1 Day	—	—	—	—	1800 (12.4)	1450 (10.0)	—	—
3 Days	1800 (12.4)	1450 (10.0)	1500 (10.3) 1000[e] (6.9)[e]	1200 (8.3) 800[e] (5.5)[e]	3500 (24.1)	2800 (19.3)	— —	2200 (15.2)
7 Days	2800 (19.3)	2250 (15.5)	2500 (17.2) 1700[f] (11.77)[f]	2000 (13.8) 1350[f] (9.3)[f]	—	—	1000 (6.9)	2200 (15.2)
28 days	—	—	—	—	—	—	2500 (17.2)	3000 (20.7)
Time of setting (alternative methods)[f]								
Gilmore test								
Initial set (min), not less than	60	60	60	60	60	60	60	60
Final set (hr), not more than	10	10	10	10	10	10	10	10
Vicat test								
Initial set (min), not less than	45	45	45	45	45	45	45	45
Final set (min), not more than	375	375	375	375	375	375	375	375

[a]Attention is called to the fact that cements conforming to the requirements for all of these types may not be carried in stock in some areas. In advance of specifying the use of other than Type I cement, it should be determined whether the proposed type of cement is or can be made available.

[b]Compliance with the requirements of this specification does not necessarily ensure that the desired air content will be obtained in concrete.

[c]Either of the two alternative fineness methods may be used at the option of the testing laboratory. However, in case of dispute, or when the sample fails to meet the requirements of the air-permeablity test, the turbidimeter test shall be used, and the requirements in this table for the turbidimeter method shall govern.

[d]The strength at any age shall be higher than the strength at any preceding age.

[e]When the optional heat of hydration on the chemical limit or the sum of the tricalcium silicate and tricalcium aluminate is specified.

[f]The purchaser should specify the type of setting-time test required. In case it is not specified or in case of dispute, the requirements of the Vicat test only shall govern.

entrained portland-pozzolan cement are used in general concrete construction. Type P portland-pozzolan cement and option Type P-A air-entrained portland-pozzolan cement are used in concrete construction where high strengths at early stages are not required. For Types IP and IP-A cement, moderate sulfate resistance or moderate heat of hydration or both may be specified by adding the suffixes (MS) or (MH) or both to the selected type designation. For Types P and PA, moderate sulfate resistance or low heat of hydration or both may be specified by adding the suffixes (MS) or (LH) or both to the selected type designation.

Portland-pozzolan cements require more water for a given consistency but exhibit greater shrinkage upon drying. Further, they exhibit less strength prior to 28 days of curing, but greater strength after 28 days of curing when compared to normal portland cement concrete.

Portland-pozzolan cements may be used for mass concrete where mass and weight are more important than strength. It also exhibits excellent sulfate resistance and hence is good for seawalls. Portland-pozzolan cements are also used in dam construction because of their low heat of hydration.

PORTLAND BLAST-FURNACE-SLAG CEMENT

Portland blast-furnace-slag cement, according to ASTM C595 (Blended Hydraulic Cements), is a hydraulic cement consisting of an intimate and uniform blend of portland cement and fine, granulated blast-furnace slag produced either by intergrinding portland cement clinker and granulated blast-furnace slag or by blending portland cement and finely ground granulated blast-furnace slag, in with the slag constituent is between 25 and 70 percent of the weight of portland blast-furnace-slag cement.

Portland blast-furnace-slag cement exists as one type, with three optional provisions. Type IS portland blast-furnace-slag cement is for use in general concrete construction. Option Type IS-A, air-entrained portland blast-furnace-slag cement, is also used in general concrete construction. Moderate sulfate resistance or moderate heat of hydration or both may be specified by adding the suffixes (MS) or (MH) or both to the selected type designation.

ALUMINA CEMENTS

Alumina cement has a high alumina content because it consists primarily of calcium aluminates. Aluminate cement is sometimes referred to as "high-alumina cement." Alumina cements have a different chemical composition and constitution from portland cement. However, they have several outstanding properties, such as high early strength (usually setting and hardening to full strength at 48 hours, compared to 28 days for portland cement), excellent refractoriness, and good resistance against chemical attacks (hence, resistance to disintegrating action of seawater).

They are somewhat limited because elevated temperatures can produce permanent strength reductions when moisture is present.

The raw materials used for the manufacture of alumina cement are limestone and bauxite. The two are ground together and then placed in a kiln until the mixture melts at 2900°F (1600°C). The clinker is cooled, ground, and gypsum is added.

Alumina cements are used where high early strength is required and moderate temperatures are to be maintained.

EXPANSIVE CEMENTS

An *expansive cement* is a hydraulic cement containing a constituent that, during the process of hydration, setting, or hardening, undergoes an expansion (increase in volume) but remains sound and eventually develops into satisfactory strength. An expansive cement is normally used in a situation where shrinkage of the concrete cannot be tolerated; hence, it compensates for the shrinkage that will take place.

There are generally three types of expansive cements: Types M (Soviet), K (Klein), and S (Portland Cement Association). Expansive cements can achieve high strength because of the high-alumina or portland cement components.

SPECIAL PORTLAND CEMENTS

A variety of *special cements* exists (white portland, colored cements, oil-well cements, regulated cements, waterproofed cements, hydrophobic cements, antibacterial cements, barium and strontium cements) that are limited for specific uses and purposes. Each type will be mentioned briefly and the reader should keep in mind that this list is by no means complete.

White portland cement is used for decorative displays. It makes an excellent base when colored aggregates are used. White portland cement is low in iron and manganese, which gives the cement its white look, compared to normal portland cement, which is gray. Hence, by reducing iron and manganese in normal portland cement, a white cement is produced.

Colored cements are made by intergrinding a chemically inert pigment such as metallic oxide in the amount of 3 to 10 percent to portland cement. Colored cements, like white portland cement, are used for decorative purposes. However, they have one disadvantage in that they have a tendency to fade over the years.

Oil-well cements are slow-setting cements that are used to seal deep wells. The cement is made in a slurry and pumped to depths within the well under high temperature and pressure before it is allowed to set. These types of cements are governed by the American Petroleum Institute for each of eight classes.

Regulated cements are rapid-setting and -hardening cements. They are used in the manufacture of blocks, pipes, prestressed and precast concrete, and, of course, for patch work. In strength, they are comparable to portland cement Types I, II, and III.

A *waterproofed cement* is a portland cement interground with a water-repellent material, such as calcium stearate. The purpose is to reduce the water permeability of the concrete.

Hydrophobic cements are similar to waterproofed cements, in that portland cement is interground with a hydrophobic (water-repellent) material. However, the purpose is to prolong the life of the cement during storage or while it is being transported long distances.

An *antibacterial cement* is a portland cement interground with an antibacterial agent with the intention of reducing harmful microorganisms. It is used in food-processing plants to minimize deterioration caused by fermentation.

Barium and strontium cements are portland cements in which the calcium oxide is replaced completely or in part by barium oxide or strontium oxide. Their purpose is to act as a concrete shield in which the barium and strontium absorb x-rays and gamma-rays.

PROBLEMS

3.1. Explain the difference between hydraulic and nonhydraulic cements.

3.2. Why is lime important in the manufacturing process of portland cement?

3.3. Discuss the uses of the following: pozzolan cements, slag cements, natural cements, portland cements.

3.4. List eight types of portland cement and explain their uses.

3.5. Explain how the following compounds affect the character of portland cement: (a) tricalcium silicate, (b) dicalcium silicate, (c) tricalcium aluminates, (d) tetracalcium aluminoferrite, and (e) alumina.

3.6. Why are portland-pozzolan cements important?

3.7. What special property of alumina cement makes it use attractive?

3.8. List several special cements and discuss their uses.

3.9 Overall, which types of cements do you view as exceptional? Explain.

3.10. After reviewing outside reference materials, discuss the manufacturing process of portland cements.

REFERENCES

AMERICAN SOCIETY FOR TESTING AND MATERIALS, *Book of Standards*, Part 14, 1979.

BAUER, E. E., *Plain Concrete*, 3rd ed., McGraw-Hill Book Company, New York, 1949.

BROWN, L. S., "Tricalcium Aluminate and the Microstructure of Portland Cement Clinker," *Proc. ASTM, 37*, Part II (1937).

DAVIS, R. E., KELLEY, J. W., TROXELL, G. E., and DAVIS, H. E., "Properties of Mortars and Concretes Containing Portland-Pozzolan Cement," *Am. Concr. Inst.*, 32 (Sept.-Oct. 1935), p. 80.

KREBS, R. D., and WALKER, R. D., *Highway Materials*, McGraw-Hill Book Company, New York, 1971.

LARSON, T. D., *Portland Cement and Asphalt Concretes*, McGraw-Hill Book Company, New York, 1963.

MILLS, A. P., HAYWARD, H. W., and RADER, L. F., *Materials of Construction*, 6th ed., McGraw-Hill Book Company, New York, 1955.

POPOVICS, S., *Concrete-making Materials*, McGraw-Hill Book Company, New York, 1979.

4 CONCRETE: STRENGTH AND BEHAVIOR

INTRODUCTION

Whether used in buildings, bridges, pavements, or any other of its numerous areas of service, concrete must have strength, the ability to resist force. The forces to be resisted may result from applied loads, from the weight of the concrete itself, or, more commonly, from a combination of these. Therefore, strength test is undoubtedly the most common type of test to evaluate the properties of hardened concrete. There are three reasons for this: (1) The strength of concrete, in compression, tension, shear, or a combination of these, has, in most cases, a direct influence on the load-carrying capacity of both plain and reinforced structures; (2) of all the properties of hardened concrete, those concerning strength can usually be determined most easily; and (3) by means of correlations with other more complicated tests, the results of strength tests can be used as qualitative indications of other properties of hardened concrete.

The results of tests on hardened concrete are usually not known until it would be very difficult to replace any concrete that is found to be faulty. These tests, however, have a policing effect on those responsible for construction and provide essential information on cases in which the concrete forms a vital structural element of any building. The results of tests on hardened concrete, even if they are known late, help to disclose any trends in concrete quality and enable adjustments to be made in production of future concrete.

UNIAXIAL BEHAVIOR

Compression Strength

The typical stress-strain relationship for concrete subjected to uniaxial compression is shown in Figure 4–1. The stress-strain curve has a linear-elastic behavior up to about 0.45

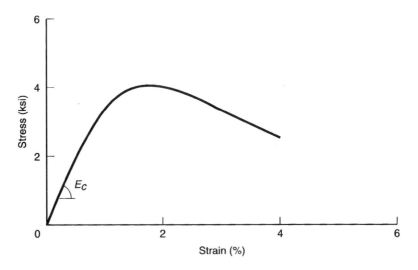

Figure 4-1 Typical Stress-Strain Curve for Concrete in Uniaxial Compression

percent of its maximum compressive strength, f_c'. At higher stresses, resulting from the propagation of a large number of microcracks, especially at interfaces between aggregates and cement paste, a nonlinear behavior of concrete is obtained. The curve in Figure 4-1 shows a gradual increase in curvature up to about 90 percent of f_c'. Beyond this stress value, a sharp bend leads to the peak stress, f_c', that corresponds to a strain value of about 0.002. The descending part of the curve follows until crushing occurs at the ultimate strain value of about 0.0035.

   If unloading concrete in the linear zone, the unloading curve demonstrates only minor nonlinearity. When reloading, a small hysteresis loop is formed, as shown in Figure 4-2. However, if unloading and reloading occur at stresses higher than about 0.45 f_c', the curves obtained indicate a strong nonlinear behavior with a significant degradation of stiffness. The stress-strain curve for monotonic loading serves as a reasonable envelope for the peak stress values for concrete under cyclic loading.

   A line joining the stress-strain curve origin to any point on the curve gives the secant modulus of elasticity. The value of the secant modulus of elasticity at 0.45 f_c', E_c, is generally used as an indicator of concrete stiffness. The value of this secant modulus of elasticity was found to be related to its unit weight and compression strength. Its value can be estimated with reasonable accuracy to form the following equation:

$$E_c = 33(w_c)^{1.5} \sqrt{f_c'}, \text{psi} \tag{4.1}$$

where (w_c) is the unit weight of concrete in pounds/cubic feet within the range of 90 to 145, and f_c' is the axial compressive strength in psi.

   Poisson's ratio, v, relating longitudinal to lateral strains for concrete under uniaxial compressive loading ranges from about 0.15 to 0.22; a typical value is 0.2.

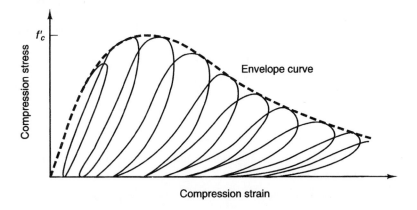

Figure 4-2   Response of Concrete to Cyclic Loading

Specimens.   Specimens to determine the compressive strength of concrete are generally obtained from four different sources: (1) cylinders made in the laboratory; (2) cylinders made in the field; (3) cores of hardened concrete cut from structures; and (4) portions of beams broken in flexure. Each type of specimen has a specific purpose(s).

ASTM C192 (Method of Making and Curing Concrete Compression and Flexure Test Specimens in the Laboratory) describes in detail methods for preparation and examination of the constituent material; proportioning and mixing of concrete; determining the consistency of the mix; and molding, curing, and capping of the specimens. Cylinders made in the laboratory constitute a large portion of the compression specimens. There are three reasons for this: (1) in research, to determine the effect of variations in the materials or conditions of manufacture, storage, or testing on the strength and other properties of concrete; (2) as control tests in conjunction with (a) tests on plain or reinforced concrete members or structures or (b) tests to determine other properties of hardened concrete; and (3) to evaluate mix designs for laboratory field use. In the making of such specimens, large variations can be introduced into the test specimen. These variations may be attributed to the character of the cement, conditions of mixing, character and grading of the aggregate, size of the aggregate, size and shape of the specimen, curing and aging, temperature, and moisture content at time of testing.

ASTM C31 (Method of Making and Curing Concrete Compression and Flexure Test Specimens in the Field) describes the detailed method of making standard cylinders in the field. When cylinders are prepared in the field, they should be made from the same concrete used on the job. In addition, the same curing process, or a process as similar as possible, should be used. The purpose of cylinders made in the field may be to check the adequacy of the laboratory mix design, to determine when a structure may be put in service, or to measure and control the quality of concrete.

Compressive test results of cored hardened concrete usually result in lower compressive strength than anticipated. ASTM C42 (Methods of Obtaining and Testing Drilled Cores and Sawed Beams of Concrete) covers the procedure for securing and testing the cylindrical cores that are most commonly used for determining compressive strength. Cores are only drilled when results of the standard cylinder test are questionable or when investigations are made of old structures.

Finally, ASTM C116 (Test for Compressive Strength of Concrete Using Portions of Beams Broken in Flexure) describes the procedure and apparatus necessary to determine the compressive strength of concrete from broken portions of beams tested in flexure. This test is extremely useful where beam specimens are made to determine the modulus of rupture, as in highway construction, of which one would like an appropriate value of the compressive strength. The method is not meant to be used as a comparison with laboratory cylinder tests. When the method is used and a correlation attempt is made, it becomes necessary to apply a correlaton factor.

Making specimens.   The compressive strength of concrete depends primarily on the water-cement ratio. However, other factors, such as character of the cement, conditions of mixing, character and grading of the aggregate, size of the aggregate, size and shape of the specimen, curing and aging, temperature, and moisture content at the time of testing also have a bearing on the compressive strength.

The characterization of the cement for a given water-cement ratio plays an important role in the early compressive strength development of concrete. All portland cements behave more or less similarly, although the gain in strength with age is not always the same. Some cements gain their strength more rapidly at first, whereas others show greater increase at later periods. This applies not only to the eight types covered by the ASTM specifications but to some extent to different cements within a single group. Tests have shown that the strength for a given water-cement ratio shows the greatest difference among cements at the early ages. For 90 days and later the differences are much less.

The importance of thorough mixing for the development of strength and for uniformity throughout the batch has been long recognized. The earliest studies in concrete showed increases in strengths with continued mixing, but the increase became slight after an initial rapid rise. The time of mixing is governed by ASTM C94.

Surface conditions, the size and shape of the particles, and the gradation are the characteristics of the aggregate that are of principal concern to the strength of the concrete. The surface conditions of the aggregate affect the adhesion of the cement paste to the aggregate particles. The presence or absence of adherent dirt or clay, the roughness, and texture affect the adhesion. These characteristics have greater effect on flexural strength than on compressive strength.

The shape of the particles influences the strength of the concrete. The shape affects the quality and the amount of paste that is required for workability with a given surface area when flat pieces of aggregate occur, especially when they happen to be combined in planes of shear and tension.

When the water-cement ratio is the same and the mixtures are plastic and workable, considerable changes in grading will affect the strength of the concrete only to a small degree. The principal effect of changing the aggregate grading is to change the amount of cement and water needed to make the mixture workable with the desired water-cement ratio.

In general, as the maximum size of the aggregate is increased, lower water-cement ratios can be used for suitable workability, and, therefore, greater strengths are obtained for a given cement content. In the high-strength range, greater than 6000 psi (41 MPa), higher compressive strengths are usually obtained at a given water-cement ratio with smaller maximum sizes of aggregates. Data from compression tests of concrete containing very large aggregates, 4 in. (10.16 cm), are conflicting because of limitations in the size of test specimens.

Compression tests of concrete are ordinarily conducted on cylindrical specimens with height equal to twice the diameter, so that surface rupture, produced on fracture, will not intersect the end bearings. The ends of the cylinders should be carefully formed to give parallel, smooth surfaces so as to obtain uniform distribution of stress. A uniformly stressed cylinder that has been properly molded will break in the shape of a double cone with vertex in the center of the cylinder.

It is generally accepted that the diameter of the specimen should be at least three times the nominal size of the coarse aggregate. A 6 in. × 12 in. (15.24 cm × 30.48 cm) cylinder is the standard for aggregate smaller than 2 in. (5.08 cm). If the aggregate is too large for the size of the mold available, the oversize aggregate may be removed by set screening. If a mold having a diameter less than three times the maximum size of the aggregate is used, the indicated compressive strength will be lowered. However, a large mold may be used; in some cases, molds as large as 36 in. (91.44 cm) in diameter have been used for concrete containing very large aggregates, such as those used in dam construction. Attention must be called to the fact that the size of the cylinder 36 in. (91.44 cm) in diameter by 72 in. (182.88 cm) high may be only about 82 percent of that of a standard 6 in. × 12 in. (15.24 cm × 30.48 cm) cylinder. A reduction in the size of the specimen below that of the standard 6 in. × 12 in. (15.24 cm × 30.48 cm) cylinder will yield a somewhat greater indicated compressive strength.

Unless the specimens are carefully molded, errant and irregular results will be obtained. Generally, a cylinder of poorly compacted concrete will have a lower strength than the one that is properly compacted. Thus, it is necessary for the standard methods for making specimens, ASTM C31 and C192, to specify procedures for compacting, rodding, or vibrating the concrete in the mold. If the specifications under which the work is being performed do not state the method of consolidation, the choice is determined by the slump. Concrete with a slump greater than 3 in. (7.62 cm) should be rodded. If the slump is between 1 and 3 in. (2.54 and 7.62 cm), the concrete may be either vibrated or rodded. When the slump is less than 1 in. (2.54 cm), the specimens must be consolidated by vibration. When the concrete is to be rodded, it should be placed in the cylinder in three layers and rodded 25 strokes per layer if the cylinder is 3 to 6 in. (7.62 and 15.24 cm) in diameter, 50 if 8 in. (20.32 cm), 75 if 10 in. (25.40 cm). When vibrating, the mold is filled and vibrated in two layers. Care must be exercised to vibrate only long enough to obtain proper consolidation. Overvibration tends to cause segregation. These methods are specified to permit reproducibility of results by different technicians.

Cylinder molds should be of nonabsorbent material and are generally steel and plastic; however, cardboard molds are often used in the field. ASTM C470 (Specification for Single-Use Molds for forming 6 × 12 in. [15.24 × 30.48 cm] Concrete Compression Test Cylinders) defines adequate paper molds as well as lightweight sheet molds. Although the cardboard is heavily paraffined, in most cases it absorbs part of the water in the concrete mixture. The use of cardboard molds may lower the observed compressive strength about 3 percent on the average; reductions as great as 9 percent have been noted.

Capping procedures are standardized under ASTM C617 (Capping Cylinder Concrete Specimens). Any material that is sufficiently strong and can be molded can be used for capping. The most common capping materials are neat portland cement paste, high-strength gypsum plaster, and sulphur compounds. The cap should be as thin as possible, and the plane surface of the cap at either end of the specimen should be truly at right angles to the axis of the cylindrical specimen. All surfaces that depart from a plane by more than 0.002 in. (0.005 cm) should be capped, and all caps should be checked for this by means of a feeler gauge and steel straightedge. Slight irregularities can be fixed by scrapping if the capping material has not set too hard. Further, according to the ASTM specification, the surface of the capping must not depart by more than 0.5° from perpendicularity with the axis of the cylinder, or a cant of 1 in 96.

If neat portland cement is used for capping, it should be mixed about 3 hours before use to minimize any harmful effect due to shrinkage. The capping procedure should be carried out at least 3 days before testing.

Gypsum plaster can, according to the ASTM specification, be used for capping, provided that it has a strength of greater than 5000 psi (31 MPa) when tested as a 2 in. (5.08 cm) cube. Suitable mixtures of sulfur (melted) and granular materials applied about 2 hours prior to testing are also recommended, but care is necessary to avoid overheating the melted mixture to prevent loss of rigidity.

Concrete can gain strength only as long as moisture is available and used for hydration. The term *curing* is used in reference to the maintaining of a favorable environment for the continuation of the chemical reactions that occur. It is through the early curing process that the internal structure of the concrete is built up to provide strength and water tightness. Although simply retaining moisture within the concrete may be sufficient for low to moderate cement contents, mixes that are rich in cement generate considerable heat hydration, which may expel moisture from the concrete in the period immediately after setting. The standard curing conditions require that the specimen be held at a temperature of 73.4 ± 3°F (23 ± 1.7°C) and in the "moist conditions" until the time of the test (ASTM C192). Any variation from this procedure may produce a specimen having a different strength from that which would be produced under standard conditions.

Cylinders to be used for quality control should be cured according to the standard conditions; however, cylinders made in the field and tested to measure the strength of the concrete in the structure should be cured in the same manner as the structure. Two methods of field curing are presently used: (1) those that interpose a source of water, in the form of ponding, or a wet material to prevent or counteract evaporation; and (2) those that minimize loss of water by interposing an impermeable medium or by other means. A third method of curing that is used in the manufacture of concrete products is the artificial application of heat while the concrete is maintained in a moist condition.

Curing and aging cannot be separated; an increase in age provides for further chemical combinations if the conditions are favorable for continued reaction. Provided that favorable conditions are met, concrete gains strength with age.

Temperature also plays an important role in the curing process of concrete. The chemical reactions proceed more rapidly at higher temperatures. Tests of specimens sealed against loss of moisture show higher early strength but lower strengths at later ages as the temperature is increased in the range of 40 to 115°F (4.4 to 46.1°C). The U.S. Bureau of Reclamation has found that for job control, specimens cured at 70°F (21.2°C) or lower temperatures at the time of casting and for a few hours thereafter give higher strengths at 1 to 3 months. The rapid stiffening in the first few hours under the higher temperature is apparently detrimental to the later development in strength.

Test procedure.   Once the specimen is made, the method by which it is tested affects the strength obtained. Two of the more important influences are the rate of loading and the eccentricity of loading.

The rate of loading has a definite effect on compressive strength, although the effect is usually fairly small over the ranges of speed used in ordinary testing. The results of tests on concrete indicate that the relationship between strength and rate of loading is approximately logarithmic; the more rapid the rate, the higher the indicated strength. A rapid rate of loading may indicate as much as a 20 percent increase in the apparent compressive strength. For this reason, ASTM C39, which applies also to testing of cores, specifies that the rate of loading for screw-powered machines shall be 0.05 in./1 minute (0.11 cm/minute) and for hydraulic machines 20 to 50 psi/second (1.41 to 3.52 kg/cm²/second).

The effect of eccentric loading is obvious and the alignment of all machines should be checked. Any eccentricity will tend to decrease the strength of the test specimen, the amount of decrease being greater for low-strength than for high-strength concrete. Therefore, to ensure that a concentric and uniformly distributed load is applied to the specimen, a spherically seated bearing block is required on one end, and the specimen should be carefully centered on this bearing block. The object of using the block is to overcome the effect of a small lack of parallelism between the head of the machine and the end face of the specimen, giving the specimen as even a distribution of initial load as possible. It is desirable that the spherically seated bearing block be at the upper end of the test specimen. In order that the resultant of the forces applied to the end of the specimen not be eccentric with the axis of the specimen, it is important that the center of the spherical surface of this block be in the flat face that bears on the specimen and that the specimen itself be carefully centered with respect to the center of this spherical surface. Because of increased frictional resistance as the load builds up, the spherically seated bearing cannot be relied on to adjust itself to bending action that may occur during the test.

Significance of results.   The results of a compression test are essentially only comparative, as the value of the compressive strength obtained cannot be regarded as equal to the strength of the concrete deposited in the work. The value will only give an indication of the quality of concrete, because of the various factors that affect the mix, which have already been discussed. This lack of knowledge regarding the relationship between the strengths of concrete in a cylinder and in a structure requires the use of a larger factor of safety than would otherwise be necessary.

Compressive strength may be used as a qualitative measure of other properties of hardened concrete. No exact relationship exists between compressive strength and flexural strength, tensile strength, modulus of elasticity, wear resistance, fire resistance, or permeability. Only an approximation can be made of these properties. Nevertheless, this approximation is very useful to the engineer.

Compressive tests further aid in the selection of ingredients that may be used in making concrete. Compressive strength is a measure of the indirect effect of admixtures that may be beneficial for one purpose but detrimental for another.

Tension Strength

Concrete is weak in tension compared with its compressive strength. The tensile strength of concrete is about 10 to 20 percent of its compressive strength. The tensile strength of concrete can be obtained directly from tension specimens. However, a direct application of a pure tension force free from eccentricity is difficult to achieve, specially with the uncertainties of secondary stresses induced by the holding devices. No standard test involving the application of direct tension exists. The direct tension strength of concrete, f'_t, is usually taken approximately as

$$f'_t = 4 * \sqrt{f'_c}, \text{psi} \tag{4.2}$$

To determine the tensile strength of concrete experimentally it is customary to test concrete in flexure or by application of a splitting tension.

Flexural test. ASTM C78 (Test Method for Flexural Strength of Concrete [using simple Beam with Third-Point Loading]) does not measure the direct tensile strength of concrete. It rather determines the modulus of rupture, f_r, of concrete prisms subjected to third-point bending loads. The value of f_r is computed from the flexural formula M/Z, where M is the bending moment at failure, and Z is the section modulus of the concrete prism's cross-section. Based on many test results a linear relationship between the modulus of rupture and the square root of the compressive strength has been adopted.

$$f_r = k \sqrt{f'_c}, \text{psi} \tag{4.3}$$

where f_r = modulus of rupture, psi
          f'_c = compressive strength, psi
          k = a constant usually taken between 8 and 10; a lower bound value of
                  7.5 is adopted by the American Concrete Institute Code (1989)

Splitting tensile test. ASTM C496 (Test Method for Splitting Tensile Strength of Cylindrical Concrete Specimens) is a more reliable measure of tensile strength than the modulus of rupture beam test. The most important advantage of this test is the approximate uniformity of tensile stress distribution over the diameter area of the cylinder. During the test, diametrically opposite compressive loads are applied to a concrete cylinder laid on its side. Splitting occurs along a diametrical plane that is subjected to tensile stresses. The tensile stress at splitting, f'_s is obtained from the following equation:

$$f'_s = 2P/(\pi h\, d) \tag{4.4}$$

where P =   the applied load at failure, lb
         h =   the length of the cylinder, in.
         d =   the diameter of the cylinder, in.

The tensile stress at splitting, f'_s, usually ranges from 50 to 75 percent of the modulus of rupture, f_r. The ACI Code (1989) has indirectly used a value $f'_s = 6.7 \sqrt{f'_c}$.

Making specimens.   ASTM C192 and C31 describe the procedures for making flexural test specimens and cylinders for the splitting tensile strength test in the laboratory and in the field. ASTM C31 stipulates that the length of the beam should be at least 2 in. (5.08 cm) longer than three times its depth and that its width should be not more than one-and-one-half times its depth. The minimum depth or width should be at least three times the maximum size of aggregate. A typical specimen used would be 6 in. $\times$ 6 in. $\times$ 21 in. (15.24 cm $\times$ 15.24 cm $\times$ 53.34 cm), and is tested under third-point loading on a span of 18 in. (45.72 cm). As the depth of the beam is increased, there is a decrease in the modulus of rupture.

Many of the parameters that affect the compressive strength of concrete also apply to the flexural strength.

Test procedure.   Flexural strength measurements are extremely sensitive to all aspects of specimen preparation and testing procedure. The principal requirements of the supporting and loading blocks of the apparatus for the flexural strength of concrete are as follows: (1) They should be of such shape that they permit use of a definite and known length of span; (2) the areas of contact with the material under test should be such that unduly high stress concentrations (which may cause localized crushing around bearing areas) do not occur; (3) there should be provision for longitudinal adjustment of the position of the supports so that longitudinal restraint will not be developed as loading progresses; (4) there should be a provision for some lateral rotational adjustment to accomodate beams that have a slight twist from end to end, so that torsional stresses will not be induced; and (5) the arrangement of parts should be stable under load.

Apparatus for measuring deflection should be so designed that crushing at the supports, settlement of the supports, and deformation of the supporting and loading blocks of parts of the machine do not introduce serious errors into the results. One method of avoiding these sources of errors is to measure deflections with reference to points on the neutral axis above the supports.

Routine flexure tests are usually simple to conduct. Ordinarily, only the modulus of rupture is required; this is determined from the load at rupture and the dimensions of the specimen. When the modulus of elasticity is required, a series of load-deflection observations are made.

The dimensions of the concrete specimen should be measured to the nearest 0.01 in. (0.025 cm). The supporting and loading blocks should be located with a reasonable degree of accuracy (0.2 percent of the span length). The supports and specimens should be placed centrally in the testing machine and checked to see that they are in proper alignment and can function as intended. Deflectometers and strianometers should be located carefully and checked to see that they operate satisfactorily and are set to operate over the range required.

The rate of load application, unless standardized, may cause considerable variation in the results of flexure tests, the variation being as much as 15 percent for the range of rates that may be obtained in the average laboratory. Concrete beams may be loaded rapidly at any desired rate up to 50 percent of the breaking load, after which loads should be applied at a rate such that the extreme fibers are stressed at 150 psi (0.93 MPa) or less per minute.

Beams may be tested under either center-point or third-point loading. Third-point loading invariably gives lower strengths than center-point loading. Tests indicate the following order of decreasing magnitude of the strength obtained: (1) center loading, with moment computed at center; (2) center loading, with moment computed at point of fracture; and (3) third-point loading. Third-point loading probably gives lower strengths because the maximum moment is distributed over a greater length of the beam; because the concrete is not homogeneous, this loading method seeks the weakest section.

Curing affects the tensile strength in much the same manner as it affects the compressive strength. A beam that has been allowed to dry before testing will yield lower flexural strength than the one tested in a saturated condition. Consequently, in tests used to determine or control the quality of concrete, uniformity of results will be assured only if the beams are cured in a standard manner and tested when wet.

The temperature of a beam at the time of testing will also affect the results. As the temperature increases, the strength decreases.

ASTM C78 is also prescribed for tests of beams sawed from hardened concrete. When such beams are used, primarily as a control of concrete quality, they should be turned on their sides before testing and will usually require capping because of the irregularity of the sawed surface (the sides are not plane and parallel).

The test for splitting tensile strength described in ASTM C496 is simple to make. The effectiveness with which the material in the bearing strips is able to conform to the irregularities of the surface of the specimen and distribute the load affects the results. For uniformity, ASTM C496 specifies that $\frac{1}{8}$-in. (0.32-cm)-thick plywood should be used for bearing strips. Care must be taken to apply the load through a diametrical plane. The load should be applied such that the stress increases between 10 and 200 psi/minute (7.03 and 14.1 kg/cm²/minute).

Significance of results.   In the case of pavement construction, it appears that the flexural strength of concrete is at least as important as the compressive strength. It has been suggested that pavements be designed on the basis of flexural strength. The design procedure would be exactly the same except that the tables would have to be prepared on the basis of flexural strength instead of compressive strength. Many agencies that are involved in pavements make only flexural tests. The results of these tests give an indication of when the concrete has gained sufficient strength so that load may be applied or the forms removed.

Splitting tension tests have been used to evaluate the bond-splitting resistance of concrete.

With increased emphasis on the control of cracking in reinforced concrete, appreciation of the importance of tensile strength has increased. However, the three general methods of estimating tensile strength give slightly different results. The splitting tensile strength test is the easiest to perform and gives the most uniform results. The tensile strength from the splitting test is about one-and-one-half times greater than that obtained in a direct tension test and about two-thirds of the modulus of rupture.

BIAXIAL BEHAVIOR

Figure 4-3a shows typical experimental stress-strain curves for concrete under biaxial compression. The results of this figure indicate that the maximum compressive strength increases for the biaxial compression state as compared with uniaxial compression state. The amount of increase depends mainly on the principal stresses ratio. Also, it can be observed that the average maximum compressive strain is about 0.003 (same as uniaxial compression state). Under biaxial compression-tension state, the compressive strength decreases linearly as the applied tensile stress is increased, as shown in Figure 4-3b. Figure 4-3c indicates that under biaxial tension the strength is nearly the same as that of uniaxial tensile strength.

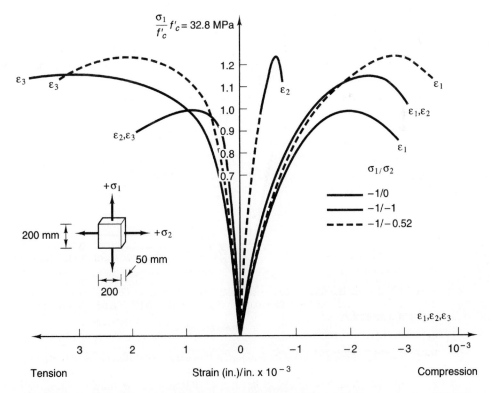

Figure 4-3a   Stress-Strain Relationship of Concrete Under Biaxial Compression State (*Source:* Kepfer, et. al., 1969, 1973)

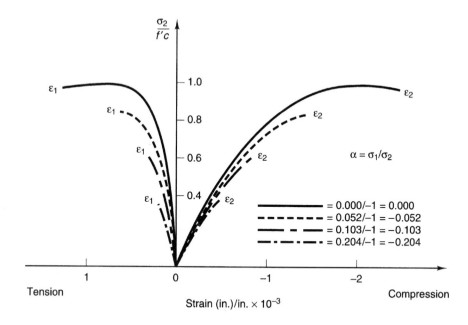

Figure 4-3b   Stress-Strain Relationship of Concrete Under Biaxial Tension-Compression State (*Source*: Kepfer, et al., 1969, 1973)

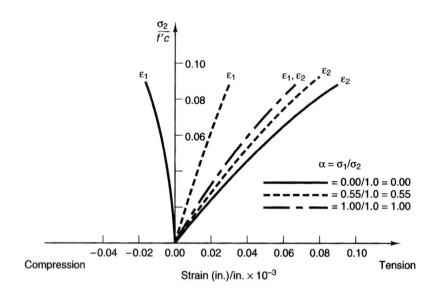

Figure 4-3c   Stress-Strain Relationship of Concrete Under Biaxial Tension State (*Source:* Kepfer, et al., 1969, 1973)

CREEP

Concrete under sustained stress undergoes a gradual increase of strain with time because of creep deformation of concrete. Figure 4-4 indicates that the final creep strain may be several times as large as the initial elastic strain. Generally, creep causes a redistribution of stress in concrete members at the service loads and can lead to an increase in expected deflection values.

The internal mechanism of creep, or as it is sometimes called "plastic flow," can be explained as the result of the combined effects of the following:

1. Crystalline flow in the aggregate and hardened cement paste.
2. Plastic flow of the cement paste surrounding the aggregate.
3. Squeezing of water out of the cement gel because of external loading.
4. Closing of internal voids.

An accurate prediction of creep effect is complicated because of the numerous variables involved, namely, fineness of cement, water-cement ratio, age at loading, magnitude of load, and so on. Several methods based on experimental observations have been proposed for the calculation of creep strain. A general prediction method developed by Branson (1977) has been adopted by the ACI Committee 209 on creep and shrinkage of concrete (1982). This method gives a standard creep coefficient of concrete, C_t, as a function of the dependent variables. The value of C_t is the ratio of creep strain to initial elastic strain. The coefficient for creep at time, t, days after application of load at age t_0, is given by

$$C_t = \left\{ \frac{(t - t_0)^{0.6}}{10 + (t - t_0)^{0.6}} \right\} C_u \qquad (4.5)$$

where
$$
\begin{aligned}
C_u &= 2.35\,\alpha_C \\
\alpha_C &= K_h^c\, K_d^c\, K_s^c\, K_f^c\, K_{ac}^c\, K_{t0}^c \\
&= 1 \text{ for standard conditions.}
\end{aligned}
$$

Graphical representation and general equations for the modification factors (K values) are given in Figure 4-5.

The effect of unloading is shown in Figure 4-6. When at a certain time, t, the load is removed, there is an immediate elastic recovery and long-time creep recovery characterized by a residual deformation.

SHRINKAGE

Because of workability requirement, concrete usually contains more water than what is necessary for the chemical hydration of cement. The excess water (free water) is lost by evaporation and other means leading to shrinkage of concrete. The resulting shrinkage strains are independent of the stress conditions in concrete. If restrained, shrinkage strains

Shrinkage

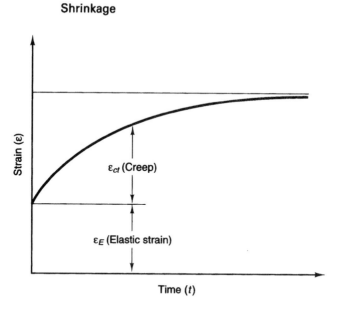

Figure 4-4 Elastic and Creep Strains

can cause cracking of concrete and can eventually increase the deflection of structural concrete members. The shrinkage strain, $\epsilon_{sh_{(t)}}$, is time dependent and increases with time toward a final maximum value called ultimate shrinkage strain, $\epsilon_{sh_{(u)}}$, as shown in Figure 4-7. In general, the same factors that influence creep have been found to also influence shrinkage strains. Branson (1977) proposed a general prediction method that has been also adopted by the ACI Committee 209 on creep and shrinkage of concrete (1982). The method gives a standard shrinkage strain in terms of time dependent variables.

For moist-cured concrete, the shrinkage strain that occurs between $t_0 = 7$ days and any time t is

$$\epsilon_{sh_{(t)}} = \left\{ \frac{t - t_0}{35 + (t - t_0)} \right\} \epsilon_{sh_{(u)}} \quad (t_0 = 7 \text{ days}) \tag{4.6}$$

For stream-cured concrete, the shrinkage strain that occurs between $t_0 = 1$ to 3 days and any time t is

$$\epsilon_{sh_{(t)}} = \left\{ \frac{t - t_0}{55 + (t - t_0)} \right\} \epsilon_{sh_{(u)}} \quad (t_0 = 1 \text{ to } 3 \text{ days}) \tag{4.7}$$

where $\epsilon_{sh_{(u)}}$ = $780 \times 10^{-6} \, \alpha_{sh}$

α_{sh} = $K_h^s K_d^s K_s^s K_f^s K_b^s K_{ac}^s$

= 1 for the standard conditions.

Graphical representation and general equations for the modification factors are given in Figure 4-8.

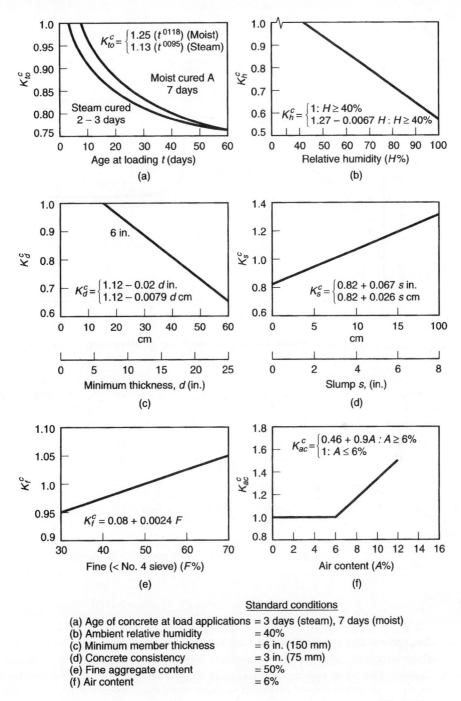

Figure 4-5   Creep Modification Factors (*Source:* ACI Committee 209, 1982)

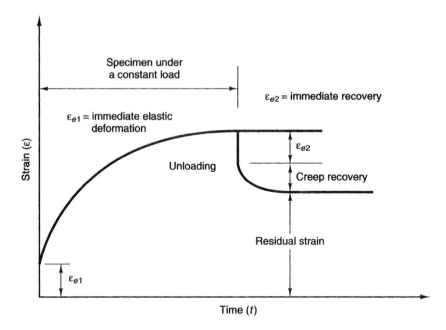

Figure 4-6   Creep Strains Due to Unloading

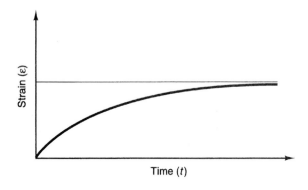

Figure 4-7   Shrinkage Strains

FATIGUE

Fatigue is the process of cumulative damage that is caused by repeated fluctuating loads. Examples of concrete structures that can be subjected to fatigue loading include bridges, offshore structures, and machine foundations. Fatigue can be considered as a special case of dynamic loading in which inertia forces do not influence the stresses. Fatigue damage can occur in concrete members subjected to elastic fluctuating stresses at regions of stress concentration where localized stresses exceed the linear limit of the material. The accumulated damage causes the initiation and the propagation of cracks in the concrete matrix. This results in an increase in deflection and crack width, and in some cases the fracture of the element.

The fatigue life, N, is defined as the total number of cycles required to cause the failure of a certain structure. The fatigue life is influenced by several variables.

1. *Stress.* State of stress, stress range, stress ratio, frequency, and stress concentration.
2. *Materials properties.* Linear and nonlinear characteristics.
3. *Surrounding environment.* Temperature, aggressive elements, and so forth.

It was early in this century when researchers observed the fatigue damage of concrete. Van Ornum (1903) reported that cementitious composites exhibited a progressive type of failure under load repetitions in a range well below the ultimate strength of the material. Researchers also noticed a variation in the shape of the concrete stress-strain curve; with the increase of number of load repetitions, the curve shape changed from concave toward the strain axis to a straight line that shifts at a decreasing rate and finally to concave toward the stress axis as shown in Figure 4-9. The degree of this latter concavity is an indication of the nearness of concrete to failure.

The detailed mechanism of fatigue failure of concrete is still not clearly understood, and a considerable research effort is currently being conducted to study its nature. However, at the present stage of knowledge the following can be concluded:

1. The fatigue of concrete is coupled with the initiation and propagation of internal microcracks at the cement paste-aggregate interface.
2. Cracks resulting from fatigue failure are more extensive than cracks initiated by static compression failure.

The concrete fatigue strength can be expressed using the *S-N* curves. *S* is a characteristic stress of the loading cycle, usually indicating a stress range or a function of the maximum and minimum stress, and *N* is the number of cycles to failure. Usually *S* is presented with a linear scale, whereas *N* is expressed using the log scale. This format results in a linear relationship as shown in Figure 4-10. A point on the curve would represent the "fatigue strength," which is the value of stress range that can cause failure at a given number of stress cycles and at a given stress ratio, and the "fatigue life," which is the number of stress cycles that will cause failure at a given stress range and stress ratio.

BOND STRENGTH

In a reinforced concrete member, several types of interface can be identified, namely: (1) interfaces between the various chemical components that make up the cement paste; (2) interfaces between cement paste and coarse aggregate; and (3) interfaces between concrete matrix and the steel reinforcement.

The cement paste is composed of individual chemical components resulting from the hydration process. The cement paste strength is derived from the chemical ionic-covalent bonds and the physical van der Waals bonds between its components. These components are very well bonded, and any loss of strength in the paste is attributed to the porosity, pore size distribution, and any existing macroscopic cracks rather than to the destruction of the bond between its components.

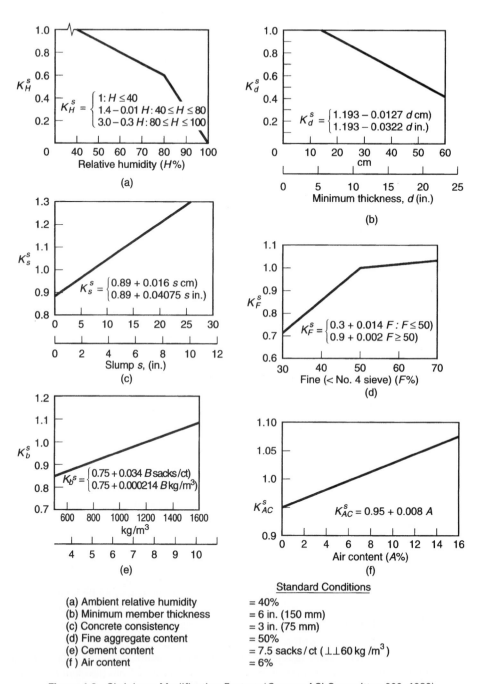

Figure 4-8 Shrinkage Modification Factors (*Source*: ACI Committee 209, 1982)

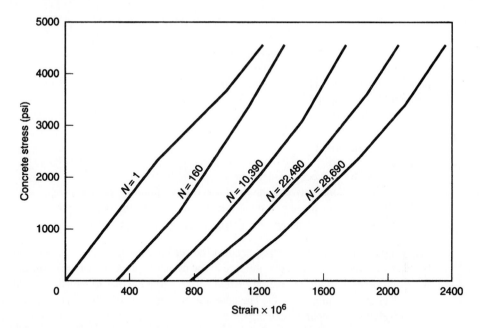

Figure 4-9 Effect of Repeated Load on Concrete Stress-Strain Curve (*Source*: ACI Committee 215, 1974)

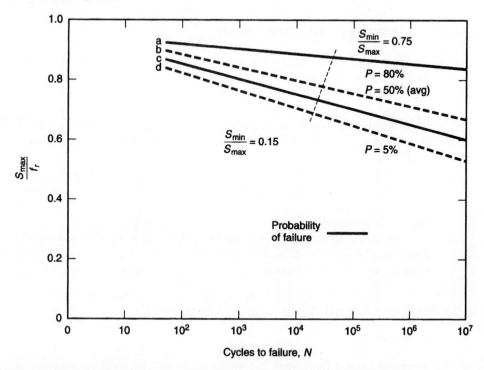

Figure 4-10 Typical *S-N* Relationship for Concrete in Compression (*Source:* ACI Committee 215, 1974)

In general, the interfaces between cement paste and coarse aggregate (transition zone) is considered the weak link in the concrete matrix. This is because bleed water accumulates at the surface of the coarse aggregate particles, creating a porous paste and planes of weakness. Addition of pozzolanic materials reduces the relative thickness of the transition zone by creating a better overall homogenous matrix. This results in an enhanced bond between the coarse aggregate particles and the cement paste. The extent of this enhancement depends on the type of pozzolanic material and its addition percentage.

Three mechanisms contributing to the bond between concrete and steel reinforcement can be identified.

1. Adhesive bond between steel and concrete matrix.
2. Frictional bond between the steel and the surrounding matrix.
3. Mechanical anchoring of steel to concrete through the bearing stresses that develop between the concrete and the deformations of the steel bars.

When a bar is pulled, the rib bears against the surrounding concrete. Friction and adhesion between concrete and steel along the face of the rib act to prevent the rebar from sliding. This force adds vectorially to the bearing stress acting perpendicular to the rib to yield the bond strength as shown in Figure 4-11. The adhesion and friction contribution to the bond strength is relatively little compared with the bearing stress. The bearing stresses are controlled by the radial pressure that the concrete cover and lateral reinforcement can resist before splitting and the effective shear strength capable of shearing the concrete surrounding the rebar.

SHEARING AND TORSIONAL STRENGTHS

The case of pure shear acting on a plane is rarely encountered in actual concrete structures. Hence, no standard tests exist to obtain the direct resistance of concrete to pure shearing stresses. Based on mechanics of materials basics, the state of pure shear stress is coupled with the existence of principal tensile stresses on another plane that are equal in magnitude to the shearing stresses. Because the strength of concrete in tension is usually lower than its shear strength, tension failure eventually occurs before the maximum shear strength is reached.

Only tests under combined stresses can be used to obtain a realistic indication of the strength of concrete in pure shear. Mohr's theory is widely used for that purpose. On the basis of these types of combined stresses tests, the value of the strength of concrete in pure shear is approximately 20 percent of the compressive strength.

Members subjected to pure torsional loads develop diagonal tensile stresses. Therefore, the strength of concrete subjected to torsion is related to its tensile strength, which is related to some function of the square root of the compressive strength.

DURABILITY

Introduction

According to ACI Committee 201 on durability of concrete, the durability of Portland Cement Concrete is defined as its ability to resist weathering action, chemical attack, abra-

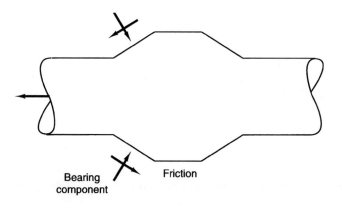

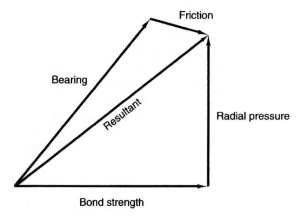

Figure 4-11 Bond Strength in Concrete (*Source:* Treece, and Jirsa, 1989)

sion, or any other form(s) of deterioration. A durable concrete should maintain its original form, quality, and its serviceability when exposed to surrounding environment for a long service life. Water is responsible for many types of physical processes of degradation. It also serves as the carrying agent of soluble aggressive ions that can be the source of chemical processes of degradation. Alkaline calcium compounds form the major part of the hydration cement paste. Hence, concrete is a basic material, and acidic waters are expected to be particularly harmful to it.

To a large extent, durability of concrete depends on the permeability of the composite. Permeability is defined as the property that governs the rate of flow of a fluid into a porous solid. For steady-state flow, the coefficient of permeability (*k*) is determined from Darcy's expression:

$$\frac{dq}{dt} = k \frac{\Delta h\, A}{L\, \mu} \tag{4.8}$$

where dq/dt = the rate of fluid flow
μ = viscosity of the fluid

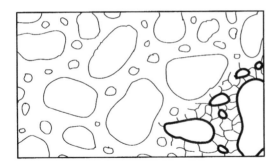

Figure 4-12   Concrete Matrix Microcracks

Tests for the measurement of the permeability are, usually, performed using water, devoid of dissolved air. The permeability coefficient of normal concrete is in the range of $1 \cdot 10^{-10}$ to $30 \cdot 10^{-10}$ cm/second. Higher or lower values could be obtained depending on water-cement ratio, presence of a pozzolanic material, and types of aggregate. Concrete permeability is higher than the permeability of its individual components (i.e., cement paste or aggregates). This is due to the presence of mirocracks in the transition zone between the aggregates and the cement paste as shown in Figure 4-12. These cracks provide the connecting path for the water in the system, thus increasing the permeability of the concrete. Permeability of concrete can be reduced by decreasing the size and number of these microcracks at the interface. This can be achieved, possibly by using an adequate cement content, low water-cement ratio, and proper compacting and curing conditions, and by avoiding early or excessive loading.

Physical Causes of Concrete Deterioration

Physical causes for concrete deterioration are due mainly to surface wear and to exposure to varying temperature extremes. Concrete surface wear, which is considered as the progressive loss of the mass from a concrete surface, can occur due to abrasion, erosion, and cavitation.

Abrasion, erosion, and cavitation.   Abrasion is the effect of dry friction resulting mainly from vehicular traffic. Erosion occurs due to abrasive action of fluids containing solid particles in suspension. This would occur in hydraulic structures and pipelines. Cavitation can also occur in hydraulic structures. Mass is lost due to the presence of high velocities and negative pressures that cause the formation, and the subsequent collapse of vapor bubbles.

Because there are several types of surface wear, no single test method has been found adequate for evaluating floor surfaces. ASTM C-779 (Test Method for Abrasion Resistance of Horizontal Concrete Surfaces) covers three different operational procedures for evaluating floor surfaces, namely, Procedure A—revolving discs; Procedure B—dressing wheels; and Procedure C—ball bearings. Each method has been used to develop information on wear resistance. Underwater surface wear presents special requirements for test procedures. ASTM C-1138 (Test Method for Abrasion Resistance of Concrete [Underwater Method]) uses agitation of steel balls in water to determine abrasion resistance. This test qualitatively simulates the behavior of swirling water containing suspended and transported waterborne particles like silt, sand, gravel, and other solids that produce abrasion of concrete.

Tests and field experience have demonstrated that compressive strength is the most important single factor controlling the abrasion resistance of concrete (abrasion resistance increases with higher compressive strength). For a given concrete mixture, the compressive strength at the surface and the surface wear resistance can be improved by:

1. Avoidance of segregation and elimination of bleeding.
2. Use of properly timed finishing and curing procedures.
3. Minimum water-cement ratio and use of minimum water to aid surface finishing.
4. Use of proper size of aggregates in the surface region.

Freezing and thawing.   It has been well established for many years that freezing and thawing action in extreme temperature regions causes severe deterioration of concrete requiring heavy expenditures for repair and replacement. The most common form of frost damage in concrete is the cracking and spalling, caused by expansion of cement paste matrix, from cyclic freezing and thawing. Typical cracking resulting from freezing and thawing of concrete is shown in Figure 4-13. In presence of moisture and deicing chemicals, scaling can also occur. Hardened cement paste and aggregate have a different behavior when exposed to repeated freeze-thaw cycles. Air entrainment has been successfully used to reduce the risk of damage to concrete by frost action.

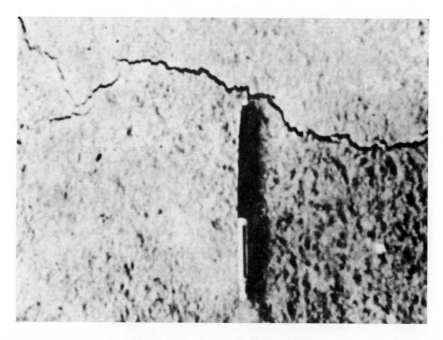

Figure 4-13   Freeze-Thaw Cracking

Frost damage in cement paste was first attributed to stresses caused by hydraulic pressure in the pores resulting from resistance to movement of water away from the regions of freezing. The positive effect of entrained air was explained in terms of shortening the flow paths to places of escape. Later some studies on the frost action of cement paste indicated that the water movement is toward, not away from, sites of freezing. It was proposed that in addition to the hydraulic pressure, the osmotic pressure resulting from partial freezing of solutions in capillaries can be another source of destructive expansion in the cement paste. The water in the cement paste is a weak alkali solution. When the temperature of the concrete drops below the freezing point, ice crystals will form in the large capillaries. This would yield an increase in the alkali content in the unfrozen portion of the solution, creating an osmotic potential that forces water in the unfrozen pore to start diffusing toward the solution in the frozen cavities. The resulting dilution of the solution in contact with the ice allows further growth of the body of ice. By the time the cavity becomes full of ice and solution, any further ice formation produces dilative pressure that can cause the paste to fail. The role of entrained air is to create air bubbles, with small distances between them. The bubbles would attract the unfrozen water from the cavities and stop ice formation in these cavities, alleviating them from the dilative pressure. In general, cement paste of adequate strength can resist freezing by means of entrained air.

Most coarse aggregates used in concrete have pore sizes much larger than those in the cement paste. There is a general agreement that the hydraulic pressure theory previously presented for cement paste is the major cause for deterioration of aggregate resulting from freezing. The role of entrained air in resisting the effect of freezing in coarse aggregate particles is minimal. The degree of permeability of the aggregate and its degree of saturation become the key parameters controlling its resistance to damage by frost action.

Freeze-thaw durability of concrete can be evaluated using ASTM C-666 (Test Method for Resistance of Concrete to Rapid Freezing and Thawing) in which the durability of concrete is measured by subjecting concrete specimens to rapid cycles of freezing and thawing in the laboratory by two different procedures. Procedure A, freezing and thawing in water; or Procedure B, freezing in air and thawing in water. A freezing-and-thawing cycle consists of alternately lowering the temperature of the specimens from 40 to 0°F (4.4 to -17.8°C) and raising it from 0 to 40°F (-17.8 to 4.4°C) in not less than 2, or more than 5, hours. The thawing period is usually higher by 25 or 20 percent of the total time for Procedures A and B, respectively.

From the previous paragraphs, it should be clear that the capability of concrete to resist freezing damage depends on the properties of both the cement paste and the aggregate. For a concrete expected to be exposed to freeze-thaw cycles the following is recommended:

1. Design the structure to minimize exposure to moisture.
2. Use low water-cement ratio.
3. Add appropriate dosage of air-entrained admixture.
4. Use high-quality aggregates.
5. Ensure adequate curing before first freezing cycle.

Chemical Causes of Concrete Deterioration

Concrete deterioration resulting from chemical processes is generally initiated by chemical reactions between aggressive agents present in the external environment and the constituents of the cement paste. Among the exceptions are alkali-aggregate reactions that occur between the alkalis in cement paste and certain reactive materials when present in aggregate and electrochemical corrosion of embedded steel in concrete.

The high pH in concrete (12.5 to 13.5) is mainly due to the large concentration of Na^+, K^+, and OH^- ions. Chemically, any environment with less than 12.5 pH may be assumed aggressive because a reduction of the alkalinity of the pore fluid would lead to destabilization of the cementitious products of hydration and the degradation of the concrete surface (see Figure 4-14). Accordingly, most industrial and natural waters can be categorized as aggressive. However, if the rate of chemical attack on concrete is slow and the permeability is low, the chemical deterioration is considered to have no serious effect. Many situations exist in which chemical deterioration cannot be ignored. The following is a presentation of the mostly known chemical deterioration processes affecting concrete.

Sulfate attack.   Naturally occurring sulfates of sodium, potassium, calcium, or magnesium that can attack concrete are found in soil or dissolved in groundwater adjacent to concrete structures. When evaporation occurs from an exposed face, the sulfates may accumulate at that face, thus increasing their concentration and potential for causing deterioration.

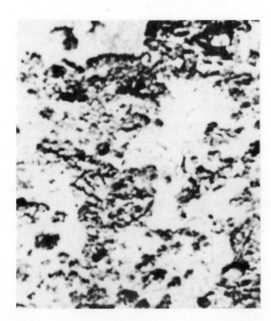

Figure 4-14   Surface Deterioration due to Chemical Attack

There are apparently two chemical reactions involved in the mechanism of sulfate attack on concrete.

1. Combination of sulfate with calcium ions liberated during the hydration of the cement to form gypsum ($CaSO_4 \cdot 2H_2O$).

$$Ca(OH)_2 + MgSO_4 \cdot 7H_2O \rightarrow CaSO_4 \cdot 2H_2O + Mg(OH)_2 + 5H_2O \qquad (4.9)$$

2. Combination of sulfate ion and hydrated calcium aluminate to form calcium sulfoaluminate ($3CaO \cdot Al_2O_3 \cdot 3CaSO_4 \cdot 31H_2O$) also known as ettringite.

$$3CaO \cdot Al_2O_3 + 3\ CaSO_4 + XH_2O \rightarrow$$
$$3CaO \cdot Al_2O_3 \cdot 3CaSO_4 \cdot 31H_2O \qquad (4.10)$$

Both of these reactions result in an increase in solid volume causing expansion and cracking of concrete.

Protection against sulfate attack is obtained by using good quality materials to produce concrete with a low water-cement ratio. There is reasonably good correlation between the sulfate resistance of cement and its tricalcium aluminate (C_3A) content. Hence, use of special cement types can provide good sulfate resistance. ASTM C-150 (Specification for Portland Cement) includes Type II cement (C_3A is limited to 8 percent), which is considered a moderately sulfate-resisting cement and Type V cement (C_3A is limited to 5 percent). Because there is also some evidence that the alumina in the aluminoferrite phase of the Portland cement may participate in the sulfate attack, ASTM C-150 (Specification for Portland Cement) provides that in Type V cement the $C_4AF + 2C_3A$ must not exceed 25 percent. Studies have also shown that some pozzolanic materials and ground granulated iron blast furnace slags, used either in blended cement or added separately to the concrete in the mixer increase the life expectancy of concrete in the sulfate exposure.

Acid attack.   In general, Portland cement does not have good resistance to acids, although some weak acids like oxalic and phosphoric acids can be tolerated, particularly if the exposure is occasional. Acids could be found in soils and industrial waters that are in contact with concrete surfaces. The deterioration of concrete by the acid attack is primarily the result of the reactions between acids and the calcium hydroxide of the hydrated Portland cement. In most cases, these chemical reactions result in the formation of water-soluble calcium compounds that are then leached away from the composite.

To increase the resistance of concrete to acid attack, a dense, low water-cement ratio concrete should be used, and exposure time to acids should be minimized. Otherwise, concrete should be protected using barrier systems.

Carbonation.   When concrete or mortar is exposed to carbon dioxide, reactions producing carbonates occur that are accompanied by shrinkage that amount to one-third of the total shrinkage. The source of the carbon dioxide can be either the atmosphere or water carrying dissolved CO_2. A development of a network of fine cracks called crazing may occur on the surface resulting from carbonation of the surface. Carbonation may reduce the pH value of the system making the reinforcing steel susceptible to corrosion.

Corrosion of reinforced concrete.   Corrosion of steel in concrete can be explained as an electrochemical process that demands the existence of an anode where electrons are liberated and ferrous ions are formed (oxidation), and a cathode where hydroxyl ions are liberated (reduction). This would occur in concrete because of differences in concentration of soluble ions. This results in parts of the reinforcing steel becoming anodic and others cathodic. The ferrous ions subsequently react with oxygen to produce various forms of rust. The transformation of metallic iron to rust is accompanied by an increase in volume, which depending on the state of oxidation may be as large as 600 percent of the original metal. This volume increase is believed to be the main reason for concrete expansion, cracking, and deterioration.

$$\text{At anode} \quad Fe = Fe^{++} + 2e^- \text{ iron dissolves}$$
$$\text{At cathode} \quad 2e^- + H_2O + \tfrac{1}{2}O_2 \rightarrow 2(OH)^-$$

then
$$Fe^{++} + 2OH \rightarrow Fe(OH)_2 \text{ (rust)} \qquad (4.11)$$

Steel in concrete is usually protected against corrosion by the high pH of the surrounding Portland cement paste (minimum pH of 12.5). Also, steel reinforcement is covered by a thin iron-oxide film that becomes impermeable and strongly bonded to the steel surface in alkaline environments. Carbonation of the Portland cement paste can lower the pH to levels of 8 to 9, destroying the passivity of steel and initiating the corrosion reactions. The presence of moisture and supply of oxygen are necessary to maintain the flow of electrons from the cathode to the anode where they are consumed.

The presence of a strong electrolyte-like soluble chloride ions can accelerate the corrosion process. The presence of chloride by more than 0.2 percent by mass of Portland cement may destroy the protective film on the steel surface at pH values even higher than 11.5, setting the stage for the corrosion process. Chloride present in the concrete when the latter is being made is called "domestic chloride," whereas that which is introduced from the surrounding environment into concrete is called "foreign chloride." Domestic chloride may be found in set-accelerating admixtures, water-reducing admixtures, aggregates, and cementitious materials. The chloride content of the Portland cement and its raw materials is typically very low. Potable water used for mixing contains insignificant levels of chloride. Aggregates can contain chloride salts, especially if they are obtained from a location in contact with seawater or groundwater containing chloride. When using admixtures to improve concrete performance, care should be taken to ensure that their contribution to the percentage of chloride ions by mass of cement is kept below the 0.1 percent level. Foreign chloride is introduced into concrete more rapidly through cracks than by the slow diffusion process.

To minimize the corrosion effect, enough concrete cover must be provided to encase the reinforcing steel fully. Crack formation can also be minimized if the concrete is made with the lowest possible water-cement ratio, and proper casting and curing specifications are adhered to.

Cost of repairing corrosion's damage are usually very high. The most successful systems that have been found to be effective are:

1. Overlays of very low water-cement ratio (less than 0.32).
2. Use of epoxy coated reinforcing steel.
3. Adding surface protective barrier systems produced from select epoxies, polyurethanes, and so forth.
4. Using of cathodic protection.
5. Using concrete carrying a corrosion inhibitor such as $NaNO_2$.

Aggregate chemical reactions.   The alkali-aggregate reactions became a noticeable problem in many countries when cement plants targeting more economical production methods began to produce cements with higher alkali contents. Other recent factors include the recycling of alkali rich flue dust, a changeover to coal firing, the widespread use of aggregates of marginal quality, the production of concrete with higher cement contents and low water-cement ratios, and the increasing use of fibers and admixtures.

The chemical reactions of aggregates in concrete can affect the performance of concrete structures. It may manifest itself as map or pattern cracking (Figure 4-15), or popouts Figure 4-16 on the exposed surface. Some reactions that improve the bond between the aggregate and the paste can be beneficial. Other reactions could yield to excessive cracking because of abnormal internal expansions.

Figure 4-15   Map Crackling due to Reactive Aggregate

Figure 4-16   Popouts due to Reactive Aggregate

Alkali-Silica Reaction.   This reaction is by far the most important of all the alkali-aggregate reactions. It occurs between alkalis and the microcrystalline phases of silica that may be found in volcanic, metamorphic, and sedimentary rocks. The expansion is caused by the formation of a gel-like material around the reactive aggregate. This material dries to a white amorphous substance around the aggregate (reactive rim). The reaction between the cement and aggregates may be explained as follows. During the initial mixing of the concrete and for several hours thereafter, the water in the concrete acquires alkalis (sodium oxide and potassium oxide) from the cement by preferential solubility. As the hydration process continues, the calcium aluminates and the silicates of the cement paste extract water. Thus, less water is left in solution, resulting in an increase in the alkalis in the remaining water. If this results and if the cement is high in alkali, a rather caustic solution is formed. During curing this caustic solution reacts chemically with susceptible aggregates to form a silica-gel reaction rim around the aggregate particle. This rim has a great affinity for water, thus attracting water from the cement paste and reducing its viscosity. Osmotic pressures are set up by the drawing of the water from the cement paste, resulting in fracture of the cement paste close to the reaction rim. As the silica gel continues to grow along the fracture, cracks develop and extend until the entire concrete surface is a series of cracks. These cracks can lead to serious problems especially if they allow water to penetrate, and then go through cycles of freezing and thawing. The alkali-silica reactions are accelerated as the temperature is increased, but beyond a certain point higher temperature actually decreases this expansion. This is possibly due to the formation of more crystallized products at higher temperatures. The maximum expansion resulting form the alkali-silica reaction seems to depend on a certain proportion of the reactive material in the aggregate, known as the pessimum content.

Alkali-Carbonate Reaction. This reaction occurs when types of carbonate rocks are used as aggregates. This reaction produces expansion in concrete in the form of pattern or map cracking. Such expansive reactions have been observed in concrete containing coarse dolomitic limestone aggregates. This type of reaction is different from the alkali-silica reaction, because of the absence of characteristic reaction product (reactive rim) normally detected by visual or microscopic methods. A typical alkali-carbonate reaction is as follows:

$$CaMg(CO_3)_2 + 2MOH \rightarrow Mg(OH)_2 + CaCO_3 + M_2CO_3 \qquad (4.12)$$

Where M represents potassium, sodium, or lithium. In concrete, the alkali carbonate produced by this reaction would react with hydration products of portland cement to regenerate alkali.

$$M_2CO_3 + Ca(OH)_2 \rightarrow 2MOH + CaCO_3 \qquad (4.13)$$

This process continues until all the dolomite has reacted or all the alkali has been used up. Another important factor in the alkali-carbonate reaction is the texture of the aggregate. If the dolomite has a coarse grain, the reaction will not occur to any significant degree. For the reaction to occur, the dolomite must consist of fine-grained particles. The alkali expansive carbonate rocks are normally gray, very fine-grained, dense, and of coarse texture. The expansion is greater with cements having a higher alkali content. Higher temperatures also promote higher expansions.

These types of unwanted reactions can be avoided by

1. Carefully choosing the aggregate type to be introduced in the concrete (avoiding reactive aggregates).
2. Not using cement with a high alkali content.
3. Using pozzolanic admixtures to mitigate the deleterious consequences of the reaction. This method of avoidance is covered by ASTM C-441 (Test Method for Effectiveness of Mineral Admixtures on Ground Blast-Furnace Slag in Preventing Excessive Expansion of Concrete Due to the Alkali-Silica Reaction).

To avoid the use of a reactive aggregate, ASTM tests could be used to evaluate the reactivity and the performance characteristics of aggregates. These tests entail a petrographic examination, a chemical test, and a proof test. ASTM C-295 (Guide for Petrographic Examination of Aggregates for Concrete) provides specific information with reference to deleterious materials. Some reactive aggregates can be recognized readily, whereas others require more testing such as ASTM C-289 (Test Method for Potential Reactivity of Aggregates [Chemical Method]) and ASTM C-227 (Test Method for Potential Alkali Reactivity of Cement-Aggregate Combinations [Mortar-bar Method]).

PROBLEMS

4.1. Why are tests to determine the strength of concrete the most frequently used to evaluate the properties of hardened concrete?

4.2. What is the significance of the compressive strength of concrete?

4.3. How are the specimens obtained for the compression test of concrete?

4.4. The compressive strength of concrete depends upon many factors. List them and discuss each in detail.

4.5. In the testing of concrete in compression, what factors can affect the results of the test? Explain.

4.6. What is the significance of flexural strength?

4.7. What is splitting tensile strength?

4.8. What factors during testing affect the final results of the flexure test?

4.9. Explain the significance of the shearing strength and the combined stresses.

4.10. Why is fatigue strength important?

REFERENCES

AAS-JAKOBSON, K., "Fatigue of Concrete Beams and Columns," *Bull. No. 70-1*, NTH Institute of Betonkonstruksjoner, Trondheim, September 1970, 148 pp.

ACI Building Code Requirements for Reinforced Concrete (ACI 318-89) and Commentary (ACI 318R-89), American Concrete Institute, Detroit, Michigan, 1989, 353 pp.

ACI Committee 209, "Prediction of Creep, Shrinkage, and Temperature Effects in Concrete Structures," *Designing for Creep and Shrinkage in Concrete Structures* (SP-76), American Concrete Institute, Detroit, Michigan 1982, pp. 193–300.

ACI Committee 201 on Durability of Concrete, "Proposed Revision of: Guide to Durable Concrete," *Mat.J.*, *88*(September-October 1991), pp. 544–582.

ACI Committee 215, "Consideration for Design of Concrete Structures Subject to Fatigue Loading," *J. Am. Concr. Inst. Proc. 71* (March 1974), pp. 97–121 (revised in 1986).

BARTOS, P. (ed.), *Bond in Concrete,* Applied Science Publishers, New York, 1982, 466 pp.

BRANSON, D. E., *Deformations of Concrete Structures,* McGraw-Hill Book Company, New York, 1977.

BROWNE, R. D., "Mechanisms of Corrosion of Steel in Concrete in Relation to Design, Inspection, and Repair of Offshore and Coastal Structures," *Performance of Concrete in Marine Environment,* SP-65, American Concrete Institute, Detroit, 1980, pp. 169–204.

CHEN, W. F., *Plasticity in Reinforced Concrete,* McGraw-Hill Book Company, New York, 1982, 474 pp.

CORDON, W. A., "Freezing and Thawing of Concrete—Mechanisms and Control," *Monograph No. 3,* American Concrete Institute/Iowa State University Press, Detroit, 1966, 99 pp.

DAHIR, S. H., "Relative Resistance of Rained-On Concrete Pavements to Abrasion, Skidding, and Scaling," *Cem., Concr. Aggreg. 3* (summer 1981), pp. 13–20.

DERUCHER, K. N., "Applications of the Scanning Electron Microscope to Fracture Studies of Concrete," *Build. Environ., 13* (1978), pp. 135–141.

DERUCHER, K. N., and HEINS, C. P., *Bridge and Pier Protective Systems and Devices,* Civil Engineering Series, Marcel Dekker, Inc., New York, 1979.

DIKEOU, J. T., "Fly Ash Increases Resistance of Concrete to Sulfate Attack," *Research Report No. 23,* U.S. Bureau of Reclamation, Denver, 1975, 17 pp.

HSU, T. C., "Fatigue of Plain Concrete," *ACI J. Proc.* (July—August 1981), pp. 292–305.

IDORNE, G. M., "Interface Reactions between Cement and Aggregate in Concrete and Mortar—bond Strength and Durability," *General Report, Theme VII, Proceedings,* 7th International Congress on the Chemistry of Cement, Paris, 1980, pp. 129–147.

KEMP, E., "Bond in Reinforced Concrete: Behavior and Design Criteria," *ACI J. Proc. 82 No. 1* (January-February 1986), pp. 49–57.

KUPFER, H., and GERSTLE, K., "Behavior of Concrete under Biaxial Stresses," *J. Eng. Mech. Div. 99* (August 1973).

KUPFER, H., HILSDORF, H. K., and RUSH, H., "Behavior of Concrete under Biaxial Stresses," *J. Am. Concr. Inst 66,* (August 1969), pp. 656–666.

LEEUWEN, J. V., and SIEMES, J. M., "Miner's Rule with Respect to Plain Concrete," *Heron, 24* (1979), 34 pp.

LUDWIG, U., "Durability of Cement Mortars and Concretes," *Durability of Building Materials and Components,* STP-691, ASTM, Philadelphia, 1980, pp. 269–281.

MEHTA, P. K., *Concrete Structure, Properties, and Materials,* Prentice Hall, Inc., Englewood Cliffs, N.J., 1986.

MINDESS, S., "Bonding in Cementitious Composites—How Important is it?", In Mindess, S., and Shah, S. P. (eds.), *Bonding in Cementitious Composites,* vol. 114, Materials Research Society, Detroit, Michigan, 1988, pp. 3–10.

MINDESS, S., and YOUNG, J. F., *Concrete.* Prentice Hall, Inc., Englewood Cliffs, N.J., 1981.

RAMACHANDRAN, V. S., FELDMAN, R. F., and BEAUDOIN, J. J., *Concrete Science: Treatise on Current Research,* Heyden & Son, Ltd., 1981, 427 pp.

RAMACHANDRAN, V.S., Editor, *Concrete Admixtures Handbook*, Noyes Publications, 1984, 626 pp.

ROGERS, C. A., "Evaluation of the Potential for Expansion and Cracking of Concrete Caused by the Alkali-Carbonate Reaction," *Cem., Concr., Aggreg. 8* (summer 1986), pp.13–23.

SIDNEY, D., "Review of Alkali-Silica Reaction and Expansion Mechanisms: I. Alkalies in Cements and in Concrete Pore Solutions," *Cem. Concr. Res., 5* (July 1975), pp. 329–346.

SIDNEY, D., "Review of Alkali-Silica Reaction and Expansion Mechanisms: II. Reactive Aggregates," *Cem. Concr. Res., 6* (July 1975), pp. 549–560.

TEPFERS, R., and KUTTI, T., "Fatigue Strength of Plain, Ordinary, and Light-weight Concrete," *ACI J.,* (May 1979), pp. 635–653.

TREECE, R. A., and JIRSA, J. O., "Bond Strength of Epoxy-Coated Reinforcing Bars," *ACI Mat. J. Proc. 86* (March-April 1989), pp. 167–174.

VAN ORNUM, J. L., "Fatigue of Cement Products," *Trans. Am. Soc. Civil Eng., 51* (1903), p. 443.

VAN ORNUM, J. L., "Fatigue of Concrete," *Trans. Am. Soc. Civil Eng., 58* (1907), pp. 294–320.

WANG, P. T., SHAH, S. P. and NAAMAN, A. E., "Stress-Strain Curves of Normal and Lightweight Concrete in Compression," *ACI J. Proc. 75* (November 1975), 603–611.

5 DESIGN PROCEDURE IN MAKING CONCRETE

Concrete is a composite material made up of inert materials of varying sizes that are bound together by a binding medium. Mortar is made up of a mixture of cement, water, air, and fine aggregate. Concrete contains coarse aggregate in addition to cement, water, air, and fine aggregate. The cement, water, and air combine to form a paste that binds the aggregates together. Thus, the strength of the concrete is dependent on the strength of the aggregate-matrix bond. The entire mass of the concrete is deposited or placed in a plastic state and almost immediately begins to develop strength (harden), a process which, under proper curing conditions, may continue for years. Because concrete is initially in a plastic state, it lends itself to all kinds of construction, regardless of size or shape. One drawback is that the concrete in a plastic condition must be placed within forms, and these forms cannot be removed until the concrete has hardened somewhat.

In types of work where concrete is used to counteract compressive stresses, it is an excellent building material. However, if the concrete must counteract tensile stresses, it must be reinforced with steel, as concrete is weak in tension.

CONCRETE MATERIALS

Cement

Usually, portland cement is specified for general concrete construction work and should conform to the standard specifications of the ASTM. From time to time special cements may be required if the project involves unusual requirements. Chapter 3 discussed the various types of portland cement as well as special cement requirements. The reader is directed to Chapter 3 for a description of the specifications established by the ASTM and also to review the importance and uses of special cements.

Water

There is usually very little trouble in obtaining water for use in concrete. Almost any water that is drinkable may be used to make concrete. Drinking water with a noticeable taste or odor should not be used until it is tested for organic impurities. Impurities in mixing water may cause any one or all of the following:

1. Abnormal setting time.
2. Decreased strength.
3. Volume changes.
4. Efflorescence.
5. Corrosion of reinforcement.

Some of the impurities in mixing water that cause these undesirable effects in the final concrete are:

1. Dissolved chemicals.
2. Seawater.
3. Sugar.
4. Algae.

Dissolved chemicals may either accelerate or retard the set and can substantially reduce the concrete strength. Further, such dissolved chemicals can actively attack the cement-sand bond, leading to early disintegration of the concrete.

Seawater containing less than 3 percent salt is generally acceptable for plain concrete but not for reinforced or prestressed concrete. The presence of salt can lead to corrosion of the reinforcing bars and prestressing of tendons.

If sugar is present in even small amounts, it can cause rapid setting and reduced concrete strength.

Algae can cause a reduction in the strength of concrete by increasing the amount of air captured in the paste and reducing the bond strength between the paste and the aggregate.

Aggregates

The requirements for fine and coarse aggregate for concrete construction were described in Chapter 2.

PRINCIPAL REQUIREMENTS FOR CONCRETE

In the design of concrete mixes, three principal requirements for concrete are of importance:

1. Quality.
2. Workability.
3. Economy.

Quality

The *quality* of concrete is measured by its strength and durability. Hardened concrete must have sufficient strength to resist the stresses from loads as well as the stresses created by its own weight. The compressive and the flexural strength of concrete are both important in the design of concrete structures. The principal factors affecting the strength of concrete, assuming sound aggregates, are the water-cement ratio and the extent to which hydration has progressed. *Hydration* is the chemical reaction that takes place between the water and cement while the concrete is hardening. The strength of concrete at the end of 28 days is the generally accepted standard for evaluating the strength properties of concrete. Laboratory compressive strengths are usually obtained by testing cylinders 6 in. (15.24 cm) in diameter and 12 in. (30.48 cm) in height under gradually increased compressive loading until failure occurs. On the other hand, flexural strengths are generally obtained by loading a suitably configured test specimen transverse to the longitudinal axis at the third points, until failure occurs. That is, the force is applied simultaneously at one-half and two-thirds of the distance between the end supports.

Durability of concrete is the ability of the concrete to resist the forces of disintegration due to freezing and thawing and chemical attack. These factors were discussed in more detail in Chapter 4.

Workability

Workability of concrete may be defined as a composite characteristic indicative of the ease with which the mass of plastic material may be deposited in its final place without segregation during placement, and its ability to conform to fine forming detail. Workability is a term for which there is no perfectly satisfactory definition. The size and gradation of the aggregate, the amount of mixing water, the time of mixing, and the size and shape of the forms are all factors that affect workability.

No satisfactory measures of workability have been defined. However, the term *consistency* is generally considered a descriptive term for workability. Consistency measures the fluidity or the lack of it. The most generally used test for the measurement of consistency is ASTM C143 (Slump of Portland Cement Concrete).

Economy

Economy takes into account effective use of materials, effective operation, and ease of handling. The cost of producing good-quality concrete is an important consideration in the overall cost of the construction project.

INFLUENCES OF INGREDIENTS ON PROPERTIES OF CONCRETE

The amount of each ingredient used in concrete must be proportioned very carefully to produce the desired effects so that the concrete may be used for its intended purposes. If the proportions are not as designed, adverse effects may result. Table 5-1 shows the influence of each principal ingredient on the properties of concrete.

TABLE 5-1   INFLUENCE OF INGREDIENTS
ON THE PROPERTIES OF CONCRETE[a]

Ingredient	Quality	Workability	Economy
Aggregate	Increases	Decreases	Increases
Portland cement	Increases	Increases	Decreases
Water	Decreases	Increases	Increases

[a]From W. A. Cordon, *Properties, Evaluation, and Control of Engineering Materials*, McGraw-Hill Book Company, New York, 1979.

Aggregates

Almost any type of aggregate can be used for the making of concrete. The most commonly used aggregates are sand, gravel, crushed stone, and air-cooled blast-furnace slag. These aggregates produce normal-weight concrete ranging from 135 to 160 lb/ft^3. Shales, clay, slag, and slate can also be used in the making of concrete. These aggregates are used to make lightweight concrete weighing between 85 and 115 lb/ft^3.

Aggregates exhibit a variety of physical and chemical characteristics; hence, their influence on concrete mixtures is varied. Physical characteristics include size and shape of the aggregate, surface texture, gradation, and top size of aggregate. Chemical characteristics of aggregates are those that may result in aggregate reactivity with the hardened concrete. In any concrete structure the maximum amount of aggregate should be used.

Water

In many concrete mix design procedures, water is considered an influential part of the total mix, as the water allows the concrete to be handled easily.

In other words, it allows the concrete mixture to be workable. Water is also important from the standpoint of hydration. Too much water added to the mixture results in poor-quality concrete. Water reacts with the cement particles, resulting in a chemical change that binds the paste to the fine and coarse aggregate particles. The addition of water to the cement forms a paste that acts as a glue. If too much water is added, the glue becomes diluted, and this leads to a weak bond between the paste (matrix) and the aggregate. In addition to this, excess water produces segregation of the aggregate particles from the paste, and this results in a nonuniform mix.

Too little water produces a dry mix that easily crumbles under its design load, resulting in failure of the structure.

A normal bag of cement requires $2\frac{1}{2}$ to 3 gallons of water to produce a mix that hydrates at the appropriate rate, resulting in a concrete mix of good quality.

Portland Cement

When water is added to portland cement a chemical reaction (hydration) takes place and a calcium silicate hydrate is produced. The amount of water needed to complete this reaction is approximately 30 percent of the cement by weight. However, if this is the only amount of water added to the mix, the mix will be stiff and unworkable. Therefore, additional water is added to the mix to make it more plastic and workable. The ratio of water to cement *(w/c)* determines the quality of the paste and controls the strength of the concrete. The *w/c* ratio is the most significant item affecting the strength of the concrete mix.

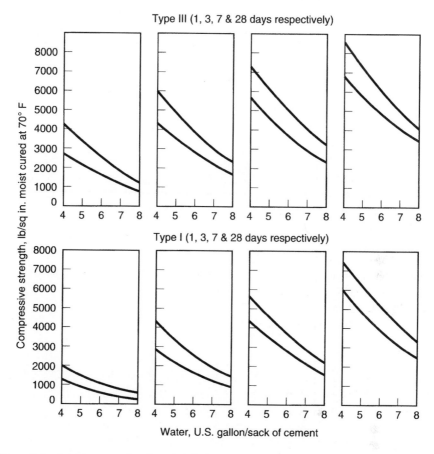

Figure 5-1 Age/Compressive Strength Relationships for Types I and III Portland Cement

A discussion of this fact was published by Duff A. Abrams in 1918 and is frequently called the *Abrams' water-cement ratio law.* Extensive research has proved this law to be valid, and graphs of such experimental work (Figure 5-1) are available for various types and combinations of materials. This type of graph is used as a beginning point in mix proportioning when an individual job design is not justifiable. Table 5-2 gives the recommended water-cement ratios for various types of structures and degrees of exposure.

PROPORTIONING CONCRETE MIXES

In proportioning concrete mixes we will briefly discuss four methods: the trial-batch method, job-curve method, the mortar-voids method, and the method of Goldbeck and Gray.

Trial-Batch Method

The purpose of the *trial-batch method* is to produce a given water-cement ratio such that the factors of quality, workability, and economy are balanced for the most desirable combination of aggregates. The size and gradation of the aggregate, surface texture of the fine

TABLE 5–2 WATER CONTENTS SUITABLE FOR VARIOUS CONDITIONS OF EXPOSURE (GALLONS PER BAG OF CEMENT)

Types or Location of Structure	Severe or Moderate Climate: Wide Range of Temperatures, Rain and Long Freezing Spells or Frequent Freezing and Thawing					Mild Climate: Rain, or Rarely Snow or Frost				
	Thin Sections		Moderate Sections		Heavy and Mass Sections	Thin Sections		Moderate Sections		Heavy and Mass Sections
	Reinforced	Plain	Reinforced	Plain		Reinforced	Plain	Reinforced	Plain	
At the water line in hydraulic or waterfront structures or portions of such structures where complete saturation or intermittent saturation is possible, but not where the structure is continuously submerged										
In seawater	5	5½		6		5	5½		6	
In fresh water	5½	6		6½		5½	6		6½	
Portions of hydraulic or waterfront structures some distance from the										

Condition						
water line, but subject to frequent wetting						
By seawater	$5\frac{1}{2}$	6	6	6	$5\frac{1}{2}$	7
By fresh water	6	$6\frac{1}{2}$	$6\frac{1}{2}$	7	6	$7\frac{1}{2}$
Ordinary exposed structures, buildings, and portions of bridges not coming under groups above	6	$6\frac{1}{2}$	7	7	6	$7\frac{1}{2}$
Complete continuous submergence						
In seawater	6	$6\frac{1}{2}$	7	$6\frac{1}{4}$	6	7
In fresh water	$6\frac{1}{4}$	7	$7\frac{1}{2}$	7	$6\frac{1}{4}$	$7\frac{1}{4}$
Concrete deposited through water	a	$5\frac{1}{2}$	$5\frac{1}{2}$	a	a	$5\frac{1}{2}$
Pavement slabs directly on ground						
Wearing slabs	$5\frac{1}{2}$	6	a	$6\frac{1}{2}$	6	a
Base slabs	$6\frac{1}{2}$	7	a	$7\frac{1}{2}$	7	a

Special cases:

(a) For concrete exposed to strong sulfate groundwaters or other corrosive liquids or salts, the maximum water content should not exceed 5 gal. per bag.

(b) For concrete not exposed to the weather, such as the interior of buildings and portions of structures entirely below ground, no exposure hazard is involved and the water content should be selected on the basis of the strength and workability requirements.

[a]These sections not practicable for the purpose indicated.

and coarse aggregate, and the proportions of fine and coarse aggregate are the most impor- tant factors in determining the combination that will give the best quality and most desir- able workability at the lowest cost.

A rough procedure for a typical trial-batch method of proportioning follows. Select the desired water-cement ratio. Weigh out a definite amount of cement and place it in the mixing apparatus, with the proper amount of water, to obtain the required water-cement ratio. Weigh out definite quantities of fine and coarse aggregate and place them in a con- tainer. Add fine and coarse aggregate from the weighed quantities to the cement-water paste until the desired consistency is obtained for the plastic state.

The remaining fine and coarse aggregate not used to make the plastic mixture is weighed and subtracted from the original quantities. The proportions of cement to sand to coarse aggregate (by weight) may be changed to volumetric proportions by dividing the weight of each material by its unit weight.

To measure the consistency of the mix, the slump-cone test is utilized. Table 5-3 shows the recommended consistency for various classes of concrete structures.

After the desired proportion has been established and the desired consistency is sat- isfactory, further experimenting may be carried out by finding a desirable ratio of fine to coarse aggregate. Concrete mixtures with a low cement-sand mortar (one that does not fill the spaces between pebbles) are hard to work with and result in a honeycombed surface. A concrete mixture with a high cement-sand mortar produces a very porous concrete.

The trial-batch method may be used for making a batch of any size. The accuracy obtained is dependent on the care exercised in producing the batch.

Job-Curve Method

The *job-curve method* utilizes Abrams' water-cement-ratio law but allows variations due to the differences caused by cements and aggregates. Trial batches are prepared utilizing dif- ferent water-cement ratios, using the cement and fine and coarse aggregate to be used on the job. Various test specimens are molded, cured, and tested. A job curve is plotted show- ing compressive strength versus gallons of water per sack of cement. If a strength of 3500 psi were required for a specific job after 28 days of curing, one would simply find 3500 psi on the job curve, read across until the job curve was intersected, and read down to the required number of gallons per sack of cement. A further trial batch would be run with the new data to refine the properties of the concrete and the necessary quantities of materials for the design mixture. Figure 5-2 shows a typical job curve in which the desired strength would be 3500 psi and the necessary water in gallons per sack of cement would be 6.

TABLE 5-3   RECOMMENDED MIX CONSISTENCIES FOR CEMENT

Type of Structure	Slump (in.)	
	Minimum	Maximum
Massive sections, pavement and floors laid on ground	1	4
Heavy slabs, beams, or walls	3	6
Thin walls and columns, ordinary slabs or beams	4	8

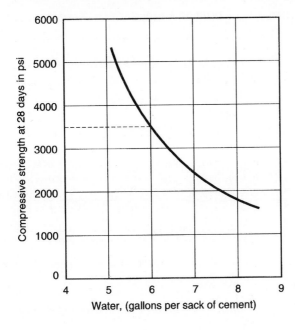

Figure 5-1 Typical Job Curve

Mortar-Voids Method

The *mortar-voids method* was developed by A. N. Talbot in 1923. His method involves the proportioning of concrete by the determination of voids in the mortar. It was his belief that the strength of concrete depends upon the composition of the cement paste (matrix). The matrix occupies all that space not occupied by the aggregate. Therefore, concrete is a composite of a matrix and an aggregate. The method is not used that much but deserves mentioning.

Method of Goldbeck and Gray

The *method of Goldbeck and Gray* has also been called the b/b_0 *method* and is sometimes referred to as the *ACI method.* This method is based on the absolute volumes of the materials in a unit volume of concrete. The method takes into account a b/b_0 ratio. The term b equals the solid volume of coarse aggregate in a unit volume concrete. The b_0 term equals the solid volume of coarse aggregate in a unit volume of dry-rodded coarse aggregate. The term b/b_0 equals the dry-rodded volume (bulk volume) of coarse aggregate in a unit of concrete. The procedure for designing concrete mixtures involves the use of data from Tables 5-4 to 5-6 which have been prepared by Goldbeck and Gray. These tables are utilized for typical materials. Table 5-7 is for pavement concrete and should be used for such.

The steps in performing a concrete mix design will be listed and then an example will be given. The steps are as follows:

1. Perform the following ASTM specifications on the fine and coarse aggregates.
   a. Determine the bulk specific gravities of the fine and coarse aggregate.
   b. Determine the dry-rodded unit weight of the coarse aggregate.

TABLE 5-4   DRY-RODDED VOLUME b/b_0 OF COARSE AGGREGATE (ANY TYPE) PER UNIT VOLUME OF CONCRETE[a]

Size of Coarse Aggregate (Square-opening Laboratory Sieves)	Fine Sand		Medium Sand				Coarse Sand	
			Fineness Modulus of Sand					
	2.4	2.5	2.6	2.7	2.8	2.9	3.0	3.1
			Values [b] for b/b_0					
No. 4 to $\frac{1}{2}$ in.	0.59	0.58	0.57	0.56	0.55	0.54	0.53	0.52
No. 4 to $\frac{3}{4}$ in.	0.66	0.65	0.64	0.63	0.62	0.61	0.60	0.59
No. 4 to 1 in.	0.71	0.70	0.69	0.68	0.67	0.66	0.65	0.64
No. 4 to 1 $\frac{1}{2}$ in.	0.75	0.74	0.73	0.72	0.71	0.70	0.69	0.68
No. 4 to 2 in.	0.78	0.77	0.76	0.75	0.74	0.73	0.72	0.71
No. 4 to 2 $\frac{1}{2}$ in.	0.80	0.79	0.78	0.77	0.76	0.75	0.74	0.73

[a]From A. T. Goldbeck and J. E. Gray, "A Method of Proportioning Concrete for Strength, Workability, and Durability," *Bull. 11*, National Crushed Stone Association, Nov. 1953.
[b]For concrete that is to be assisted in place by internal vibration under very rigid inspection, increase tabulated values of b/b_0 approximately 10%.

    c. Determine the gradation and fineness modulus of the fine aggregate and the gradation of the coarse aggregate.

2. Compute the solid weights per cubic foot of the cement and the fine and coarse aggregates. This is computed by multiplying the bulk specific gravity by 62.4 lb.

3. Pick the size of coarse aggregate desired along with the determined fineness modulus of sand and enter Table 5-4 to determine b/b_0.

4. Calculate the solid volume of coarse aggregate per cubic foot of dry-rodded materials.

$$b_0 = \frac{\text{dry} - \text{rodded weight per cubic foot}}{\text{solid weight per cubic foot}}$$

5. Calculate b, which is equal to $b/b_0 \times b_0$.

6. Knowing the kind and size of the coarse aggregate desired as well as the 28-day compressive strength desired and the chosen slump, determine from Table 5-5 (if non-air-entrained concrete is to be used) or Table 5-6 (if air-entrained concrete is to be used) the cement factor and the water content. Also select from the appropriate tables the percentage of entrapped air or optimum entrapped air and calculate its solid volume per cubic yard of concrete.

7. Determine and sum up the solid volumes of cement, coarse aggregate, water, and air.

8. Determine the solid volume of sand in a cubic yard of concrete by subtracting from 27 ft³ from item 7.

9. Convert solid volumes to pounds per cubic yard of concrete.

10. Calculate the additional water requirement.

The following example should illustrate the procedure that could be used to design a concrete mix and to calculate the batching weight.

TABLE 5-5 NON-AIR-ENTRAINING STRUCTURAL CONCRETE CEMENT FACTORS REQUIRED FOR 28-DAY COMPRESSIVE STRENGTHS LISTED[a, b]

Size of Coarse Aggregate (Square-opening Laboratory Sieves):	No. 4 to $\frac{1}{2}$ in.		No. 4 to $\frac{3}{4}$ in.		No. 4 to 1 in.		No. 4 to $1\frac{1}{2}$ in.		No. 4 to 2 in.		No. 4 to $2\frac{1}{2}$ in.	
Slump (in.):	3	6	3	6	3	6	3	6	3	6	3	6
Water[c] (gal yd³ of concrete) Angular coarse aggregate	42	44	40	42	38	40	36	38	35	37	34	36
Rounded coarse aggregate	38	40	36	38	34	36	32	34	31	33	30	32
28-Day Compressive Strength (psi)	Cement (Sacks/Yd³ of Concrete)											
2000	4.6	4.8	4.4	4.6	4.2	4.4	4.0	4.2	3.9	4.0	3.8	3.9
2500	5.0	5.2	4.8	5.0	4.5	4.8	4.2	4.5	4.1	4.3	4.0	4.2
3000	5.4	5.7	5.2	5.4	4.9	5.2	4.6	4.9	4.4	4.7	4.3	4.6
3500	5.9	6.3	5.6	5.9	5.3	5.6	5.0	5.3	4.9	5.2	4.8	5.0
4000	6.5	6.9	6.2	6.5	5.8	6.2	5.5	5.8	5.4	5.7	5.2	5.5
4500	7.2	7.5	6.8	7.1	6.4	6.8	6.1	6.4	5.9	6.3	5.7	6.1
5000	8.1	8.5	7.7	8.1	7.3	7.7	6.9	7.3	6.7	7.1	6.5	6.9
Entrapped air (approx. %)	2.5		2		1.5		1		1		1	

[a]From A. T. Goldbeck and J. E. Gray, "A Method of Proportioning Concrete for Strength, Workability, and Durability," *Bull. 11*, National Crushed Stone Association, Nov. 1953.

[b]For concrete to be assisted in place by internal vibration, use 3-in. slump and decrease tabulated water contents by approximately 4 gal. No reduction in cement factor is suggested.

[c]This is water actually effective as mixing water.

Example 5.1:

Determine the batch quantities for a concrete mix design to have a 28-day compressive strength of 3000 psi using angular aggregate from a No. 4 sieve opening to $\frac{1}{2}$ in. in size and a medium fine aggregate (sand) with a fineness modulus of 2.70. A slump of 3 in. is desired.

The material descriptions are as follows:

Coarse Aggregate (CA)

Size, No. 4 to $\frac{1}{2}$ in., angular aggregate

Absorption (%) = 0.1

TABLE 5-6 AIR-ENTRAINING STRUCTURAL CONCRETE CEMENT FACTORS REQUIRED FOR 28-DAY COMPRESSIVE STRENGTHS LISTED[A, B]

Size of Coarse Aggregate Square-opening Laboratory Sieves)	No. 4 to $\frac{1}{2}$ in.		No. 4 to $\frac{3}{4}$ in.		No. 4 to 1 in.		No. 4 to $1\frac{1}{2}$ in.		No. 4 to 2 in.		No. 4 to $2\frac{1}{2}$ in.	
Slump (in.):	3	6	3	6	3	6	3	6	3	6	3	6
Water[c] (gal yd³ of concrete) Angular coarse aggregate	38	40	36	38	34	36	32	34	31	33	30	32
Rounded coarse aggregate	35	37	33	35	31	33	29	31	28	30	27	29
28-Day Compressive Strength (psi)	Cement (Sacks/Yd³ of Concrete)											
2000	4.4	4.7	4.2	4.4	3.9	4.2	3.7	3.9	3.6	3.8	3.5	3.7
2500	4.9	5.2	4.6	4.9	4.4	4.7	4.2	4.4	4.0	4.3	3.9	4.2
3000	5.6	5.9	5.3	5.6	5.0	5.3	4.7	5.0	4.5	4.8	4.3	4.7
3500	6.3	6.7	6.0	6.3	5.6	6.0	5.3	5.6	5.1	5.4	4.9	5.3
4000	7.2	7.5	6.8	7.2	6.4	6.8	6.0	6.4	5.8	6.2	5.6	6.0
4500	8.1	8.5	7.6	8.1	7.2	7.6	6.8	7.2	6.6	7.0	6.4	6.8
5000	9.2	9.7	8.7	9.2	8.2	8.7	7.7	8.2	7.4	8.0	7.2	7.7
Optimum entrained-air content (%)	6.0		6.0		5.5		5.0		5.0		4.5	

[a]From A. T. Goldbeck and J. E. Gray, "A Method of Proportioning Concrete for Strength, Workability, and Durability," *Bull. 11*, National Crushed Stone Association, Nov. 1953.
[b]This table should always be used to proportion concrete subject to freezing.
[c]This is water actually effective as mixing water.

SG (bulk dry) = 2.7

SG (bulk saturated surface dry) = 2.73

Solid weight (lb/ft³) = SG (bulk dry) × unit weight of water
$$= 2.7 \times 62.4 = 168.48$$

Dry-rodded unit weight (lb/ft³) = 101.5

b_0 = dry-rodded unit weight ÷ solid weight = 101.5 ÷ 168.48
$$= 0.60$$

Fine Aggregate (FA)
Fineness modulus (FM) = 2.7

Absorption (%) = 0.8

SG (bulk dry) = 2.5

SG (bulk saturated surface dry) = 2.53

Solid weight (lb/ft³) = SG (bulk dry) × unit weight of water
$$= 2.5 \times 62.4 = 156.00$$

Cement
SG = 3.14

Solid weight (lb/ft³) = SG × unit weight of water
$$= 3.14 \times 62.4 = 195.9$$

Weight per sack (lb) = 94

Solid volume per sack (ft³) = weight per sack ÷ solid weight
$$= 94 \div 195.9 = 0.48$$

Calculations of Proportions

b/b_0 (from Table 5-4) = 0.56

$b = b/b_0 \times b_0 = 0.56 \times 0.60 = 0.34$

Designing for a 3-in. slump and 3000-psi compressive strength for non-air-entrained concrete (2.5%) using Table 5-5.

Cements (sacks/yd³) = 5.4

Water (gal/yd³) = 42

	Solid Volume (ft³/yd³ of concrete)	Quantities (lb/yd³ of concrete)
Cement: 5.4 × 0.48 =	2.59 × 195.9 =	507.38
CA: 0.34 × 27 =	9.18 × 168.5 =	1546.83
Water: 42 ÷ 7.5 =	5.60 × 62.4 =	349.44
Air: 0.025 × 27 =	0.68 × 0	0
	Σ 18.05	
FA: 27 − 18.05 =	8.96 × 156.00 =	1397.76

Notice that the weights for fine and coarse aggregate are dry, and thus the water content or amount of water must be increased. This increase in water is equivalent to the amount absorbed by the aggregates. In this example the increase would be as follows:

$$\text{water absorbed by CA} = 0.001 \times 1546.83 = 1.547 \text{ lb}$$
$$\text{water absorbed by FA} = 0.008 \times 1397.76 = 11.180 \text{ lb}$$
$$\text{total} = 12.73 \text{ lb}$$

Thus, the actual amount of mixing water would be increased by 12.73 lb, for a total of 362.17 lb. In most cases the amount of water added is insignificant, but in large jobs it may be significant. In situations in which the aggregate is wet, the mixing water is decreased by the amount of free water.

TABLE 5-7 PAVEMENT CONCRETE USE: DRY-RODDED VOLUME OF COARSE AGGREGATE AND MIXING WATER REQUIRED[a]

Size of Coarse Aggregate (Square-opening Laboratory Sieve)	Fine Sand		Medium Sand			Coarse Sand			Water (gal/yd³ of Concrete for 2-in. Slump)			
	Fineness Modulus of Sand								Non-air-entrained[c]		Air-entrained	
									Coarse Aggregate			
	2.4	2.5	2.6	2.7	2.8	2.9	3.0	3.1	Angular	Rounded	Angular	Rounded
	Values for b/b₀											
No. 4 to 1 in.	0.75	0.74	0.73	0.72	0.71	0.70	0.69	0.68	35	31	31	28
No. 4 to 1½ in.	0.79	0.78	0.77	0.76	0.75	0.74	0.73	0.72	33	29	29	26
No. 4 to 2 in.	0.82	0.81	0.80	0.79	0.78	0.77	0.76	0.75	32	28	28	25
No. 4 to 2½ in.	0.84	0.83	0.82	0.81	0.80	0.79	0.78	0.77	31	27	27	24

[a]From A. T. Goldbeck and J. E. Gray, "A Method of Proportioning Concrete for Strength, Workability, and Durability," *Bull. 11*, National Crushed Stone Association, Nov. 1953.
[b]This is to water actually effective as mixing water.
[c]Since the maximum size of coarse aggregate used in highways is 1½ in. or larger, constant value of 1% may be taken for entrapped air, or 27 ft³/yd³ of concrete.

Keep in mind that pavement concrete is a less workable mix and therefore requires different b/b_0 values and slumps from those given here. Use Table 5-7.

PROPORTIONING MATERIALS

Cement

Cement is usually purchased in bags or sacks. A sack contains 94 lb and is assumed to be 1 ft³ in volume. Cement may also be purchased in the form of a barrel, which is equal to four sacks weighing 376 lb. Cement may also be purchased by bulk weight. The reason for measuring cement by weight rather than volume is that if cement were purchased by the cubic foot, the weight might vary due to different amounts of compaction.

Water

The accurate weighing of water is important due to the fact that the cement and water form a paste that binds the aggregate and thus influences its strength. In stationary mixing plants, the water is usually weighed out. Volumetric measurements are usually made with modern concrete mixers.

Aggregates

In modern proportioning plants aggregates are weighed, thus eliminating the errors of volumetric measurement. Volumetric measurements are common but are likely to be inaccurate, due to the differences in compaction.

MIXING AND DEPOSITING

Mixing

When the ingredients have been measured in the proper proportion, the next step in the manufacturing of concrete is to mix the ingredients until the aggregate particles are coated with cement and a homogeneous mixture is obtained. Mixing may be done by hand or by a power-driven machine.

Most concrete, regardless of the size of the job, is machine-mixed in batch mixers of 3 to 4 yd³ capacity. Most batch mixers operate with a revolving drain fitted with blades projecting inward. The drum revolves and the ingredients are carried part way around and turned over as they drop to the bottom, thus producing a thorough mix. The time of mixing varies depending on the consistency of the mix.

Transporting Concrete

After being mixed, concrete is transported to the place of deposit in such a way as to prevent segregation of the ingredients. The means of transportation is a question of economy. The only important requirement is that the concrete arrives at the place of deposit and is of good quality and uniformity. On small projects, wheelbarrels or two-wheeled carts are used for transportation. On large projects, industrial cars, cable cars, conveyor belts, or towers may be used. Sometimes, concrete is pumped through pipelines from the mixer to the forms. Concrete may also be transported and placed by trucks.

Depositing Concrete

The proper placement of concrete in the forms is an important factor in obtaining durable concrete structures. Before the placement of the concrete, the forms should be free of debris and accumulated water. To obtain uniform concrete it should be placed in the forms evenly and without segregation of materials. If the concrete is of a dry consistency, it should be placed in layers 6 to 8 in. thick. If the concrete is of a wet consistency, it may be placed in layers 12 to 15 in. thick. After placement the concrete is spaded to remove any entrapped air, thus producing a smooth surface. The same results may be obtained by vibration, but care should be taken not to segregate the ingredients.

Depositing Concrete under Water

When one places concrete under water, care should be taken to prevent the cement from being washed away or cements from segregating. The common method of depositing concrete under water is by use of a *tremie,* a long pipe about 1 ft (30.5 cm) in diameter with a hopper at the top and a slightly flaring bottom. It is plugged at the bottom, filled, then lowered to position and kept filled at all times. The only problem is that one never knows if the concrete is being placed uniformly.

Vibrated Concrete

Vibration is a mechanical method (high-frequency electric or pneumatic) of puddling concrete that is used extensively for large masses. Vibration is usually recommended for mixes that are stiff in consistency and may need help to effectively consolidate. Vibration may be internal, or may be achieved by vibration of forms, or by vibrating floats on top of the concrete.

Bonding New Concrete to Old Concrete

In most construction projects it is best to try to avoid the bonding of new concrete to old concrete, and thus a continuous pour is recommended. However, it is sometimes impossible to avoid joints, and so precautions should be taken to make the bond between the new concrete and the old concrete as strong as possible.

In massive work with horizontal joints it will probably be sufficient to wet the old concrete and continue with the new concrete. When the walls are thin, the old concrete is roughened and cleaned of foreign matter and laitance and slushed with grout of neat cement or mortar.

Finishing Concrete

After the concrete has been deposited, it may be finished off by several methods: steel trowel, spading, canvas belt, wooden float, burlap, or a grinding tool.

Curing

Concrete sets and hardens as a result of a chemical reaction that takes place between the cement and water. This process continues as long as water is present. Thus, a strong concrete results if water is present to carry on the reaction. One other factor that is important to curing is temperature. If the temperature is high, undue cracking will occur. If the temperature is too low, the curing process will not continue. Chapter 4 discusses some of the curing techniques.

COMPRESSIVE, TENSILE, AND FLEXURAL STRENGTH

The major factors that affect concrete strength have been discussed in Chapter 4. Only a comparison table is shown here (Table 5-8).

PROBLEMS

5.1. Define the term "concrete."

5.2. Impurities in mixing water may cause undesirable effects in the final concrete. List five undesirable effects.

5.3. List four undesirable impurities in mixing water and discuss each in detail.

5.4. In the design of concrete mixes, three principal requirements for concrete are of importance. Name them and discuss each in detail.

5.5. For the three requirements requested in Problem 5.4, list the ways in which the ingredients of a concrete mix affect each. Explain.

5.6. List four methods of proportioning concrete mixes and discuss each in detail.

TABLE 5-8 COMPARISON OF COMPRESSIVE, FLEXURAL, AND TENSILE STRENGTH

Strength of Plain Concrete (psi)			Ratio (%)		
Compressive	Modulus of Rupture	Tensile	Modulus of Rupture to Compressive Strength	Tensile Strength to Compressive Strength	Tensile Strength to Modulus of Rupture
1000	230	110	23	11	48
2000	375	200	19	10	53
3000	485	275	16	9	57
4000	580	340	14	8	59
5000	675	400	14	8	60
6000	765	460	13	8	61
7000	855	520	12	7	61
8000	930	580	12	7	62
9000	1010	630	11	7	63

5.7. What is meant by the term b/b_0?

5.8. List the steps in performing a concrete mix design according to the method of Goldbeck and Gray.

5.9. In the method of Goldbeck and Gray, how is the b/b_0 ratio affected if the concrete is to be assisted by internal vibration?

5.10. Determine the batch quantities for a concrete mix design that is to have a 28-day compressive strength of 2500 psi using rounded aggregate from a No. 4 sieve opening to 1 in. in size and a medium-fine aggregate with a fineness modulus of 2.65. A slump of 3 in. is desirable. The following descriptions apply to the coarse and fine aggregates.

Coarse Aggregate

Size, No. 4 to 1 in., rounded
Absorption (%) = 0.1
SG (bulk dry) = 2.7
SG (bulk saturated surface dry) = 2.74
Dry-rodded unit weight (lb/ft^3) = 101.5

Fine Aggregate

Fineness modulus = 2.65
Absorption (%) = 0.8
SG (bulk dry) = 2.4
SG (bulk saturated surface dry) = 2.45

REFERENCES

ABRAMS, D.A., Design of Concrete Mixtures, *Lewis Inst. Struct. Mater. Res. Lab Bull. 1*, Chicago, 1918.

AMERICAN SOCIETY FOR TESTING AND MATERIALS, "Significance of Tests and Properties of Concrete and Concrete-making Material," *ASTM Spec. Tech. Publ. 169B*, 1979.

BANER, E. E., *Plain Concrete*, 3rd ed., McGraw-Hill Book Company, New York, 1949.

CORDON, W. A., *Properties, Evaluation, and Control of Engineering Materials*, McGraw-Hill Book Company, New York, 1979.

CORDON, W. A., and MERRILL, D., "Requirements for Freezing and Thawing Durability for Concrete," *Proc. ASTM, 63* (1963), pp. 1026–1035.

FULLER, W. B., and THOMPSON, S. E., "The Laws of Proportioning Concrete," *Trans. Am. Soc. Civil Eng.* (1907), p. 67.

GOLDBECK, A. T., and GRAY, J. E., "A Method of Proportioning Concrete for Strength, Workability and Durability," *Bull. 11*, National Crushed Stone Association, Nov. 1953.

KREBS, R. D., and WALKER, R. D., *Highway Materials*, McGraw-Hill Book Company, New York, 1971.

MILLS, A. P., HAYWARD, H. W., and RADER, L. F., *Materials of Construction*, 6th ed., John Wiley & Sons, Inc., New York, 1955.

POWERS, T. C., "The Air Requirements in Frost-resistant Concrete," *Proc. Highway Res. Board* (1949), p. 184.

POWERS, T. C., "The Mechanism of Frost Action in Concrete," Stanton Walker Lecture Series on the Material Sciences, Lecture No. 3, National Ready-Mixed Concrete Association, Silver Spring, Md., 1965.

TALBOT, A. N., "Proposed Method of Estimating Density and Strength of Concrete and Proportioning Materials by Experimental and Analytical Considerations of Voids in Mortar and Concrete," *Proc. ASTM* (1921).

WALKER, S., BLOEM, D. L., and GAYNOR, R. D., "Relationships of Concrete Strength to Maximum Size of Aggregate," *Proc. Highway Res. Board* (1959).

6

PROPORTIONING STRUCTURAL CONCRETE MIXTURES WITH FLY ASH AND OTHER POZZOLANS*

No history of concrete as it is made and used today would be complete without reference to the expanding use of pozzolans as an ingredient in concrete. It has been stated that pozzolans, along with other finely divided mineral admixtures, are used in mass concrete, structural concrete, pavements, dam locks, canal linings, tunnels, sewage works, waterworks, high-rise residential and commercial structures, and residential concrete, including sidewalks, driveways, and parking areas. In short, concrete containing pozzolans as an ingredient is being used in virtually every application for which concrete is used. Literature provides much information on the desirable effects of pozzolans in concrete. The history of the use of pozzolans parallels the history of the recorded works of man. About 2000 years ago Greeks, Romans, and other Mediterranean peoples discovered the value of using pozzolans (fine volcanic ash) with burned lime to build historic structures—some are still in use today. The earliest methods of proportioning with pozzolan in the United States consisted primarily of substituting pozzolans on a pound-for-pound or volume-for-volume basis with portland cement. Modern building practices have required, however, that a new method of proportioning with pozzolans be developed whereby rapid construction can be coupled with long-range owner satisfaction. These procedures are in accord with ACI 211.1-70.

BACKGROUND

The American Society for Testing and Materials (ASTM) defines pozzolan as "a siliceous or siliceous and aluminous material which in itself possesses little or no cementitious value

*This chapter was reproduced with permission from the American Concrete Institute, where it appeared in their SP 46-8 publication. The authors of this article are C. E. Lovewell and Edward J. Hyland.

146

but will, in finely divided form and in the presence of moisture, chemically react with cal-cium hydroxide at ordinary temperatures to form compounds possessing cementitious properties." In brief, pozzolans are powders that will harden when combined with lime and water at ordinary temperatures. In concrete the lime is provided as a by-product of the hydration of the portland cement in the mixture. A commonly accepted amount of lime produced by hydration is 15 percent, by weight, of the portland cement.

There are in general two types of pozzolans: natural and synthetic. Natural poz-zolans are found in the earth's crust and include volcanic ashes, pumicites, tuffs, diatoma-ceous earths, opaline cherts and shales and certain processed clays, shales, and diatomites. Some natural pozzolans are capable of being used in concrete with minimal preparation such as screening and grinding whereas others require calcination and finish grinding to provide satisfactory properties.

Synthetic pozzolan refers almost exclusively to fly ash. Fly ash is defined as "the fine-ly divided residue that results from the combustion of ground or powdered coal and is trans-ported from the combustion chamber by exhaust gasses." Some factors affecting the chemical and physical characteristics of fly ash are the chemical composition or type of coal, duration of time in open-air storage, fineness of grind, and other factors relating to combustion.

Most recognized specifications for pozzolans in the United States are written to cover both natural pozzolans and fly ash, with the properties of each arranged in conven-ient tables. Some of these specifications are ASTM C-618 "Fly Ash and Raw or Calcined Natural Pozzolans For Use in Portland Cement Concrete," CRD C262-63 "Corps of Engineers Specifications For Pozzolan For Use in Portland Cement Concrete" and SS-P-570b "Federal Specifications for Pozzolans for Use in Portland Cement Concrete." The characteristics of the pozzolan intended for use must be known before any attempt can be made to proportion concrete with it. For example, a favorable particle shape and fineness of the pozzolan are necessary if low-water demand is to be achieved. Coarse pozzolan of unfavorable shape (such as volcanic glass) may cause an increase in mixing water require-ments with consequent excessive bleeding and segregation. Other characteristics, such as the tendency of high-carbon fly ash to reduce entrained air should also be known prior to proportioning. Highway Research Board Special Report No. 119 "Admixtures in Con-crete" contains much of this type of information helpful to the proportioner. Selected data from that report are shown in Table 6-1.

Early proportioning of concrete with pozzolans in this country was concerned almost exclusively with mass concrete. Since the proportioner of mass concrete is concerned with producing a mixture having a low heat of hydration and since the structure itself will prob-ably not sustain design loads for many months or years and most of the concrete used in the mass will not be exposed to freezing and thawing, certain procedures were developed whereby the quantity of portland cement was held to an absolute minimum. The pozzolan used therefore comprised a very high percentage of the total cementitious ingredients. However, in proportioning the concrete for use in structures, an entirely different set of cri-teria has to be applied. For example, most structures have a required design strength at 28 days. Further, floors in many high-rise buildings receive higher loadings during the con-struction period than they ever do when they are turned over to the owner. A cube of brick or concrete block resting on concrete that is seven days old causes more damaging stress than a stenographer seated at a typewriter in a two-year-old building. Concrete in the struc-tural market is also invariably designed for higher strength than mass concrete. Further,

TABLE 6-1   EFFECT OF TYPE, FINENESS, AND AMOUNT OF POZZOLAN ON
WATER CONTENT AND AIR-ENTRAINING ADMIXTURE REQUIREMENT

| Cement (lb/cu yd) | Pozzolan | | | Surface (sq cm/g) | Water (lb/cu yd) | Neutralized Vinsol Resin (ml/cu yd) |
	(lb/cu yd)	Percent	Designation			
532	0	0	None	3,550	265	485
346	116	30	Fly Ash I	3,565	247	379
416	103	25	Pumicite I	4,410	277	529
404	156	35	Tuff	10,460	310	800
346	139	35	Obsidian	3,415	266	378
458	91	20	Calcined Shale I	13,685	286	670
423	141	30	Calcined diatomite	10,450	302	1,018
462	64	16	Uncalcined diatomite	12,125	274	540

Source: "Admixtures in Concrete," Highway Research Board, *Special Report 119*, 1971.

since many of these structures are designed in areas that experience freezing and thawing, the exposed concrete must be air-entrained. Even in those parts of the building not exposed to the elements air entrainment may be required by the designer because of the volume of reinforcing steel in the thin members and the need to improve the workability of concrete.

Under these conditions it becomes imperative that a new method of proportioning concrete for the structural market be developed.

Lovewell and Washa reported the development of a method of proportioning concrete mixtures using fly ash. Basically, the method showed that with relatively few tests a designer could develop data that provided guidelines for proportioning concrete and fly ash with predictable results. This was further elaborated by Cannon.

The method that follows is based on the references previously cited and includes additional data developed for mixtures containing fly ash and two types of chemical admixtures. The authors have found that the procedures outlined have proved valid with fly ashes conforming to ASTM C-618 and with a great variety of aggregates and brands of portland cement. The procedure has been proven to be satisfactory by the excellent performance of many millions of cubic yards of structural concrete made with fly ash, with and without chemical admixtures. With a minimum number of check tests the procedure can also be applied to natural pozzolans.

PROCEDURE

As is pointed out in ACI 211.1-70, compressive strength is not the only important characteristic of concrete. Workability, durability, wear resistance and other characteristics may be equally or even more important. The ACI "Guide for Use of Admixtures in Concrete" discusses the effects of admixtures in concrete. Assume that based upon this guide and other data the designer requires that a pozzolan and perhaps other admixtures be used. Other requirements are also specified such as strength, slump, air content, maximum size of aggregate, type of portland cement, and type of chemical admixtures. The following steps should be followed:

1. *Selection of slump.* If the designer has not specified a minimum and maximum slump choose values according to ACI.
2. *Selection of aggregate.* The designer may have specified a certain aggregate size and perhaps even shape. However, if this has been left open, the proportioner should in general select the maximum size of coarse aggregate that is economically available and one that can be placed readily in the structure. Other considerations should be made, such as whether the concrete is intended to produce a very high strength, or whether it is intended for pumping through small lines, in which case a smaller aggregate size may be selected and a specific particle shape may be desired. In no case should the maximum size of the aggregate exceed one-fifth of the narrowest dimension of the size of the form or one-third of the depth of slabs or three-fourths of the minimum clear spacing between individual reinforcing bars, bundles of bars, or pretensioning strands.
3. *Estimate the mixing water and air content.* Table 6-2 gives an estimate of the mixing water required with a given size and shape of aggregate to produce a given slump. The table also gives an estimate of the amount of entrapped and entrained air normally present in non-air-entrained concrete. This table is for concrete that does not contain pozzolan or a chemical admixture.

TABLE 6-2 APPROXIMATE MIXING WATER AND AIR CONTENT REQUIREMENTS FOR DIFFERENT SLUMPS AND MAXIMUM SIZES OF AGGREGATES[a]

Slump (in.)[b]	Water, lb/yd³ of Concrete for Indicated Maximum Sizes of Aggregate						
	$\frac{3}{8}$ in.	$\frac{1}{2}$ in.	$\frac{3}{4}$ in.	1 in.	$1\frac{1}{2}$ in.	2 in.	3 in.
Non-air-entrained Concrete							
1 to 2	350	335	315	300	275	260	240
3 to 4	385	365	340	325	300	285	265
6 to 7	410	385	360	340	315	300	285
Approximate amount of entrapped air in non-air-entrained concrete, %	3	2.5	2	1.5	1	0.5	0.3
Air-entrained Concrete							
1 to 2	305	295	280	270	250	240	225
3 to 4	340	325	305	295	275	265	250
6 to 7	365	345	325	310	290	280	270
Recommended average total air content, %	8	7	6	5	4.5	4	3.5

[a]These quantities of mixing water are for use in computing cement factors for trial batches. They are maxima for reasonably well-shaped angular coarse aggregates graded within limits of accepted specifications.
[b]The slump values for concrete containing aggregates larger than $1\frac{1}{2}$ in. are based on slump tests made after removal of particles larger than $1\frac{1}{2}$ in. by wet-screening.
Note: This is a reproduction of Table 5.2.3 of ACI 211.1-70.

If no other data is available the proportioner can use Table 6-2 and apply a correction factor based on other information, such as HRB Report 119, Wallace and Ore, or other available sources. However, the producer or vendor of the pozzolan proposed for use may have useful information of this type. As an example of what may be available to the proportioner, Figure 6-1 and Figure 6-2 give the approximate mixing water requirements of concrete containing fly ash at different slumps made with rounded or angular coarse aggregate of the maximum sizes commonly used in structural concrete. These curves, while not absolute, are sufficiently accurate to provide the proportioner with a figure for developing a first trial mixture. The curves have been developed over many years and much field work.

The use of chemical admixtures, specifically water-reducers or water-reducing-retarders, has become widespread. These admixtures are used with great frequency in all concrete, including mixtures containing pozzolans. The chemical composition of the water-reducing admixture affects the amount of mixing water required, and its characteristics must be understood. Figure 6-3 and Figure 6-4 give estimates of mixing water required for mixtures of cement plus fly ash and a chemical admixture with the salt of a lignosulfonic acid (lignin) as its base. Figure 6-5 and Figure 6-6 give similar information, but in this case the chemical admixture has a carbohydrate base.

4. *Selection of compressive strength and air-entrainment.* The designer will ordinarily specify the required average compressive strength and whether or not the concrete is to be air-entrained, along with the required limits of air entrainment. If the designer fails to specify average strength and entrained-air content the proportioner may turn to Table 6-3 and Table 6-4.

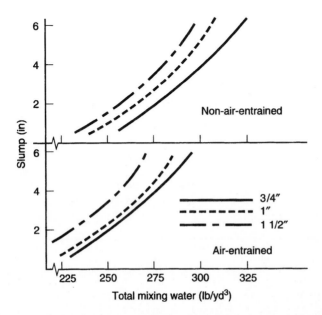

Figure 6-1 Approximate Mixing Water Requirements for Different Slumps and Maximum Sizes of Typical Structural Concrete Aggregates
Materials: Portland Cement—Type I
                Pozzolan—Fly Ash
                Chemical Admixture—None
                Course Aggregate—Rounded

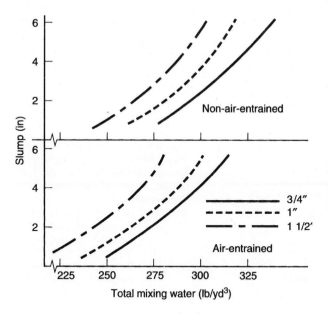

Figure 6-2 Approximate Mixing Water Requirements for Different Slumps and Maximum Sizes of Typical Structural Concrete Aggregates
Materials Cement—Type I
                Pozzolan—Fly Ash
                Chemical Admixture—None
                Course Aggregate—Angular

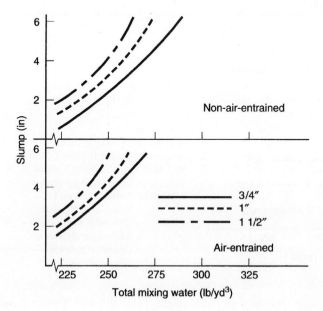

Figure 6-3 Approximate Mixing Water Requirements for Different Slumps and Maximum Sizes of Typical Structural Concrete Aggregates
Materials: Cement—Type I
                 Pozzolan—Fly Ash
                 Chemical Admixture—Lignin Base
                 Course Aggregate—Rounded

The durability of concrete containing fly ash or other pozzolans is equal to that of concrete mixtures not containing these materials, provided that the mixtures have

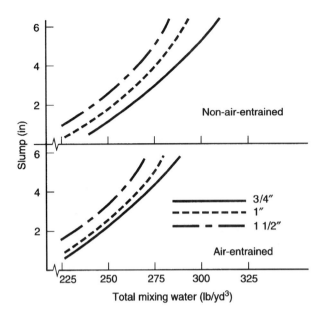

Figure 6-4 Approximate Mixing Water Requirements for Different Slumps and Maximum Sizes of Typical Structural Concrete Aggregates
Materials: Cement—Type I
            Pozzolan—Fly Ash
            Chemical Admixture—
            Lignin Base
            Course Aggregate—
            Angular

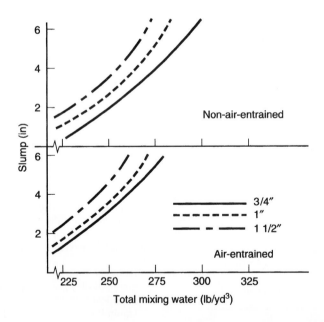

Figure 6-5
Approximate Mixing Water Requirements for Different Slumps and Maximum Sizes of Typical Structural Concrete Aggregates
Materials: Cement—Type I
            Pozzolan—Fly Ash
            Chemical Admixture—
            Carbohydrate Base
            Course Aggregate—
            Rounded

equal compressive strength and equal air entrainment. Thus, although Table 6-3 and Table 6-4 were developed for concrete mixtures not containing fly ash or other pozzolans, they may be used safely to determine the required strength and air entrainment of mixtures containing fly ash and other pozzolans as well.

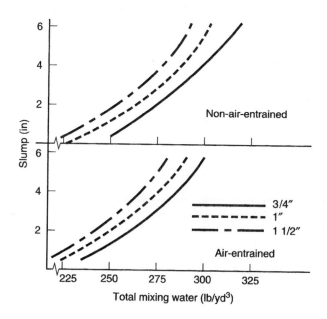

Figure 6-6
Approximate Mixing Water
Requirements for Different Slumps
and Maximum Sizes of Typical
Structural Concrete Aggregates
Materials: Cement—Type I
Pozzolan—Fly Ash
Chemical Admixture—
Carbohydrate Base
Course Aggregate—
Angular

TABLE 6-3 RELATIONSHIPS BETWEEN WATER-CEMENT RATIO AND COMPRESSIVE STRENGTH OF CONCRETE

Compressive Strength at 28 Days (psi)[a]	Non-air-entrained Concrete	Air-entrained Concrete
6000	0.41	
5000	0.48	0.40
4000	0.57	0.48
3000	0.68	0.59
2000	0.82	0.74

[a]Values are estimated average strengths for concrete containing not more than the percentage of air shown in Table 5.2.3. For a constant water-cement ratio, the strength of concrete is reduced as the air content is increased.
Strength is based on 6 in. × 12 in. cylinders moist-cured 28 days at 73.4 ± 3°F (23 ± 1.7°C) in accordance with Section 9(b) of ASTM C 31 for Making and Curing Concrete Compression and Flexural Test Specimens in the Field.
Relationship assumes maximum size of aggregate about $\frac{3}{4}$ to 1 in.; for a given source, strength produced for a given water-cement ratio will increase as maximum size of aggregate decreases; see Sections 3.4 and 5.2.2
Note: This is a reproduction of Table 5.2.4 (a) of ACI 211.1-70.

5. *Calculation of cement content.* Knowing the required average compressive strength, the proportioner now turns to curves that have been previously established or that can be established with relatively few mixtures. An example of such a family of curves is shown in Figures 6-7, 6-8, and 6-9. These figures are averages of hundreds of tests of structural concrete made either in commercial testing laboratories or in laboratories owned and operated by suppliers of ready-mixed concrete. These 3-, 7-, and 28-day

TABLE 6-4   MAXIMUM PERMISSABLE WATER-CEMENT
RATIOS FOR CONCRETE IN SEVERE EXPOSURES[a]

Type of Structure	Structure Wet Continuously or Frequently and Exposed to Freezing and Thawing[b]	Structure Exposed to Seawater or Sulfates
Thin sections (railings, curbs, sills, ledges, ornamental work) and sections with less than 1 in. cover over steel	0.45	0.40[c]
All other structures	0.50	0.45[c]

[a]Based on report of ACI Committee 201, "Durability of Concrete in Service," previously cited.

[b]Concrete should also be air entrained.

[c]If sulfate resisting cement (Type II or Type V of ASTM C150) is used, permissable water-cement ratio may be increased by 0.05.

Note: This is a reproduction of Table 5.2.4 (b) of ACI 211.1-70.

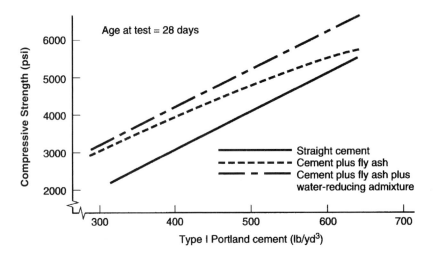

Figure 6-7   Average Compressive Strength of Laboratory-Made, Air-Entrained Concrete Cylinders Containing Maximum Sizes of $\frac{3}{4}$- to $1\frac{1}{2}$-in. Aggregates at 3- to 5-in. Slump

curves are averages of results gained with many different brands of Type I Portland cement and various sources of three-quarter inch to one and one-half inch coarse aggregate, both angular and rounded. The water-reducing admixtures conform to ASTM C 494-71 (Type A) and had either lignin or carbohydrate bases. The fly ash came from eight different sources, the loss on ignition ranging from 5 percent to less than 1 percent. Concrete cylinders used to obtain data were cast in either waxed cardboard molds or single-use metal molds. Testing procedures were standard. The great many variables in materials, molds and laboratories naturally cause some

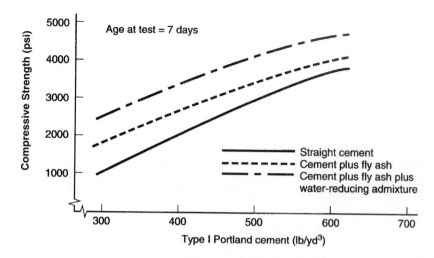

Figure 6-8 Average Compressive Strength of Laboratory-Made, Air-Entrained Concrete Cylinders Containing Maximum Sizes of $\frac{3}{4}$- to $1\frac{1}{2}$-in. Aggregates at 3- to 5-in. Slump

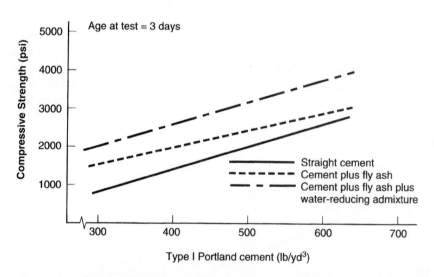

Figure 6-9 Average Compressive Strength of Laboratory-Made, Air-Entrained Concrete Cylinders Containing Maximum Sizes of $\frac{3}{4}$- to $1\frac{1}{2}$-in. Aggregates at 3- to 5-in. Slump

scatter in the points used to plot the curves, yet the authors feel that the curves are reasonably conservative and represent a good usable average strength of laboratory-fabricated mixtures of structural concrete.

Structural concrete is ordinarily specified on the 28 day compressive strength basis. Therefore selecting the required average 28 day compressive strength of cylinders made, cured, and tested in the laboratory on the ordinate scale and proceeding horizontally until intersecting the appropriate curve, and proceeding downward to the abscissa scale, the estimated amount of Type I portland cement required to produce air-entrained concrete can be read.

Figures 6-8 and 6-9 are similar curves developed to show the relationship of the various mixtures at 7 days and at 3 days. Such figures can be invaluable in helping the field engineer determine at an early age whether concrete specimens will actually approach the desired strengths. They are also helpful in determining when loads may be applied to "green" concrete.

Figures such as Figures 6-7, 6-8, and 6-9 can also be established for non-air-entrained concrete.

6. *Determination of pozzolan content.* The optimum amount of the pozzolan can be determined by relatively few mixtures and a curve plotted similar to that shown in Figure 6-10, or the producer or vendor of the pozzolan may have such data available. Lovewell and Washa established the curve shown in Figure 6-10 for fly ash, and experience over the years with this material has shown that the curve is valid with only minor modifications. Experience in the field has shown that it is beneficial to increase the value shown in Figure 6-10 by about 20 percent when the sand available for use is extremely coarse or where the concrete is to be pumped. When the sand selected for use is extremely high in 50 and 100 mesh particles it is also beneficial to decrease the amount shown in Figure 6-10 by about 20 percent to avoid stickiness. In all cases it should be remembered that the use of figures and tables in any suggested method of proportioning is for the establishment of a trial mix only and the numbers obtained from such references should be considered as estimates and modified when appropriate with good judgment by the proportioner.

7. *Estimation of coarse and fine aggregate contents.* Having determined the cement and pozzolan contents, an estimate of the approximate mixing water requirements, and the entrained-air content (if any), the proportioner is now ready to complete the design by selecting a coarse aggregate by means of Table 6-5. The proportioner may

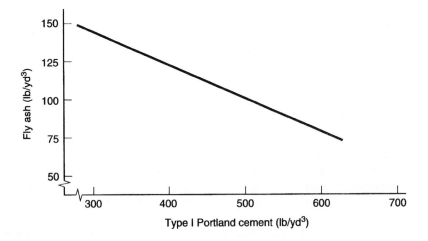

Figure 6–10   Optimum Amount of Fly Ash at Varying Cement Contents of Air-Entrained Concrete with $\frac{3}{4}$- to 1 $\frac{1}{2}$-in. Maximum Size Aggregate

TABLE 6-5   VOLUME OF COARSE AGGREGATE
PER UNIT OF VOLUME OF CONCRETE

Maximum size of Aggregate (in.)	Volume of Dry-rodded Coarse Aggregate[a] per Unit Volume of Concrete for Different Fineness Moduli of Sand			
	2.40	2.60	2.80	3.00
$\frac{3}{8}$	0.50	0.48	0.46	0.44
$\frac{1}{2}$	0.59	0.57	0.55	0.53
$\frac{3}{4}$	0.66	0.64	0.62	0.60
1	0.71	0.69	0.67	0.65
$1\frac{1}{2}$	0.75	0.73	0.71	0.69
2	0.78	0.76	0.74	0.72
3	0.82	0.80	0.78	0.76
6	0.87	0.85	0.83	0.81

[a]Volumes are based on aggregates in dry-rodding condition as described in ASTM C 29 for Unit Weight of Aggregate. These volumes are selected from empirical relationships to produce concrete with a degree of workability suitable for usual reinforced construction. For less workable concrete such as required for concrete pavement construction they may be increased about 10 percent. For more workable concrete, such as may sometimes be required when placement is to be by pumping, they may be reduced up to 10 percent.
Note: This is a reproduction of Table 5.2.6 of ACI 211.1-70.

have sufficient data available on the unit weight of fresh concrete containing the pozzolan with or without a chemical admixture to allow solving for the fine aggregate content by the "weight" method, but the probability is that sufficient background information will not be available and so it is suggested that sand content be calculated by the absolute volume method. To do this the specific gravity of the pozzolan in question must be known. Because the specific gravity of one pozzolan can be widely different from another, no assumption as to specific gravity should be made. Exact information should be furnished by the producer of the pozzolan or it should be determined by laboratory tests.

8. *Adjustment.* ACI 211.1-70 shows the methods employed to adjust the first trial mix. The methods of the standard are equally valid when applied to mixtures containing pozzolans with or without chemical admixtures.

APPLICABILITY TO LIGHTWEIGHT CONCRETE

In general, the method of selecting the proportions of cement and pozzolan with or without the chemical admixture previously described is also applicable to lightweight concrete mixtures. Because the amount of water required in a lightweight concrete mixture is so dependent upon the sizes, shape, and texture of the lightweight aggregate and the fineness and amount of natural sand, if that is to be included, it is impossible to generalize on an estimate of mixing water. This must be determined by actual tests although a good estimate of it may perhaps be furnished by the vendor of the pozzolan or by the producer of the lightweight aggregate. Curves showing the relationship of cement, cement plus pozzolan, and cement plus pozzolan plus water-reducing or water-reducing–retarding admix-

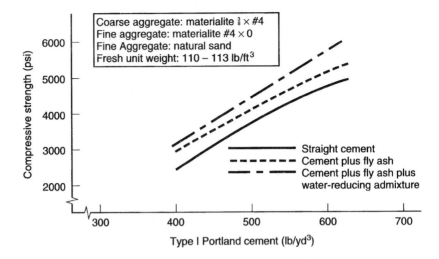

Figure 6-11   Average Compressive Strength of Laboratory-Made Air-Entrained Lightweight Concrete Cylinders at 28 Days

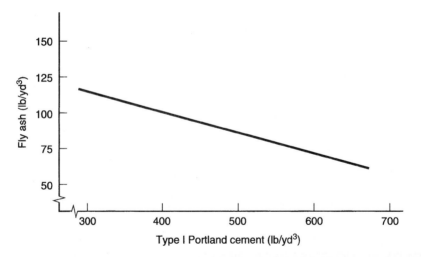

Figure 6–12   Optimum Amounts of Fly Ash at Varying Cement Contents with Materialite Aggregate

ture can then be developed. A further curve showing the optimum amount of pozzolan to be used in cement with the aggregates in question should also be drawn. With these two curves in hand it then becomes easy for the proportioner of lightweight concrete to use ACI 211.2-69 to determine a first trial mixture. An example of a family of curves developed for concrete containing an expanded shale lightweight aggregate and natural sand is shown in Figure 6-11. The relationship of fly ash to cement content with this particular combination of aggregate is shown in Figure 6-12.

SAMPLE COMPUTATIONS

Example 1:

Normal weight concrete is to be proportioned for a structure, part of which will be exposed to frequent freezing and thawing. Structural demands require an average compressive strength of 3500 psi for cylinders fabricated, cured, and tested in the laboratory. It is expected that specimens from the field will average over 2500 psi. Job conditions indicate that a slump of 4 in. $\pm \frac{1}{2}$ in. should be used and the coarse aggregate will be crushed stone with a maximum size of 1 in. The dry rodded unit weight of the coarse aggregate is 96.3 lb/ft^3 and the bulk specific gravity is 2.64. The fine aggregate is natural sand of 2.66 specific gravity and fineness modulus of 2.70. The designer has required that fly ash be used and its specific gravity is 2.44. The concrete shall be air entrained 5% $\pm \frac{1}{2}$ percent.

(a) As indicated, the desired slump is 4 in.

(b) The designer has required 5% entrained air plus fly ash with angular coarse aggregate. Using Figure 6-2 it appears that the approximate amount of mixing water to produce a 4 in. slump concrete with 1 in. maximum size aggregate is 285 lbs.

(c) From Table 6-4 a water-cement ratio of 0.50 would be required for the structure if cement alone were the cementing material. However, Table 6-3 estimates that a water-cement ratio of 0.50 for air-entrained concrete will produce approximately 3800 psi compressive strength at 28 days. Although the designer has required only 3500 psi, the 3800 is required for durability, so this higher figure will govern and the concrete should be designed accordingly.

(d) Using Figure 6-7, entering at 3800 psi on the ordinate axis and proceeding laterally to intersect the curve marked "Fly Ash Mixture" and then dropping down to the abscissa axis, 390 lb/yd^3 of Type I Portland cement is estimated as necessary to fulfill the requirements.

(e) To determine the optimum amount of fly ash for use in the mixture Figure 6-10 is used. Entering with 390 lb/yd^3 of Type I Portland cement, proceeding upward until intersecting the curve, then laterally to the ordinate scale, a value of 125 lb/yd^3 of fly ash combines with the 390 lb/yd^3 of Type I Portland cement previously determined to give the total weight of the cementing material.

(f) The amount of coarse aggregate is estimated from Table 6-5. With a fineness modulus of the fine aggregate of 2.70 and 1 in. maximum size of coarse aggregate, the table indicates that 0.68 ft^3 of coarse aggregate on a dry rodded basis may be used in each cubic foot of concrete. Therefore, the coarse aggregate content is 27 $\times$ 0.68 = 18.36 ft^3/yd^3. Because it weighs 96.3 lb/ft^3, the dry weight of coarse aggregate is 18.36 $\times$ 96.3 or 1768 lb/yd^3.

(g) The quantities of cement, fly ash, coarse aggregate, entrained air, and water having been established, the following absolute volumes of ingredients are calculated:

$$\text{Cement:} \qquad \frac{390}{3.15 \times 62.4} = 1.98 \text{ ft}^3$$

$$\text{Fly ash:} \qquad \frac{125}{2.44 \times 62.4} = 0.82 \text{ ft}^3$$

$$\text{Coarse Aggregate:} \qquad \frac{1768}{2.64 \times 62.4} = 10.73 \text{ ft}^3$$

$$\text{Entrained air:}\qquad 0.05 \times 27.00 = 1.35 \text{ ft}^3$$

$$\text{Water:}\qquad \frac{285}{62.4} = 4.57 \text{ ft}^3$$

Total solid volume of ingredients except sand = 19.45 ft^3. Solid volume of sand required = 27 – 19.45 = 7.55 ft^3. Required weight of dry sand = 7.55 × 2.66 × 62.4 = 1253 lb/yd^3.

(h) We will not proceed further on this example as the method of making adjustments and compensations is well covered in ACI 211.1-70.

Example 2:

Concrete is required for a high-rise structure that will not be exposed to freezing and thawing, but because of the heavy reinforcement the designer has required that concrete shall be air entrained 5% ± $\frac{1}{2}$ percent. Crushed stone aggregate shall be used with a top size of $\frac{3}{4}$ inch and the slump shall be 3 in. ± $\frac{1}{2}$ in. The designer has further required that fly ash with an average specific gravity of 2.50 and a lignin-based water-reducer shall be used in the concrete. Structural requirements are such that laboratory-made specimens shall average 6000 psi at 28 days in compression.

(a) Slump and aggregate size and shape were given previously. Dry rodded weight of coarse aggregate is 100 lb/ft^3 and its specific gravity is 2.67. The fineness modulus of sand is 2.80 and its specific gravity is 2.66.

(b) Since the designer has required air-entrained concrete with a lignin-based water-reducer and fly ash we proceed to Figure 6-4, which gives us an approximation of 260 lb/yd^3 of water required for 3 in. slump.

(c) Turning next to Figure 6-7, we enter at the 6000 psi ordinate and proceed horizontally to where we intersect the curve "Fly Ash plus Water-Reducing Mixture." Moving downward to the abscissa scale we find that we require approximately 580 lb/yd^3 of Type I Portland cement to fulfill requirements.

(d) We then turn to Figure 6-10, where we obtain an estimated 85 lb/yd^3 of fly ash which will be required for the job. Because the manufacturer of the water reducing admixture has recommended that 4.0 fluid ounces of the water reducer be used per 100 lbs of portland cement, we determine that we will require 23.2 fl oz of this admixture per cubic yard.

(e) The quantity of coarse aggregate is estimated from Table 6-5 to be 0.62 cubic feet on a dry rodded basis per cubic foot of concrete. For a cubic yard therefore, the coarse aggregate will be 16.74 ft^3, and since the coarse aggregate weighs 100 lb per cubic foot the dry weight of coarse aggregate is 1674 lb/yd^3.

(f) With quantities of cement, fly ash, coarse aggregate, water, and air known, the following calculations are made:

$$\text{Cement:}\qquad \frac{580}{3.15 \times 62.4} = 2.95 \text{ ft}^3$$

$$\text{Fly ash:}\qquad \frac{85}{2.5 \times 62.4} = 0.55 \text{ ft}^3$$

$$\text{Coarse aggregate:} \qquad \frac{1674}{2.67 \times 62.4} = 10.05 \text{ ft}^3$$

$$\text{Water:} \qquad \frac{260}{62.4} = 4.16 \text{ ft}^3$$

$$\text{Entrained air:} \qquad 0.05 \times 27.00 = 1.35 \text{ ft}^3$$

Total solid volume of ingredients except sand = 19.06 ft^3. Solid volume of sand required = 27.00 −19.06 = 7.94 ft^3. Required weight of dry sand = 7.94×2.66×62.4 = 1316 lb/yd^3.

(g) Adjustments may be required and they can be performed as shown in ACI 211.1-70.

CONCLUSIONS

There are many methods that can be used to proportion concrete containing pozzolans. This method has given good results and is applicable to many different materials and situations. Certain preliminary work must be done before the method can be employed. It is of utmost importance that the type of pozzolan and other admixtures used are known and their characteristics well established before using the method. As is always the case, the proportioner must use good judgment and experience in applying the method.

REFERENCES

ACI Committee 211, "Recommended Practice for Selecting Proportions for Normal Weight Concrete," (ACI 211.1-70). American Concrete Institute, Detroit, Michigan, 1970.

"Standard Specifications for Fly Ash and Raw or Calcined Natural Pozzolans for Use in Portland Cement Concrete" (ASTM C 618–71), *1971 Book of ASTM Standards*, Part 10, ASTM, Philadelphia, Pennsylvania, pp. 345–349.

"Corps of Engineers Specifications for Pozzolan for Use in Portland-Cement Concrete" (CRD-C262-63), *Handbook for Concrete and Cement,* U.S. Army Engineer Waterways Experiment Station, Vicksburg, Miss., 1949, plus quarterly supplements.

"Specifications for Pozzolans for Use in Portland Cement Concrete," Federal Specifications SS-P-570b, General Services Administration, Washington, D.C.

"Admixtures in Concrete," Highway Research Board, Special Report 119, 1971.

LOVEWELL, C. E. and WASHA, G. W., "Proportioning Concrete Mixtures Using Fly Ash," *ACI J. Proc. 54* (June 1958), pp. 1093–1102.

CANNON, ROBERT W., "Proportioning Fly Ash Concrete Mixes for Strength and Economy," *ACI J. Proc. 65* (November 1968), pp. 969–979.

WALLACE, GEORGE, B. and ORE, ELWOOD L., "Structural and Lean Mass Concrete as Affected by Water-Reducing, Set-Retarding Agents," *ASTM Special Technical Publication No. 26*, ASTM, Philadelphia, 1959.

7 ADVANCES IN CONCRETE TECHNOLOGY

ADMIXTURES

General

According to ASTM C125 (Terminology Relating to Concrete and Concrete Aggregates), an admixture is defined as a material other then water, aggregates, hydraulic cements, or fiber reinforcement, that is used as an ingredient of concrete or mortar and added to the batch immediately before or during mixing. Admixtures are generally added to the basic ingredients that make up concrete to modify various properties of the concrete to make it more suitable for use. They are used only when the desired properties cannot be achieved by altering the design mix. Currently, most used concrete contains at least one admixture. The proportion of concrete in which admixtures are used is reported to be 88, 85, and 71 percent, in Canada, Australia, and United States, respectively. Admixtures consist broadly of

1. Air-entraining admixtures.
2. Water reducing admixtures.
3. Set controlling admixtures (retarders-accelerators).
4. Mineral admixtures.

Air-entraining Admixtures

Fresh concrete contains some air bubbles because of the difficulty of eliminating all air during mixing. Some of these bubbles are lost shortly after mixing. Others are entrapped in concrete. They are usually large in size and could form continuous channels in the concrete that may increase the permeability and reduce durability. The air-entraining agents

162

stabilize the air bubbles that are originally formed during mixing. By stabilizing the air bubbles, they introduce minute voids in concrete that are discontinuous, and on the average less then 0.05 mm in diameter and are spaced less than 0.20 mm. The most common way of incorporating air in concrete is by means of surface active agents that reduce surface tension. The main entraining agents are natural wood resins and their soaps, animal vegetable fats and oils, and alkali salts of sulphonated or sulfated organic compounds.

The most important application of air-entraining chemicals is to increase the resistance of concrete to freezing and thawing. Research studies as well as practical experiences have shown that the entrapment of air in concrete increases very considerably the resistance of concrete to freezing and thawing. This durability enhancement is due mainly to the relief, caused by the minute dispersed air bubbles that act as expansion chambers of pressure caused by temperature and moisture variations in concrete.

ASTM C233 (Test Method for Air-entraining Admixtures for Concrete) covers the testing of materials proposed for use as air-entraining admixtures for concrete mixtures in the field. This test is based on arbitrary stipulations permitting highly standardized testing in the laboratory and are not intended to simulate actual job conditions. ASTM C260 (Specifications for Air-entraining Admixtures for Concrete) covers the specifications of chemicals proposed for use as air-entraining admixtures.

Water-reducing and Set Controlling Admixtures

ASTM C494 (Standards Specifications for Chemical Admixtures for Concrete) divides the water-reducing and set-controlling chemicals into seven types.

1. Type A: Water-reducing admixtures.
2. Type B: Retarding admixtures.
3. Type C: Accelerating admixtures.
4. Type D: Water-reducing and set-retarding admixtures.
5. Type E: Water-reducing and set-accelerating admixtures.
6. Type F: High-range water-reducing admixtures.
7. Type G: High-range water-reducing and set-retarding admixtures.

Water-reducing admixtures.   Water-reducing admixtures reduce the quantity of mixing water required to produce concrete of a given consistency. Usually, water-reducing admixtures are based on lignosulphonic acids, hydroxy carboxylic acids, and processed carbohydrates. For a given workability, they can reduce the water requirement of concrete by 5 to 15 percent.

High-range water-reducing admixtures (superplasticizers) were introduced in Japan in 1964, and later in 1970 in Europe and the United States. They are chemically different from the normal water reducers and capable of reducing water requirement by about 30 percent for the same consistency. The superplasticizers are broadly classified into four groups.

1. Sulphonated melamine-formaldehyde condensates.
2. Sulphonated naphtaline formaldehyde condensates.

3. Modified lignosulphonates.
4. Sulphonic acid esters, carbohydrate esters, and so on.

The advantage of water reducers and superplasticizers can be categorized in three ways.

1. *Obtaining higher compressive strength.* The addition of the admixture reduces the water-cement ratio without altering the workability producing a denser and stronger concrete.
2. *Obtaining better workability.* The inclusion of the admixture to concrete without reducing the water-cement ratio will improve the workability of concrete without compromising its strength.
3. *Reducing the cement ratio.* The addition of the admixture to a concrete mixture would yield a concrete with the same strength and workability characteristics of another concrete mixture with higher cement content.

The effect of water-reducing admixtures on fresh concrete can be described as follows. When water is added to the cement the cement particles tend to cluster together. This is attributed to the attractive forces that exist between positively and negatively charged surfaces. To break these clusters and improve the workability a larger amount of water is needed (higher water-cement ratio). This could result in lower strength for the hardened concrete. When water-reducing additives are added, they are absorbed by the cement particles, causing them to repel each other. Hence, a well-dispersed system is obtained, and less water is required.

Retarding admixtures.   Retarding admixtures are used to lengthen the setting and workability times of a concrete mixture. The main known retarders are calcium sulfates, sugars, hydroxycarboxylic acids, and organic compounds based on phosphate, borax, and magnesium salts. The most important application of retarders is in hot weather climate especially if delays in transportation and placing of concrete are expected. The retardation of cement hydration is believed to be caused by the absorption of retarders on the surface of the cement particles, and reduction of their contact with water and hence the formation of the hydration products.

Accelerating admixtures.   An accelerating admixture accelerates the setting and early-strength development of concrete. Accomplishing this results in better scheduling of the workload, earlier removal of construction forms, and compensation effects of low temperature on early-strength development. The mechanism of the action of accelerators on cement hydrate can be explained by the formation of complex compounds that accelerate setting by providing nuclei for the hydration of the silicate phase. The most common accelerator is calcium chloride ($CaCl_2$).

In addition to the increased early strength, calcium chloride has the following secondary effects:

1. Increases workability by a small amount.
2. Increases air content and air-bubble size in air-entrained concrete.
3. Reduces bleeding.
4. Increases drying shrinkage slightly.
5. Lowers resistance to sulfate attack.

Water-reducing and set-retarding admixtures.   Water-reducing and set-retarding admixtures reduce the quantity of mixing water required to produce concrete of a given consistency and retard the setting of concrete.

Water-reducing and set-accelerating admixtures.   Water-reducing and set-accelerating admixtures reduce the quantity of mixing water required to produce concrete of a given consistency and accelerate the setting and early-strength development of concrete.

MINERAL ADMIXTURES

Of the many admixtures available for concrete, an important group falls into the category of finely divided mineral admixtures, which are divided into three classifications:

1. Those that are chemically inert.
2. Those that are pozzolanic.
3. Those that are cementitious.

The inert finely divided material includes quartz, limestone, bentonite, and hydrated lime. These finely divided materials are often substituted for portland cement in concrete to save on portland cement. To have the necessary workability and plasticity, most concrete must have large amounts of cement. Substituting finely divided material for portland cement saves a considerable amount of portland cement.

Pozzolanic or cementitious finely divided material contributes to the strength development of concrete and reduces the amount of portland cement required for a given strength. A pozzolan is a siliceous, or siliceous and aluminous material that itself has little or no cementitious value but that will, in finely divided form and in the presence of moisture, chemically react with calcium hydroxide at ordinary temperatures to form compounds that do have cementitious properties. Pozzolans include diatomaceous earth, opaline cherts, shales, volcanic ashes, fly ash, and clays. The most common pozzolan is the fly ash, a finely divided residue that results from the combustion of ground or powdered coal. ASTM C618 (Fly Ash and Raw or Calcined Natural Pozzolan for Use as a Mineral Admixture in Portland Cement Concrete) covers fly ash and raw or calcined natural pozzolan for use as a mineral admixture in concrete. Originally, the main justification for using mineral admixtures was cost reduction and workability improvement. However, later research studies have clearly indicated that they can be included in concrete as well to enhance its durability, strength, and cracking performances.

The following classifications exist under ASTM C618:

1. *Class N.* Class N comprises raw or calcined natural pozzolans. They include diatomaceous earths; opaline cherts and shales; tuffs and volcanic ashes or pumices, any of which may or may not be processed by calcination; and various materials requiring calcination to induce satisfactory properties, such as some clays and shales.
2. *Class F.* Class F is fly ash normally produced from burning anthracite or bituminous coal that meets the applicable requirements for this class. This class of fly ash has pozzolanic properties.

3. *Class C.* Class C is fly ash normally produced from lignite or subbituminous coal that meets the applicable requirements for this class. This class of fly ash, in addition to having pozzolanic properties, has some cementitious properties. Some Class C fly ashes may have a lime content exceeding 10 percent.

HIGH-STRENGTH CONCRETE

General

High-strength concrete refers to concrete that has a uniaxial compressive strength greater than the usual strength obtained in a particular region. This definition does not include a compressive strength value indicating a transfer from a normal-strength concrete to a high-strength one. This can be attributed to the fact that, as the development of concrete technology has continued, the strength that can be achieved has increased. In the middle of this present century, concrete with a compressive strength of 5000 psi was considered high-strength concrete. Currently in the 1990s compressive strength greater than 16000 psi has been used in cast-in-situ buildings and prestressed concrete members. However, this compressive strength could be considerably lower depending on the characteristics of the local materials used for these concrete products.

Because the definition of high-strength concrete has changed over the years, and because the definition of high-strength concrete varies regularly depending on the local materials used, ACI Committee 363 on High-Strength Concrete had to define an applicable range of concrete strength for its activities. In 1979, the committee selected 6000 psi as the lower limit for defining high-strength concrete. Two reasons could be presented to justify this definition:

1. To produce concrete above 6000 psi, more quality control is needed for the selection and proportioning of the materials.
2. Most empirical equations used to predict properties of concrete or to design structural members are based on tests using concrete with a compressive strength less than 6000 psi. If using higher-strength concrete, the designers should be aware of the applicability of these equations for this higher range.

Materials

Conventional production techniques have been used in producing high-strength concrete. However, more stringent quality control, better quality materials, and increased care in the proportioning of mortar and choosing admixtures must be maintained continuously.

Proper selection of the cement is an important criteria in the production of high-strength concrete. The cement suppliers should report uniformity in accordance with ASTM C917 (Standard Method of Evaluation of Cement Strength Uniformity from a Single Source). Also ASTM C109 (Test Method for Compressive Strength of Hydraulic Cement Mortars [using 2-in. or 50-mm specimens]) mortar cube test should be used to check the cement strength characteristics. Higher cement content is generally used for high-strength concrete production (650 to 950 lb/yd^3). However, using cement content beyond an optimum level would not necessarily result in an increase in strength. In evaluating an optimum cement content, trial mixes are usually proportioned to equal consistencies, allowing the water content to vary according to the water demand in the mixture. The most important variable in achieving higher-strength concrete is the water-cement ratio. When mineral admixtures are used in concrete, a water/(cement+mineral) ratio by weight is usually considered in lieu of the traditional water-cement ratio. This ratio typically has ranged from 0.27 to 0.50. The lower value has been successfully used with the introduction of high-range water reducers. The requirements for water quality for higher-strength concrete are no more stringent than those for conventional concrete. Potable water is usually specified.

Finely divided mineral admixtures have been widely used in high-strength concrete, among which, silica fume, Class C fly ash and Class F fly ash are the most common. They have been used to supplement the portland cement from 10 to 40 percent by weight of the cement content. Optimum desirable quantities of these materials should be evaluated using laboratory trial batches.

Chemical admixtures and air-entraining agents have been commonly used in high-strength concrete production to attain optimum performance at lower cost. Superplasticizers, or high-range water reducers, are the chemical admixtures mostly used because of their capability in reducing the water-cement ratio without loss in strength or workability.

Fine and coarse aggregates used for high-strength concrete should meet the requirement of ASTM C33 (Specifications for Concrete Aggregates) as a minimum. To obtain a better workability and compressive strength, it is often recommended to use sand with a fineness modulus of about 3.0. The maximum size of the coarse aggregate should be kept to a minimum ($\frac{1}{2}$ or $\frac{3}{8}$ in.). Studies have shown that crushed stones produce higher strengths than rounded gravel. However, excessive angularity should be avoided, not to permit the use of high-water requirements. Also, because bond strength is the limiting factor in the development of high-strength concrete, the mineralogy of the aggregates should be such as to promote chemical bonding. Typical high-strength concrete mixtures are presented in Table 7-1.

Material Properties

Extensive studies have been conducted during the last 15 years on the properties and the microstructural behavior of high-strength concrete. Axial compression stress versus strain curves for concrete, with a compressive strength up to 12000 psi are shown in Figure 7-1. From these curves, the following observations can be made:

1. Stiffness of concrete increases with the increase of the compressive strength.
2. Higher compressive strength results in higher strain at ultimate loading.
3. The fracture strain (maximum sustained strain) seems to decrease with increase in strength leading to a more brittle and explosive type of failure. Special testing procedures and equipment are usually needed to obtain the complete stress-strain curve of high-strength concrete.

The tensile behavior of high-strength concrete is usually evaluated using two types of testing: (1) The split cylinder test (ASTM C496 [Test Method for Splitting Tensile Strength of Cylindrical Concrete Specimens]), and (2) the modulus of rupture or bending test (ASTM

TABLE 7-1 TYPICAL HIGH-CONCRETE MIXTURES

(a) Using Fly Ash

| Concrete Design Strength (psi) | | Quantities per yd³ | | | | |
	Cement (lb)	Fine (lb)	Coarse (lb)	Water (lb)	Fly Ash	Pozzolith (fl. oz.)
9000	846	1025	1800	330	100	25.4
7500	729	1250	1695	315	100	21.9
6000	705	1190	1825	303	100	21.1
5000	541	1385	1850	268	100	16.2
4000	447	1485	1860	257	100	13.4

Metric equivalent: 1 lb/yd³ = 0.59 kg/m³, 1 Ksi = 1000 psi = 6.9 MPa.
Source: Russell and Corley, 1978.

(b) Using Silica Fume

Contents	Mix Proportion (kg/m³)	Average Slump (mm.)	Compressive Strength (MPa.) (90 days)
Type I cement	593		
Silica fume admixture	119		
Elgin sand	537	119	115.1
Thorton limestone	997		
Water	159		

Metric equivalent: 1000 psi = 6.895 MPa.
Source: Wolsiefer, 1984.

C78 [Test Method for Flexural Strength of Concrete (Using Simple Beam with Third-Point Loading)]). Experimental investigations have shown that the tensile strength of high-strength concrete can be expressed as a function of $\sqrt{f'_c}$ as is the case of normal-strength concrete. However, the rate of increase of tensile strength seems to decrease with higher compressive strength. Figures 7-2 and 7-3 indicate the linear relationship between $\sqrt{f'_c}$ and the tensile splitting strength f'_t, and the modulus of rupture f_r, respectively.

For compressive strength range of 3000 to 12000 psi, the tensile splitting strength f'_t can be expressed as

$$f'_t = 7.4\sqrt{f'_c}, \text{psi} \tag{7.1}$$

For the same compressive strength range, the modulus of rupture, f_r, can be expressed as

$$f_r = 11.7\sqrt{f'_c}, \text{psi} \tag{7.2}$$

The modulus of elasticity, E_c, of concrete increases with the compressive strength. The following equation for a compressive strength range of 3000 to 12000 psi seems to express the relation satisfactorily:

$$E_c = [40,000\sqrt{f'_c} + 10^6], \text{psi} \tag{7.3}$$

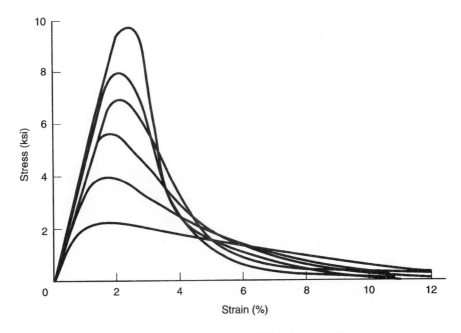

Figure 7-1 Stress-Strain Curves of High-Strength Concrete

Poisson's ratio for high-strength concrete was found to be close to 0.2, as in the case of normal-strength concrete.

Little information is available on the shrinkage behavior of high-strength concrete. A relatively high initial rate of shrinkage has been reported, but after drying for 180 days there is little difference between the shrinkage of high-strength concrete and lower-strength concrete. A significant difference was reported for the ultimate value of creep strain under sustained axial compression to initial elastic strain. High-strength concrete may be as low as one-half the value generally associated with low-strength concrete.

Applications of High-strength Concrete

Using high-strength concrete in structures today would result in both technical and economical advantages. Most applications of high strength to date have been in high-rise buildings, long-span bridges, and some special structures.

Tall structures whose construction using normal strength would not have been feasible in the past have now been completed using high-strength concrete. Major applications of high-strength concrete in tall structures have been in columns and shear walls, which resulted in decreased dead weight of the structure and an increase in the amount of the rental floor space, in the lower stories. High-strength concrete application in high-rise buildings have been reported in New York City, Houston, Dallas, Chicago, Minneapolis, Toronto (Canada), and Melbourne (Australia). Table 7-2 includes some of these applications.

The benefits of using high-strength concrete in bridges have been summarized as follows:

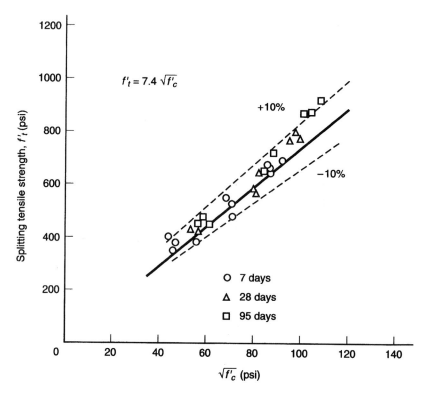

Figure 7–2   Compressive Strength versus Tensile Splitting Strength
(*Source*: ACI 363, 1984)

1. Greater compressive strength per unit cost, per unit weight, and per unit volume.
2. Increased tensile strength that is a controlling parameter in the design of prestressed concrete members under service loads.
3. Increased modulus of elasticity that would result in lower short-term deflection.
4. Improved long-term deflection properties and reduced losses in prestressing forces.

Table 7-3 presents a list of bridges in which high-strength concrete has been used in United States and Japan. Other applications of high-strength concrete include marine foundations, offshore structures, and industrial chemical-resistant structures.

FIBER-REINFORCED CEMENTITIOUS COMPOSITES

General

The technology of fiber-reinforced building materials goes back to ancient times. Approximately 3500 years ago, sun-baked bricks reinforced with straws were used in

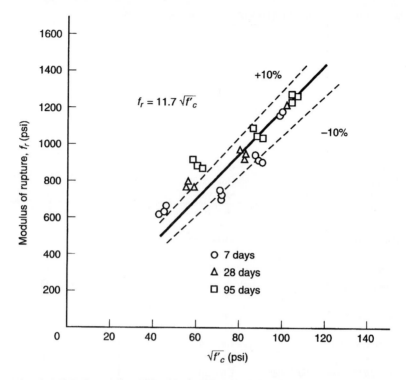

Figure 7–3 Compressive Strength versus Modulus of Rupture
(*Source*: ACI 363, 1984)

buildings, and animal hair was used to strengthen plaster. In the past 30 years, there has been a renewed interest in the science and applications of fiber-reinforced construction materials. This section includes a presentation on the use of different fiber types in cementitious composites and their construction applications.

Plain concrete possesses low-tensile-strength and tensile-strain capacities, making it a brittle material. To use concrete economically as a construction material, reinforcement must be provided. Traditionally, reinforcement has been in the form of continuous reinforcing bars placed at select locations to withstand resulting tensile stresses. Fibers added to concrete are short, discontinuous, and randomly distributed throughout the matrix. Experimental studies showed that with volumes and sizes that could be conveniently incorporated into conventional cementitious composites without handling difficulties, fibers were not able to offer a substantial improvement in strength over corresponding mixtures without fibers. Therefore, fibers are not generally considered highly efficient in resisting tensile stresses. However, because they are closely spaced, they are very effective in providing better crack control and improving the composite toughness.

Despite the limited contribution of fibers to the tensile strength of cementitious composites, there are certain applications in which fiber reinforcement is better than conventional reinforcing bars.

TABLE 7-2   BUILDINGS WITH HIGH-STRENGTH CONCRETE

Building	*Location*	*Year*[a]	*Total Stories*	*Maximum Concrete Design Strength psi*
Pacific Park Plaza	Emeryville, Calif.	1983	30	6500
S. E. Financial Center	Miami	1982	53	7000
Petrocanada Building	Calgary	1982	34	7250
Lake Point Tower	Chicago	1965	70	7500
1130 S. Michigan Avenue	Chicago			7500
Texas Commerce Tower	Houston	1981	75	7500
Helmsley Palace Hotel	New York	1978	53	8000
Trump Tower	New York		68	8000
City Center Project	Minneapolis	1981	52	8000
Collins Place	Melbourne		44	8000
Larimer Place Condominiums	Denver	1980	31	8000
499 Park Avenue	New York		27	8500
Royal Bank Plaza	Toronto	1975	43	8800
Richmond—Adelaide Centre	Toronto	1978	33	8800
Midcontinental Plaza	Chicago	1972	50	9000
Frontier Towers	Chicago	1973	55	9000
Water Tower Place	Chicago	1975	79	9000
River Plaza	Chicago	1976	56	9000[b]
Chicago Merchantile Exchange	Chicago	1982	40	9000[c]

[a] Year in which high-strength concrete was cast.
[b] Two experimental columns of 11,000 psi strength were included.
[c] Two experimental columns of 14,000 psi strength were included.
Metric equivalent: 1000 psi = 6.895 MPa.
ISource: ACI Committee 363, 1984.

1. Thin sheet materials, in which traditional reinforcing bars cannot be used because of space limitations and in which the fibers therefore constitute the primary reinforcement. In thin sheet materials, fiber concentrations are relatively high, typically exceeding 5 percent by volume. In these applications, the fibers act to increase both the strength and the toughness. Toughness is defined as the energy absorbed by the material when subjected to loading. It is estimated by the area under the load-deflection curve of the composite as shown in Figure 7-4.
2. Structural components in which fibers' main role is to control cracking, to improve the postfailure behavior, and to enhance the energy absorption capacity are shown in Figure 7-5. In this application, fibers are not a substitute for conventional reinforcement and hence are not used to improve strength.

Similar to all construction materials, the evaluation of the properties of fiber-reinforced cementitious composites is of major importance. Most of these properties are matrix dependent and can be measured by the methods commonly used for conventional concrete (examples: compressive strength, freeze-thaw durability). However, some other properties are highly affected by the addition of fibers and therefore must be evaluated by test meth-

TABLE 7-3 BRIDGES WITH HIGH-STRENGTH CONCRETE

Bridge	Location	Year	Maximum Span (ft)	Maximum Concrete Design Strength psi
Willows Bridge	Toronto	1967	158	6000
Houston Ship Canal	Texas	1981	750	6000
San Diego to Coronado	California	1969	140	6000 L[a]
Linn Cove Viaduct	North Carolina	1979	180	6000
Pasco-Kennewick Intercity	Washington	1978	981	6000
Coweman River Bridges	Washington		146	7000
Huntington to Proctorville	West Virginia	1984	900	8000
Nitta Highway Bridge	Japan	1968	98	8500
Kaminoshima Highway Bridge	Japan	1970	282	8500
Fukamitsu Highway Bridge	Japan	1974	85	10,000
Ootanabe Railway Bridge	Japan	1973	79	11,400
Akkagawa Railway Bridge	Japan	1976	150	11,400

[a]Lightweight concrete.
Metric equivalent: 1000 psi = 6.895 MPa.
Source: ACI Committee 363, 1984.

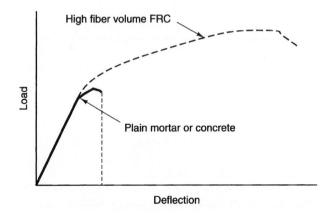

Figure 7-4 Typical Load Deflection Curve of High Volume Fiber-Reinforced Cementitious Composite

ods not used in normal concrete practice. ASTM C995 (Time of Flow of Fiber Reinforced Concrete through Inverted Slump Cone) is used to measure the workability of fiber reinforced composites. In this test, the slump cone is inverted over a yield bucket and is then filled with concrete without compaction. An immersion type vibrator is placed vertically into the concrete at the center of the cone, and the time required for it to drop through the concrete until it touches the bottom is determined. The shorter is the time, the more workable is the concrete. Another test, namely, ASTM C1018 (Flexural Toughness and First-crack Strength of Fiber-reinforced Concrete [Using Beam with Third-point Loading]) is used to measure the flexural properties and postfailure performance of fiber-reinforced

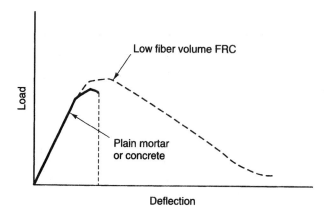

Figure 7-5 Typical Load Deflection
Curve of Low Volume Fiber-
Reinforced Cementitious Composite

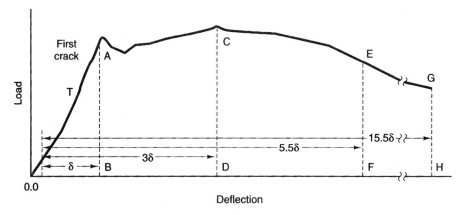

Figure 7-6   Interpretation of Load-Deflection Curve According to ASTM C1018

cementitious composites. A prism is loaded at two points and the complete load-deflection
curve is obtained. Figures 7-4 and 7-5 show that plain concrete fails suddenly once the
deflection corresponding to the ultimate flexural strength is reached. Conversely, fiber-
reinforced cementitious prism continues to sustain loads even at deflections considerably
in excess of the fracture deflection of the plain concrete. Toughness indexes are defined in
ASTM C1018 and are used to evaluate the postfailure performance. These indexes are
obtained by dividing the area under the load-deflection curve, determined at a deflection
that is a multiple of the first crack deflection, by the area under the curve up to the first
crack. I_5 is determined at a deflection 3 times, I_{10} is determined at 5.5 times, and I_{30} at 15.5
times the first-crack deflection. Figure 7-6 indicates how to obtain these indexes. The postfailure behavior depends primarily on the type, content, and aspect ratio (length-diameter
ratio) of the fibers.

  There are numerous types of fibers used in cementitious composites, namely, steel
fibers, glass fibers, asbestos fibers, synthetic fibers, and natural fibers. These types of

TABLE 7-4 TYPICAL PROPERTIES OF FIBERS

Fiber	Diameter (μm)	Specific Gravity	Modulus of Elasticity (GPa)	Tensile Strength (GPa)	Elongation at Break (%)
Steel	5.0–500	7.84	200	0.5–2.0	0.5–3.5
Glass	9.0–15	2.60	70–80	2.0–4.0	2.0–3.5
Asbestos					
Crocidolite	0.02–0.4	3.40	196	3.5	2.0–3.0
Chrysotile	0.02–0.4	2.60	164	3.1	2.0–3.0
Fibrillated polypropylene	20.0 200	0.90	5.0–77	0.5–0.75	8.0
Aramid (Kevlar)	10.0	1.45	65–133	3.6	2.1–4.0
Carbon (high strength)	9.0	1.90	230	2.6	1.0
Nylon	—	1.10	4.0	0.9	13.0–15.0
Cellulose	—	1.20	10.0	0.3–0.5	—
Acrylic	18.0	1.18	14–19.5	0.4–1.0	3.0
Polyethylene	—	0.95	0.3	0.7×10^{-3}	10.0
Wood fiber	—	1.50	71	0.9	—
Sisal	10.0–50	1.50	—	0.8	3.0
Cement matrix (for comparison)	—	2.50	10.0–45	3.7×10^{-3}	0.02

Source: Bentur and Mindess, 1990.

fibers vary considerably in properties and effectiveness. Table 7-4 includes some common fibers and their physical properties.

Steel Fibers

Steel fiber-reinforced cementitious composites are made with hydraulic cements, aggregates along with discontinuous discrete steel fibers. With the exception of asbestos, steel is considered the most commonly used fiber for cementitious composites' reinforcement. During the last three decades, steel fiber-reinforced cementitious composites have continued to attract more research than any other fiber cementitious composite. The following are some applications of steel fiber cementitious composites:

Hydraulic structures such as dams, spillways, and sluice ways as new or replacement slabs to resist cavitation.

Airport and highway paving and overlay.

Industrial floors for impact resistance and resistance to thermal shock.

Refractory concrete, using high-alumina cement in both castable and shot-creted applications.

Bridge decks, as an overlay.

Explosion-resistant structures, usually in combination with reinforcing bars.

ASTM A820 (Standard Specifications for Steel Fibers for Fiber-reinforced Concrete) covers steel fibers that could be used for steel fiber-reinforced concrete. When specifying a fiber type, it is necessary to specify the fiber length and diameter (aspect ratio l/d), minimum ultimate tensile strength, and deformations.

Currently, the steels used for making steel fibers are

1. Carbon steels for standard applications.
2. Alloy steels for corrosion resistant applications (refractories and marine structures).

Steel fibers may have different shapes.

1. Round, straight fibers produced by cutting or chopping wires (typical diameter from 0.01 to 0.039 in.).
2. Flat, straight fibers produced by shearing sheets or flattening wire (typical cross section ranging from 0.006- to 0.025-in. thickness by 0.01- to 0.08-in. width).
3. Crimped and deformed fibers produced with full-length crimping or bent at the ends only. The deformations are used to increase mechanical bonding.

To facilitate handling and mixing, some fibers are collated into bundles using a water-soluble glue. The glue dissolves during the mixing process, and the bundles separate into individual fibers. The amount of steel fibers added to a cementitious composite ranges from 0.25 percent by volume (33 lb/yd^3) to 2 percent by volume (265 lb/yd^3).

Fiber-reinforced cementitious composites, in general, may be viewed as composites with improved strain capacity, impact resistance, energy absorption, and possible tensile strength. The increase in these properties could vary from substantial to marginal depending on the quantity and type of fibers used.

Glass Fibers

The process of producing glass fibers consists of extracting molten glass in the form of filaments through the bottom of a heated platinum tank. These filaments, 10 to 20 μm in diameter, solidify after cooling outside the heated tank. The filaments are then collected into a strand consisting of about 200 filaments. Several strands are wound together to form a roving. The glass fibers are marketed either as continuous or chopped roving.

Glass fibers are mainly used for the production of thin sheet components with cement paste or cement mortar and about 5 percent fiber content. The constituents of a typical glass fiber reinforced cementitious mixture are shown in Table 7-5. Admixtures and polymers are not always used in mixtures, but, when added, are intended to improve specific properties of the mixture. Another application is the making of reinforcing bars with continuous glass fibers joined together and impregnated with plastics.

TABLE 7-5 TYPICAL COMPOSITION
OF GLASS FIBER-REINFORCED
CEMENTITIOUS COMPOSITES

Component	Weight (%)
Cement	39
Sand	39
Water	13
Fiber	5
Admixtures and polymer	4

Source: Swamy, 1984.

TABLE 7-6   CHEMICAL COMPOSITION
OF GLASS FIBERS

Component	E-glass (%)	AR-glass (%)
SiO_2	55	71
Al_2O_3	15	1
B_2O_3	7	—
CaO	21	—
Alkalis	2	11
ZnO_2	—	16
LiO_2	—	1

Source: Swamy, 1984.

The main problem of glass fibers (E-glass fibers) is their low alkali resistivity making them not durable to chemical attack by portland cement paste. To overcome this problem, alkali-resistant glass fibers with better durability have been developed (AR-glass fibers). The composition of E-glass and AR-glass is given in Table 7-6.

Asbestos Fibers

Asbestos fibers are made of natural fibrous minerals. Depending on their mineralogical composition, they can be divided into two groups, namely, amphibole and chrysotile. Amphibole has a layered structure with its external surface being a silicate layer. Chrysotile consists of parallel sheets of $Mg(OH)_2$ and silicates. Asbestos fibers are added to a cement-water slurry. Then, a thin lamina is formed of the dewatered slurry. The thickness of the final asbestos-cement product is obtained by piling wet laminae over each other to the desired thickness.

Asbestos fiber-reinforced cement is used to produce pipe, flat sheets, and corrugated roofing sheets. Some asbestos fibers, when inhaled, can be a health hazard. The health risks are greatest during the production process, but are minimal during the use of the asbestos fiber-reinforced cement product. However, as a result of such health risks and strict environmental regulations, current research is conducted to obtain replacement for asbestos fibers while maintaining the same efficient production process.

Synthetic Fibers

Synthetic fibers have been used in recent years for the reinforcement of cementitious composites. These composites are primarily used to produce cladding panels, roof tiles, and sheets. Synthetic fibers can be broadly divided into two groups.

1. *Fibers with low modulus of elasticity.* When added to cementitious composites, considerable improvements can be obtained with respect to the strain capacity, toughness, impact resistance, and crack control. This group of fibers includes polypropylene, nylon, polyethylene, saran, and orlon. All of these fibers showed effective contribution to the properties of cementitious composites when mixed as short fibers (25 to 75 mm long) and contents of up to about 5 percent by volume.

2. *Fibers with high modulus of elasticity.* The effect of mixing short high modulus of elasticity fibers to cement mixtures is usually to increase both the flexural strength and the postcracking behavior. The fibers' contribution depends on the fiber type, aspect ratio, and content. This group of fibers includes carbon fibers, aramid (kevlar) fibers, acrylic fibers, and polyvinyl alcohol fibers.

Natural Fibers

In recent years, a great deal of interest has been created on the potential applications of natural fiber-reinforced cementitious composites. Natural fibers can be grouped, based on their morphology, into four groups.

1. Stem or bast fibers, obtained from the stalks of plants.
2. Leaf fibers, obtained from the leaves of plants.
3. Surface fibers, found on the surface of stems, fruits, and seeds of plants.
4. Wood fibers, obtained from wood chips.

Research investigations and site applications have shown that the uniform dispersal of short discontinuous natural fibers in cementitious composites enhances resistance to cracking, impact, and shock loading. It would also improve ductility for better energy absorption. Because of the relatively simple production processes required to obtain these fibers, they are highly suitable for low-cost housing applications.

WASTE AND BY-PRODUCTS USE IN CONCRETE

General

Concrete is and should continue to be a dominant construction material in the foreseeable future. With the increasing awareness toward our environment and natural resources, progress in concrete technology should consider proper use of resources and energy. Consequently, it can be expected that efforts will be made to use waste and by-products in concrete technology. Currently, the concrete conducted research and applications in this area can be broadly divided into two groups, namely: use of by-products as cement replacement and use of by-products as aggregate replacement.

Generally, the use of any by-product should not create an interference mechanism that would delay the setting and curing of concrete or excessively reduce its strength and durability. Bizock (1967) presented two lists of industrial wastes that are harmless or deemed to be aggressive to cementitious products. The list of wastes that can be tolerated includes the following:

1. Brines containing bases but not sulfates.
2. Potassium permanganate, occurring at fermenting and purification installations.
3. Sodium carbonate and potassium carbonate.
4. Bases provided that their concentrations is not excessively high.
5. Oxalic acid occurring at tanneries.

6. Mineral oils and petroleum products (benzene, kerosene, cut-back oil, naphta, paraffin, tar) as long as these contain no acids that can continue to remain in the products after chemical treatment.

The list of wastes considered aggressive to cementitious products includes the following:

1. Water containing gypsum.
2. Ammonia salts.
3. Hydrochloric acid, nitric acid, acetic acid, and sulfuric acid.
4. Chlorine and bromine.
5. All sulfur and magnesium salts.
6. Salts of strong acids.

By-Products Use as Cement Replacement

Power plants using coal or rice husks as fuel, and metallurgical furnaces producing pig iron, steel, and silicon alloys are major sources of by-products. Traditionally, disposing of these by-products has been by dumping into landfills. Most of these by-products have pozzolanic properties. According to ASTM C595 (Specifications for Blended Hydraulic Cements), a pozzolan is defined as "a siliceous or siliceous and aluminous material which in itself possesses little or no cementitious value but will, in finely divided form and in the presence of moisture, chemically react with calcium hydroxide at ordinary temperature to form compounds possessing cementitious properties." Therefore, a great interest has been shown in using them as cement replacement. Because cement production requires substantial amounts of energy, a partial replacement of cement with these by-products, available at lower prices, would result in substantial energy and cost savings. Furthermore, there is evidence that most of these by-products enhance the properties of concrete.

The major by-products used as cement replacement follow.

Coal ashes. During the coal combustion in power plants, the volatile materials and coals particles are burned off. The mineral impurities fuse and remain in suspension in the flue gas. After cooling, the fused minerals solidify as spherical particles and most of it *flies* out with flue gas stream. This "fly ash" is removed from the flue gas by mechanical separators and special filters. Based on their mineralogical composition and properties, fly ashes can be divided into two categories; fly ash containing less than 5 percent CaO (i.e., Class F) obtained from combustion of anthracite and bituminous coals, and fly ash containing between 15 to 35 percent CaO (i.e., Class C) obtained from combustion of lignite and subbituminous coals.

Rice husk ash. Rice husks are the shells produced during the dehusking operation of paddy rice. When used as fuel, the ash formed after their combustion consists mainly of crystalline silica minerals. When ground to fine particle size, this ash can develop pozzolanic properties.

Silica fume. Silica fume is produced by electric arc furnaces as a by-product of the production of metallic silicon or ferrosilicon alloys. Silica fume is extremely fine and

TABLE 7-7   CLASSIFICATION OF WASTE MATERIALS

Group I	Group II	Group III
Blast-furnace slag	Steel slag	Alumina muds
Fly ash	Bituminous coal refuse	Phosphate slimes
Bottom ash	Phosphate slag	Sulphate sludge
Boiler slag	Slate mining waste	Scrubber sludge
Reclaimed concrete	Foundry waste	Copper tailings
Anthracite coal refuse	Taconite tailings	Dredge spoil
	Incinerator residue	Feldspar tailings
	Waste glass	Iron ore tailings
	Zinc smelter waste	Lead/zinc tailings
	Building rubble	Nickel tailings
		Rubber tires
		Battery casings
		Residual contaminated sands

Source: Ramchandran et al., 1981.

is removed by filtering the outgoing gases in a bag filter. It possesses good pozzolanic and cementitious properties.

By-Products Use as Aggregates Replacement

There are many wastes and by-products that have been used or have potential applications as concrete aggregate replacement. These materials can be broadly categorized into three groups as shown in Table 7-7.

The materials included in group I present the best potential for use as aggregates. Most of them are currently in use. They generally have desirable properties such as soundness, strength, shape, abrasion resistance, and gradation.

The materials of group II could be considered as aggregate replacement in concrete. However, because their properties are not compatible to concrete as group I materials, they are only used to a limited extent after processing.

The materials of group III require more research and development work before being economically used as concrete aggregate replacement. Most of them need extensive processing to end up with uniform characteristics. Their potential is for the production of low-strength concrete.

PROBLEMS

7.1. Discuss the effect of air-entraining admixtures on concrete.

7.2. What is the major application of water-reducing admixtures?

7.3. Why are retarders and accelerators admixtures used in concrete?

7.4. Discuss the effect of adding mineral admixtures in concrete.

7.5. Discuss the main ingredients used in concrete mixtures to obtain a compressive strength higher than 6000 and 12,000 psi.

7.6. Discuss the major applications of high-strength concrete.

7.7. Why are fibers added to concrete?

7.8. Discuss the major applications of steel and synthetic fibers in cementitious composites.

7.9. What is the major by-product used as cement replacement?

7.10. Discuss the different groups of by-products used as aggregate replacement.

REFERENCES

ACI COMMITTEE 363, "State-of-the-art Report on High-Strength Concrete," *ACI J.* (July–August 1984), pp. 364–411.

ACI COMMITTEE 363, "Research Needs for High-Strength Concrete," *ACI J.* (November–December 1987), pp. 559–561.

BALAGURU, P. N., and SHAH, S. P., *Fiber-Reinforced Cement Composites*, McGraw-Hill Book Company, New York, 1992.

BENTUR, A., and MINDESS, S., *Fibre Reinforced Cementitious Composites*, Elsevier Applied Science Publishers, New York, 1990, 449 pp.

BIZOCK, I., *Concrete Corrosion and Concrete Protection*, Chemical Publishing, New York, 1967.

EZELDIN, A. S., BALAGURU, P. N., and SHAH, S., "High Strength Concrete: Proportioning, Behavior and Applications," *Proceedings of Structural Congress 89*, ASCE, Structural Division, San Francisco, Calif., 1989, pp. 21–30.

HANNANT, D. J., *Fibre Cements and Fibre Concretes*, John Wiley & Sons, New York, 1978.

HESTER, W., (ed.), *Utilization of High-Strength Concrete,* Special Publications, SP-121, American Concrete Institute, Detroit, Michigan, 1990, 794 pp.

MALHOTRA, V. M. (ed.), *Fly Ash, Silica Fume, Slag and Natural Pozzolans in Concrete*, 2 vols., Special Publications, SP-132, American Concrete Institute, Detroit, Michigan, 1993.

RAMACHANDRAN, V. S. (ed.), *Concrete Admixtures Handbook*, Noyes Publications, 1984, 626 pp.

RAMACHANDRAN, V. S., FELDMAN, R. F., and BEAUDOIN, J. J., *Concrete Science: Treatise on Current Research*, Heyden & Son, Ltd., 1981, 427 pp.

RUSSELL, H. (ed.), *High-Strength Concrete*, Special Publications, SP-87, American Concrete Institute, Detroit, Michigan, 1985, 278 pp.

RUSSELL H. G., and CORLEY, W. G., *Time Dependent Behavior of Columns in Water Tower Place*, Special Publications, SP-55, ACI 1978, pp. 347–373.

SHAH, S. P. (ed.), "High Strength Concrete," *Proceedings of a Workshop Held at the University of Illinois at Chicago Circle*, December 2–4, 1979, 226 pp.

SWAMY, R. N. (ed.), *New Reinforced Concrete*, United Press, 1984, 200 pp.

WOLSIEFER, J., "Ultra High-Strength Field Placeable Concrete with Silica Fume Admixtures," *Concr. Int. 6* (April 1984), pp. 32–34.

8 ■ TIMBER

Timber has been one of the basic materials of construction since the earliest days of humankind. Today it has been largely superseded by concrete and steel. However, the use of timber remains quite extensive.

In the United States 600 different tree species exist, of which 15 are used in the building trades (ash, birch, cedar, cypress, fir, gum, hemlock, hickory, maple, oak, pine, poplar, redwood, spruce, and walnut). A number of these include several varieties, each showing different characteristics. One variety of walnut, for example, might make a better building product than another.

Therefore, the engineer should have some knowledge of the classification of trees, as well as of their growth and structural ability, to understand the physical and mechanical properties of each.

CLASSES OF TREES

Exogens and Endogens

Botanists classify trees into two main categories: exogens and endogens. These terms refer to the pattern of growth of a particular species. *Exogenous trees* are those that grow diametrically, by adding new cells in a layer between the existing wood and the bark. Almost all wood that is of commercial use falls in this category. The exogenous growth pattern forms a characteristic transverse section with concentric annual rings that correspond to each year's growth. At the center of a cross-section of a tree is pith, a dark area of small diameter that does not increase in size after the first year of growth. Encircling the outermost annual ring is a layer of microscopic thickness called the cambium ring, and it is here that new growth is formed. The outmost layer is known as the bark.

Endogenous trees add new living fiber to the old by allowing new fiber to intermingle with the old, thus producing growth both diametrically and longitudinally. Most endogens are fairly small plants, such as corn, sugar cane, palm, and bamboo. The latter two plants are used extensively throughout the Orient as a structural material.

Hardwoods and Softwoods

Exogenous trees are further classified as hardwoods or softwoods. These are some what misleading terms, because they refer not to a quality of the wood, but to the types of tree the wood comes from. Hardwoods comprise the broadleafed trees, mostly deciduous, and include the many varieties of oak, maple, ash, hickory, cherry, poplar, gum, walnut, elm, beech, birch, basswood, whitewood or tulip, and ironwood. These trees are in general slow growing, forming heavy and hard wood that is used more extensively for furniture, interior finishing, and cabinetwork than for structural purposes. The softwoods are the conifers, such as pine, spruce, fir, hemlock, cedar, cypress, and redwood.

WOOD STRUCTURE

Many of the qualities that are unique to wood for use as a building material arise from the structure of wood. Wood is formed mostly of filamentous cells, fairly tubular in shape, varying in length from 1 to 3 mm and having a diameter that is about $\frac{1}{100}$ of their length. The shape of the cell in transverse section is approximately rectangular with rounded corners. Perforated pits in the cells make possible the flow of sap between cells.

Wood cells are composed primarily of the polymers cellulose and hemicellulose. These are similar to the sugar molecule, but form long chains that run in bundles helically. They are cemented together by a third polymer, lignin. Wood is typically 50 to 60 percent cellulose material and 20 to 35 percent lignin, with small amounts of other carboxydrates, such as resin and gum. It is the cellulose that gives wood its axial strength and elastic properties, and the lignin that is responsible for its compressive strength. These carboxydrates, although indigestible by man, are a favorite food for some other forms of life, such as termites, marine borers, and fungi, and measures must often be taken to protect wooden structures from these species.

Most fibers run parallel to the axis of the tree, and these cells, which are more or less empty in wood that is ready for use, act as a collection of empty tubes under stress. Since the tree develops from a sapling by adding new longitudinal cells in layers around the existing wood, differences in periods of growth provide rings of cells with different characteristics. Growth is rapid in the spring and slows down in the summer, stopping altogether in the fall and winter. During the spring growth period, cells that are formed have relatively thin walls and an open texture, producing what is known as spring wood. The cross-sectional area of spring wood may form as little as 10 percent of the cell wall. The summer growth is slower and the wood formed is denser and stronger, with thicker cell walls. This summer wood may have cell walls constituting up to 90 percent of the cross-sectional area. The pattern of a layer of spring wood followed by increasingly dense wood that is formed over each year gives the wood its annual rings. These rings vary in size with the type of tree involved and the growth conditions during a particular year. A typical annual thickness is approximately $\frac{1}{10}$ in., with $\frac{1}{2}$ in. being about the maximum. In tropical trees, the growth rings correspond to the rainy and dry seasons.

Over a period of several years the wood ceases to function as part of the living tree and has only structural value for the tree. The living part is called *sapwood*: the older wood is *heartwood.* Heartwood is easily distinguished from sapwood because it is darker in color. It is more highly valued because it is more resistant to decay than sapwood, although the strength of each is about the same.

Also parallel to the axis of the tree and running the length of the tree are vessels formed by the fusion of chains of cells. These vessels are characteristic of hardwoods, and are sometimes visible in cross-section without a microscope. Other groups of cells lie in radial arrangements, and these form what are known as radial, pith, or medullary rays. Medullary rays are present in all trees, but are especially prominent in woods such as oak. In the living tree they carry food from the inner bark to the cambrium layer and add strength in a radial direction.

The terminology of the tree is best illuminated by a cross-section such as the one shown in Figure 8-1.

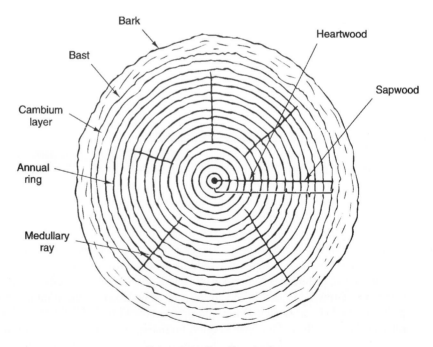

Figure 8-1   Tree Terminology

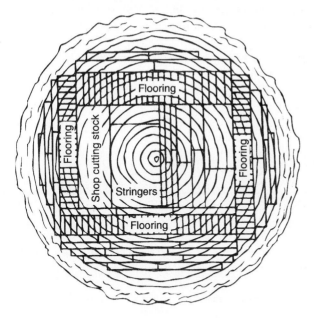

Figure 8-2 Log Cutting into Standard Size

PHYSICAL CHARACTERISTICS OF WOOD

Season of Cutting; Slash and Rift Cutting

The first step in the manufacturing process of lumber is to select and fell the trees. This is normally done in the fall and winter when the flow of sap is minimal and destructive fungi and insects are least active. The logs are cut into standard-size boards at sawmills, as illustrated in Figure 8-2. The lumber must then be dried or seasoned.

The angle of the plane of the cut of the boards will have an effect on the resulting lumber. Wood that is quartersawed or *rift-cut* is cut parallel to the axis of the tree and radially across the annual rings. When the wood is cut tangential to the annual rings, it is called *slash-cut*. Rift-cut lumber generally warps less in the drying process and wears more evenly, but in the interest of practicality, more wood is slash-cut. A typical sawing pattern is shown in Figure 8-3.

Seasoning of Timber, Moisture Content

Wood must be seasoned before it is put to use so that its moisture content will become stabilized. Unseasoned wood readily takes up and retains moisture, and this moisture will adversely affect the wood in a number of ways. One is that changes in moisture content will produce dimensional changes in the timber. Also, a high moisture content diminishes the strength of the wood. Wood that is properly dried may be up to two and one-half times stronger than a comparable piece of green wood. Other problems, such as the presence of

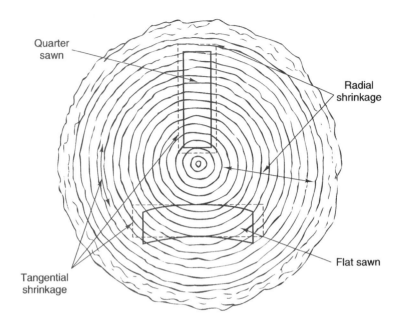

Figure 8-3 Sawing Pattern

harmful fungi, occur when wood is cut. The moisture in wood is in two forms: free and combined. *Free moisture* is moisture outside the cell, and *combined moisture* is water that has been incorporated into the cell walls. In green wood, moisture will constitute from 25 to 200 percent of the oven-dry weight of the lumber, and most of this must be evaporated from the wood in the seasoning process.

The timber can be either air-dried or kiln-dried. As the wood dries, the free moisture is evaporated from the surface and the moisture from the interior of the porous lumber equalizes with the drier surface. When the moisture content stabilizes, the drying process is complete.

$$\text{moisture content } (\%) = \frac{\text{original amount of moist wood—oven-dry weight}}{\text{oven-dry weight}}$$

Air drying will reduce the moisture content to about 15 percent, a level that is acceptable for most construction purposes. Lumber that is to be used for purposes such as furniture, flooring, and cabinetwork is kiln-dried in a few days using temperatures of 155 to 180°F (68 to 82°C). A 1-in. pine board, for example, can be dried in 4 days. Hardwoods warp more easily, so the drying process is more gradual. They are normally air-dried for several months before they are kiln-dried. Temperatures used need to be lower, 100 to 120°F (38 to 49°C) for those woods that warp most easily. At the hotter kiln temperatures, a 1-in. hardwood board may take 6 to 10 days to dry.

All wood has the ability to reabsorb moisture from the atmosphere, depending on the humidity of its surroundings. This can cause a certain amount of dimensional instability, which may lead to problems in certain applications. Wood that has been subjected to temperatures of 212°F (100°C) for a length of time during boiling, steaming, or soaking will, with time, become less able to absorb humidity from the atmosphere.

Shrinkage, Warping, and Checking

Shrinkage will occur in the drying of wood when the combined moisture begins to be evaporated from the cell walls. The contraction will be across the cell walls rather than longitudinal, cells shrinking proportional to the thickness of their cell walls. The heavy-celled summer wood thus shrinks far more than the thin-walled cells of spring wood. In a tangential direction, the bands of heavy summer wood force the lighter spring wood to contract with them, and for this reason there is a variation in the amount of shrinkage that a piece of wood will exhibit in different directions. The usual limits in shrinkage between a green condition and an over-dry condition are:

Volumetric shrinkage	7–21%
Longitudinal shrinkage	0.1–0.3%
Radial shrinkage	2–8%
Tangential shrinkage	4–14%

When shrinkage occurs unequally for any reason, warping results. If drying proceeds at unequal rates over different portions of a piece of lumber, warping will occur. When wood is cut tangentially, the dried piece will be convex because of unequal tangential shrinkage. Irregularities in the grain of the wood will produce more pronounced irregularities in the dried boards.

Checks are longitudinal cracks across the growth rings that can occur during drying of timber. Unequal shrinkage produces strains within the wood, and checks can result. They can be temporary, occurring when the outer portion of a piece of wood dries too rapidly and contracts over the inner portion. These checks that appear in the outer layer of the wood may close up and become imperceptible as the wood of the inner part of the lumber dries and contracts. Other checks of a more serious nature can occur when the difference in tangential shrinkage is too great relative to the radial shrinkage to be accommodated. Large radial checks such as those sometimes seen in posts can then occur.

Defects in Lumber

Anything that adversely affects the strength, durability, or utility of a piece of wood is a defect, and these may be within the wood itself or be produced by warping.

Shakes are longitudinal cracks in the wood that follow the growth rings and develop prior to the lumber's being cut. They are sometimes a result of heavy winds.

Checks, as described earlier, are longitudinal splits across the growth rings resulting from uneven drying.

Knots are formed at the base of branches where they extend into the wood of the tree. Only wood that comes from the base of the tree where there are no branches is free of knots. If the branch was dead, a *loose knot* is formed. A *spike knot* occurs when the cut is longitudinal to the branch. The effect of knots depends on the use of the wood and the location and size of the knots. They have little effect on wood in shear or in compression members. For beams in bending, however, a knot in the part of the beam subject to tension can significantly reduce its maximum load.

Pitch pockets are accumulations of resins in openings between the annual rings.

Bark pockets are formed when bark is wholly or partially encased in wood.

Waynes are areas where the lumber has been cut too close to the edge of the log and there is bark on the boards.

Compression wood is formed on the lower side of branches or leaning tree trunks. It is darker than normal wood, has a high lignin content, higher specific gravity, greater longitudinal shrinkage, and is not as tough as normal wood. The strength of compression wood is not predictable, and many failures in wood members have been found to be caused by compression wood. Tension wood, like compression wood, has higher specific gravity and greater longitudinal shrinkage. It is sometimes stronger and sometimes weaker than normal wood.

The effects of fungus or insect attack can be considered defects and are addressed later in the chapter.

Unequal shrinkage during the drying process produces warped lumber. The different types of warping are bow, crook, twist, and cup, as illustrated in Figure 8-4.

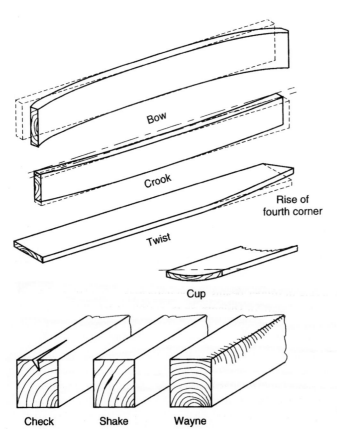

Figure 8-4 Unequal Shrinkage and Various Defects

MECHANICAL PROPERTIES OF TIMBER

Strength of Wood

Because of the structure of wood, the fibers act like so many cylindrical tubes firmly bound together in a way that they withstand stress. Loads applied parallel to the grain are carried by the strongest of the fibers; in loads perpendicular to the grain, the weakest fibers have to take their share of the load.

Tensile Strength

Tensile strength in wood parallel to the grain is much higher (three times as much) than compressive strength. Thus, the limiting factor in tension members is usually compression or shear at the points of concentration. Thus far, no means have been developed to connect a tension member without it somehow compressing and eventually resulting in failure.

Tensile loads perpendicular to the grain of wood fibers cause the fibers to split apart. The tensile loads that can be carried are only one-tenth or less of the tensile load that can be withstood by the wood in a parallel direction with the grain. An example of the differences is shown in Figure 8-5, which shows Douglas fir and its appropriate values for various loading conditions.

Compression Strength

Compression loads parallel to the grain can be carried by the strongest fibers, whereas compression loads perpendicular to the grain are carried by both weak and strong fibers. Wood in compression parallel to the grain can carry three to four times the load that wood in compression perpendicular to the grain can carry.

Compression failure of wood perpendicular to the grain involves the complete crushing of the wood fiber. Compression failure of wood parallel to the grain involves the bending or buckling of the wood fibers.

Flexural Strength and Stiffness

For timber beams in flexure, the critical factors in evaluating the load that can be carried are the compressive strength parallel to the grain and the shear strength parallel to the grain. The resistance of shear strength parallel to the grain is very low. However, if the wood is free of defects, the initial failure will be compressive.

Stiffness also plays a role in timber beams, since deflection is usually limited to $\frac{1}{360}$ of the span. Stiffness can be a measure of strength, in that stiffer timber beams are usually more dense.

Elastic Properties of Timber

The modulus of elasticity in wood is high relative to its compressive strength, compared with other building materials. Wood does not, however, exhibit a well-defined yield point, and therefore the proportional limit is used to measure elastic strength. Therefore, for a specific

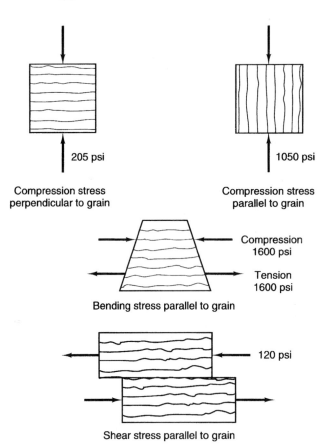

Figure 8-5   Allowable Stresses for
Douglas Fir

load requirement and sufficient wood to carry such a load, there is a high degree of elastic strength. For this reason, wood members are considered to have good elastic properties.

Factors Affecting Timber Strength

Many factors affect the strength of timber, such as direction of the wood fibers, moisture, weight, and rate of growth. The direction of the wood fibers is called the *grain*. The strength of timber depends so heavily on the direction of the loads with respect to the grain that it is an important consideration in design. Diagonal grain is measured by the tangent of the angle it makes with the cut edge of a piece of lumber. The diagonal grain or direction of the wood fibers will affect the strength ratios of lumber as shown in Table 8-1.

As indicated earlier, wood that is properly dried is far stronger than wood that has a high moisture content. When all the free water is evaporated and all the moisture is removed from the cell walls, the fiber saturation point is said to have been reached. The presence or absence of free moisture does not, in general, affect the strength properties. However, if drying occurs beyond the fiber saturation point, there will be an increase in strength. The increase in strength beyond the fiber saturation point may be two or three times the value of wet wood.

TABLE 8-1   STRENGTH RATIOS CORRESPONDING TO VARIOUS
SLOPES OF GRAIN[a]

	Maximum Strength Ratio (%)	
Slope of Grain	*For Stress in Extreme Fiber in Bending (Beams and Stringers of Joists and Planks)*	*For Stress in Compression Parallel to Grain (Posts and Timbers)*
1 in 6	—	56
1 in 8	53	66
1 in 10	61	74
1 in 12	69	82
1 in 14	74	87
1 in 15	76	100
1 in 16	80	—
1 in 18	85	—
1 in 20	100	—

[a] ASTM D245-49T.

TABLE 8-2   GROWTH RINGS FOR
VARIOUS SPECIES

Species	*Rings per Inch*
Douglas fir	24
Shortleaf pine	12
Loblolly pine	6
Western hemlock	18
Redwood	30

All other things being equal, the denser the wood, the stronger it will be. This is so true that a formula is based on it:

$$\text{modulus of rupture} = 26{,}200 \times (\text{specific gravity})^{1.25} \qquad \text{(air-dry timber)}$$

$$\text{modulus of rupture} = 18{,}500 \times (\text{specific gravity})^{1.25} \qquad \text{(green timber)}$$

These formulas were developed by the Forest Products Laboratory and are very accurate, regardless of species.

Finally, rate of growth greatly affects the strength of timber. In general, most timber used for construction will exhibit an optimal number of annual rings per inch of wood (Table 8-2). Usually, that with the greater number of rings per inch will be the stronger one.

Tabulation of the Mechanical Properties for Wood

Tables 8-3 and 8-4 show the mechanical properties for wood. The tables are based upon clear wood (wood that is free of defects) and a 100 percent strength ratio. The tables also have a factor of safety built into them for design purposes. Further, various grades of lum-

TABLE 8-3 MECHANICAL PROPERTIES OF A FEW IMPORTANT WOODS GROWN IN THE UNITED STATES[a]

Common and Botanical Name	Weight (lb/ft³)		Shrinkage from Green to Oven-dried (% of Green Volume)	Modulus of Rupture (psi)		Modulus of Elasticity (1000 psi)	
	Green	Air-dried[b]		Green	Air-dried	Green	Air-dried
Hardwoods							
Ash, white (*Fraxinius* sp.)	48	41	12.8	9,500	14,600	1,410	1,680
Elm, American (*Ulmus americana*)	54	35	14.6	7,200	11,800	1,110	1,340
Hickory, true (*Carya* sp.)	63	51	17.9	11,300	19,700	1,570	2,190
Maple, red (*Acer rubrum*)	50	38	13.1	7,700	13,400	1,390	1,640
Maple, sugar (*Acer saccharum*)	56	44	14.9	9,400	15,800	1,550	1,830
Oak, red (*Quercus* sp.)	64	44	14.8	8,500	14,400	1,360	1,810
Oak, white (*Quercus* sp.)	63	47	16.0	8,100	13,900	1,200	1,620
Walnut, black (*Juglans nigra*)	58	38	11.3	9,500	14,600	1,420	1,680
Conifers							
Cedar, western red (*Thuja plicata*)	27	23	7.7	5,100	7,700	920	1,120
Cypress, bald (*Taxodium distichum*)	51	32	10.5	6,600	10,600	1,180	1,440
Douglas fir, coast type (*Pseudotsugu taxifolia*)	38	34	11.8	7,600	12,700	1,570	1,950
Fir, white (*Abies* sp.)	46	27	9.8	5,900	9,800	1,150	1,490
Hemlock, eastern (*Tsuga canadensis*)	50	28	9.7	6,400	8,900	1,070	1,200
Pine, longleaf (*Pinus palustris*)	55	41	12.2	8,700	14,700	1,600	1,990
Pine, shortleaf (*Pinus echinata*)	52	36	12.3	7,300	12,800	1,390	1,760
Pine, western white (*Pinus monticola*)	35	27	11.8	5,200	9,500	1,170	1,510
Redwood (old growth) (*Sequoia sempervirens*)	50	28	6.8	7,500	10,000	1,180	1,340
Spruce, Sitka (*Picea sitchensis*)	33	28	11.5	5,700	10,200	1,230	1,570
Spruce, eastern (*Picea* sp.)	34	28	12.6	5,600	10,100	1,120	1,450
Tamarack (*Larix laricina*)	47	37	13.6	7,200	11,600	1,240	1,640

(*Continued*)

TABLE 8-3 *(Continued)*

Common and Botanical Name	Maximum Crushing Strength Parallel to Grain (psi)		Compression Perpendicular to Grain, Proportional Limit (psi)		Maximum Shearing Strength, Parallel to Grain (psi)	
	Green	Air-dried	Green	Air-dried	Green	Air-dried
Hardwoods						
Ash, white (*Fraxinius* sp.)	4,060	7,280	860	1,510	1,350	1,920
Elm, American (*Ulmus americana*)	2,910	5,520	440	850	1,000	1,510
Hickory, true (*Carya* sp.)	4,570	8,970	1,080	2,310	1,360	2,130
Maple, red (*Acer rubrum*)	3,280	6,540	500	1,240	1,150	1,850
Maple, sugar (*Acer saccharum*)	4,020	7,830	800	1,810	1,460	2,330
Oak, red (*Quercus* sp.)	3,520	6,920	800	1,260	1,220	1,830
Oak, white (*Quercus* sp.)	3,520	7,040	850	1,410	1,270	1,890
Walnut, black (*Juglans nigra*)	4,300	7,580	600	1,250	1,220	1,370
Conifers						
Cedar, western red (*Thuja plicata*)	2,750	5,020	340	610	710	860
Cypress, bald (*Taxodium distichum*)	3,580	6,360	500	900	810	1,000
Douglas fir, coast type (*Pseudotsuga taxifolia*)	3,860	7,430	440	870	930	1,160
Fir, white (*Abies* sp.)	2,830	5,480	360	620	770	990
Hemlock, eastern (*Tsuga canadensis*)	3,080	5,410	440	800	850	1,060
Pine, longleaf (*Pinus palustris*)	4,300	8,440	590	1,190	1,040	1,500
Pine, shortleaf (*Pinus echinata*)	3,430	7,070	440	1,000	850	1,310
Pine, western white (*Pinus monticola*)	2,650	5,620	290	540	640	850
Redwood (old growth) (*Sequoia sempervirens*)	4,200	6,150	520	860	800	940
Spruce, Sitka (*Picea sitchensis*)	2,670	5,610	340	710	760	1,150
Spruce, eastern (*Picea* sp.)	2,600	5,620	290	590	710	1,070
Tamarack (*Larix laricina*)	3,480	7,100	480	990	860	1,280

[a]Compiled from R1903-10, June, 1952, Forest Products Laboratory, U.S. Dept. of Agriculture.
[b]Air-dried lumber contained 12% moisture.

TABLE 8-4   BASIC STRESSES FOR CLEAR MATERIAL[a,b]

Species	Extreme Fiber in Bending	Modulus of Elasticity	Compression Parallel to Grain[c] (L/d = 11 or less)	Compression Perpendicular to Grain	Maximum Horizontal Shear
Hardwoods					
Ash, commercial white	2,050	1,500,000	1,450	500	185
Elm, white	1,600	1,200,000	1,050	250	150
Hickory, true and pecan	2,800	1,800,000	2,000	600	205
Maple, sugar and black	2,200	1,600,000	1,600	500	185
Oak, commercial red and white	2,050	1,500,000	1,350	500	185
Conifers					
Cedar, western red	1,300	1,000,000	950	200	120
Cypress, southern	1,900	1,200,000	1,450	300	150
Douglas fir, coast region	2,200	1,600,000	1,450	320	130
Fir, commercial white	1,600	1,100,000	950	300	100
Hemlock, eastern	1,600	1,100,000	950	300	100
Pine, western white, eastern white, ponderosa, and sugar	1,300	1,000,000	1,000	250	120
Pine, Norway	1,600	1,200,000	1,050	220	120
Pine, southern yellow (longleaf or shortleaf)	2,200	1,600,000	1,450	320	160
Redwood	1,750	1,200,000	1,350	250	100
Spruce, red, white, and Sitka	1,600	1,200,000	1,050	250	120
Tamarack	1,750	1,300,000	1,350	300	140

[a]From Forest Products Laboratory, U.S. Dept. of Agriculture. See also ASTM D245-49T.
[b]All values in pounds per square inch and for material under long-time service conditions at maximum design load.
[c]L, unsupported length; d, least dimension of cross-section.

ber have a strength ratio stamped on them, and by multiplying this number times that in the tables, basic stresses for that particular grade of lumber can be obtained.

DECAY, DURABILITY, AND PRESERVATION OF TIMBER

Mechanism of Decay

Various factors that cause damage to a wooden structure, the durability of the wooden structure, and its preservation will be discussed in detail. Molds, stains, and decay in timber are caused by fungi, microscopic plants that require organic material to live on. Reproduction occurs through thousands of small windblown particles called *spores.* Spores send out small arms that destroy timber through the action of enzymes. Fungi require a temperature of between 50 and 90°F (10 and 32°C), food, moisture, and air. Timber that is soaked or dried normally will not decay. Molds cause little direct staining because the color caused is largely superficial. The cottony or powdery surface growths

Figure 8-6 Early Stages of Decay at Base of
Timber Member

Figure 8-7 Advanced Stage of Decay

range in color from white to black and are easily brushed or planed off the wood. Stains, on the other hand, cannot be removed by surface techniques. They appear as specks, spots, streaks, or patches of varying shades and colors, depending upon the organism that is infecting the timber. Although stains and molds should not be considered as stages of decay since the fungi do not attack the wood substance to any great degree, timber infected with mold and stain fungi has a greater capacity to absorb water and is therefore more susceptible to decay fungi. Decay fungi may attack any part of the timber, causing a fluffy surface condition indicative of decay or rot. Early stages of decay may show signs of discoloration or mushroom-type growths (Figure 8-6).

In some types the color differs only slightly from the normal color, giving the appearance of being water-soaked. Later stages of decay are easily recognized because of the definite change in color and physical properties of the timber (Figure 8-7). The surface becomes spongy, stringy, or crumbly, weak, and highly absorbent. It is further characterized by a lack of resonance when struck with a hammer. Brown, crumbly rot in a dry condition is known as *dry rot*. Serious decay problems are indicative of faulty design or construction or a lack of reasonable care in the handling of the timber. Principles that assure long service life and avoid decay hazards in construction include building with dry timber, using designs that will keep the wood dry and accelerate rain runoff, and using preservative-treated wood where the wood must be in contact with water.

Insect damage may result from infestation of the timber members with any of a number of insects, among them powder-post beetles, termites, and marine borers. Powder-post beetles are reddish brown to black, hard-shelled insects from $\frac{1}{8}$ to $\frac{1}{2}$ in. long. The life cycle of the beetle includes four distinct stages: egg, larva, transformation, and adult. The adult bores into the timber, producing a cylindrical tunnel just under the surface in which the eggs are laid. The larvae burrow through the wood, leaving tunnels packed with a fine powder $\frac{1}{16}$ to $\frac{1}{8}$ in. in diameter. Powder-post damage is indicated by this fine powder, either fallen from the timber or packed into tunnels within the wood and by the tunnels and holes in the timber. The beetles attack both sound and decayed wood but are not active in decayed wood that is water-soaked. They may also cause damage by transmitting destructive fungi from one site to another, thus spreading decay.

Termites resemble ants in size and general appearance and live in similarly organized colonies. Destruction is done by the workers only, not by the soldiers or winged sexual adults. Subterranean termites are responsible for most of the damage done in the United States. They are more prevalent in the southern states but are found in varying numbers throughout the states. Termites build dark, damp tunnels well below the surface of the ground, with some of these tunnels leading to the wood, which is the termites' source of food. Termites also require a constant water supply in order to survive.

Subterranean termites do not infest structures by being carried to the construction site. They must establish a colony in the soil before they are able to attack the timber. Telltale signs are the tunnels in the earth leading to unprotected timber and swarms of male and female winged adults in the early spring and fall. When termites successfully enter the wood, they make tunnels that follow the wood grain, leaving a shell of sound wood to conceal the tunnels (Figure 8-8). Methods of controlling termites include breaking the path from the timber to the ground, although the best method is to treat the timber with a preservative.

Wood-inhabiting termites are found in a narrow strip around the southern boundary of the United States. They are most common in southern California and southern Florida. They are fewer in number than the subterranean termites, do not multiply as fast, and do not cause as much damage. But because they can live without contact with the ground and in either damp or dry wood, they are a definite problem and do considerable damage in the coastal states. They are carried to a building site in timber that has been infested prior to delivery, thereby making inspection prior to delivery a necessity. Full-length treatment with a good wood preservative is recommended.

Carpenter ants are usually found in stumps, trees, or logs, but are sometimes found in structural timbers. They range in color from brown to black and in size from large to small. Although carpenter ants are often confused with termites, they can be distinguished by comparison of the wings and thorax (waist) sizes of the insects. The carpenter ant has short wings and a narrow waist, whereas the termite has long wings and a thicker waist. Carpenter ants use wood as shelter, not food. They prefer naturally soft or decayed wood and construct tunnels that are very smooth and free of dust. Carpenter ants tunnel across the wood grain and cut small exterior openings for access to the food supplies (Figure 8-9). A large colony takes from 3 to 6 years to develop. Prevention of ant infestation can be accomplished by the use of preservatives along the length of the timber member.

Figure 8-8 Characteristic Termite Tunnel

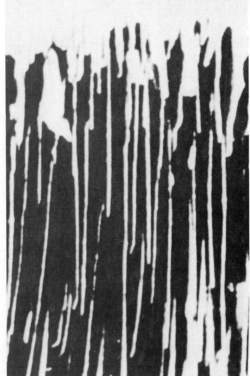

Figure 8-9 Carpenter Ant Tunnels

By far the most damaging insect pest is the marine borer (Figure 8-10). Borers have been known to ruin piles and framing within a few months. No ocean is completely free of borers and, although some waters may be relatively free, the status of an area may change drastically within a relatively short period of time. The main point of attack is generally between the high tide level and the mud line. Submergence often hides tell-tale signs of infestation, so that the first sign of attack may be the failure of the structure. The borers that do the most damage are the mollusks, related to clams and oysters, and the crustaceans, related to lobsters and crabs.

The mollusk borers consist of the "shipworm" teredo, the "shipworm" bankia, and the pholadidae. There are other types of shipworms throughout the world, but they all live and survive in much the same way, although size and environmental requirements may vary. The teredo has a wormlike, slimy gray body with two shells attached to the head that are used for boring. Two tubes that resemble a forked tail and normally remain outside the burrow are the only external indications that a shipworm is present. The shipworm can seal its entrance and thereby protect the burrow from intruders or foreign substances. The size of the teredo varies from $\frac{3}{8}$ to 1 in. in diameter and from 6 in. to 6 ft in length. The size of the bankia is generally larger than the teredo, but its other characteristics are the same. Both types bore tiny holes when they are young and grow to maturity inside the wood.

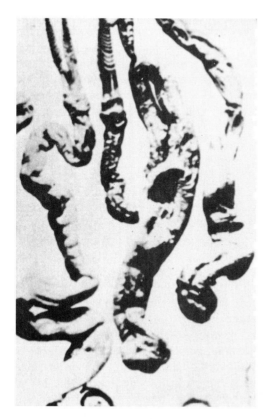

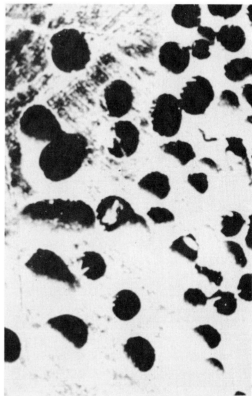

Figure 8-10 Marine Borers

Figure 8-11 Cross-Cut Timber Member Showing Extensive Damage

Once the young shipworm has entered the wood, it normally turns down and expands its burrow to its full diameter. Extremely careful observation with a hand lens is required to detect the entrance. The only way to detect the extent of the damage is to cut the wood (Figure 8-11) or take borings by some other method. Because the first sign of marine borer infestation may be the failure of timber members, the shipworm should be considered extremely dangerous. Pholadidae resembles a clam with its body entirely enclosed in a two-part shell. It is of particular danger because it can bore holes up to $1\frac{1}{2}$ in. deep into the hardest timber.

The most common crustacean borer is the limnoria or "wood louse." Its body is slipper-shaped, from $\frac{1}{8}$ to $\frac{1}{4}$ in. long and from $\frac{1}{16}$ to $\frac{1}{8}$ in. wide. It has a hard boring mouth, two sets of antennae, and seven sets of legs with sharp claws. The limnoria is able to roll itself into a ball, to swim, and to crawl. It will gnaw interlacing branching holes in the surface of the wood with as many as 200 to 300 holes per square inch. These holes follow the softer wooden rings and are 0.05 to 0.025 in. in diameter and seldom over $\frac{3}{4}$ in. long. As a result, the wood becomes a mass of thin walls between burrows that break away, exposing new areas to attack. In this manner the pile is slowly reduced in diameter. In soft woods such as pine and spruce, the diameter may be reduced as much as 2 in. per year. The chief area of attack is between the low-water level and the mud line, with occasional activity up to the high-water level. The limnoria does not seem to be affected by small environmental changes and may be found in salt or brackish water of any temperature, polluted or clean.

Complete protection from borer attack is essential. Metal armoring is falling into disuse, as is concrete casing. The very best method may prove to be jacketing creosoted piles. The most practical method involves heavy treatment with high-quality coal-tar creosote by the full-cell method to the point of near saturation. Although shipworms will generally not attack a creosoted member, they may attack any area that has been damaged, and any untreated area. Limnoria attack the creosoted timber directly but at a decelerated rate. Other types of wood preservatives in use include a plastic outer wrap, which is successful as long as the plastic is not damaged, and cyanide treatment, which is messy to apply, leaches out with time, and may cause damage to the environment.

Fire damage to wood is a real problem on bridge and wharf structures. Treatment of wood with preservatives may protect the wood from fungi and insect attack, but it generally results in making the wood more susceptible to fire. Damage resulting from fire is readily apparent because of the charred appearance and burnt odor of fire-damaged wood.

Chemical damage to wood is very difficult to determine because it often resembles damage done by other factors. Fungi attack employs a chemical reaction between the enzymes particular to each fungus and the wood fiber. Fire damage involves the oxidation of the wood fiber, which is again a chemical process. It is therefore important to report any apparent damage to the timber as soon as possible.

Impact damage may occur in the event of a high-energy collision. Because wood has good impact characteristics, it may show only limited signs of external damage and must therefore be inspected carefully after an accident or suspected accident. The timbers may have a shattered appearance as opposed to the sagging appearance caused by overload. Compression failure will resemble wrinkled skin; tension failure looks as if the fibers have been pulled apart.

Another method of protecting timber is to inject preservatives into the timber. There are two methods utilized: the pressure process and the nonpressure process. There are two types of *pressure processes:* the full cell and the empty cell. In the *full-cell process* it is intended that the cells remain filled with the preservative. The initial treatment removes most of the air and water before impregnation. The timber is then covered with preservatives under pressure. In the *empty-cell process,* the preservative merely forms a film over the cell walls. In the *nonpressure process* (hot and cold), the timber is immersed in the preservative in an open tank and heated to 200°F (93°C). The heat of the preserving fluid expands the air within the cells. The cooling of the bath or the immersing of the timber in a cold medium causes a contraction of the air still remaining in the cells, which tends to produce infiltration by the preservative.

PROBLEMS

8.1. List the two principal categories of trees and describe each.

8.2. How do hardwood trees differ from softwood trees?

8.3. Discuss the role of the following polymers: cellulose, hemicellulose, and lignin.

8.4. Explain the differences between slash cut and rift cut.

8.5. What role does moisture content play in timber used for structural purposes?

8.6. What are some defects in lumber? Describe each.

8.7. How does compressive strength parallel to the grain differ from compressive strength perpendicular to the grain?

8.8. Explain the mechanism of decay in lumber.

8.9. How do marine borers damage timber?

8.10. Explain the methods utilized for protecting timber.

R E F E R E N C E S

American Society for Testing Materials, *Book of Standards*, Part 22, 1979.

BROWN, H. P., PANSHIN, A. J., and FORSAITH, C. C., *Textbook of Wood Technology*, McGraw-Hill Book Company, New York, 1949.

BRUST, A. W., and BERKLEY, E. E., "The Distributions and Variations of Certain Strength and Elastic Properties of Clear Southern Yellow Pine Wood," *Proc. ASTM*, *35*, Part II (1935), pp. 643–693.

CLAPP, W. F., "Recent Increases in Marine Borer Activity," *Civil Eng.*, *7*, No. 12 (December 1937), pp. 836–838; and *17*, No. 6 (June 1947), pp. 324–327.

"Defects in Timber Caused by Insects," *U.S. Dept. Agr. Farmers Bull. 1490*, 1927.

DESCH, H. E. *Timber, Its Structure and Properties*, 2nd ed., Macmillan Publishing Co., Inc., New York, 1947.

DIETZ, A. G. H., *Materials of Construction, Wood, Plastics, Fabrics*, D. Van Nostrand Company, New York, 1949, pp. 1–189.

DIETZ, A. G. H., and GRINSFELDER, A., "Fatigue Tests on Compressed and Laminated Wood," *ASTM Bull. 129*, Aug. 1944.

ELMENDORF, A., "The Uses and Properties of Water-resistant Plywood," *Proc. ASTM*, *20*, Part II (1920), p. 324.

Forest Products Laboratory, *Wood Handbook*, rev. ed., Superintendent of Documents, Government Printing Office, Washington, D.C., 1940.

FREAS, A. D., "Studies of the Strength of Glued Laminated Wood Construction," *ASTM Bull. 70*, Dec. 1950; and *Bull. 73*, Apr. 1953.

GAY, C. M., and PARKER, H., *Materials and Methods of Architectural Construction*, 2nd ed., John Wiley & Sons, Inc., New York, 1943, pp. 39–50, pp. 304–371.

HANSEN, H. J., *Timber Engineers Handbook*, John Wiley & Sons, Inc., New York, 1948.

HOLTMAN, D. F., *Wood Construction*, McGraw-Hill Book Company, New York, 1929.

HUNT, G. M., and GARRATT, G. A., *Wood Preservation*, McGraw-Hill Book Company, New York, 1938.

KOFOID, C. A., *Termites and Termite Control*, University of California Press, Berkeley, Calif., 1934.

KUERRZI, E. Q., "Testing of Sandwich Construction at the Forest Products Laboratory," *ASTM Bull. 164*, Feb. 1950.

LEWIS, W. C., "Fatigue of Wood and Glued Wood Construction," *Proc. ASTM*, *46* (1946), p. 814.

MARKWARDT, L. J., and WILSON, T. R. C., "Strength and Related Properties of Woods Grown in the United States," *U.S. Dept. Agr., Tech. Bull. 479*, 1935.

MILLS, A. P., HAYWORD, H. W., and RADER, L. F., *Materials of Construction*, 6th ed., John

Wiley & Sons, Inc., New York, 1955, pp. 508–543.

"Preventing Damage by Termites and White Ants," *U.S. Dept. Agr. Farmer's Bull. 1972*, 1927.

TRUAX, T. R., "The Gluing of Wood," *U.S. Dept. Agr. Bull. 1500*, 1929.

United States Coast Guard, *The State of the Art: Bridge Protective Systems and Devices*, A Report by the U.S.C.G. Bridge Modification Branch, R. T. Mancill, Editor, 1979.

9 ASPHALT CEMENTS

Asphalt materials have been utilized since 3500 B.C. in building and road construction. Their main uses have been as adhesives, waterproofing agents, and as mortars for brick walls. These early asphalt materials were native asphalts. *Native asphalts* occur when petroleum rises to the earth's crust and the volatile oils are evaporated. These native asphalts were found in pools and asphalt lakes. One of the more well-known deposits of native asphalts is the "Trinidad Lake" deposit on the island of Trinidad off the north coast of South America. Before the development of the processes for producing asphalt cement from crude petroleum products, native asphalts were the only sources of supply for early pavement projects. The first asphaltic pavement was built in 1869 in London, England. A year later construction of road and street pavements began in the United States in Newark, New Jersey.

With the invention of the automobile, which required smooth all-weather pavements, the demand for asphaltic products for pavements grew. Thus, the processes for producing asphalt cement from crude petroleum products grew, which led to the development of modern asphalt cement.

ASPHALT CEMENTS

Asphalt cement, according to ASTM D8 (Materials for Roads and Pavements), is fluxed or unfluxed asphalt specially prepared as to quality and consistency for direct use in the manufacture of bituminous pavements, and having a penetration at 77°F (25°C) of between 5 and 300, under a load of 0.2 lb (100 g) applied for 5 seconds. Asphalt cements fall within a broad category known as bitumens. *Bitumen,* according to ASTM D8, is a class of black or dark-colored (solid, semisolid, or viscous) cementitious substances, natural or manufactured, composed principally of high-molecular-weight hydrocarbons, of which asphalts, tars, pitches, and asphaltites are typical.

Bitumen by definition is soluble in carbon disulfide. The hydrocarbons that make up bitumen can generally be made up of the following:

1. Asphaltenes.
2. Resins.
3. Oils.

Asphaltenes are large, high-molecular-weight hydrocarbon fractions precipitated from asphalt by a designated paraffinic naphtha solvent at a specified solvent-asphalt ratio. Asphaltenes have a carbon-to-hydrogen ratio of 0.8. Asphaltenes constitute the body of the asphalt. *Resins* are hydrocarbon molecules with a carbon-to-hydrogen ratio of more than 0.6 but less than 0.8. Resins affect the adhesiveness and ductility properties of asphalt. *Oils* are hydrocarbon molecules with a carbon-to-hydrogen ratio of less than 0.6. Oils influence the viscosity and flow of the asphalt.

Ductility and adhesiveness are two properties that make asphalt cement attractive as a highway material. Oxidation of an asphalt cement causes a loss of ductility and adhesiveness, resulting in an asphalt cement that is harder and less ductile and adhesive. Oxidation results in the creation of more asphaltenes at the expense of resins. Thus, oxidation is a serious problem.

Production and Distillation

Asphalt cement is a valuable by-product obtained when petroleums are processed to obtain gasoline, kerosene, fuel oil, motor oil (diesel and lubricating), and other asphalt products. There are three types of petroleum found in the earth's crust:

1. Asphaltic-base crude oils.
2. Paraffin-base crude oils.
3. Mixed-base crude oils.

Asphalt cement is easily obtained from asphaltic-base crude oils by a straight-run distillation process. Bituminous products may be obtained from paraffin-base crude oils by a destructive distillation process involving chemical changes. These bituminous materials should not be classified as asphalt. Asphalt cement may also be obtained from mixed-base crude oils but the process is complicated.

All distillation of asphalt-base petroleum is fractional. During the distillation of petroleum several fractions are separated from the petroleum as given in Table 9-1.

Figure 9-1 shows a flowchart of the refining process necessary for the production of asphalt cement. When the process is controlled to prevent overheating and eventual chemical changes, the asphalt cement that remains as a residue is a straight-run asphalt. In the first operation the petroleum is pumped through the tube heater, in which the petroleum is placed under pressure and heated to a temperature of about 550°F (288°C). This crude product, released through the bottom of the tube heater, enters an atmospheric fractionating column

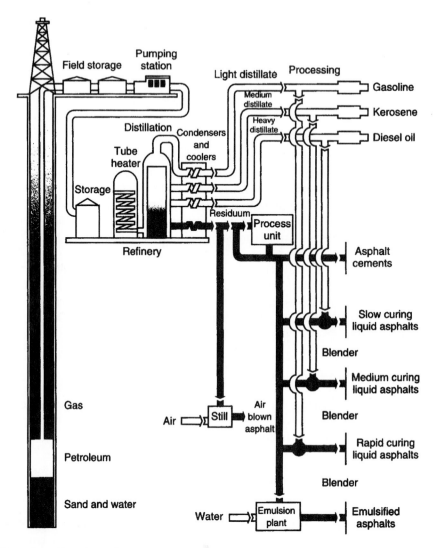

Figure 9-1   Flowchart of the Refining Process for Asphalt Cement (Courtesy of Asphalt Institute)

TABLE 9-1   FRACTIONS OF PETROLEUM

Fraction	Product Type	Boiling-Point Range (°F)
Light distillate	Gasoline	100–400
Medium distillate	Kerosene	350–575
Heavy distillate	Diesel oil	425–700
Very heavy	Lubricating oil	Over 650
Residue	Asphalt	

(tower distillation). On being exposed to atmospheric pressure within the column the more volatile fractions rise to the top of the column, where the vapors pass through condensers and coolers. Traps are arranged in the column at different levels where each trap is just below the boiling point of the liquid at which it is to collect. Once the vapor passes through the appropriate traps, the vapors are condensed and cooled. These condensed vapors are referred as distillates (light, medium, and heavy). After further processing, gasoline, kerosene, and diesel oil result. The product that settles down in the atmospheric fractionating column is *residuum,* sometimes called *hot topped crude.*

The crude then enters a second fractionation column. This column allows steam to be introduced at the bottom of the fractionation column in such a way as to become mixed with the hot topped crude. The boiling point of the various fractions becomes a combination of the boiling points of the oils being vaporized plus water. A second procedure (which is an aid for the use of lower temperatures) is the application of a partial vacuum in the fractionation column. The greater the vacuum applied, the lower the boiling point of the fractions to be separated from the crude. By controlling the steam, the temperature, and the partial vacuum, the quality of asphalt cement is maintained. The residue that results from the second fractionation column is asphalt cement. The volatiles are light vacuum distillate, nonvolatile oils, and heavy vacuum distillate.

Air is sometimes used to improve an asphalt material. The air is blown through an asphalt stock at temperatures of 400 to 500°F (205 to 288°C). A reaction results in which the oxygen from the air combines with the hydrogen in the hydrocarbon molecule to form water, which is emitted as steam. As a result, polymerization of the hydrocarbons takes place to form heavier and harder materials. This chemical reaction may be speeded up by the addition of ferric chloride or phosphorous pentoxide. The main advantage of air blowing is that the final product is less susceptible to temperature fluctuations than an unheated asphalt.

METHODS OF TESTING

Most present-day standards were developed in the early 1900s. Very few new tests or standards have been developed in recent years. Equipment has changed, becoming electric, and better control can be accomplished during the testing period of the product. We shall look at some of these standards in greater detail.

Penetration of Bituminous Materials

ASTM D5 (Penetration of Bituminous Materials) covers determination of the penetration of semisolid and solid bituminous materials. In other words, it measures the hardness or softness of the material. It is a consistency test for bituminous material expressed as the distance in tenths of a millimeter that a standard needle vertically penetrates a sample of the material under known conditions of loading, time, and temperature.

The procedure for the standard test, which is implied unless other conditions are stated, is for pressure to be applied to a load of 100 grams for a period of 5 seconds at a temperature of 77°F (25°C).

In running the test at least three determinations should be made on the surface of the sample at points not less than 10 mm from the side of the container and not less than 10 mm apart.

The loss on weight with today's manufactured products is quite low, and thus the test has lost much of its meaning. The test back in the early 1900s was extremely valuable in characterizing early steam-refined asphalts, which had low flash points and high losses on heating.

Lower penetration grades are generally needed in warmer climates to avoid softening in the summer. Obviously, higher penetration grades will be used in colder climates so that excessive brittleness does not occur during cold winter weather.

The penetration test has provided long experience and service records associated with asphalts classified in this manner. However, the test is empirical and many engineers would like to replace it with ASTM D2171 (Viscosity of Asphalts by Vacuum Capillary Viscometer).

Specific Gravity

ASTM D70 (Specific Gravity and Density of Semi-solid Bituminous Materials) covers the determination of the specific gravity and density of semisolid bituminous materials, asphalt cements, and soft tars pitches by use of a pycnometer.

In running this test, the sample is heated until it can be poured. The material is placed in a pycnometer. The asphalt volume is determined by taking the difference between the total volume of the bottle and the volume of water required to complete the filling. From this information the specific gravity can be expressed as the ratio of the weight of a given volume of the material at 77°F (25°C) or at 60°F (15.6°C) to that of an equal volume of water at the same temperature. The following formula may be used to calculate the specific gravity:

$$\text{specific gravity} = \frac{C - A}{(B - A) - (D - C)}$$

where A = weight of pycnometer (plus stopper)
       B = weight of pycnometer filled with water
       C = weight of pycnometer partially filled with asphalt
       D = weight of pycnometer plus asphalt plus water

The density is determined by multiplying the specific gravity by the density of water at test temperature in desired units where the specific gravity is calculated as in the given equation.

The specific gravity and density is reported to the nearest third decimal place at 77°F (25°C) or 60°F (15.6°C).

Specific gravity is usually not a requirement of asphalt specifications but is of value in the design of bituminous mixtures as well as for determining costs.

Ductility

ASTM D113 (Ductility of Bituminous Materials) covers the procedure for determining ductility. The ductility of a bituminous material is measured by the distance to which it will elongate before breaking when two ends of a briquette specimen of the material are pulled apart at a specified speed and temperature.

The results of the ductility test are controversial. The test is believed to measure the adhesiveness and elasticity of the asphalt. The property of adhesiveness is most important, since asphalt cement is used to bind stone, sand, and filler to make bituminous concrete. Also of concern is the temperature at which the product is run, 77°F (25°C). It is believed that this temperature is too high and that a much lower temperature should be used, because lack of ductility or brittleness, is more serious in cold weather.

Float Test

ASTM D139 (Float Test for Bituminous Materials) is a consistency test used for materials that are too soft to undergo the standard penetration test and too hard for use with the Saybolt-Furol viscosity test.

The test is performed by filling a brass collar with the asphalt or asphalt product to be tested, cooling the holder with its contents to 41°F (5°C), screwing the collar to a float, and floating the apparatus in water at a specified temperature. The float-test results are reported as the time in seconds from placing the float in water until the water breaks through the material in the collar.

Sampling Bituminous Materials

ASTM D140 (Sampling Bituminous Materials) covers the method used to sample bituminous materials at points of manufacture, storage, or delivery. The purpose of this procedure is to determine the true nature and condition of the material. Samples are taken as a representation of the bulk of the material and to ascertain the maximum variation in characteristics that the material possesses. This procedure covers sampling of semisolid or uncrushed solid material and of crushed or powdered material at place of manufacture, from containers, tankcars, vehicle tanks, distributor trucks, recirculating storage tanks, tankers, barges, pipelines, drums, and barrels.

Viscosity

Viscosity of asphalt materials can be determined by one of two methods: ASTM D2170 [Kinematic Viscosity of Asphalt (Bitumens)] and ASTM D2171 (Viscosity of Asphalts by Vacuum Capillary Viscometer).

ASTM D2170 [Kinematic Viscosity of Asphalt (Bitumens)] covers determination of the kinematic viscosity of liquid asphalts (bitumens), road oils, and distillation residues of liquid asphalts (bitumens), all at 140°F (60°C), and of asphalt cements at 275°F (135°C) in the range of 6 to 100,000 centistokes. Kinematic viscosity is the ratio of the viscosity to the density of a liquid. It is a measure of the resistance to flow of a liquid under gravity. The SI unit of kinematic viscosity is m^2/s; for practical use, a submultiple (mm^2/s) is more convenient. The centistoke (cSt) is 1 mm^2/s and is customarily used. *Viscosity* as referred to herein is the ratio between the applied shear stress and the rate of shear, called the *coefficient of dynamic viscosity*. This coefficient is a measure of the resistance to flow of a liquid. The SI unit of viscosity is the pascal-second; for practical use, a submultiple (mPa·s) is more convenient. The centipoise, 1 mPa·s, is customarily used. Density as referred to herein is the mass per unit volume of liquid. The cgs unit of density is 1 g/cm^3 and the SI unit of density is 1 kg/m^3.

The method includes measuring the time for a fixed volume of the liquid to flow through the capillary of a calibrated glass capillary viscometer under an accurately reproducible head and at a closely controlled temperature. The kinematic viscosity is then calculated by multiplying the efflux time in seconds by the viscometer calibration factor.

The results of this test can be used to calculate viscosity when the density of the test material at the test temperature is known or can be determined.

The second method, ASTM D2171 (Viscosity of Asphalts by Vacuum Capillary Viscometer), covers procedures for the determination of viscosity of asphalt (bitumen) by vacuum capillary viscometers at 140°F (60°C). It is applicable to materials having viscosities in the range from 0.036 to over 200,000 poises (P).

In this procedure the time is measured for a fixed volume of the liquid to be drawn up through a capillary tube by means of vacuum, under closely controlled conditions of vacuum and temperature. The viscosity, in poise, is calculated by multiplying the flow time in seconds by the viscometer calibration factor.

MODIFICATIONS OF ASPHALT CEMENTS

Asphalt cements may be modified into several different products to make it easier for distribution and use. Asphalt cement is generally hard and relatively solid. It must therefore be heated or treated before it can be mixed with aggregates to produce an asphaltic concrete (pavement). Asphalt cement can be made into a liquid at lower temperatures by mixing it with a volatile oil, or it can be emulsified with water to produce liquids at normal temperatures. When the volatile oils evaporate or the emulsion breaks, a hard stable asphalt cement remains which, with the aggregate particles, makes up the pavement.

Liquid Asphalts (Cutbacks)

When volatile oils are mixed with asphalt cement to make a liquid product, the product is referred to as a *cutback*. The purpose of the cutback is to allow relatively easy placement of the asphalt product without the use of high temperatures. After the material has been placed, the product reverts to its natural penetration value through the evaporation of the volatile oils. In general, there are three types of liquid asphalt in the cutback category:

1. Rapid-curing (RC), (ASTM D2028)
2. Medium-curing (MC), (ASTM D2027)
3. Slow-curing (SC; road oils), (ASTM D2026)

Rapid-curing (RC) cutbacks are made by diluting gasoline or naphtha with asphalt cement. The four grades of RC cutbacks range from RC 70 to RC 3000 depending on the amount of gasoline or naphtha used. Careful control must be maintained when utilizing gasoline or naphtha to keep the flash point of the material above 80°F (27°C). The percentage of the total volatile oils driven off at 500°F (260°C) varies from 70 percent for RC 70 to 25 percent for RC 3000. Obviously, the viscosity increases from RC 70 to RC 3000.

Medium-curing (MC) cutbacks are similar to rapid-curing cutbacks, with the exception that kerosene is used rather than gasoline or naphtha to liquefy the asphalt cement. The asphalt cement used to make medium-curing cutbacks should have a penetration of 70 to 250 at 77°F (25°C). There are five grades of MC cutbacks that range from MC 30 to MC 3000.

Comparing rapid-curing cutbacks with medium-curing cutbacks, it can be shown that rapid-curing cutbacks have a harder base asphalt and that the gasoline or naphtha will evaporate at lower temperatures, resulting in a material that is believed to cure rapidly. The medium-curing cutbacks have a softer base asphalt and a less volatile solvent (kerosene), and as a result will cure much more slowly than will a rapid-curing cutback.

Therefore, if one is constructing a pavement in a northern region, one might well choose the medium-curing cutback, as it has a softer base, and after curing the material would be less brittle in the winter and subject to less cracking. In a southern region, one might well choose the rapid-curing cutback, to obtain a harder asphalt base. This harder base may well provide a much more stable pavement under the hot sun.

Many state specifications for rapid-curing and medium-curing cutbacks are based upon viscosity tests run at 140°F (60°C). In this system RC and MC asphalt cutbacks have the following grades: 70, 250, 800, and 3000 for RC and 30, 70, 250, 800, and 3000 for MC. These numbers refer to minimum allowable range of viscosity as determined by the kinematic viscosity test, in units of centistokes (Table 9-2).

Cutback liquid asphalts are governed by ASTM D2027 [Cutback Asphalt (Medium-Curing Type)] and ASTM D2028 [Cutback Asphalt (Rapid-Curing Type)].

Road Oils

Slow-curing liquid asphalts (*road oils*) were originally manufactured by a straight-run distillation process and were really liquid asphalt cements. In other words, they were manufactured like asphalt cements, but the distillation process was cut off earlier and many of the volatile oils remained as part of the asphalt. Today, slow-curing (SC) road oils are cutback asphalts. However, they are not cutback with gasoline, naphtha, or kerosene but are fluxed with nonvolatile oils.

The grading system for SCs is the same as that for RC and MC materials. Their main applications are for dust control (dust binding). *Dust binding* is a light application of bituminous material for the express purpose of laying and bonding loose dust. Slow-curing road oils are governed by ASTM D2026 (Cutback Asphalt [Slow-Curing Type]).

Asphalt Emulsions

An *asphalt emulsion* is a suspension of minute globules of water, or an aqueous solution in a liquid bituminous material. It is a mixture of asphalt cement and water in which water is

TABLE 9–2 CUTBACK GRADES

Viscosity	Grade (MC)	Grade (RC)
30–60	MC-30	
70–140	MC-70	RC-70
250–500	MC-250	RC-250
800–1600	MC-800	RC-800
3000–6000	MC-3000	RC-3000

the continuous phase and asphalt cement is the dispersed phase. Asphalt emulsions are another way of liquefying asphalt cements. The two most commonly used types of emulsified asphalts are anionic emulsions and cationic emulsions. *Anionic emulsions* are a type of emulsion such that a particular emulsifying agent establishes a predominance of negative charges on the discontinuous phase. *Cationic emulsions* are a type of emulsion such that a particular emulsifying agent establishes a predominance of positive charges on the discontinuous phase. Originally, all asphalt emulsions were of the anionic type. The type of asphalt emulsion is determined by the kind of emulsifying agent or soap used. An anionic emulsion works well with aggregates that have a positive charge (thus, opposites attract). Cationic emulsions work well with all types of aggregate. Cationic emulsions can extend the paving season, as they can better handle cold weather and are not damaged by sudden rain.

The setting time of each of the three general grades of asphalt emulsions (rapid setting, medium setting and slow setting) is dependent upon the breaking time of the emulsion and the evaporation of the water. Each type of emulsion has two grades, 1 and 2. There also exists an MS-2h grade, which is suitable for cold and hot plant mix as well as four HFMS grades. Most types are used cold, although in special cases they may be used hot.

The advantages of asphalt emulsions are as follows:

1. Can be used with cold or hot aggregate.
2. Can be used with aggregate that is dry, damp, or wet.
3. Eliminates the fire and toxicity hazards of cutback liquid asphalt.

Testing of Asphalt Cements Emulsions

Most of the ASTM test procedures previously discussed for asphalt cements may be used for emulsions. However, ASTM D244 (Emulsified Asphalt) covers specific testing for emulsified asphalts. The method covers the compositions, consistency, stability, and examination of residue of asphalt emulsions composed principally of a semisolid or liquid asphaltic base, water, and an emulsifying agent. The composition test includes water content, residue by distillation, identification of oil distillation by microdistillation, residue by evaporation, and particle charge of emulsified asphalts. The consistency test is the Saybolt viscosity test. The stability test includes demulsibility, settlement, cement mixing, sieve test, coating, miscibility with water, modified miscibility with water, freezing, coating ability and water resistance, and storage stability of asphalt emulsions. The final test is examination of residue.

ROAD TARS

Road tars were the most common paving materials until the development of the petroleum industry. With the widespread use of gasoline for automobile consumption, the asphalt by-product of the refining process has become so readily available that it has put tars out of use. However, in England, road tars are still in use because of the large bituminous coal industry.

Manufacture

Tars are brown or black bituminous materials, liquid or semisolid in consistency, in which the predominating constituents are bitumens obtained as condensates in the destructive distillation of coal, petroleum, oil shale, wood, or other organic materials, and which yield substantial quantities of pitch when distilled. *Road tars* are the product of straight-run distillation of crude tars. The two most common methods of production are the coke-oven process and the water-gas method.

In the *coke-oven method,* bituminous coal is processed to produce coke with crude coke-oven tar as a by-product. Approximately 10 tons of bituminous coal are heated to 2500°F (1370°C) in a brick-lined oven. The crude tar is part of the volatile product that is removed in the heating process, the residue being coke. With this method the properties of the tar vary with the makeup of the coal, the kind of oven, the temperature, the length of time the temperature is applied, and the pressure of the system. When the vapors are removed from the oven, condensed, and cooled, crude tar is formed. The crude tar is then placed through a straight-run distillation process to form tar.

Although most of the road tars are produced by destructive distillation of bituminous coal in the coke-oven tar procedure, a very small amount is produced by the *water-gas operation* of cracking petroleum. In this method, the crude tar is a by-product of the second stage of producing heating gas. Initially, steam is passed over a bed of incandescent coke that decomposes the steam; carbon monoxide and hydrogen gases are formed, which are known as *water gas.* This water-gas tar is too low in Btu for heating purposes, and hydrocarbon gases must be added for enrichment. These enriching hydrocarbon gases result from the passing of petroleum oils over hot firebrick in a carburetor. The high temperature in this process causes cracking (destructive distillation) of the petroleum oils, and in addition to the hydrocarbon gases, which enrich the water gas, a heavy residue of hydrocarbons is condensed to form crude tar. The tar is referred to as water-gas tar if a light petroleum fraction is used, and as residuum tar if a heavier residue is used.

In the manufacture of tar, the crude tar is stored until the water settles out. At this point the crude tar is transported to the refining plant, where the small amount of water that remains is heated out. In the tar refining process, many other products are produced besides road tars. The crude tar is treated by fractional distillation in much the same manner as petroleum is treated to produce asphalt. The viscosity of tars is produced by straight-run distillation. The distillation is carried to that extent required for the grade desired. Road tars are graded from RT 1 up to RT 12. RT 1 is a light fraction of tar and RT 12 is as hard in consistency as the No. 200 penetration asphalt. Cutback tars also exist in two grades, RTCB 5 and RTCB 6. These two grades are made by using light or middle oils with RT 10, 11, or 12 tar.

RT 1 is used for dust control. RT 2 and 3 are light prime coats. RT 4 is for prime coats and sometimes surface treatments. RT 5, 6, and 7 are used for surface treatments and road mixes. RT 8 and 9 are for use as surface treatments, seal coats, and road mixes. RT 10 and 11 are used in seal coats, tar concrete, and hot repairs. RT 12 is used in penetration macadam work, tar concrete, and hot repairs.

The RT 1-6 and RTCB 5-6 can be used at temperatures up to 150°F (66°C). RT 7 and above may be used at higher temperatures.

Methods of Test

Road tars are tested in much the same manner as asphalt cements, but the tests differ somewhat, primarily because the tests were developed at different times. In general, the tars are many years older.

PROPORTIONING ASPHALTIC MIXES

The most suitable asphaltic concrete is one that produces a stable, durable, flexible, and skid-resistant pavement at a minimum cost with adequate aggregate. It is not possible to optimize all four properties; thus, compromises result. We will now look at the compromises. One was made in the design of bituminous-concrete mix design.

Properties

The term *stability* is related to strength and refers to the ability of a pavement to resist deformation under application of loads. The stability depends on the distribution of loads by point-to-point contact of the aggregate particles. The forces of this distribution are affected by aggregate interlocking developed among aggregate particles and the cohesiveness supplied by the asphalt cement.

In maximizing stability, the aggregate would have to be of crushed angular particle shape with a rough surface texture. It should also be hard and dense, its grade approaching the Fuller curve. Further, there should be just enough asphalt to coat the aggregate particles such that all particles benefit from the adhesiveness.

Rounded particles would result in low stability and load distribution. Rounded particles tend to slide over one another, whereas angular particles interlock with one another. If soft aggregate particles were used, they would break and wear under the impact of vehicular loads and reduce stability greatly.

Durability refers to the resistance of the pavement to disintegration under traffic loads. In other words, the pavement will remain smooth and serviceable during summer heat and will not crack or ravel during the winter cold. It will further resist the forces of freezing and thawing. In general, the greater protection of the aggregate provided by the asphalt cement, the more durable the pavement. Thus, in maximizing durability, one would not only wish to coat all the aggregate particles with asphalt but also to fill the voids within it. In this matter, the aggregate would be immune to the forces of freezing and thawing, for no water could enter the mix. Further stripping would not occur, because the water could not get between the aggregate and the asphalt. In addition, oxygen would not penetrate the pavement; thus, oxidation cannot take place and durability and adhesiveness would last longer.

In maximizing durability, stability and skid resistance are compromised. In a thick coating of asphalt over the aggregate particles, the aggregates tend to float in the asphalt. The result is that no interlocking of the aggregate particles takes place, and stability is lost.

As traffic tends to compact a pavement in which there is more asphalt than that needed to cover the aggregate, skid resistance is compromised. As the pavement compacts, bleeding of the asphalt occurs (asphalt comes to the surface), making a very slippery pavement.

Flexibility refers to the ability of a pavement to withstand deflections and bending without cracking. To maximize flexibility, one would use an open-graded aggregate mixture. However, in this case, stability is compromised. The stability might be improved by increasing the pavement thickness or by increasing such flexibility decreases.

Skid resistance is a form of stability, and there are two categories that cause slippery pavements:

1. Pavement bleeding.
2. Aggregate polishing.

In *pavement bleeding* too few voids are left in the mix and compaction occurs under traffic loads on a hot day, forcing asphalt to the surface of the roadway. This covers the exposed rough aggregate and results in a slippery pavement. With voids in pavements as low as $\frac{1}{2}$ or 1 percent, slippery pavements will not occur; however, most specifications require 2 or 3 percent air voids to prevent surface bleeding.

Aggregate polishing results after a pavement receives continuous wear from traffic and the surface aggregates polish, forming a slippery pavement. One method to prevent aggregate polishing is to use relatively hard aggregates. Another is to use a mixture of varying aggregates of different hardnesses. In this way, as one type of aggregate polishes, the other, harder aggregates do not, so that the road does not totally polish.

Mix Design

A good asphalt pavement is one that compromises among stability, durability, flexibility, and skid resistance for a given gradation of aggregate with just enough asphalt to cover the aggregate particles for good adhesive properties. In proportioning asphaltic concrete, seven steps should be followed:

1. Evaluate and select the aggregates to be used.
2. Select the aggregate gradation.
3. Make trial mixes with various amounts of asphalt cement.
4. Measure the relative stability of each mix.
5. Compute the void content of each compacted mix.
6. Compute the voids in the mineral aggregate and voids in the aggregate filled with asphalt.
7. Select the optimum design.

In step 1, the contractor is allowed to select the most convenient and economical aggregate source that meets the general specifications. These specifications usually demand an aggregate of the highest quality that is economically feasible.

For step 2, aggregate gradation, many highway and government agencies have specifications that define aggregate gradings for various types of mixes and sizes of aggregates. For reference, the Asphalt Institute publishes various manuals that give minimum recommendations on grading and sizes.

Step 3, the making of the trial mixes with different amounts of asphalt cements (2 to 6 percent), is probably the most important step. As previously discussed, the optimum mix should be one in which a compromise is made among stability, durability, flexibility, and skid resistance with sufficient asphalt to coat the aggregate particles of a given size and gradation.

Step 4, which allows for the measurement of stability, may use either of two laboratory tests that measure the deformation under load of compacted laboratory specimens. These tests are both governed by ASTM specifications and are as follows:

1. Marshall, ASTM D1559.
2. Hveem Stabilometer, ASTM D1560.

The most common test of these two is the Marshall test, ASTM D1559.

In step 5, the void content of the compacted specimen is computed. The percentage of voids in a compacted asphaltic concrete influences the stability and can be used as a specification to prevent pavement bleeding and thus slippery roadways.

The void content is computed by comparing the volume of the compacted mix with the volume occupied by each of the ingredients. The volume of any amount of material can be computed by knowing its specific gravity or density.

$$\text{specific gravity} = \frac{\text{density of the material}}{\text{density of water}} \tag{9.1}$$

$$\frac{\text{solid volume of}}{\text{each ingredient}} = \frac{\text{weight of the ingredient in the specimen}}{\text{specific gravity of the ingredient}} \tag{9.2}$$

$$\text{percent voids} = 100 \, \frac{V_S - V_1}{V_S} \tag{9.3}$$

where V_S =   volume of specimen
        V_1 =   combined solid volumes of all ingredients in the specimen

In step 6, we compute the voids in the mineral aggregate. When one discusses the total voids within the compacted aggregate particles, it becomes clear that asphalt occupies most of the void space. As indicated, if the voids are completely filled with asphalt, the pavement loses its stability. Therefore, it is useful to compute the voids in the mineral aggregate (VMA) and the percentage of voids filled with asphalt. The VMA and the voids filled with asphalt are usually controlled by specifications. The VMA may be computed as follows:

$$\text{VMA} = \text{volume of specimen} - \frac{\text{volume occupied}}{\text{by the aggregate}} \tag{9.4a}$$

Therefore, $$\text{VMA} = V_S - V_{agg} \tag{9.4b}$$

$$\text{percentage VMA filled with asphalt} = 100 \, \frac{V_a}{\text{VMA}} \tag{9.5}$$

where V_a is the volume of asphalt in the aggregate voids.

The final step, 7, is to select the optimum asphalt content according to the design specifications.

Example 9.1:

Calculate the percent voids and VMA in a compacted mixture with a unit weight of 144 lb/ft³ with the following criteria:

Ingredients	Percent of Total Aggregate	Effective Specific Gravity
Coarse aggregate	70	2.70
Fine aggregate	25	2.65
Filler aggregate	5	2.60

Percent asphalt=5 with an effective specific gravity=1.00

Solution:

Weight			Volume
	0	Air	$0.06 = 1 - 0.94$
$0.05 \times 144 =$	7.2	Asphalt	$0.12 = \dfrac{7.2}{1 \times 62.4}$
$0.047 \times 144 =$	6.7	Filler	$0.04 = \dfrac{6.8}{2.60 \times 62.4}$
$0.24 \times 144 =$	34.1	Fine aggregate	$0.21 = \dfrac{34.1}{2.65 \times 62.4}$
$0.67 \times 144 =$	95.9	Coarse aggregate	$0.57 = \dfrac{95.9}{2.70 \times 62.4}$
Σ	144 lbs.		Σ 1 ft³

1. Assume a 1-ft³ volume and determine the total weight of the mix.
2. Compute the percent of aggregate based upon 95 percent of total ingredients.
3. Determine the weights of each ingredient.
4. Determine the volume of each ingredient by Equation 9.2.
5. Use Equation 9.3 to determine the volume of the air.
6. Use Equation 9.4b to determine the VMA.

Answer: 6 percent air and 18 percent VMA.

In Example 9.1 the term "effective specific gravity" was used, and needs an explanation. The specific gravity of aggregates is an important consideration in the determination of void contents in compacted bituminous materials. There are three types of specific gravity values used in the field: apparent, bulk, and effective.

The *apparent specific gravity* is the ratio of the weight of an aggregate particle to the weight of a volume of water equal to the volume of solid aggregate and pores impermeable to water. The diagram below illustrates the term of apparent specific gravity:

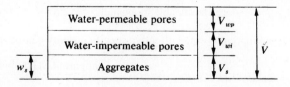

The apparent specific gravity is as follows:

$$\text{apparent specific gravity} = \frac{W_s}{(V_s + V_{wi})\gamma_w} \tag{9.6}$$

where W_s = weight of the aggregate particles (dry)
   V_s = volume of solid aggregate
   V_{wi} = volume of water-impermeable pores
   γ_w = unit of water
   V_{wp} = volume of water-permeable pores

ASTM C127 and C128 give the procedure for the determination of the specific gravity (apparent) for coarse aggregate and fine aggregate, respectively.

The *bulk specific gravity* is the ratio of the weight of aggregate particles to the weight of a volume of water equal to the volume of solid aggregate, pores impermeable to water, and pores permeable to water. In viewing the previous diagram, the bulk specific gravity is as follows:

$$\text{bulk specific gravity} = \frac{W_s}{(V_{wp} + V_{wi} + V_s)\gamma_w} \text{ or } \frac{W_s}{V} \tag{9.7}$$

where V is the total volume of aggregate. The same ASTM procedures, ASTM C127 and C128, may be used to determine the bulk specific gravity.

In the determination of the air-void content in bituminous concrete mixtures, the use of the apparent or bulk specific gravity is incorrect. If the apparent specific gravity is used, it is incorrect because this assumes that the permeable voids are filled with bitumen to the same extent as with water. On the other hand, if the bulk specific gravity is used, the asphalt is assumed not to penetrate into the permeable voids. Therefore, effective specific gravity is used.

The *effective specific gravity* is the ratio of the weight of aggregate particles to the weight of a volume of water equal to the volume of solid aggregate and pores impermeable to asphalt. Again using a diagram, the effective specific gravity is illustrated.

$$\text{effective specific gravity} = \frac{W_s}{(V_s + V_{pi})\gamma_w} \tag{9.8}$$

where V_{pi} = pores impermeable to asphalt
      V_{pa} = pores permeable to asphalt

Generally speaking, an average effective specific gravity can be obtained as follows:

$$\text{effective specific gravity} = \frac{\text{apparent SG} + \text{bulk SG}}{2} \tag{9.9}$$

Marshall Mix Design

The Marshall mix design procedure is probably the most widely used method of bituminous mix design. It is governed by ASTM C1559 (Resistance to Plastic Flow of Bituminous Mixtures Using Marshall Apparatus). This method covers the measurement of the resistance to plastic flow of cylindrical specimens of bituminous paving mixtures loaded on the lateral surface by means of the Marshall apparatus. This method is for use with mixtures containing asphalt cement, asphalt cutback or tar, and aggregate up to 1 in. (2.54 cm) maximum size.

TABLE 9-3 MARSHALL DESIGN CRITERIA

Traffic Category:	Heavy and Very Heavy		Medium		Light	
Number of Compaction Flows Each End of Specimen:	75		50		35	
Test Property	Min.	Max.	Min.	Max.	Min.	Max.
Stability, all mixtures	750	—	500	—	500	—
Flow, all mixtures	8	16	8	18	8	20
Percentage air voids						
Surfacing or leveling	3	5	3	5	3	5
Sand or stone sheet	3	5	3	5	3	5
Sand asphalt	5	8	5	8	5	8
Binder or base	3	8	3	8	3	8
Percentage voids in mineral aggregate			Varies with particle size			

In this procedure, a 4-in. (10.16-cm) diameter by 2.5-in. (6.35-cm)-high specimen is prepared by compacting in a mold with a compacting hammer that weighs 10 lb (4.54 kg) and has a free fall of 18 in. (45.72 cm). Depending upon the expected design traffic, either 35, 50, or 75 blows of the hammer are applied to each side of the specimen. After 24 hours of curing, the density and voids are determined and the specimen is heated to 140°F (60°C) for the Marshall stability and flow tests. The specimen is situated in a cylindrical half-split breaking head and the specimen is loaded at a rate of 2 in./min (5.08 cm/min). The maximum load registered, in pounds (kilograms), is referred to as the Marshall stability of the specimen. The amount of movement between no load and maximum load is referred to as *strain* and is measured in units of 0.01 in.; it is generally referred to as the *flow*.

Figure 9-2 illustrates test property curves for hot mix-design data by the Marshall method. As shown, five curves are presented: unit weight, percent air voids, Marshall stability, percent voids filled with bitumen, and flow. The stability value obtained for each test specimen is modified if the thickness is greater or less than a height of 2 in. (5.08 cm).

Table 9-3 illustrates the Marshall design criteria. This information is based upon the Asphalt Institute Marshall design criteria. These criteria are applied to the curves and a suitable asphalt percentage is determined. The most common procedure for doing this is to take the most desirable asphalt percentages for stability, unit weight, and percentage of voids and to average them. This average value should satisfy the required criteria.

PROBLEMS

9.1. Define asphalt cement.

9.2. List and explain the purpose of the hydrocarbons that make up bitumen.

9.3. What is the best type of petroleum found in the earth's crust? Why?

9.4. Explain the distillation process of crude oil.

9.5. Discuss the specific ASTM standards for asphalt cements. Give the purpose and procedure for each test.

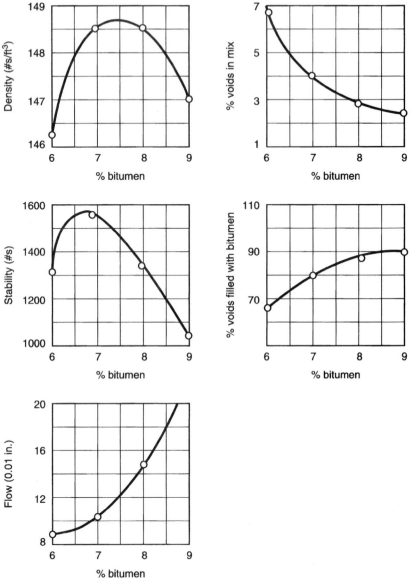

Figure 9-2 Typical Test Data for Marshall Mix

9.6. Explain liquid asphalts and discuss in detail.
9.7. Describe the manufacturing process for road tars.
9.8. List and describe the four essential properties of a good bituminous mix design.
9.9. What are the seven basic steps in proportioning asphaltic concrete? Explain.
9.10. Calculate the percent voids and VMA in a compacted mixture with a unit weight of 135 lb/ft³ with the following criteria:

Ingredient	Percentage of Total Aggregate	Effective Specific Gravity
Coarse aggregate	76	2.67
Fine aggregate	20	2.65
Filler aggregate	4	2.55

Percent asphalt=6 with an effective specific gravity=1.00

REFERENCES

Asphalt as a Material, *Asphalt Inst. Inform. Ser. 93*, 1965.

Asphalt Institute, *Asphalt Paving Manual*, 1962.

Asphalt Institute, "Brief Introduction to Asphalt," *Manual MS-5*, 1974.

BROOME, D. C., "Native Bitumens," in Arnold J. Holberg (ed.), *Bituminous Materials: Asphalts, Tars and Pitches*, vol. 2, part 1, Interscience Publishers, John Wiley & Sons, Inc., New York, 1965.

GOETZ, W. H., and WOOD, L. E., "Bituminous Materials and Mixtures," in K. B. Woods (ed.), *Highway Engineering Handbook*, sec. 18, McGraw-Hill Book Company, New York, 1960.

KREBS, R. D., and WALKER, R. D., *Highway Materials*, McGraw-Hill Book Company, New York, 1971.

WHITEHURST, E. A., and GOODWIN, W. A., *Bituminous Materials and Bitumen Aggregate Mixes*, Pitman Publishing Corp., New York, 1958.

10 *METALLIC STATE*

To understand the macroscopic properties of metals and alloys in any detail, it is necessary to have some knowledge of the metallic state at the atomic level. All of the large-scale effects metals exhibit under stress may be accurately explained by the interactions occurring microscopically. Indeed, it is knowledge of the physics involved between individual atoms and groups of atoms that allows predictions of large-scale behavior to be made. The degree to which one can correlate actual test results with known atomic structure will mark the extent to which one can make correct decisions in the selection of metals for a given design. Such ability is only possible through an understanding of the manner in which metallic properties derive from the nature of the metallic state. To facilitate this understanding, we will begin with a discussion of the crystalline structure of metals.

CRYSTALLINE NATURE OF METALS

The Metallic Bond

In a pure metal or alloy, each atom is independent of its neighbors; that is, there are no definable molecules. The valence electrons of each atom exist in an unconfined state, free to move randomly among the other atoms at very high velocity. This freedom of movement is due to the nearly equal energy levels existing on the valence shells of the atoms in the lattice. Thus, electrons may move from one atom to the next without disrupting the electronic equilibrium of the structure. The aggregation of all free valence electrons is termed the *electron cloud,* a singular property of the metallic state. As we shall see, the freedom of electrons from individual atoms is responsible for the unique elastic behavior of metals.

When the metal atoms give up their valence electrons to the electron cloud in metallic bonding, they become positive ions, held in position through their attraction to the electron cloud and their mutual repulsion. For the structure to exist in equilibrium, these forces must balance in such a way that the resultant is zero. At some specific atomic spacing a level of lowest energy is reached for which the internal forces cancel each other. The structure will seek to return to this equilibrium position if displaced, giving rise to the elastic behavior of metals in tension and compression. Figure 10-1 is a graph of force versus distance for two ions. The points above the horizontal axis correspond to repulsive force, those below represent the action of attraction, and the point where the curve crosses the horizontal axis is the equilibrium spacing or mean interatomic distance. If the structure is placed in tension (i.e., an attempt is made to increase the atomic spacing), the attractive force predominates, resisting the deformation and restoring the equilibrium position after load is removed. Similarly, when compressive forces are applied, the ions are pushed closer together and the repulsive force due to their like charges resists the deformation, again restoring the lattice to its equilibrium spacing when load is removed.

For equilibrium to exist throughout the structure (in the absence of any external or internal forces), the interatomic distance must be the same for all atoms in the lattice. Each metal or alloy is associated with a unique interatomic distance which allows individual metals or alloys to be identified through their characteristic x-ray diffraction pattern. This pattern is a result of the crystal structure and atomic spacing, thus possessing the same uniqueness as the interatomic distance of an individual metal.

Unit Cells

The absence of any bonding requirement beyond a uniform spacing results in a very closely packed structure resembling spheres stacked in a box. The majority of metals have as

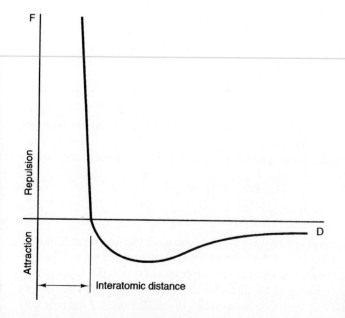

Figure 10-1   Force versus Distance

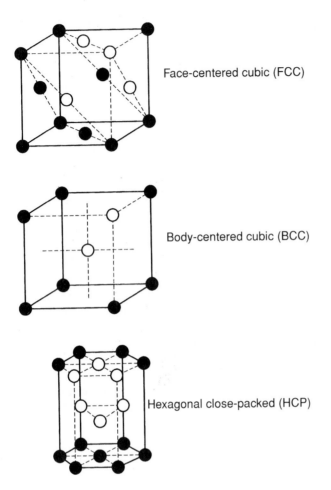

Face-centered cubic (FCC)

Body-centered cubic (BCC)

Hexagonal close-packed (HCP)

Figure 10-2 Lattice Structure

lattice structure one of the following three types, illustrated in Figure 10-2: body-centered cubic (BCC), face-centered cubic (FCC), or hexagonal close-packed (HCP). These geometrical figures, called *unit cells,* represent the smallest configuration of atoms that preserves the characteristic arrangement of the lattice. Table 10-1 lists some of the important metallic elements with their interatomic spacing and unit cell configuration in order of increasing interatomic distance.

Allotropic Behavior

Inspection of Table 10-1 will reveal several elements that have more than one crystalline configuration. The different crystal formations characteristic of an element are called its allotropic forms or *allotropes.* The property itself is termed *polymorphism* and occurs due to the almost identical interatomic distances existing in the allotropes of elements that exhibit this behavior. Examination of the interatomic spacing corresponding to the allotropes of the polymorphous elements cobalt, iron, and titanium listed in Table 10-1 will reveal this similarity.

TABLE 10-1 INTERATOMIC SPACING

Element	Unit Cell	Interatomic Spacing (Å)
Beryllium	HCP	2.23
Iron	BCC	2.48
	FCC	2.58
Nickel	FCC	2.49
Chromium	BCC	2.50
Cobalt	HCP	2.506
	FCC	2.511
Copper	FCC	2.56
Vanadium	BCC	2.63
Zinc	BCC	2.66
Molybdenum	BCC	2.73
Tungsten	BCC	2.74
Platinum	FCC	2.78
Aluminum	FCC	2.86
Tantalum	BCC	2.86
Gold	FCC	2.88
Titanium	HCP	2.89
	BCC	2.89
Silver	FCC	2.89
Magnesium	HCP	3.20
Lead	FCC	3.50

The interatomic spacing, and hence the crystal structure of an allotrope, is determined by the energy level of the lattice. Iron, for example, has an FCC structure with an interatomic distance of 2.58 Å at temperatures above 3038°F (1670°C). This is a higher energy state than that of the BCC structure. It occurs at lower temperatures with the closer spacing of 2.48 Å.

Grain Formation and Growth

The majority of metals used in industry are polycrystalline; they are composed of a great many small crystals, called *grains,* each of which is made up of metallic ions arranged in the particular space lattice characteristic of that metal. Grain size is determined by the rate of cooling from the liquid state. As the temperature of the liquid metal falls below its melting point, a phase change begins to occur; low-energy atoms precipitate out of the liquid state first, forming the nuclei upon which further growth takes place. The rate at which nuclei precipitate and growth progresses is a direct function of the cooling rate of the melt. If cooling is slow, few nuclei precipitate and growth is slow. This tends to produce a matrix of large, well-developed crystals. If cooling is rapid, as when the metal is quenched (immersed in a medium such as soil, water, or brine at much lower temperature), many nuclei form almost instantaneously and the structure produced is very finely grained.

Regardless of the rate of cooling, individual grains form in the dendritic pattern illustrated in Figure 10-3, with growth occurring predominantly at the ends of the dendrites. Eventually, the growth of an individual dendrite progresses to the point that other dendrites

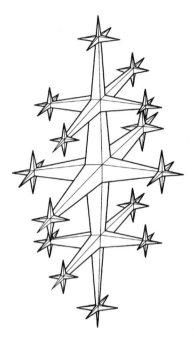

Figure 10-3 Density Pattern

restrict its expansion. When this happens, growth continues on the existing arms until all internal liquid is frozen. This is a random process, determined by the charge positions that nuclei occupy when they precipitate. Thus, the final shape of a grain is irregular and bears little resemblance to its highly ordered internal structure.

The space separating individual grains, typically one or two atoms wide, is a region of higher energy (less order) than the grains themselves. It is a transition zone from the ordered crystalline direction of one grain to the differing orientation of adjacent grains. We shall see that grain boundaries play a significant role in influencing the characteristic behavior of a metal when stresses exceed its elastic limit. A diagram of a grain boundary is shown in Figure 10-4.

Alloying

There are certain additional elements, predominantly other metals, which will combine with a pure metal in such a way that the resulting solid solution, called an *alloy,* is also metallic. This is possible only if the alloying element lies within specific boundaries of size and valence. The size factor divides alloys into two distinct groups: substitutional and interstitial alloys. For simplicity, we consider only binary alloys, those composed of two constituent elements. Alloys composed of three or more elements are very common. However, the complexity of dealing with them increases rapidly as their number of constituent elements increases. We limit the scope of our examination accordingly.

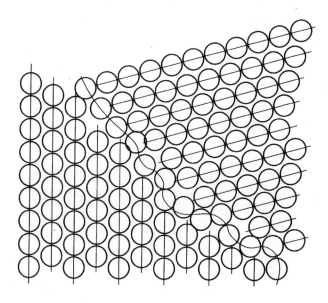

Figure 10-4 Grain Boundary

In the *substitutional* alloy, the alloying element replaces the primary element in random positions in its lattice. To accomplish this without destroying the structure, the alloying element must have an atomic radius within 15 percent of that of the primary element. If the radial difference is greater than this, the energy required to place the alloying element in position is more than the structure can accommodate and still retain its crystalline form.

As the name implies, an *interstitial* alloying element fits into the spaces or interstices of the primary element's lattice. There is a definite geometrical restriction on the size of the interstitial atom; to fit into the lattice spaces, the alloying element's radius must be less than approximately six-tenths that of the primary element. This corresponds to the maximum-size sphere that will fit into the spaces existing in a lose-packed lattice of larger spheres. The transition metals are the main group of elements that act as primary elements in interstitial alloys. There are only four elements small enough to meet the size requirement for interstitial alloying to occur when the primary elements are of this group. They are hydrogen, boron, carbon, and nitrogen and are termed *metalloid elements*.

Table 10-2 is a list of the metallic elements in a form known as the *electromotive series*. As one moves down the series, elements contain increasingly more complete valence shells and thus tend to lose electrons less easily than do the elements above them.

For two metallic elements to form a substitutional alloy over a wide range of composition, they must lie near each other in the series; that is, they must be of similar valence. If their valences differ by a sufficient amount, they will tend to exchange electrons and form a compound rather than a metallic bond when their relative proportions approach the necessary ratio. This is also true of interstitial alloys where the metalloid elements tend to form ionic compounds with nontransition metals that are relatively much more electronegative.

Alloys containing more than two elements are often both substitutional and interstitial. An important class of alloys in this category is the nickel steels; nickel and carbon are substitutional and interstitial elements, respectively, in iron, the primary element.

TABLE 10-2 METALLIC
                ELEMENTS

Lithium	Cobalt
Potassium	Nickel
Sodium	Tin
Barium	Lead
Calcium	Antimony
Magnesium	Bismuth
Aluminum	Copper
Manganese	Mercury
Zinc	Silver
Chromium	Platinum
Iron	Gold

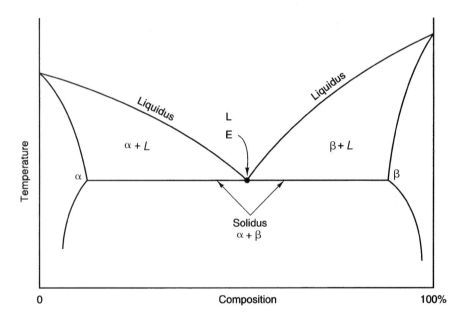

Figure 10-5 Phase Diagram of Binary Alloy

Phase Diagram

The *phase diagram* is a very useful tool in illustrating the characteristics of an alloy in relation to its temperature and composition. Also called an *equilibrium diagram,* this is a
graph of the phases existing in an alloy for any combination of temperature and composition. Figure 10-5 is a typical phase diagram for substitional binary alloys. In the diagram,
relative percentages of one of the two constituent elements are plotted on the horizontal
axis with the vertical scale corresponding to temperature. Any two distinct phases are separated by a region in which both exist simultaneously. The curves separating the different
regions are the specific points of temperature and composition where one phase or combi-

nation of phases can exist in equilibrium with another. These are the loci of saturated equilibrium states.

Three different phases are shown in Figure 10-5: α, β, and L, where α and β are solid phases and L is a liquid phase. Also shown are the three two-phase regions separating the single-phase regions, $\alpha+\beta$, $\alpha+L$, and $\beta+L$. The liquidus and the solidus are the boundaries between the liquid and solid phases, respectively, and the mixed regions. They converge to point E, the intersection of three regions, which describes the singular temperature and composition at which three phases may exist simultaneously. Such a point, which has a single liquid phase above two solid phases, is called a *eutectic*; one with a single solid phase above two other solid phases is termed a *eutectoid*. Many binary alloys exhibit phase relationships similar to Figure 10-5, with the sizes of the regions and position of the eutectic varying substantially from one alloy to the next. There are also binary alloys with a completely different form from that of Figure 10-5, and eutectoids. An example of an alloy with a phase diagram similar to Figure 10-5 is the copper-tin system. One differing markedly from this is the aluminum-nickel binary system.

For alloys with three or more elements, graphic illustration of phase relationships becomes increasingly difficult. The ternary or three-element alloys, for example, require a three-dimensional graph for complete description. Two-dimensional plots of the system may be made by taking a horizontal cross-section of the three-dimensional graph, resulting in the phase relationships of the three constituents for varying compositions at a constant temperature.

MECHANICAL BEHAVIOR OF METALS

In this section we discuss the mechanical behavior of metals. Included in our discussion will be the resistance of metals to failure caused by slip, fatigue, impact, flow or creep, stress rupture, and corrosion and wear.

Slip (Inelastic Action)

Slip is a phenomenon of movement along gliding planes of crystals due to the application of stress beyond the material's elastic limit. When a material undergoes slip, one will observe slip lines or slip bends (fine lines) across the faces of a number of the crystalline grains of the overstressed metal.

Slip is an inelastic action that does not occur to any appreciable extent in any of the ordinary materials of construction until a fairly well defined limiting stress has been applied; and if slip occurs, it does not continue indefinitely, under load, but ceases after a short time.

The most satisfactory explanation of the fact that, under a load above the elastic strength of a metal, slip soon ceases, and of the observed increase of elastic strength after such overload, was furnished by the slip interference theory of Jeffries and Archer. As slip takes place, the crystalline grains of metal are fragmented into thin plates and the surfaces of the sliding plates tend to "dig into" each other, causing increasing resistance as slip proceeds. In addition to this, adjacent grains, in general, have slip planes lying in different directions, and as slip proceeds, the slip in one grain tends to hinder the slip in the adjacent grain.

Moreover, in metals composed of two or more different kinds of crystalline grains, the stronger grains act as "keys," tending to stop or to hinder slip in weaker grains. Slip interference theory further holds that a metal may contain definite, very small grain sizes that will be more effective as brakes on slip than larger, more widely scattered grains, and also more effective than extremely small grains, grains so small that slip can proceed right by them without much deviation from its direct path. In the case of steel, such an "optimum" size of grain may be produced by heat treatment.

Fatigue

For structural parts (axles, machine parts, shafts, etc.) under infrequent variation of load, the outstanding danger of failure is failure by inelastic distortion (slip).

Examination under the microscope and the scanning electron microscope of the behavior of the crystalline constituents of metals under repeated stress has shown that failure of crystals is caused by a succession of shear slips on parallel planes of vast strength. The failure of the piece as a whole is due to the successive failures of individual crystals and the development of cracks. Failure very seldom occurs at crystalline boundaries.

Fatigue was once considered a phenomenon of the cohesion between crystalline constituents of the metal subjected to many repetitions or reversals under comparatively low stresses. However, the failure of a material under repeated stress is a process of gradual or progressive fracture of the crystals themselves. Thus, the term "fatigue" is improperly used and the appropriate term should be "progressive fracture."

In a general way, the difference in the behavior of material under repeated loads and under a single load is shown by the tendency toward gradual cracking of the material under repeated loads. Under a single application of load the material of a structure either withstands the load or fails. Under a repeated load, the load may be applied many thousands of times and the material may withstand the load for a while, and then fail by the gradual spread of cracks in the material. Under repeated loads, local strains (which would be of little importance in a single load) may form a nucleus for damage that gradually spreads until the whole member fails. Slip in crystals is likely to occur at points where high localized stresses are set up. These stresses depend upon the distribution of stress between crystals, which in turn depends upon the homogeneity of the structure. Thus, flaws and cracks in the internal structure tend to cause internal stresses, which were formed by cooling, heat treatment, or mechanical working of the metal. These stresses would not normally affect the static structural strength but tend to produce slip in crystals when the metal is subjected to repeated stresses.

Impact

Impact is the resistance of a material to failure due to brittleness under service conditions in a structure. In selecting materials for members in a structure that must resist impact, two factors must be considered:

1. Total stress allowable.
2. Total strain allowable.

Resistance to impact is a function of both of these factors and both must be taken into consideration in design.

For materials that must withstand heavy accidental impact without actual rupture, toughness of the material is of primary concern. Toughness of a material may be measured by the area under the stress-strain curve of the material. According to Moore, a striking illustration of the resistance of materials to rupture under impact is furnished by comparing the action of oak with that of cast iron. Under static load cast iron is about three times as strong as oak, but the strain that oak will stand before rupture is about nine times the strain that cast iron will stand. The area under the stress-strain diagram for oak is about three times that for cast iron, and under impact loads oak requires about three times as much energy for fracture as cast iron.

The ability of a material to resist impact without a permanent distortion is measured by the area under the stress-strain diagram up to the elastic limit: in other words, the area under the straight-line portion of the stress-strain curve. If elastic resistance to impact is desired, a material with a high elastic strength or a low modulus of elasticity should be used.

Flow or Creep

Creep is the very slow flow of a material at elevated temperatures under sustained stress. The rate of flow of a given metal depends on the magnitude of the stress and the temperature. Lead and zinc, for example, will creep under normal temperatures at relatively low stresses. However, most metals require a relatively high temperature before they will begin to flow under low stresses. Creep may continue for an indefinite amount of time under a sustained load tending to distort the material and eventually cause failure of the material by rupture. Unlike repeated loads (fatigue), creep affects the entire body of the material under stress instead of producing a localized rupture.

Stress Rupture

With the development of jet propulsion and aircraft gas turbines, greater attention has been placed on metals to withstand stresses at high temperature. In the aircraft industry this temperature may be as high as 982°C (1800°F). Various laboratory tests have been developed and proven reliable in indicating the performance of the metal at high temperatures under stress. These laboratory tests eliminate time-consuming engine tests. In determining the material's life under continued high temperature and loading, stress necessary to cause rupture at a given time and temperature is the property of greatest importance. This property is sometimes referred to as *stress rupture.*

In jet engines and aircraft gas turbines operations, severe local overheating and overstressing may occur in a very short time. Thus, designs must provide for short-term strengths of alloys at temperatures somewhat higher than expected operating levels as well as sufficient stress rupture strengths at the anticipated operating temperature levels.

Corrosion and Wear

Each year millions of dollars worth of damage to iron and steel structures is caused by corrosion. Further, millions of dollars are spent to replace worn-out parts of machine structures. Corrosion and wear damage take place gradually. They seldom cause sudden or dramatic structural disasters, as parts damaged by corrosion or wear can be replaced or repaired before failure occurs.

Most metals associated with construction materials come in contact with water that contains dissolved oxygen or with moist air and enter into solution readily. The rate of solution is usually retarded by a film of hydrogen forming on the metal or by coating the metal with a protective coating. However, oxygen will combine with the hydrogen and over a period of time will strip it away from the metal, and thus further corrosion will result.

Five classifications of corrosion for metals exist:

1. Atmospheric
2. Water immersion
3. Soil
4. Chemicals other than water
5. Electrolytic

In atmospheric corrosion a large excess of oxygen is available and the rate of corrosion is largely determined by the quantity of moisture in the air and the length of time in contact with the metal.

When metals are immersed in water, the amount of oxygen dissolved in the water is an important factor. If the water does not contain any dissolved oxygen, the metal will not corrode. If the water is acidic, the corrosion rate is increased, whereas water that is alkaline has very little corrosion activity unless the solution is highly concentrated.

In soil corrosion and in corrosion by chemicals other than water, the most important item is the ingredient coming in contact with the iron or steel.

Corrosion by electrolysis due to stray currents from power circuits may be disastrous, but in nearly all cases it can be prevented by suitable electrical precautions.

The most common protective coating against corrosion for iron and steel is paint. The paint coating is usually mechanically weak and it cracks and wears out. Thus, to do a satisfactory job, the paint must be renewed every 2 or 3 years. Before the structure is painted, it should first be cleaned and the rust removed.

If the structure is to be immersed in water or if it comes in contact with water, paint provides little protection. Thus, the portion that is in contact with water might require a coating of asphalt or coal tar to protect it.

Another excellent method of preventing corrosion is to encase the iron or steel in concrete. Although concrete is porous, it will provide adequate protection for years. However, if the concrete becomes cracked, it loses most of its protecting ability and should be replaced if possible, or patched.

Metals under stress, especially those beyond their elastic strength, corrode more rapidly than do unstressed metals.

In nearly all cases the failure of materials by mechanical wear under abrasion occurs gradually, the progress of wear is evident, and the failure is not a definitely defined event but one whose occurrence is a matter of judgment on the part of the user of the material. Failure by wear rarely leads to disaster, and usually involves repair or replacement of a part.

TESTING AND EVALUATION OF METALS

Tests are conducted on materials of construction in order to determine their quality and their suitability for specific uses in machines and structures. It is necessary for the producer, consumer, and the general public to have tests for the determination of quantitative properties of materials such that the material may be properly selected, specified, and designed. Tests are further needed to duplicate materials and to check upon the uniformity of different shipments.

Testing and evaluation of the many various metals and alloys requires hundreds of tests and specifications. The ASTM specifications dealing with metals and their alloys comprise several volumes. ASTM, Section 3, which covers various tests for steel products, is a primary reference for civil and highway engineers.

The most common mechanical tests (for metals and metallic products) are listed in ASTM A370 as follows:

1. ASTM E8: Tension Testing of Metallic Materials.
2. ASTM E10: Test for Brinell Hardness of Metallic Materials.
3. ASTM E18: Test for Rockwell Hardness and Rockwell Superficial Hardness of Metallic Materials.
4. ASTM E23: Notched-Bar Impact Testing of Metallic Materials.

ASTM E8: Tension Testing of Metallic Materials

Most commercial specifications for metals have requirements for physical properties as determined by the tensile strength test. The tension test is one of the most important tests for determining the structural and mechanical properties of metals. These properties include the ultimate strength (tensile), ductility (elongation and reduction of area), modulus of elasticity, yield point, offset yield strength, proportional limit, and the elastic and inelastic range, among others.

The *tensile test* is performed by gripping the opposite ends of a test specimen called a "coupon" and pulling it apart. Various standard ASTM coupons are shown in Figure 10-6. Standard test specimens are obtained from sheared, blanked, sawed, trepanned, or oxygen-cut materials that are being tested. In all cases, special attention should be taken to ensure the removal by machining of all distorted cold-worked or heat-affected areas. Test specimens should be machined such that they have a reduced cross-section at the midpoint to localize the zone of fracture. In addition, the specimen should be gaged—marked such that the percent elongation may be determined.

When loading a metal specimen, the rate of loading is usually unimportant but should not exceed 690 MPa/minute. Most modern testing machines are equipped with electronic strain gages; otherwise, extensometers or foil gages (in conjunction with strain indicators and switching and balancing units) may be employed. When the coupon is fastened in the machine and the strain gage is fastened, the test begins. As the specimen is pulled apart, a load-strain (stress-strain) diagram may be plotted (automatically if electronic strain gages are used) as shown in Figure 10-7. From this diagram the various physical and/or mechanical properties may be obtained.

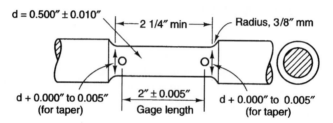

(a) Standard round specimen with 2" gage length

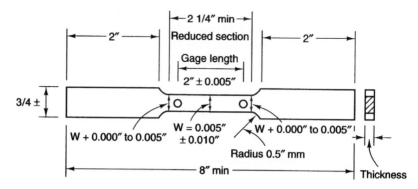

(b) Standard rectangular speciment with 2"
gage length for testing metals in form of
plate, sheet, etc. having thickness from
0.005" to 5/8".

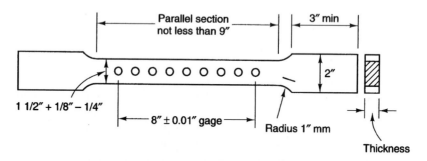

Figure 10-6   Various Coupons (Courtesy of ASTM)

In a tensile test it is customary to give the strength of a material in terms of unit stress or internal force per unit of area. Also, the point at which yielding starts is expressed as unit stress.

Further, in a tensile test, strain is measured as unit strain. *Unit strain* in any direction is the deformation per unit of length in that direction. If electronic strain gages are employed in a tensile test, the plot on the stress-strain diagram is a function of the force per unit area versus the unit strain.

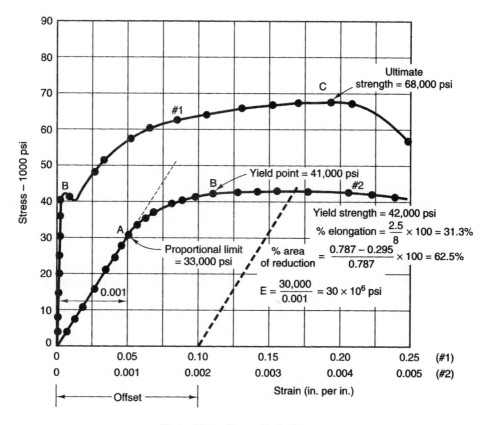

Figure 10-7   Stress-Strain Diagram

However, in some cases the unit strain cannot be obtained directly (as in the use of an extensometer), so the total strain or deformation is measured. *Deformation* in any direction is the total change in the dimension of a member in that direction. When the loading is such that the unit strain is constant over the length of a member, it may be computed by dividing the deformation by the original length of the member (gage length). Thus, unit strain equals total strain divided by the gage length.

Tensile strength.   The *tensile strength* of metals is the maximum axial load (ultimate load) observed in a tension test divided by the original cross-sectional area. The strength increases and reaches a maximum in mild steel, after extensive elongation and necking. As indicated, it is characterized by the beginning of necking down, a decrease in cross-sectional area of the specimen, or local instability. The tensile strength is the ultimate strength expressed in units of pounds per square inch (Newtons per square meters) and as shown in Figure 10-7 is 68,000 psi.

In design, one cannot base the working stress on the ultimate strength because excessive strain cannot be tolerated. Thus, one must find the ultimate unit stress produced by the design loads in the member and reduce it by a factor of safety such that one works with an allowable unit stress.

Ductility.   *Ductility* is the ability of a material to undergo large deformations without fracture. Thus, the material will deform in the inelastic (plastic) range. Ductility is measured by the elongation and reduction of area in a tension test and is expressed as a percentage:

$$\frac{\text{percentage}}{\text{elongation}} = \frac{\text{final length} \; - \; \text{original length}}{\text{original length}} \times 100 \qquad (10.1)$$

and

$$\begin{array}{c}\text{percentage reduction}\\ \text{in area}\end{array} = \frac{\text{original length} \; - \; \text{area after fracture}}{\text{original area}} \times 100 \qquad (10.2)$$

For our example (as given in Figure 10-7) the percentage elongation is 31.3 and the percentage reduction in area is 62.5.

Modulus of elasticity.   The *modulus of elasticity (E)* is given by the slope of the straight-line portion of the stress-strain curve. It is a measure of the inherent rigidity or stiffness of a material. For a given geometric configuration, a material with a large E deforms less under the same stress.

The modulus of elasticity (Young's modulus) is the ratio of unit stress to unit strain in the elastic range of the stress-strain curve as follows:

$$E = \frac{\text{stress (psi)}}{\text{strain (in./in.)}} \qquad (10.3)$$

or in SI units,

$$E = \frac{\text{stress (MPa)}}{\text{strain (cm/cm)}} \qquad (10.4)$$

In Figure 10-7 the modulus of elasticity is 30,000,000 psi.

Yield point.   At the termination of the linear portion of the stress-strain curve, some materials (such as a low-carbon steel) develop a yield point. The *yield point* is the first load at which there is a marked increase in strain without an increase in stress. This behavior may be a consequence of inertia due to the effects of the testing machine and the deformation characteristics of the test specimen. The yield point is sometimes taken as the proportional limit and elastic limit, which is an incorrect practice. Most metals do not have a yield point, and thus an offset method is used. In Figure 10-7 the yield point would correspond to 41,000 psi.

Offset yield strength.   The *offset yield strength* is defined as the stress corresponding to a permanent deformation, usually 0.10 or 0.20 percent (0.001 or 0.002 in./in.). The offset method is usually used with materials that have a definite straight-line portion to their stress-strain curve. One measures the corresponding offset percentage on the stress-strain curve and projects upward a straight line parallel with the straight-line portion of the stress-strain curve. Where the line intersects the stress-strain curve, the value is read off as the offset yield strength. In Figure 10-7 this value would be 42,000 psi.

In situations when the stress-strain curve does not exhibit a straight-line portion, the secant modulus or tangent modulus method may be used.

Proportional limit.   The *proportional limit* is the greatest stress that a material is capable of without deviating from the law of proportionally of stress to strain (Hooke's law). *Hooke's law* is defined as

$$f = E\epsilon \tag{10.5}$$

where f =   unit stress
        ϵ =   unit strain
        E =   Young's modulus of elasticity

Metals are elastic within the proportional limit and thus the proportional limit has significance in the elastic stability of columns and shells. In our example, as shown by Figure 10-7, the proportional limit is 33,000 psi.

Elastic limit and inelastic limit.   The *elastic limit* is the largest unit stress that can be developed without a permanent set remaining after the load is removed. In most cases the elastic limit is difficult to determine, and many materials do not have a well-defined proportional limit, or any at all; thus, the offset yield strength is used to measure the beginning of plastic deformation (inelastic limit).

Modulus of rigidity.   The *modulus of rigidity* or the shearing modulus of elasticity, as it is sometimes called, is defined as

$$G = \frac{v}{\gamma} \tag{10.6}$$

where G =   modulus of rigidity
        v =   unit shearing stress
        γ =   unit shearing strain

The modulus of rigidity may also be rewritten and related to the modulus of elasticity by

$$G = \frac{E}{2(1 + \mu)} \tag{10.7}$$

where μ is a constant known as *Poisson's ratio* and for structural steel equals 0.3.

Modulus of toughness.   The *toughness* of a material is the ability of a material to absorb large amounts of energy. The *modulus of toughness* can be related to the area under the entire stress-strain curve, and it depends on both strength and ductility. Because of the difficulty of determining toughness analytically, toughness is often measured by the energy required to fracture a specimen, usually notched and sometimes at low temperatures, in an impact test such as the Charpy or the Izod.

Modulus of resilience. The *resilience* of a material is that property of an elastic body by which energy can be stored up in the body by loads applied to it and given up in recovering its original shape when the loads are removed. Thus, the *modulus of resilience* is equal to the area under the straight-line portion of the stress-strain curve (a triangle).

ASTM E10: Brinell Hardness of Metallic Materials

The *hardness* of metals is usually determined by measuring the resistance to penetration of a ball, cone, or pyramid. The results of a hardness test can be directly related to the tensile strength of a material. Thus, it provides a quick check on the tensile strength of a material without going through a time-consuming tensile test.

The *Brinell hardness method* is based upon determining the resistance offered to indentation by a hard ball of specific diameter that is subjected to a given pressure. The pressure used in testing steel is usually 6600 lb (3000 kg) and the diameter of the ball is 0.4 in. (10 mm). When softer metals are utilized, a pressure of 1100 lb (500 kg) is used. The Brinell hardness number can be computed by the following formula:

$$BH = \frac{2P}{\pi D(D - \sqrt{D^2 - d^2})} \tag{10.8}$$

where P = pressure, kg
        D = diameter of the steel ball, mm
        d = average diameter of indentation, mm

As is obvious from the equation, the smaller the indentation, the greater the Brinell hardness number.

ASTM E18: Test for Rockwell Hardness

ASTM E18 specifies the test for the determination of the Rockwell hardness and Rockwell superficial hardness of metallic materials. The *Rockwell hardness method* employs either a ball or a diamond cone in a precision testing instrument that is designed to measure depth of penetration accurately. Two superimposed impressions are made. The depth to which the major load drives the ball or cone below that depth to which the minor load has previously driven it is a measure of the hardness. For the harder steels, greater accuracy is obtained by use of a diamond cone.

Rockwell superficial hardness machines are used for the testing of very thin shells or thin surface layers of materials. In this case the minor load is 6.6 lb (3 kg) and the major load varies from 33 to 99 lb (15 to 45 kg).

In all cases the Rockwell hardness number is read directly from the scales on the machine. The relationship of the Brinell and Rockwell hardness numbers to tensile strength is shown in Table 10-3.

ASTM E23: Notched-Bar Impact Testing of Metallic Materials

Impact tests are performed primarily for two reasons:

TABLE 10-3   RELATIONSHIP OF BRINELL AND ROCKWELL HARDNESS NUMBERS
TO TENSILE STRENGTH

Brinell Indentation Diameter (mm)	Brinell Hardness Number		Rockwell Hardness Number		Rockwell Superficial Hardness Number, Superficial Diamond Penetrator			(Approximate) Tensile Strength (MPa)
	Standard Ball	Tungsten Carbide Ball	B Scale	C Scale	15– N Scale	30– N Scale	45– N Scale	
2.50		601		57.3	89.0	75.1	63.5	2262
2.60		555		54.7	87.8	72.7	60.6	2055
2.70		514		52.1	86.5	70.3	47.6	1890
2.80		477		49.5	85.3	68.2	54.5	1738
2.90		444		47.1	84.0	65.8	51.5	1586
3.00	415	415		44.5	82.8	63.5	48.4	1462
3.10	388	388		41.8	81.4	61.1	45.3	1331
3.20	363	363		39.1	80.0	58.7	42.0	1220
3.30	341	341		36.6	78.6	56.4	39.1	1131
3.40	321	321		34.3	77.3	54.3	36.4	1055
3.50	302	302		32.1	76.1	52.2	33.8	1007
3.60	285	285		29.9	75.0	50.3	31.2	952
3.70	269	269		27.6	73.7	48.3	28.5	897
3.80	255	255		25.4	72.5	46.2	26.0	855
3.90	241	241	100.0	22.8	70.9	43.9	22.8	800
4.00	229	229	98.2	20.5	69.7	41.9	20.1	766
4.10	217	217	96.4					710
4.20	207	207	94.6					682
4.30	197	197	92.8					648
4.40	187	187	90.7					621
4.50	179	179	89.0					607
4.60	170	170	86.8					579
4.70	163	163	85.0					566
4.80	156	156	82.9					552
4.90	149	149	80.8					503
5.00	143	143	78.7					490
5.10	137	137	76.4					462
5.20	131	131	74.0					448
5.30	126	126	72.0					434
5.40	121	121	69.0					414
5.50	116	116	67.6					400
5.60	111	111	65.7					386

1. To determine the ability of the material to resist impact under service conditions.
2. To determine the quality of the metal from a metallurgical standpoint.

As previously indicated, two types of impact tests are usually performed, the Charpy and the Izod tests. Both apply a dynamic load by use of a pendulum that has enough kinetic energy to rupture a specimen in its path.

P R O B L E M S

10.1. Explain the crystalline nature of metals.

10.2. What is meant by allotropic behavior?

10.3. Why is grain formation and growth important in metals?

10.4. Discuss the phenomenon of slip.

10.5. Why is "fatigue" an improper term as it applies to metals?

10.6. Explain the process of flow or creep.

10.7. What are the five classifications of corrosion? Discuss.

10.8. Explain by use of your own diagram the tension testing of steel. Show all physical and mechanical properties and explain each.

10.9. Explain the two most common hardness tests.

10.10. Using the library and ASTM standards, explain the Charpy impact test.

R E F E R E N C E S

BRICK, R. M., GORDEN, R. B., and PHILLIPS, A., *The Structure and Properties of Alloys*, McGraw-Hill Book Company, New York, 1964.

CORDON, W. A., *Properties, Evaluation, and Control of Engineering Materials*, McGraw-Hill Book Company, New York, 1979.

COTTRELL, A. H., *An Introduction to Metallurgy*, Edward Arnold Publishers, Ltd., London, 1967.

DAVIS, H. E., TROXELL, G. E., and WISKOCIL, C. T., *The Testing and Inspection of Engineering Materials*, 3rd ed., McGraw-Hill Book Company, New York, 1964.

DIETER, G. E., *Mechanical Metallurgy*, McGraw-Hill Book Company, New York, 1961.

DOAN, G. E., and MAHLA, E. M., *The Principles of Physical Metallurgy*, McGraw-Hill Book Company, New York, 1941.

GILLET, H. W., *The Behavior of Engineering Metals*, John Wiley & Sons, Inc., New York, 1951.

JASTRZEBSKI, Z. D., *The Nature and Properties of Engineering Materials*, John Wiley & Sons, Inc., New York, 1959.

JEFFRIES, Z., and ARCHER, R. S., *The Science of Metals*, McGraw-Hill Book Company, New York, 1924.

MOORE, H. F. *Materials of Engineering*, 7th ed., McGraw-Hill Book Company, New York, 1947.

SEITZ, F. B., *The Physics of Metals*, McGraw-Hill Book Company, New York, 1943.

VAN VLACK, L. H., *Elements of Material Science*, 2nd ed., Addison-Wesley Publishing Company, Inc., Reading, Mass., 1964.

11

FERROUS METALS

In general, metals can be classified into two major groups: ferrous and nonferrous. A *ferrous metal* is one in which the principal element is iron, as in cast iron, wrought iron, and steel. A *nonferrous metal* is one in which the principal element is not iron, as in copper, tin, lead, nickel, aluminum, and refractory metals.

In this chapter we discuss ferrous metals. Special attention will be placed on the production techniques of steel.

GENERAL

Sources of Metals

In general, more than 45 metals of industrial importance are found within the earth's crust. With the exception of aluminum, iron, magnesium, and titanium, which occur in appreciable percentages within the earth's crust, all other metals comprise less than 1 percent of the earth's crust. Thus, most metals occur in the form of ore, from which the metal has to be extracted. An ore is usually referred to as a *mineral*, which is a chemical compound or mechanical mixture. The material associated with the ore which has no commercial use is referred to as *gangue*. Basically, six classifications of ore exist:

1. Native metals
2. Oxides
3. Sulfides
4. Carbonates

5. Chlorides
6. Silicates

The native metals consist of copper and precious metals. Oxides are the most important ore source, in that iron, aluminum, and copper can be extracted from them. Sulfides include ores of copper, lead, zinc, and nickel. Carbonates include ores of iron, copper, and zinc. The chlorides include ores of magnesium, and the silicates include ores of copper, zinc, and beryllium.

Production of Metals

Four operations are required for the production of most metals:

1. Mining the ore
2. Preparing the ore
3. Extracting the metal from the ore
4. Refining the metal

In the mining operation, the methods of open-pit borrowing and underground mining are both utilized. The most famous examples of an open-pit operation are the iron mines in the Mesabi Range in Minnesota. An example of underground mines are the copper mines in upper Michigan. Most underground mines are of the room-and-pillar type for working horizontal veins or of the stepping type for working vertical veins by cutting a series of steps.

In the preparation process the ore is crushed and large quantities of gangue are removed by a heavy-media-separation method. In some cases, the preparation of the ore may involve roasting or calcining. In roasting, the ore of sulfide is heated to remove the sulfur and in calcining the carbonate ores are heated to remove carbon dioxide and water.

The extraction of the metal from the ore is accomplished through chemical processes. These chemical processes reduce the compounds, such as oxides, by releasing the oxygen from chemical combinations and thus freeing the metal.

Basically, three types of processes of extraction are used:

1. Pyrometallurgy
2. Electrometallurgy
3. Hydrometallurgy

In the *pyrometallurgy* process (generally referred to as *smelting*) the ore is heated in a furnace producing a molten solution, from which the metal can be obtained by chemical separation. The blast furnace or reverberatory furnace is used in this process.

In the *electrometallurgy* process metals are obtained from ores by electrical processes utilizing an electric furnace or an electrolytic process.

Hydrometallurgy or *leaching* involves subjecting the ore to an aqueous solution from which the metal is dissolved and recovered.

As a result of the extraction process, the metals will contain impurities, which must be removed by a refining process. If the metal was extracted by the pyrometallurgy process, the most common method of refining is by oxidizing the impurities in a furnace (steel from pig iron). However, other methods are utilized, such as liquidation (tin), distillation (zinc), electrolysis (copper), and the addition of a chemical reagent (manganese to molten steel).

FERROUS METALS

Ferrous metals comprise three general classes of materials of construction:

1. Cast iron
2. Wrought iron
3. Steel

All of these classes are produced by the reduction of iron ores to pig iron and the subsequent treatment of the pig iron to various metallurgical processes. Both cast iron and wrought iron have fallen in production with the advent of steel, as steel tends to exhibit better engineering properties than do cast and wrought iron. The application of steel and steel alloys is so widespread it has been estimated that there are over a million uses. In construction, steel has three principal uses.

1. Structural steel
2. Reinforcing steel
3. Forms and pans

Classification of Iron and Steel

Iron products may be grouped under six headings:

1. Pig iron
2. Cast iron
3. Malleable cast iron
4. Wrought iron
5. Ingot iron
6. Steel

Pig iron is obtained by reducing the iron ore in a blast furnace. This is accomplished by charging alternate layers of iron, ore, coke, and limestone in a continuously operating *blast furnace*. Blasts of hot air are forced up through the charge to accelerate the combustion of coke while raising the temperature sufficiently to reduce the iron ore to molten iron. The limestone is a flux that unites with impurities in the iron ore to form slag. The blast furnace accomplishes three functions:

1. Reduction of iron ore
2. Absorption of carbon
3. Separation of impurities

The amount of carbon present in pig iron is usually greater than 2.5 percent but less than 4.5 percent. The iron may be cast into bars, referred to as *pigs*.

Cast iron is pig iron remelted after being cast into pigs, or about to be cast in final form. It does not differ from pig iron in composition and it is not in a malleable form.

Malleable cast iron is cast iron that has undergone special annealing treatment after casting and has been made malleable or semimalleable.

Wrought iron is a form of iron that contains slag, is initially malleable but normally possesses little to no carbon, and will harden quickly when rapidly cooled.

Ingot iron is a form of iron (or a low-carbon steel) that has been cast from a molten condition.

Steel is an iron–carbon alloy that is cast from a molten mass whose composition is such that it is malleable in some temperature range. Carbon steel is steel that has a carbon content of less than 2 percent and generally of less than 1.5 percent; its properties are dependent on the amount of carbon it contains. Alloy steels are steels in which the properties are due to elements other than carbon.

MANUFACTURE OF STEEL

As previously stated, the first process in the manufacture of steel is the reduction of iron ore to pig iron by use of a blast furnace. This is followed by the removal of impurities, and four principal methods are used to refine the pig iron and scrap metal:

1. Open-hearth furnace
2. Bessemer furnace
3. Electric furnace
4. Basic oxygen furnace

In this section we discuss in detail the blast furnace, the open-hearth furnace, the Bessemer converter, the electric furnace, and the basic oxygen furnace. Most of the discussion is taken directly from the *Steel Products Manual* of the American Iron and Steel Institute.

Blast Furnace

Blast furnaces (Figure 11-1) are so named because of the continuous blast of air required to bring about the necessary heat and chemical reactions in the raw materials in the stack. As much as 4.5 tons of air may be needed to make 1 ton of pig iron.

Most blast furnaces, and all of the more modern installations, are served by *turboblowers.* These enormously powerful steam engines are designed to receive steam at 700 psi (492,100 kg/m^2) at a temperature of 750°F (399°C).

The blowers force air from the atmosphere through piping into *stoves,* which are the most prominent auxiliaries serving a blast furnace. Each furnace must have at least two stoves and may have three.

Essentially, a stove consists of two parts: a *combustion chamber,* which is a vertical passageway wherein cleaned blast furnace gas is burned, and *brick checkerwork,* which contains many small passageways for heating the air as it passes over the hot masonry.

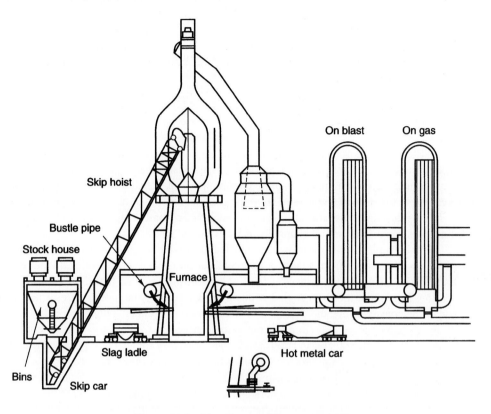

Figure 11-1 Blast Furnace (Courtesy of AISI)

The design of the stoves varies in complexity, but it is important to remember that each stove serves two alternating functions. A stove will receive cold air from the blowers and pass it on through the heated brick checkerwork to the blast furnace; when the atmospheric air from the blower has cooled the brick checkerwork, another stove will take over the function and hot air from the top of the blast furnace will burn in the combustion chamber to reheat the brick checkerwork in the furnace stove.

Modern stoves for large furnaces are somewhat like farm silos in appearance (cylindrical in shape with a domed top). They may be up to 28 ft (8.5 m) in diameter and about 120 ft (36 m) high from the bottom to the top of the dome. Depending upon the type of brick checkerwork used, the stoves may contain upwards of 250,000 ft^2 (22,500 in.^{2}) of heating surface.

It is common for steel columns to support steel grids inside the stove, and these grids bear the brick checkerwork. Insulation between the brick and the steel shells prevents distortion of the stove.

Special valves control the flow of air from the blower to the heated stove, which is referred to as being "on blast." The stove or stoves in the process of having their checkerwork heated are referred to as being "on gas." In a three-stove arrangement serving one blast furnace, each stove is on gas twice as long as it is on blast.

The air from an on-blast stove speeds toward the furnace at a temperature that ranges from 1400 to 2100°F (760 to 1150°C). Air enters a bustle pipe made of heavy steel plate which completely encircles the stack of the blast furnace. From this bustle pipe smaller pipes extend at an angle downward and enter the stack of the furnace. These entry pipes, called *tuyeres,* enter the furnace at a level considerably above the hearth, in which lies a pool of molten iron at approximately 2700°F (1482°C) with the layer of impurities, called *slag,* floating on top. In modern practice it is not uncommon for fuels, such as oil or a coal sledge, to be injected through the tuyeres along with the air for higher output.

The term "pig iron" may seem obscure to modern city dwellers, but its origin was perfectly clear in an earlier agricultural age when small blast furnaces largely confined their services to their own communities. In those days molten iron from a furnace ran down a channel prepared in the ground and flowed from the channel into small holes dug on either side of it. To neighboring farmers the arrangement resembled a familiar sign in their lives (newborn pigs, sucklings). The central channel was known as a "sow."

Today's typical pig casting machines feature twin conveyor lines of shallow molds. Molten iron is poured from a ladle or hot metal car into a short channel that divides to feed the molds passing on the conveyor lines at a controlled rate so that there is little or no spilling. Iron in the molds is cooled with water. By the time the mold is overturned at the end of the conveyor line, the iron is solid and each piece is called a *pig.* It probably weighs 40 lb (18 kg) or more and simply drops into a railway car. There it is further cooled by spraying. Samples are taken and transported to a laboratory for analysis. Finally, the carload of cool iron is taken to a numbered pile and stored until customers require it.

Most of the iron from pig casting machines is made to rigid specifications that vary widely according to end use. The generic name for this product is *merchant pig iron,* because it is intended to be sold as an end product, although some small percentage of it may be used in steel making.

The customers for merchant pig iron are foundries that cast pipe for water and waste, for oil burner parts, and for automobile engine blocks, to name a few of a hundred uses. Among others are the dutch ovens used in many household kitchens, intricately shaped pipe fittings, heavy bases of machine tools, and major elements in a magnet developed at the Department of Energy's Argonne Laboratories.

A merchant pig iron producer may have several hundred piles of different grades of iron, each manufactured to meet the specifications of a given customer. A merchant pig iron producer may blend a dozen or more grades of iron ore to arrive at a desired chemical composition. Thus, the different products may be considered as iron alloys.

Most merchant pig iron furnaces are considerably smaller than the machines producing iron for steelmaking purposes. Even so, it is worthy of note that while, for example, 30,000 ft^3 (900 m^3) of solids weighing just under 1000 tons is being processed in a furnace stack, the chemical composition of the pigs is being controlled within fractions of 1 percent relative to carbon, silicon, sulfur, phosphorus, and manganese.

Foundries buy and melt scrap to supplement merchant pig iron in making their products. Some of them also buy imported iron. In a recent year about 20 percent of the available supply of merchant pig iron in the United States was of foreign origin. But domestic pig iron remains the standard for quality control in making castings because foundries are dependent on the chemistry of the iron they buy from various producers in the United States.

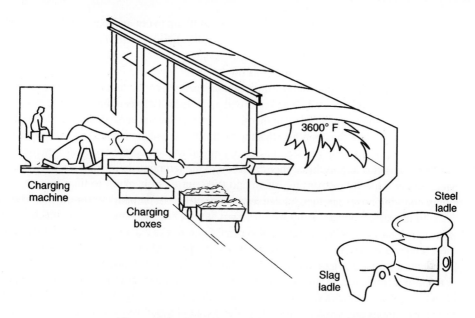

Figure 11-2 Open-Hearth Furnace (Courtesy of AISI)

Serious changes in chemistry occur when too much poorly graded scrap and iron are consumed. The precise specifications to which American merchant iron is made provide a standard by which foundries can control the quality of their product.

Open-Hearth Furnace

Open-hearth furnaces (Figure 11-2) are so named because limestone, scrap steel, and molten iron are charged into a shallow steelmaking area called a "hearth" and are exposed, or "open," to the sweep of flames that emanate alternately from opposite ends of the furnace. The first open hearths for steelmaking were built in the United States during the late 1800s, and they have since been developed to a high degree of sophistication. In the beginning, as now, one of their most attractive features concerns the use of steel scrap.

Theoretically, an open-hearth furnace can operate using either blast furnace iron alone or steel scrap alone, but most are designed to operate using both in approximately equal proportions. These proportions vary according to such economic factors as the price of scrap or the availability of molten iron. For more than half a century open-hearth furnaces were the outstanding high-tonnage producers of steel for America's expanding industry and are still major factors in the steel industry. A furnace that will produce a fairly typical 350 tons (318,000 kg) of steel in 5 to 8 hours may be about 90 ft (27 m) long and 30 ft (9 m) wide. It is very likely to be one of several furnaces contained in a single building referred to as an *open-hearth shop*. These very large furnaces are installed close together so that they may be serviced by the same units of highly specialized auxiliary equipment.

Because open-hearth shops are so large, it is difficult to obtain meaningful photographs. The arrangement can be likened to that of a split-level house with the entrance on the top floor and with a bottom level exit and heating system.

The charging side of the furnace is on the upper level and faces a charging floor full of highly specialized equipment. The most notable piece of equipment is a charging machine, which is electrically operated on a very broad gage track. Between the door side of the furnace and the charging machine is another track system. Special cars containing boxes of raw materials are run into the open-hearth shop on this track and are parked in position near the furnace doors. The charging machines are equipped to pick up the boxes of raw materials from their buggies and to thrust them in through the temporarily opened doors of the furnace, empty them by overturning them and then put them back on to their buggies. The empty charging boxes are then hauled away and refilled in preparation for the next charge.

In the rafters above the tracks for raw material, buggies, and charging machines are enormously strong overhead cranes, also running on tracks extending the full length of the shop. This crane system is used in many ways, but primarily to carry ladles of molten iron from a blast furnace or a mixer to the open hearths. Special troughs are wheeled into place in front of a furnace door and the huge ladles of molten iron are tilted so that a stream of the white-hot metal pours from the lip of the ladle through the trough and into the furnace. These paragraphs are devoted primarily to the layout of the shop; the actual processes of operation will be described later.

The only other major elements on the charging-floor side of the open-hearth furnace are the control areas, which contain panels permitting operators to control most of the processes at a safe distance from the heat and moving equipment.

The dish-shaped hearth of an open-hearth steelmaking furnace is quite shallow. The walls containing the charging doors are vertical and each unit is covered by an arched refractory roof. On the side of the furnace opposite the charging floor is a *taphole,* which is so arranged that molten steel can rush by gravity through a spout into a large ladle on the pouring floor or pit side of the open-hearth shop, which is at a considerably lower level than the charging floor.

A very considerable portion of an open-hearth facility is not visible at all. Brick *checker chambers* are located at both ends of each furnace below the level of the charging floor. The bricks in these chambers are arranged to leave a great number of passages through which the hot waste gases from the furnace pass and heat the brickwork prior to going through the gas cleaning equipment. Later, the flow of gas is reversed, and the atmospheric air supporting combustion in the furnace passes through the heated bricks and is heated on its way to the hearth.

Today, practically all open-hearth furnaces have been converted to use oxygen. The gas is fed into the open hearth through the roof by means of retractable lances. The use of gaseous oxygen in open hearth increases temperatures and thereby speeds up the melting process.

Each open-hearth shop is supervised by a melter supervisor. This person is in charge of all furnaces in the shop and their crews. He or she operates the controls, directs any repairs to the furnace in operation, and reports any variations from standard procedure. There are many variables in open-hearth steelmaking practice, the most significant being the ratio of pig iron to scrap. For the purpose of this book, the rather common "fifty-fifty" practice is chosen for description. In this practice the limits of molten pig iron are usually 45 to 55 percent and the charge may include relatively small amounts of solid pig iron or iron scrap generated within the plant. Limestone and some ore agglomerates are also charged.

Assuming that the raw steel has been tapped from an open-hearth furnace, the first step in making a new heat of steel is to examine the interior of the furnace carefully to determine if any damage has been done to the hearth or roof. If so, a special patching gun may be wheeled into position and the damage patched by literally shooting a refractory material into the holes or cracks.

As the furnace is being prepared, buggies containing boxes of raw materials are rolled into position next to the furnace. The long-armed charging machine picks up boxes of limestone and fairly light steel scrap. One by one these boxes are thrust through the furnace doors and dumped. The flame of burning fuel oil, tar, or gases shoots from one end of the furnace across the solid materials in the hearth, partially melting them. (Most of the combustion gases are collected at the other end of the furnace and used to heat the checker brick.) Molten iron is then poured into the furnace.

All of this is much more complex than it sounds, because timing is extremely important. It is necessary to have the solid materials at such a temperature and degree of oxidation that, on the one hand, the molten pig iron will not be chilled by the scrap and, on the other hand, the oxidation of the metalloids of the pig iron will not be delayed by insufficient oxygen support from the oxidized scrap. The present-day usage of oxygen roof lances has greatly aided this critical phase of charging.

The chemical reactions for making steel in an open-hearth furnace relate to the removal of carbon, manganese, phosphorus, sulfur, and silicon from the metallic bath. First, silicon and manganese are oxidized and become part of the slag, which is a layer of molten limestone and other materials floating on top of the molten steel. Next, the oxidation of carbon speeds up, creating carbon monoxide gas, which causes agitation as it leaves the metal. Eventually, phosphorus and sulfur are transferred to the slag.

The agitation of the metal bath caused by carbon monoxide is called the *ore boil*. The more violent turbulence caused by calcination of limestone is called the *lime boil*. After both "boils" have subsided, the refining phase of open-hearth steelmaking begins, with the aim of lowering the phosphorus and sulfur contents to meet end-product specifications, controlling the carbon content, and generally bringing the composition and grade of the steel to predetermined levels.

The manufacture of steel in open-hearth furnaces takes much more time per ton than it does by the basic oxygen process, but it affords very good control of chemical analysis and is capable, in some instances, of making very large batches—sometimes 600 tons (540,000 kg) of steel of a given analysis in a single heat.

When a heat of steel is ready to be tapped, a member of the furnace crew, working from the tapping side, breaks out the clay plug and refractory that has closed the taphole since before the furnace was charged. The operation may involve the use of a special explosive charge and is undertaken following the best safety practice.

The highest level of the taphole is located at the lowest part of the hearth and slopes down to a spout. Most of the steel surges out of the furnace into a large ladle before any of the slag floating on top of the molten bath appears. The late appearance of slag permits alloy and recarburized and deoxidized materials to be added to the steel in the ladle. The ladle is so placed beneath the spout that the stream of molten steel is given a swirling motion that tends to mix and make more homogeneous the metal and the additions.

The capacity of the ladle is matched to the amount of steel in the furnace. When the slag comes through the taphole, a layer of it is permitted to lie on top of the molten metal as a covering while the remainder overflows through a special notch into a *slag thimble,* a small vessel placed next to the steel ladle.

Some very large open-hearth furnaces are provided with two tapholes, therefore requiring two ladles on the pit-side floor.

Bessemer Process

In the Bessemer process (Figure 11-3) the impurities in pig iron are removed by oxidation, by finely divided air currents blowing through a bath of molten iron contained in a vessel called a *converter.*

The Bessemer converter consists of a heavy steel pear-shaped shell with refractory brick. It is supported by two trunnions upon which it can rotate 180°. The upper portion

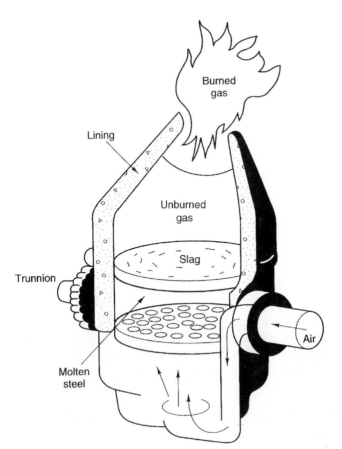

Figure 11-3 Bessemer Process
(courtesy of AISI)

can be either concentric or eccentric. The bottom of the converter is pierced with a large number of small holes, called tuyeres, through which the air blast is forced by means of a blower from the windbox up through the molten metal, oxidizing the impurities.

Converters generally have a capacity of 1 to 40 tons, with the average around 25 tons. For a small converter of 15-ton capacity, the clear-mouth opening is 2 to 2.5 ft. (0.6 to 0.75 m) in diameter with the inside cylindrical diameter at the bottom opening 8 ft. (2.4 m) and the height of the converter approximately 15 ft. (4.5 m).

The lining of the converter is usually 12 to 15 in. (0.3 to 0.33 m) thick and is made of a refractory material of strongly acidic character, with silica being the primary constituent. This lining lasts for several months before it has to be replaced.

The bottom is lined with 24 to 30 in. (0.6 to 0.75 m) of damp siliceous material bound together with clay in which the tuyere bricks are set. Each tuyere brick is about 30 in. long and has about 10 blast holes, each approximately 0.25 in. (0.6 cm) in diameter. Tuyere bricks have a useful life of about 1 month.

In the operation of the Bessemer process, the converter is tilted to a horizontal position to receive a charge. The blast is turned on after charging and before righting, to prevent the metal from entering the tuyere.

As soon as the blow is on, the silicon and manganese begin to burn to form oxides and thus are reduced to traces before the oxidation of the carbon becomes appreciable. As soon as the carbon is burned out, the converter is turned down, the blasts are shut off, and the re-carburizer is added.

This method of making steel is of little importance today because of the advent of better production facilities.

Electric Furnace

Electric arc furnaces (Figure 11-4) have a long history of producing alloy, stainless, tool, and other specialty steels. More recently, operators have also learned to make larger heats of

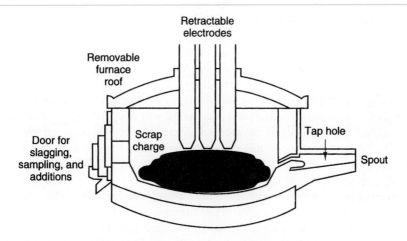

Figure 11-4   Electric Arc Furnace

carbon steels in these furnaces. Therefore, the electric steelmaking process is presently becoming a high-tonnage producer.

Electric arc furnaces are shallow steel cylinders lined with refractory brick. In the first half of the twentieth century most electric furnaces were loaded, or changed, with scrap iron and steel through a door in the side of the cylinder. However, today, the entire roof of an electric furnace is mounted on cantilevered steel beams so that it can be lifted and swung to one side. Thus, electric furnaces are now charged in one operation from buckets or other containers brought in by overhead cranes.

The roof of an electric furnace is pierced so that three carbon or graphite electrodes can be lowered into the furnace, and these electrodes give the electric arc furnace its name, because, in operation, the current arcs from one electrode to the metallic charge and then from the charge to the next electrode. This provides intense heat.

Opposite one of the doors is a tapping spout through which the molten steel is poured into a ladle. The entire furnace is mounted on "rockers" so that it can be tilted to permit the molten steel to emerge through the spout.

From this description and the accompanying diagram it can be seen that the charge for electric furnace steelmaking is entirely solid. The principal metallic charge is steel scrap. When a complex specialty steel is to be made, the scrap is carefully selected so that no unwanted elements are present. Pig iron is charged into some electric furnaces and, more recently, prereduced iron ore, which may contain up to 98 percent of the element iron. Such ore, under these conditions, is handled as a primary metallic source.

One of the major factors in making electric furnaces economically competitive as high-tonnage producers of high-carbon steel is the increased size of the units. Most experienced operators remember when a furnace capable of producing 50 tons of steel was considered large. Today, some furnaces average 300 tons per heat.

A relatively few years ago, new electric arc furnaces customarily operated on 25,000 to 35,000 kilovolt-amperes, compared with up to 80,000 kVA for today's units. Electric furnace steelmakers figure their electricity costs on the basis of kilowatt-hours per ton of raw steel. As power input and furnace size increase, the time required to produce a heat of steel decreases. Therefore, the kilowatt-hours per ton also decrease. As indicated, this economy has thrust electric arc furnaces into the high-tonnage production picture, particularly in areas where molten iron is not readily available.

The charge to most electric furnaces is largely scrap with small amounts of burned lime and mill scale. The scrap is segregated (separated) into stockpiles of identified grades, a procedure that is necessary for several reasons. Primary among these is the matter of economics. It would be wasteful to use valuable alloy steel scrap as a material for making common grades of carbon steel. The most economical way of operating an electric furnace is to assure, in advance, that only the elements desired for a special heat are introduced to the furnace. This means that the preliminary operations in collecting, sorting, and preparing scrap are extremely important. Steelmakers are very concerned with the recycling aspect of environmental quality control. However, this does not mean that any classification of steel scrap can be put back into a furnace without further treatment, and this is the case with electric arc furnaces. The fragmentation of old automobiles and the recycling of "tin cans" and other steel-base containers cannot make up too large a portion of the charge. These items contain contaminants such as copper and tin that might spoil a new heat of steel.

Once the sorted scrap has been placed into the hearth of the furnace, ferroalloys and other elements that are not easily oxidized may be charged in the furnace prior to melting down. If the metallic charge is too low in carbon, additions of coke or of scrap electrodes are included with the scrap. In all instances, the amount of carbon charged into the electric arc furnace is higher than is required to produce the desired chemical analysis in the finished steel.

The top of the furnace is now swung into position, the electrodes are lowered through the roof, and the electric power is turned on. The heat resulting from the arcing between electrodes and metal begins to melt the solid scrap. The electrodes, in some cases 2 ft thick and 24 ft long, turn white-hot to a considerable distance above the arc. As the charge melts, they may rise and fall vertically, a process called *searching*.

At a given point, iron ore is added to reduce the carbon content and cause a "bubbling" or boiling action. This is one of the most important factors in the production of high-quality steel.

Limestone and flux are charged on top of the molten bath. Through a chemical interaction, impurities in the steel rise in the molten slag, which floats on top of the melt. When the desired product is a carbon steel, a *single-slag* practice is used. This means that the furnace is tilted slightly and much of the slag is raked off through the charge door. In the *double-slag* method, an oxidizing slag is first formed, then raked off, and an additional slag is formed for reducing purposes.

The direct use of oxygen gas is extremely important in modern practice. It is of great value in the rapid removal of carbon from the bath.

In the double-slag method, the original slag is removed from the surface of the bath by cutting off the power to the electrodes, back-tilting the furnace slightly, and then pouring the slag out through the charge door into a slag thimble. The original slag must be removed thoroughly to prevent delay in making up the second slag. This might cause the reversion of some elements from slag to metal. Once the first slag is removed, additional lime and iron ore can be added.

The steel should not be held under the second slag any longer than is absolutely necessary. As soon as the results of the last analysis are reported, additions are made to the bath to adjust the carbon and alloying element content. When all of the additions are in solution, ferrosilicon may be added and aluminum thrust through the slag into the bath to control the grain size of the steel.

When the chemical composition of the steel meets specifications, the furnace (roof, electrodes, and all) is tilted forward so that molten metal may pour out the taphole through the spout. The slag comes after the steel and serves as an insulating blanket during tapping into a ladle.

Basic Oxygen Process

The basic oxygen steelmaking process (Figure 11-5) uses as its principal raw material molten pig iron from a blast furnace. The other source of metal is scrap. Lime, rather than limestone, is the fluxing agent. As the name implies, heat is provided by the use of oxygen.

Although the basic oxygen steel production in the United States was first reported by the American Iron and Steel Institute in 1954, the modern technique was pioneered in Austria shortly after World War II, and the concept of using oxygen in a pneumatic steelmaking

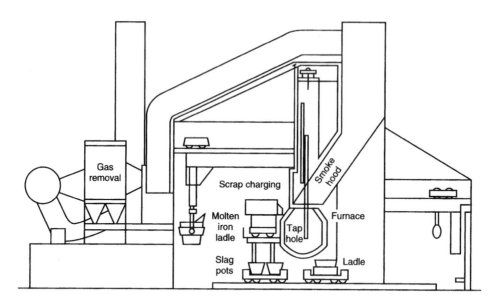

Figure 11-5 Basic Oxygen Steelmaking Process

process originated with Sir Henry Bessemer in the mid-1800s, at the time when he was developing the process now bearing his name. Bessemer recognized that blowing pure oxygen into his converter would be advantageous, but the technology for the bulk production of oxygen had not yet been conceived.

The first European basic oxygen furnaces were too small to be of commercial value in this country. Much of the developmental work in adapting the processes to big furnaces, containing over 300 tons of molten metal, was done in the United States.

The basic oxygen furnace is a steel shell lined with refractory materials. The body of the furnace is cylindrical, the bottom is slightly cupped, and the top is shaped like a truncated cone with its open base set on top of the cylinder. The narrow part of the truncated cone is open to receive raw materials and a jet of oxygen and also to permit dirty waste gases from the steelmaking process to escape into extensive air treatment facilities. The entire furnace is supported on horizontal trunnions so that it can be tilted.

Usually, these furnaces are installed in pairs so that one of them can be making steel while the other is being filled with raw materials. A basic oxygen furnace (BOF) may produce batches of over 300 tons in 45 minutes as against 5 to 8 hours for the older open-hearth process. Most grades of steel can now be made in BOFs, although this was not true in their early history. Even today, many foreign BOFs concentrate on making only a few uncomplicated high-tonnage steels, whereas in the United States it is the practice to produce a wide range of compositions.

The first step for making a heat of steel in a BOF is to tilt the furnace and charge it with steel scrap. This is done by swinging the furnace on its trunnions through an arc so that the open top can be reached by a charger that runs on rails at an appropriate level. The charger looks like a combination small railroad car and dump truck. The boxlike part of the charger is filled with carefully graded steel scrap and is carried by the railcar to a position where it can dump its contents into the tilted furnace. Pistons, or jointed arms, then lift and tilt the box so that the material slides out the open end into the preheated furnace. The furnace is rotated forward to distribute the scrap over the furnace bottom.

Immediately following the scrap charge, an overhead crane presents a ladle of molten iron from a blast furnace or from a holding device called a *mixer*. The iron, accounting for 65 to 80 percent of the charge, is also poured into the top of the tilted furnace.

As soon as the furnace with its charge of scrap and molten iron is in a vertical position, the oxygen lance is lowered and the oxygen is turned on to a flow of up to 6000 ft^3 (170 m^3) per minute at a pressure of up to 160 psi (480 kg/cm^2). The tip of the water-cooled lance may be about 6 ft above the metal surface when it is locked into place.

In a very short period of time, ignition causes increased heat and provides correct conditions for adding lime, fluorspar, and sometimes scale via a retractable chute to the metallic charge.

From that point on, the blowing procedure is uninterrupted. Oxygen combines with carbon and other unwanted elements, eliminating those impurities from the molten charge and converting it to steel. The lime and fluorspar help to carry off the impurities as a floating layer of slag on top of the metal, which is now entirely molten. In this function, lime is usually consumed at a rate of about 150 lb (680 kg) per ton of raw steel produced.

A BOF shop must be designed so that it is possible for the oxygen lance to be lowered into the vessel through the open top while that aperture is also hooded by the open end of the ductwork which carries away the waste gases. These gases are conducted to air treatment facilities, which are among the most complex and effective of the many pollution control systems employed by the iron and steel industry. In fact, the gas cleaning equipment is so complex that it may account for one-third or more of the overall structure of a BOF shop.

Experienced operators of BOFs can identify the completion of the steelmaking process by a variety of signs, including a decrease in the flame, a change in the wound level, a reading of the amount of oxygen used as indicated by the composition of gases emitted from the furnace, and a consideration of the blowing time for iron of a given composition.

When the batch of steel is complete, the clamps on the lance are released and the lance is retracted through the hood. The furnace is then rotated back toward the charging floor until the slag floating on top of the metal is even with the lip at the top of the furnace. Special equipment is then used to determine the temperature of the bath, which may be 3000°F (1649°C) for many of the more common types of steel. Once a correct temperature reading has been obtained, it may be necessary to test for carbon content. When both temperature and carbon content are acceptable, the furnace is rotated away from the charging platform past the vertical position and onto the opposite tilt. In this position the taphole, set into the cone-shaped portion of the furnace top, is aimed at a ladle that receives the molten steel. The slag, which floats on top of the steel, is caused to stay above the taphole by the progressive tilt of the furnace. Alloys are added to the ladle of steel, often by chutes extended from above the teeming floor.

The ladle into which the metallic contents of the furnace have been teemed is usually mounted on a railcar that is removed to a position where an overhead crane can lift it. The overhead crane carries the ladle of molten steel to a position where it can be poured into ingot molds or into a strand casting machine for solidification.

Meanwhile, the molten slag remaining in the furnace is emptied into a receptacle which is carried away for disposal. Except for possible minor refractory lining repair, the BOF is then ready to be recharged immediately. At approximately the same time, the other furnace in a pair should be ready to begin the steelmaking process.

STRUCTURE OF IRON AND STEEL

Carbon Steel

Carbon steel is an alloy of iron and carbon. The carbon atoms actually replace or enter into solution among the lattice structure of the iron atoms and limit the slip planes in the lattice structure. The amount of carbon within the lattice determines the properties of the steel.

Alloys containing less than 0.008 percent carbon are classed as irons. Steel is an iron–carbon alloy in which the carbon content is less than 2.0 percent. These steel products, including structural steel and reinforcing steel, can be rolled and molded into a shape. However, as the carbon content goes above 2.0 percent, the material becomes increasingly hard and brittle. Thus, cast iron has a carbon content above 2.0 percent. *High-strength steels* are alloys containing less than 0.8 percent carbon (the eutectoid composition) and are sometimes referred to as *hypoeutectoid steels. Structural steels* are alloys containing less than 2.0 percent but more than 0.8 percent carbon and are referred to as *hypereutectoid steels.* Wrought iron is a combination of iron and slag.

Phase Diagrams

A typical phase diagram of iron–carbon is shown in Figure 11-6. This equilibrium diagram is plotted for amounts of carbon up to 5 percent, which is sufficient for practical purposes. The diagram at times may be extended up to 6.67 percent of carbon, as this corresponds to 100 percent cementite. *Cementite,* a compound of iron and carbon, is a high-carbon steel that exists in a stable phase as iron carbide (Fe_3C). Cementite contains 6.67 percent carbon and 93.33 percent iron and is very hard and brittle. Other terms of importance shown on the phase diagram of Figure 11-6 are austenite, eutectic, eutectoid, ferrite, graphite, and pearlite. These terms are used to define the various stages or phases of iron–carbon alloys. *Austenite* is gamma iron with carbon in solution. The *eutectic* of the carbon–steel alloy is the combination that melts at the lowest temperature. Thus, according to Figure 11-6, the eutectic at 4.3 percent carbon melts at 2060°F (1130°C) and will contain a solid solution of austenite and a solid cementite. Below the 1330°F (723°C) mark, the solution changes to cementite and pearlite. The *eutectoid* in an equilibrium diagram for a solid solution is the point at which the solution on cooling is converted to a mixture of solids. Pearlite is a eutectoid and changes to a solid at 1330°F (723°C). *Ferrite* is iron that has not combined with carbon in pig iron as steel. This fact allows the steel to be cold-worked. *Graphite* (black lead) occurs as small flakes of carbon that become mixed with the steel. *Pearlite* is a lamellar aggregate of ferrite and cementite often occurring in carbon steels and in cast irons.

Properties of Cementite, Ferrite, and Pearlite

Cementite is a hard and very brittle material with a Brinell hardness of 650 and a diamond hardness of 760. Ferrite is a soft and ductile material with a Brinell hardness of 90 and a diamond hardness of 170. It has a 40 percent elongation in 2-in. (5.08-cm) specimen. Pearlite is harder and less ductile than ferrite but is softer and less brittle than cementite. Pearlite has

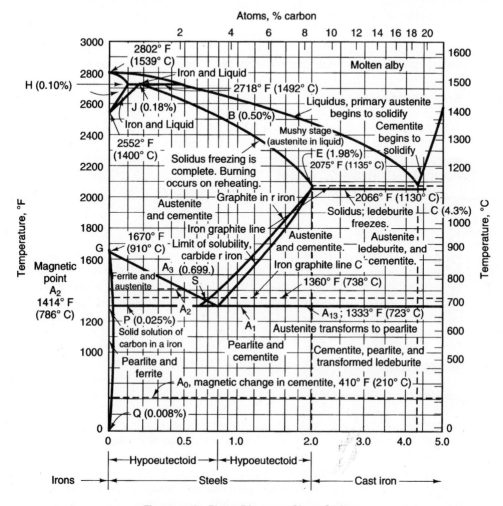

Figure 11-6   Phase Diagram of Iron–Carbon

a Brinell hardness of 275 and a diamond hardness of 300. It has an elongation of 15 percent in a 2-in. (5.08-cm) specimen.

Impurities in Steel

The principal impurities in steel are silicon, phosphorus, sulfur, and manganese.

The amount of silicon in structural steel is less than 1 percent and forms a solid solution with iron. This small amount of silicon increases both the ultimate strength and the elastic limit of steel with no appreciable change in its ductility. Silicon may further prevent the solution of carbon in iron.

The phosphorus in steel is in the form of iron phosphide (Fe_3P). For low-grade structural steel the amount of phosphorus is about 0.1 percent and decreasing to 0.05 percent for high-grade structural steel. Tool steel is approximately 0.02 percent phosphorus.

Sulfur in steel combines with the iron to form iron sulfate (FeS). This compound has a low melting point and segregation may take place.

Manganese has an affinity for sulfur and combines with such as well as with other impurities to form slag. In other words, manganese acts like a cleanser. Manganese is used to harden steels.

Structural Steel

Table 11-1 illustrates the uses of structural steels. It becomes obvious that the steel must have strength, toughness, and, above all, durability. The requirements for structural steel are presented in Table 11-2. In most cases the maximum percent of carbon is less than 0.27, but most structural steels average 0.2 percent.

Reinforcing Steel

As was explained in Chapter 4, concrete exhibits great compressive strength but little tensile or flexural strength. Thus, deformed bars of structural steel are embedded in the concrete to take up the tensile or flexural forces. These deformed bars have been developed in such a way as to force the concrete between the deformations such that failure in shear will occur before slippage. Table 11-3 lists the ASTM standard-size reinforcing bars as to their weights and dimensions. ASTM has also standardized reinforcing steel according to its yield point, as shown in Table 11-4.

HEAT TREATMENTS

Hardening or Quenching

Whenever a solid solution, such as steel, decomposes because of a falling temperature into the eutectoid, the decomposition may be more or less completed, depending on the cooling rate. This process is utilized in the hardening of steel. If the steel is cooled slowly, the changes just discussed will occur; however, if the steel is cooled too quickly, decomposition into the eutectoid will be prevented and a structure called *martensite* is produced rather than steel. Martensite is a hard structure with little ductility (a necessary property in steels).

The successful hardening of steel may be achieved by the application of three general principles:

1. Steel should always be annealed before hardening, to remove forging or cooling stains.
2. Heating for hardening should be slow.
3. Steel should be quenched on a rising, not on a falling temperature.

Quenching media vary, but are basically of three types:

1. Brine for maximum hardness.
2. Water for rapid cooling of the common steels.
3. Oils (light, medium, or heavy) for use with common steel parts of irregular shapes or for alloy steels.

TABLE 11-1 STRUCTURAL STEEL[a]

ASTM Designation	Product	Use
A36	Carbon-steel shapes, plates, and bars	Welded, riveted, and bolted construction; bridges, buildings, towers, and general structural purposes
A53	Welded or seamless pipe, black or galvanized	Welded, riveted, and bolted construction; primary use in buildings, particularly columns and truss members
A242	High-strength, low-alloy shapes, plates, and bars	Welded, riveted, and bolted construction; bridges, buildings, and general structural purposes; atmospheric-corrosion resistance about four times that of carbon steel; a weathering steel
A245	Carbon-steel sheets, cold- or hot-rolled	Cold-formed structural members for buildings, especially standardized buildings; welded, cold-riveted, bolted, and metal-screw construction
A374	High-strength, low-alloy, cold-rolled sheets and strip	Cold-formed structural members for buildings, especially standardized buildings; welded, cold-riveted, bolted, and metal-screw construction
A440	High-strength shapes, plates, and bars	Riveted or bolted construction; bridges, buildings, towers, and other structures; atmospheric-corrosion resistance double that of carbon steel
A441	High-strength, low-alloy manganese–vanadium steel shapes, plates, and bars	Welded, riveted, or bolted construction but intended primarily for welded construction; bridges, buildings, and other structures; atmospheric-corrosion resistance double that of carbon steel

(Continues)

TABLE 11-1   STRUCTURAL STEEL[a] (CONTINUED)

ASTM Designation	Product	Use
A446	Zinc-coated (galvanized) sheets in coils or cut lengths	Cold-formed structural members for buildings, especially standardized buildings; welded, cold-riveted, bolted, and metal-screw construction
A500	Cold-formed welded or seamless tubing in round, square, rectangular, or special shapes	Welded, riveted, or bolted construction; bridges, buildings, and general structural purposes
A501	Hot-formed welded or seamless tubing in round, square, rectangular, or special shapes	Welded, riveted, or bolted construction; bridges, buildings, and general structural purposes
A514	Quenched and tempered plates of high yield strength	Intended primarily for welded bridges and other structures; welding technique must not affect properties of the plate, especially in heat-affected zone
A529	Carbon-steel plates and bars to 1/2 in. thick	Buildings, especially standardized buildings; welded, riveted, or bolted construction
A570	Hot-rolled carbon-steel sheets and strip in coils or cut lengths	Cold-formed structural members for buildings, especially standardized buildings; welded, cold-riveted, bolted, and metal-screw construction
A572	High-strength, low-alloy columbian-vanadium steel shapes, plates, sheet piling, and bars	Welded, riveted, or bolted construction of buildings in all grades; welded bridges in grades 42, 45, and 50 only
A588	High-strength, low-alloy steel shapes, plates, and bars	Intended primarily for welded bridges and buildings; atmospheric-corrosion resistance about four times that of carbon steel; a weathering steel
A606	High-strength, low-alloy hot- and cold-rolled sheet and strip	Intended for structural and miscellaneous purposes where savings in weight or added durability are important

After E. H. Gaylord, Jr., and C. N. Gaylord, *Design of Steel Structures*, McGraw-Hill Book Company, New York, 1972.

TABLE 11-2 ASTM STRUCTURAL STEEL REQUIREMENTS

Product:	Shapes[a]	Plates					Bars			
Thickness [in. (mm)]:	All	To $\frac{3}{4}$ (19), incl	Over $\frac{3}{4}$ to $1\frac{1}{2}$ (19 to 38), incl.	Over $1\frac{1}{2}$ to $2\frac{1}{2}$ (38 to 64), incl.	Over $2\frac{1}{2}$ to 4 (64 to 102), incl.	Over 4 (102)	To $\frac{3}{4}$ (19), incl.	Over $\frac{3}{4}$ to $1\frac{1}{2}$ (19 to 38), incl.	Over $1\frac{1}{2}$ to 4 (102), incl.	Over 4 (102)
Carbon, maximum (%)	0.26	0.25	0.25	0.26	0.27	0.29	0.26	0.27	0.28	0.29
Manganese, (%)	—	—	0.80–1.20	0.80–1.20	0.85–0.20	0.85–0.20	—	0.60–0.90	0.60–0.90	0.60–0.90
Phosphorus, max. (%)	0.04	0.04	0.04	0.04	0.04	0.04	0.04	0.04	0.04	0.04
Sulfur, max. (%)	0.05	0.05	0.05	0.05	0.05	0.05	0.05	0.05	0.05	0.05
Silicon (%)	—	—	—	0.15–0.30	0.15–0.30	0.15–0.30	—	—	—	—
Copper when copper steel is specified, min. (%)	0.20	0.20	0.20	0.20	0.20	0.20	0.20	0.20	0.20	0.20

(Continues)

TABLE 11-2 ASTM STRUCTURAL STEEL REQUIREMENTS

		Minimum Tensile Properties of Structural Steels	
ASTM Designation	Yield (ksi)	Strength (ksi)	Elongation (8 in. Unless Noted) (%)
Carbon steels			
A36	36	58–80	20
A529	42	60–85	19
High strength steels			
A242, A440, A441			
To $\frac{3}{4}$ in. thick	50	70	18
Over $\frac{3}{4}$ in. to $1\frac{1}{2}$ in.	46	67	19
Over $1\frac{1}{2}$ in. to 4 in.	42	63	16
A572			
Grade 42, to 4 in. incl.	42	60	20
Grade 45, to $1\frac{1}{2}$ in. incl.	45	60	19
Grade 50, to $1\frac{1}{2}$ in. incl.	50	65	18
Grade 55, to $1\frac{1}{2}$ in. incl.	55	70	17
Grade 60, to 1 in. incl.	60	75	16
Grade 65, to $\frac{1}{2}$ in. incl.	65	80	15
A588			
To 4 in. thick	50	70	19–21[b]
Over 4 in. to 5 in.	46	67	19–21[b]
Over 5 in. to 8 in.	42	63	19–21[b]
Quenched and tempered steels			
A514			
To $2\frac{1}{2}$ in. thick	100	115–135	18[b]
Over $2\frac{1}{2}$ in. to 4 in.	90	105–135	17[b]

[a] Manganese content of 0.85–1.35% and silicon content of 0.15–0.30% is required for shapes over 426 lb/ft.
[b] In 2 in.

TABLE 11-3   ASTM STANDARD-SIZE REINFORCING BARS[a]

Bar Designation Number[c]	Nominal Weight (kg/m)	Nominal Dimensions[b]			Deformation Requirements (mm)		
		Diameter (cm²)	Cross-Sectional Area (mm)	Perimeter (mm)	Maximum Average Spacing	Minimum Average Height	Maximum Gap (Chord $12\frac{1}{2}$ % of Nominal Perimeter)
3	0.560	9.52	0.71	29.9	6.7	0.38	3.5
4	0.994	12.70	1.29	39.9	8.9	0.51	4.9
5	1.552	15.88	2.00	49.9	11.1	0.71	6.1
6	2.235	19.05	2.84	59.8	13.3	0.96	7.3
7	3.042	22.22	3.87	69.8	15.5	1.11	8.5
8	3.973	25.40	5.10	79.8	17.8	1.27	9.7
9	5.059	28.65	6.45	90.00	20.1	1.42	10.9
10	6.403	32.26	8.19	101.4	22.6	1.62	11.4
11	7.906	35.81	10.06	112.5	25.1	1.80	13.6
14[d]	11.384	43.00	14.52	135.1	30.1	2.16	16.5
18[d]	20.238	57.33	25.81	180.1	40.1	2.59	21.9

[a]Based on ASTM 6615, 6616, 6617.
[b]The nominal dimensions of a deformed bar are equivalent to those of a plain round bar having the same weight per foot as the deformed bar.
[c]Bar numbers are based on the number of eighths of an inch included in the nominal diameter of the bars.
[d]Available in billet steel only.

All hardened steel is in a state of strain, and steel pieces with sharp angles or grooves sometimes crack immediately after hardening. For this reason, tempering must follow the quenching operation as soon as possible.

Tempering

Tempering of steel is defined as the process of reheating a hardened steel to a definite temperature below the critical temperature, holding it at that point for a time, and cooling it, usually by quenching for the purpose of obtaining toughness and ductility in the steel.

Annealing

Annealing has basically the opposite objective of hardening. *Annealing* has the process of heating a metal above the critical temperature range, holding it at that temperature for the proper period, and then slowly cooling. During the cooling process, pearlite, ferrite, or cementite form. The objectives of annealing are the following:

1. To refine the grain.
2. To soften the steel to meet definite specifications.
3. To remove internal stresses caused by quenching, forging, and cold working.

TABLE 11-4 KINDS AND GRADES OF REINFORCING BARS

Type of Steel and ASTM Specification Number	Grade Designation	Size Numbers	Yield Point Minimum (Mpa)	Tensile Strength Minimum (MPa)	Elongation in 203-mm Minimum (%)	Diameter Bend Test Pin[a]
Billet steel	40	3	276	483	11	4d
A615		4, 5			12	4d
		6			12	5d
		7			11	5d
		8			10	5d
		9			9	5d
		10			8	5d
		11			7	5d
		14, 18			—	None
	60	3, 4, 5	415	621	9	4d
		6			9	5d
		7, 8			8	6d
		9, 10, 11			7	8d
		14, 18			7	None
Rail Steel	50	3	345	550	6	6d
A616		4, 5, 6			7	6d
		7			6	6d
		8			5	6d
		9, 10, 11[b]			5	8d
	60	3	415	620	6	6d
		4, 5, 6			6	6d
		7			5	6d
		8			4.5	6d
		9, 10, 11[b]			4.5	8d
Axle steel	40	3	275	480	11	4d
A617		4, 5			12	4d
		6			12	5d
		7			11	5d
		8			10	5d
		9			9	5d
		10			8	5d
		11			7	5d
	60	3, 4, 5			8	4d
		6			8	5d
		7			7	6d
		8	415	620	8	6d
		9, 10, 11			7	8d

[a]d, diameter of specimen.
[b]Number 11, 90° bend; all others 180°.

4. To change ductility, toughness, electrical, and magnetic properties.
5. To remove gases.

PHYSICAL PROPERTIES OF STEELS

In general, and as previously mentioned, three principal factors influence the strength, ductility, and elastic properties of steel:

1. The carbon content.
2. The percentages of silicon, sulfur, phosphorus, manganese, and other alloying elements.
3. The heat treatment and mechanical working.

Carbon

The various properties of different grades of steel are due more to variations in the carbon content of the steel than to any other single factor. Carbon acts as both a hardener and a strengthener, but at the same time it reduces the ductility.

Silicon, Sulfur, Phosphorus, and Manganese

The effect of silicon on strength and ductility in ordinary proportions (less than 0.2 percent) is very slight. If the silicon content is increased to 0.3 or 0.4 percent, the elastic limit and ultimate strength of the steel are raised without reducing the ductility. This is a procedure used for steel castings.

Sulfur within ordinary limits (0.02 to 0.10 percent) has no appreciable effect upon the strength or ductility of steels. It does, however, have a very injurious effect upon the properties of the hot metal, lessening its malleability and weldability, thus causing difficulty in rolling, called "red-shortness."

Phosphorus is the most undesirable impurity found in steels. It is detrimental to toughness and shock-resistance properties, and often detrimental to ductility under static load.

Manganese improves the strength of plain carbon steels. If the manganese content is less than 0.3 percent, the steel will be impregnated with oxides that are injurious to the steel. With a manganese content of between 0.3 and 1.0 percent, the beneficial effect depends upon the amount of carbon content. As the manganese content rises above 1.5 percent, the metal becomes brittle and worthless.

Effect of Heat Treatment upon Physical Properties

The effects of various heat treatments upon the mechanical properties of wrought or rolled carbon steels of various compositions are shown in Table 11-5.

Tensile Strength

The tensile strength and properties of various carbon steels as given by the ASTM are shown in Table 11-6. The modulus of elasticity of all grades varies from 28,000,000 to 30,000,000 psi (1.9×10^{11} to 2.1×10^{11} N/m^2).

TABLE 11-5   MECHANICAL PROPERTIES OF HEAT-TREATED WROUGHT OR ROLLED CARBON STEELS[a]

Composition (%)			Yield Point (psi)	Tensile Strength (psi)	Elongation in 2 in. (%)	Reduction of Area (%)	Brinell Hardness
Carbon	Manganese	Heat Treatment					
0.14	0.45	Hot-rolled	45,000	59,500	37.5	67.0	112
		Annealed	31,000	54,500	39.5	67.0	107
		Quenched in water	—	90,000	21.0	67.0	170
		Quenched in oil	56,500	71,500	34.0	75.5	134
0.32	0.5	Hot-rolled	49,500	75,500	30.0	51.9	144
		Annealed	41,000	70,000	30.5	51.9	131
		Quenched in water	—	135,000	8.0	16.9	255
		Quenched in oil	67,500	101,000	23.5	62.3	207
		Quenched in oil, tempered at 650°F	61,500	84,000	30.0	71.4	163
0.46	0.40	Hot-rolled	52,500	86,500	22.5	30.7	160
		Annealed4	8,000	79,500	28.5	46.2	153
		Quenched in water	—	220,000	1.0	0.0	600
		Quenched in oil	87,500	126,500	20.5	51.9	225
		Quenched in oil, tempered at 560°F	81,500	111,500	24.0	57.2	—
		Quenched in oil, tempered at 650°F	73,000	98,000	25.5	59.8	192
0.57	0.65	Hot-rolled	57,000	106,500	19.0	27.4	220
		Annealed	50,000	95,000	25.0	40.3	183
		Quenched in water	—	215,000	0.0	0.0	578
		Quenched in oil	105,000	152,000	16.5	40.3	311
		Quenched in oil, tempered at 460°F	97,500	145,000	16.0	46.2	293
		Quenched in oil, tempered at 650°F	79,500	113,000	24.0	62.3	228
0.71	0.67	Hot-rolled	66,000	128,000	15.0	20.5	240
		Annealed	46,500	111,500	16.5	24.0	217
		Quenched in oil	100,000	184,500	1.5	0.0	364
		Quenched in oil, tempered at 460°F	115,500	177,000	10.0	34.0	340
		Quenched in oil, tempered at 560°F	106,000	148,500	17.0	43.0	311
		Quenched in oil, tempered at 650°F	91,000	125,500	19.5	57.2	269

[a] Data from the American Society for Steel Treating.

TABLE 11-6 TENSILE PROPERTIES OF VARIOUS STEELS

Kind and Use of Steel	Tensile Strength (psi)	Yield Point, Min. (psi)	Elongation Min. in 8 in. (%)	Elongation Min. in 2 in. (%)	Reduction of Area, Min. (%)
Structural steel for bridges and buildings					
Structural	60,000–72,000	33,000	21	22	
Rivet	52,000–62,000	28,000	24		
Structural steel for ships					
Structural	58,000–71,000	32,000	21	22	
Rivet	55,000–65,000	30,000	23		
Carbon steel forgings for locomotives and cars					
Annealed or normalized	75,000 min.	37,500		20	33
Normalized, quenched, and tempered	115,000 min.	75,000		16	35
Carbon steel bolting material, heat-treated					
Grade BO	100,000 min.	75,000		16	45
Boiler and firebox steel					
Flange	55,000–65,000	0.5 u.t.s.[a]	1,500,000/u.t.s.	1,750,000/u.t.s.	
Firebox, grade A	55,000–65,000	0.5 u.t.s.	1,550,000/u.t.s.	1,750,000/u.t.s.	
Firebox, grade B	48,000–58,000	0.5 u.t.s.	1,550,000/u.t.s.	1,750,000/u.t.s.	
Boiler-rivet steel					
Grade A	45,000–55,000	0.5 u.t.s.	1,500,000/u.t.s.		
Grade B	58,000–68,000	0.5 u.t.s.	1,500,000/u.t.s.		
Billet-steel-concrete reinforcing bars					
Plain bars					
Structural grade	55,000–70,000	33,000	1,400,000/u.t.s.		
Intermediate grade	70,000–90,000	40,000	1,300,000/u.t.s.		
Hard grade	80,000 min.	50,000	1,100,000/u.t.s.		
Deformed bars					
Structural grade	55,000–75,000	33,000	1,200,000/u.t.s.		
Intermediate grade	70,000–90,000	40,000	1,100,000/u.t.s.		
Hard grade	80,000 min.	50,000	1,000,000/u.t.s.		
Concrete reinforcing bars from rerolled steel rails					
Plain bars	80,000 min.	50,000	1,000,100/u.t.s.		
Deformed bars	80,000 min.	50,000	1,000,000/u.t.s.		

[a] u.t.s., ultimate tensile strength.

Structural Steel

Notice that the carbon content is not specified. The reason for this is that the manufacturer is permitted to vary the carbon percent age to meet the tensile strength requirements of Table 11-6. The chemical compositions for structural steel used in bridges and buildings as given by the ASTM are listed in Table 11-7.

Torsional Shear

The strength of steel in torsional shear is shown in Table 11-8. The modulus of elasticity for shear is about 12,000,000 psi (8.2×10^7 N/m^2).

ALLOY STEELS

Alloy steels are steels that owe their distinctive properties to elements other than carbon. Common alloys include chromium, nickel, manganese, molybdenum, silicon, copper, vanadium, and tungsten.

These alloys can be classified into two groups: those that combine with the carbon to form carbides, such as nickel, silicon, and copper, and those that do not combine with carbon to form carbides, such as manganese, chromium, tungsten, molybdenum, and vanadium.

TABLE 11-7 CHEMICAL COMPOSITION OF STRUCTURAL STEEL FOR BRIDGES AND BUILDINGS, ASTM SPECIFICATIONS

Element	Ladle Analysis, Max. (%)	Check Analysis, Max. (%)
Open-hearth and electric-furnace structural and structural-rivet steel		
Phosphorus		
Acid process	0.06	0.075
Basic process	0.04	0.05
Sulfur	0.05	0.063
Copper (when specified)	0.020	0.18
Acid Bessemer structural steel[a]		
Phosphorus	0.11	0.138

[a]Not permitted in bridges or in building members subject to dynamic loads.

TABLE 11-8 STRENGTH OF STEEL IN TORSIONAL SHEAR

Class of Steel	Computed Extreme Fiber Stress (psi)	Shearing Modulus of Elasticity (psi)
Mild Bessemer	64,200	11,320,000
Medium Bessemer	68,300	11,570,000
Hard Bessemer	74,000	11,700,000
Cold-rolled	79,900	11,950,000

Alloys are added to steel for three principal reasons:

1. To increase hardness.
2. To increase the strength.
3. To add special properties, such as
   a. Toughness
   b. Improved magnetic and electrical properties
   c. Corrosion resistance
   d. Machinability

Chromium

Chromium is primarily a hardening agent and is generally added to steel in amounts of 0.70 to 1.20 percent, with a variation in carbon content of 0.17 to 0.55 percent. Its value is due principally to its property of combining intense hardness after quenching with very high strength and elastic limit. Thus, it is well suited to withstand abrasion, cutting, or shock. It does lack ductility, but this is unimportant in view of its high elastic limit. Chromium steels corrode less rapidly than do carbon steels.

Nickel–Chromium

Nickel–chromium steels, when properly heat-treated, have a very high tensile strength and elastic limit, with considerable toughness and ductility. The nickel content is usually 3.5 percent, with a carbon content ranging from 0.15 to 0.50 percent.

One very important property of nickel–chromium steels is that by adding aluminum, cobalt, copper, manganese, silicon, silver, or tungsten, stainless steel results.

Manganese

As previously indicated, manganese is present in all steels as a result of the manufacturing process. When the manganese is 1.0 percent or greater in solution with steel, it is considered an alloy. Manganese will add hardness to steel if used within the proper range.

Molybdenum

Molybdenum provides strength and hardness in steel. It inhibits grain growth on heating as a result of its slow solubility of austenite. When in solution in the austenite, it decreases the cooling rate and, therefore, increases the depth of hardening.

Silicon

Silicon is added to carbon steel for the purpose of deoxidizing. For this reason, silicon may be added in amounts of up to 0.25 percent. Silicon does not form carbides but does dissolve in the ferrite up to about 15 percent. Silicon decreases hysteresis and eddy-current losses, and thus is valuable for electrical machinery.

Vanadium

Vanadium is a powerful element for alloying in steel. It forms stable carbides and improves the hardenability of steels. Vanadium promotes a fine-grained structure and promotes hardness at high temperatures. The amount of vanadium present is 0.10 to 0.30 percent when used.

Copper

Copper increases the yield strength, tensile strength, and hardness of steel. However, ductility may be decreased by about 2 percent. The most important use of copper is to increase the resistance of steel to atmospheric corrosion.

Tungsten

Tungsten increases the strength, hardness, and toughness of steel. After moderately rapid cooling from high temperatures, tungsten steel exhibits remarkable hardness, which is still retained upon heating to temperatures considerably above the ordinary tempering heats of carbon steels. It is this property of tungsten that makes it a valuable alloy, in conjunction with chromium or manganese, for the production of high-speed tool steel.

NONFERROUS METALS

In this section, various nonferrous metals will be listed with only a brief statement; specific details will be omitted.

Basically, three groups of nonferrous metals exist. In the first group, those of greatest industrial importance, are aluminum, copper, lead, magnesium, nickel, tin, and zinc. The second group includes antimony, bismuth, cadmium, mercury, and titanium. The third and final group, important in that they are used to form alloy steels, includes chromium, cobalt, molybdenum, tungsten, and vanadium.

The nonferrous alloys of the greatest importance are alloys of copper with tin (bronzes), alloys of copper with zinc (brasses), and alloys of aluminum, magnesium, nickel, and titanium.

PROBLEMS

11.1. Explain the purpose and use of the blast furnace.

11.2. Discuss four manufacturing processes of steel.

11.3. Why is carbon so important in steel?

11.4. Define austenite, cementite, eutectic, eutectoid, ferrite, graphite, and pearlite.

11.5. Discuss the various heat-treatment processes.

11.6. What is tempering?

11.7. What factors influence the strength, ductility, and elastic properties of steel?

11.8. Discuss the importance of silicon, sulfur, phosphorus, and manganese in steel.

11.9. List and discuss the various alloy steels.

11.10. Discuss briefly the use of nonferrous metals that are alloyed with steel.

REFERENCES

American Iron and Steel Institute, *The Making of Steel*, Washington, D.C., 1970.

American Society for Metals, *Metals Handbook*, Cleveland, Ohio, 1948.

American Society for Testing and Materials, "Specifications for Structural Steel," *ASTM A-36, C1020.*

BULLENS, D. K., *Steel and Its Heat Treatment*, 5th ed., 3 vols., John Wiley & Sons, Inc., New York, 1948. See especially vol. 1, *Principles*, and vol. 2, *Tools, Processes, and Control.*

DIGGES, T. G., "Effect of Carbon on the Hardening of High Purity Iron Carbon Alloys," *Trans. Am. Soc. Met.*, 25 (1938).

GAYLORD, E. H., JR., and GAYLORD, C. N., *Design of Steel Structures*, McGraw-Hill Book Company, New York, 1972.

LIPSON, H., and PARKER, A. M. B., "The Structure of Martensite," *J. Iron Steel Inst. (Lond.)*, *149* (1944).

SISCO, F. T. *Modern Metallurgy for Engineers*, 2nd ed., Pitman Publishing Corp., New York, 1948.

TEICHERT, E. J., *Ferrous Metallurgy*, 2nd ed., McGraw-Hill Book Company, New York, 1944.

12 PLASTICS

A textbook on materials for civil and highway engineers would not be complete without a chapter on plastics, as plastics comprise an important group of materials of construction. Many plastic substances have combinations of properties that cannot be duplicated by other materials. In general, plastics exhibit a number of outstanding characteristics:

1. Lightness in weight (generally half as light as aluminum).
2. High dielectric strength (electrical insulation).
3. Low heat conductivity (heat insulation).
4. Special properties toward lights (colorability).
5. Extremely resistant toward chemicals.
6. Metal inserts may be molded into the plastic (since plastics are inert toward such materials).
7. Many high-quality products can be developed by using a lathe, sawing, punching, and drilling.

CLASSIFICATION

A *plastic* is a polymeric material (usually organic) of high molecular weight which can be shaped by flow. The term usually refers to the final product, with fillers, plasticizers, pigments, and stabilizers included (as opposed to the resin, the homogeneous polymeric starting material). Most plastics are synthetic organic compounds that derive their coherence and strength from large chain-linked macromolecules formed from one or more single molecules (monomers) to a macromolecule (polymer) by a chemical reaction called *polymerization*. Examples include polyvinyl chloride, polyethylene, and urea-formaldehyde. Most organic plastics contain a *binder,* which imparts the plastic properties to the composition. In addition, many plastics contain a *filler,* an inert extender that adds hardness, strength, and other desirable properties. In general, organic plastics can be divided into three general classifications:

1. Thermoplastics.
2. Thermosetting plastics.
3. Chemically setting plastics.

Table 12-1 shows the classification of organic plastics.

Thermoplastics

Thermoplastics are organic plastics, either natural or synthetic, which remain permanently soft at elevated temperatures. Upon cooling, they again become hard. These materials can be shaped and reshaped any number of times by repeated heating and cooling. Natural thermoplastics include asphalts, bitumen, pitches, and resin, to name some of the most familiar.

Thermosetting Plastics

Thermosetting plastics are organic plastics that were originally soft or soften at once upon heating, but upon further heating, they harden permanently. Thermosetting plastics are hardened by chemical changes due to heat, a catalyst, or to both. Thermosetting plastics remain hardened without cooling and do not soften appreciably when reheated. The most common thermosetting plastic is polyester.

Chemically Setting Plastics

Chemically setting plastics are those that harden by the addition of a suitable chemical to the composition just before molding or by subsequent chemical treatment following fabrication.

TYPES OF PLASTICS

Polymerization and Condensation

Polymerization involves unsaturated molecules that contain double or triple bonds between carbon atoms that are weaker than single bonds. Unsaturated molecules are unstable and they react in such a way as to break the multiple bond. The mechanism by which polymerization takes place is grouped into two categories:

1. Addition polymerides.
2. Condensation polymerides.

Addition polymerides are mixtures of polymers that have been formed by addition of like molecules. The molecules are added to increase the average molecular size and weight. This

TABLE 12-1   CLASSIFICATION OF ORGANIC PLASTICS

Plastic Binder

Thermoplastics

- **Natural**
  - **Resin binders**
    1. Rosin
    2. Copals
  - **Asphaltic binders**
    1. Asphalts
    2. Bitumens
    3. Gilsonite
    4. Glance pitch
    5. Pitches
- **Synthetic**
  - **Cellulose derivatives**
    1. Pyroxylin
    2. Cellulose
    3. Benzyl
    4. Ethyl
    5. Carboxymethyl
    6. Hydroxyethyl
    7. Butyrate
  - **Addition polymerides**
    1. Vinyl chloride
    2. Chlorovinyl
    3. Vinyl acetate
    4. Vinyl ethers
    5. Styrenes
    6. Acrylics
    7. Acrolein
    8. Methyl
  - **Condensation polymerides**
    1. Glycol-modified
    2. Acid-catalyzed
    3. Polyamide

Thermosetting plastics

- **Addition polymeride**
  1. Polyvinyl
- **Condensation polymeride**
  1. Phenol–formaldehyde
  2. Cast phenolics
  3. Transparent molded phenolics
  4. Resorcinol
  5. Acroite
  6. Phenolfurfural
  7. Urea
  8. Casein
  9. Solfonamide
  10. Polyester

Chemically setting plastics

1. Harvel resins
2. Cold
3. Casein

reaction occurs by the breaking of double bonds between atoms in monomers and the forming of two single bonds in their place.

Condensation polymerides are formed by chemical reactions in which two or more different molecules combine with the separation of water or other simple substances in the formation of resins. Condensation polymerides are produced by a by-product as well as the growing polymer molecules. The nonpolymerizable molecule is usually water or some simple molecule.

Copolymers are produced by mixing monomers and then polymerizing. In other words, copolymers are produced by the simultaneous polymerization of two or more chemically different monomers.

Thermoplastics

Properties and uses of common plastics of the thermoplastic type are shown in Table 12-2. Thermoplastic types of binders are divided into two groups: natural and synthetic. Of the *natural* type two categories exist: resin and asphaltic binders. The resin binders at this point in materials of construction are of little importance. The asphaltic binders were thoroughly discussed in Chapter 9.

The *synthetic* group consists of three categories: cellulose derivatives, addition polymerides, and condensation polymerides. *Cellulose* plastic is the term employed for plastics made from derivatives of cellulose, such as pyroxylin, cellulose acetate, benzyl cellulose, ethyl cellulose, carboxymethyl cellulose, hydroxyethyl cellulose, and cellulose acetate butyrate. Cellulose plastics are true thermoplastics and exhibit the greatest toughness and resilience of any of the plastics. They are used for objects having thin-walled sections where other plastics may be too brittle.

Addition and condensation polymerides are grouped together and referred to as *noncellulose* type. This group consist of vinyl chloride, chlorovinyl chloride, vinyl acetate, vinyl ethers, styrene, acrylics, acrolein, methyl methacrylate, cumarone–indene, glycolmodified alkyds, acid-catalyzed phenol–formaldehyde, and polyamide (nylon). The noncellulose plastics are all odorless, tasteless, nontoxic, and transparent. These plastics are strong, tough, and chemically inert with little water absorption. The noncellulose plastics are suited for electrical insulation or cable coatings, safety glass, nylon, and Saran (covering to prevent corrosion and chemical attack on pipes and tubing).

Thermosetting Plastics

Thermosetting plastics can be divided into two groups: addition polymerides and condensation polymerides. Addition polymerides consist of any polyvinyl types of resins; condensation polymerides, the largest group, consist of phenolformaldehyde, cast phenolics, transparent molded phenolics, resorcinol–formaldehyde, acroite, phenolfurfural, urea, casein–formaldehyde, sulfonamide resins, and polyesters.

Plastics of the thermosetting group have excellent mechanical and electrical properties and are highly resistant to heat (Table 12-3). These plastics are also highly resistant to

TABLE 12-2 PROPERTIES AND USES OF COMMON PLASTICS (THERMOPLASTICS)

Property	Shellac	Polyethylene	Polymono-chloro-trifluoro-ethylene	Vinylidene Chloride Molding	Polystyrene	Methyl methacryl-ate, Cast	Polyamide (Nylon) Molding	Cellulose Acetate Molding	Cellulose Nitrate (Pyroxylin)
Injection molding pressure (1000 psi)	1.00–1.20	8–15	20–60	10–30	10–30	—	10–25	8–32	
Specific gravity	1.1–2.7	0.92	2.10	1.65–1.72	1.05–1.07	1.18–1.20	1.14	1.27–1.37	1.35–1.40
Tensile strength (1000 psi)	0.9–2.0	1.5–1.8	5.7	3–5	5–9	6–7	7–9	1.9–8.5	7–8
Elongation (% in 2 in.)	—	50–400	28–36	20–250	0.1–3.6	2–7	40–100	6–50	40–45
Modulus of elasticity in tension (100,000 psi)	5–6	0.19	1.9	0.5–0.8	4–6	3.5–5	2.6–4.0	0.86–4.0	1.9–2.2
Compressive strength, (1000 psi)	10–17	—	32–80	7.5–8.5	11.5–16.0	11–19	7.2–13.	13–36	22–35
Impacts strength, Izod test on $\frac{1}{2}$-in. notched bar (ft-lb/in. width of notch)	2.6–2.9	Less than 16	3.6	0.3–1.0	0.26–0.50	0.4–0.5	1.0	0.4–5.2	5–7
Hardness, Rockwell[a]	—	R11	R110–115	M50–65	M65–90	M90–100	M111–118	R50–125	R95–115

Highest usable temperature continuous (°F)	150–190	212	390	160–200	150–205	140–200	270–300	140–220	140
Thermal conductivity 10^{-4} cal/ (sec) (cm^3) (°C)	—	8	1.4	3	2.4–3.3	4–6	5.2–5.5	4–8	3.1–5.5
Thermal expansion (10^{-5} in./in. °C)	—	16–18	4.5–7.0	19	6–8	9	10–15	8–16	8–12
Dielectric strength, short time, $\frac{1}{8}$ in. thickness (V/mil)	200–600	400	2500	350	500–700	450–500	385–470	250–365	300–600
Water absorption 24 hr, $\frac{1}{8}$ in. thick (%)	0–0.1	Less than 0.01	0.00	0–0.1	0.03–0.05	0.3–0.4	0.4–1.5	1.9–6.5	1.0–2.0
Effect of strong acids	Deteriorated	Attacked by oxidizing acids	None	Highly resistant acids	Attacked by oxidizing acids	Attacked by oxidizing	Attacked	Decomposed	Decomposed
Color possibilities	Limited	Unlimited	Unlimited	Extensive	Unlimited	Unlimited	Unlimited	Unlimited	Unlimited
Common uses	Phonograph records, electrical insulation	Bottle stoppers, flexible bottles, wire insulation, textiles, tablewear	Filter disks, insulators, gaskets	Screening, chemical tubing, auto seat covers	Electrical insulators, battery boxes, lenses, toys, boxes,	Windows, furniture, dentures, picture frames	Bearings, cups, fabrics, bristles	Fountain pens, tools, toys, spectacles, packaging	Packaging, foils, glazing, materials, photographic film

[a] Rockwell scales: M, $\frac{1}{4}$-in.-diameter ball, 100-kg major load; R, $\frac{1}{2}$-in.-diameter ball: 60-kg major load.

TABLE 12-3 PROPERTIES AND USES OF COMMON PLASTICS (THERMOSETTING)

Property	Phenol–Formaldehyde Resin		Urea-Form-aldehyde α-Cellulose, Molded	Melamine–Formaldehyde, Asbestos, Paper or Fabric Laminate	Polyester, Glass Fiber, Mat, Laminate	Silicone Glass Fabric Laminate	Cold-Molded[a] Cement Binder Asbestos-filled	Hard Rubber[b] No Filler
	Macerated Cotton Fabric or Cord Filler, Molded	Mechanical Grade, No Filler, Cast						
Compression molding pressure (1000 psi)	2.00–8.00	0	2.00–8.00	1.00–1.80	0.01–0.15	1.00–2.00	1.00–10.00	1.20–1.80
Specific gravity	1.34–1.47	1.25–1.30	1.45–1.55	1.75–1.85	1.5–1.8	1.6–1.8	1.6–2.2	1.4
Tensile strength (1000 psi)	2–9	4–7	6–13	6.5–12.	10–20	10–25	1.6–2.2	8–10
Elongation (% in 2 in.)	0.4–0.6	Very small	0.5–1.0	Very small	Very small	Very small	Very small	5–7.5
Modulus of elasticity in tension (100,000 psi)	9–13	5–7	12–15	16–39	10–19	20	—	3.0
Compressive strength (1000 psi)	15–30	15–20	25–35	27–50	30–50	35–46	16	8–12
Impact strength, Izod test on $\frac{1}{2} \times \frac{1}{2}$ in. notched bar (ft-lb/in. width of notch)	1–8	0.3–0.4	0.24–0.36	0.7–5.0	11–25	5–22	0.4	0.5
Hardness, Rockwell[c]	M110–120	M70–110	M115–120	M110–115	M90–100	M100	M75–95	HR95
Highest usable temperature, continuous (°F)	250	250	170	225–245	300–400	400–480	900–1300	—

Property								
Thermal conductivity [10^{-4} cal (sec)(cm³)(°C)]	4–7	3–5	7–10	10–17	8–12	3.5	—	2.9
Thermal expansion (10^{-5} in./in. C)	1–4	8–11	2.5–4.5	2.0–4.8	1.0–3.0	0.5	—	7.7
Dielectric strength, short time $\frac{1}{8}$ in. thickness (V/mil)	200–400	—	300–400	40–150	250–400	200–480	45	470
Water absorption, 24 hr, $\frac{1}{8}$ in. thick (%)	0.04–1.8	0.2–0.4	0.4–0.8	1–5	0.3–1.0	0.2–0.7	0.5–15	0.02
Effect of strong acids	Decomposed by oxidizing acids	Decomposed by oxidizing acids	Decomposed	Decomposed	Some attack	Very slight	Decomposed	Attacked by oxidizing acids
Color possibilities	Limited	Limited	Unlimited	Limited	Unlimited	Limited	Gray and black	Limited
Common uses	Serving trays, radio cabinets, electrical parts	Punches and dies	Tablewear, electrical controls, housings	Aircraft structural parts, high-strength electrical parts, stove switches	Aviation and automotive structures, decorative applications	High-temperature resisting electrical insulation	Arc-shield terminal insulators, electric heater elements	Beakers, funnels, etc., for chemicals, combs

[a] The cement binder is not strictly a thermosetting material but is set by chemical combination with water from steam (hydration).
[b] Hard rubber is not usually classified as thermosetting but as vulcanizing.
[c] Rockwell scales: M, $\frac{1}{4}$-in.-diameter ball, 100-kg major load; HR (hard rubber), $\frac{1}{4}$-in.-diameter ball, 60-kg major load.

water, oil, alkalies, and acids. They also exhibit very little shrinkage. Thus, they are excellent where high first degree of precision is required. In this group the phenol–formaldehyde resins are probably the most important resins. The phenol–formaldehyde group can be subdivided into four general classes:

1. Cellulose-filled compositions.
2. Mineral-filled compositions.
3. Molding sheet.
4. Impact-resistant materials.

The *cellulose-filled composition* utilizes wood flour as the filler and the final product has high dielectric strength, mechanical strength, and it is very light. Uses have been found in the aerospace and automotive industries.

The *mineral-filled composition* utilizes asbestos as the filler. Thus, articles made of these plastics resist chemicals and heat. Uses are for insulation of high-voltage transmissions.

Molding sheets are made by impregnating paper with phenolic resins. These sheets can be softened by heat and become brittle in the cold weather. These sheets are used for molding applications where the material must be able to give.

In the final group, *impact-resistant materials,* the plastic is made by impregnating paper and fabric fillers, built up in layers. This product is called laminated plastic.

Chemically Setting Plastics

In the chemically setting group three categories exist: harvel resins, cold-molded plastics, and casein. These plastics are generally used where resistance to heat and arcing are of primary importance. Thus, they are utilized for electrical insulating parts.

MANUFACTURE OF ORGANIC PLASTICS

In general, four main steps are required in the manufacturing of articles made from organic plastics:

1. The production of intermediate materials (chemical) from the raw materials of coal, petroleum, and cotton.
2. The manufacture of synthetic resins from the above.
3. The preparation of molding powders, fillers, rods, and sheets from step 2.
4. Molding the articles from the powders, fillers, rods, and sheets.

Methods of Forming and Fabricating Plastics

Most of the methods utilized in the forming and fabrication of plastic have their counterpart in the processing of other materials of engineering. Examples include the following:

1. Casting.
2. Compression molding.
3. Injection molding.

4. Transfer molding.
5. Extruding.
6. Blowing.
7. Laminating.

Casting is the simplest molding method available. Casting utilizes low-cost lead-antimony dies (molds) where the melted resins are poured into open molds and cured. The cast resins are baked in an oven at about 167°F (75°C) for several days, until a permanent set is obtained. Both thermoplastics and thermosetting plastics may be cast.

Compression molding is the most widespread molding operation for thermosetting plastics. Compression molding requires the use of pressure on the molding compound placed in the mold of heated platens of the press. Thermoplastics may be compression-molded, but the mold must be cooled after each molding before the article is removed from the press. This is time-consuming and uneconomical, so other methods are utilized for thermoplastic plastics.

Injection molding is one of the most widely used and most rapid methods of producing articles of intricate shape. Injection molding consists of forcing softened plastic materials into a closed mold maintained at a temperature below the softening point of the compression. This method is widely used for thermoplastic materials. The thermoplastic material is fed into a pressure chamber and a plunger compresses the material and forces it into a heating chamber, where it is progressively heated to a uniform temperature. The thermoplastic material emerges through a nozzle in a thoroughly softened condition and is then forced into a cold mold, where pressure causes it to take the shape of the cavity. The plastic is cooled, the mold is opened, and the part removed.

Transfer molding is used for thermosetting materials when the article is to include delicate metal inserts for any reason. In this method the molding compound is heated under pressure until it is soft enough to flow. The compound is then placed into a cavity in which the part is formed and allowed to cool.

Extruding is basically the opposite of injection molding. Extruding of thermoplastic plastics is a widely used practice for producing rods, tubes, and other cylindrical shapes.

Blowing of thermoplastics is a process by which hollow objects are made. It is accomplished in much the same manner as the blowing of glass.

Laminating is used to produce hard boards or sheets of resin-impregnated papers, wood veneers, or fabrics. Laminates are made into stock sizes of flat sheets under medium to high pressure in a press. The process involves building up individual resin-coated fabric or veneer over a form, inserting the assembly into a large rubber bag (called bag-molding process), withdrawing the air from the bag, and subjecting the bag and contents to pressure in an autoclave.

PROPERTIES OF ORGANIC PLASTICS

Table 12-4 shows the mechanical properties of molded compositions. Table 12-5 shows the electrical properties of molded compositions for which most plastics are used. Table 12-6 illustrates the strength-weight relationships for plastics and some common structural materials.

TABLE 12-4 MECHANICAL PROPERTIES OF MOLDED COMPOSITIONS

Type of Material	Tensile Strength (psi)	Impact Strength Notched Bar, (ft-lb./in.)	Modulus of Rupture (psi)	Modulus of Elasticity (psi)	Specific Gravity	Heat Distortion (°C)	Hardness	Coefficient of Linear Expansion per °C
Phenolic laminated, paper base	6,000–13,000	0.8–2.4C[a]	13,000–20,000	$1.0–2.0 \times 10^6$	1.34–1.55	100–140	85–125R[b]	$20–50 \times 10^{-6}$
Phenolic laminated, canvas base	8,000–12,000	1.6–10.4C	12,000–19,000	$1.0–2.0 \times 10^6$	1.34–1.55	100–140	95–115R	$30–70 \times 10^{-6}$
Phenolic molded, woodflour filled	6,000–11,000	0.26–0.50C	8,000–15,000	$0.8–1.5 \times 10^6$	1.25–1.52	120–140	95–120R	$35–80 \times 10^{-6}$
Phenolic molded, cellulose filled	6,000–11,000	0.40–0.80C	8,000–15,000	$0.8–1.5 \times 10^6$	1.32–1.48	120–140	95–115R	$35–80 \times 10^{-6}$
Phenolic molded, fabric filled	6,000–8,000	0.80–6.0C	8,000–13,000	$0.8–1.5 \times 10^6$	1.35–1.40	120–140	90–115R	$35–80 \times 10^{-6}$
Phenolic molded, mineral filled	5,000–9,000	0.26–1.0C	8,000–18,000	$1.0–5.0 \times 10^6$	1.70–2.05	120–150	100–120R	$25–50 \times 10^{-6}$
Cast phenolics	3,000–7,000	0.30–0.50C	3,000–14,000	$0.25–0.75 \times 10^6$	1.26–1.70	40–80	70–110R	$70–160 \times 10^{-6}$
Phenolic molded, transparent	8,000	—	1600	1.27	107	—	—	
Furfuryl-phenol molded, woodflour filled	5,000–12,000	1.0–6.51[c]	10,000–16,000	—	1.3–1.4	131	35–40B[d]	30×10^{-6}
Furfuryl-phenol molded, asbestos filled	4,000–12,000	1.0–6.0I	8,000–14,000	—	1.6–2.0	136	44–46B	20×10^{-6}

Material							
Furfuryl-phenol molded, fabric filled	5,000–10,000	20–39I	10,000–16,000	—	1.3–1.4	—	30–35B	—
Furfuryl-phenol laminated, paper base	10,000–20,000	5.0–20I	20,000–30,000	—	1.3–1.4	—	—	—
Furfuryl-phenol laminated, cloth base	9,000–12,000	10.0–50I	—	—	1.3–1.4	—	—	—
Urea molded	9,000–12,000	0.28–0.36C	10,000–14,000	$1.2–1.9 \times 10^6$	1.45–1.55	95–130	110–125R	—
Polystyrene molded	5,000–7,000	0.40–0.60C	6,000–8,000	$0.40–0.60 \times 10^6$	10.5–1.07	75–80	82–92R	$65–75 \times 10^{-6}$
Acrylate molded	4,000–8,000	0.3–4.0C	9,000–16,000	$0.4–0.6 \times 10^6$	1.18–1.19	51–60	18–20B	80×10^{-6}
Methyl methacrylate molded	9,000–12,000	0.2–3.0C	12,000–14,000	—	1.18–1.20	60–13	17–20B	$70–90 \times 10^{-6}$
Cellulose nitrate molded	4,900–8,500	10–11.5I	5,000–8,000	$0.2–0.4 \times 10^6$	1.35–1.60	71–91	7–12B	$120–160 \times 10^{-6}$
Cellulose acetate molded	4,000–5,000	1.7–3.0C	5,000–7,000	$0.20–0.40 \times 10^6$	1.27–1.63	50–80	85–120R	$140–160 \times 10^{-6}$
Ethyl cellulose	5,000–9,000	3.2–9.6I	—	0.3×10^6	1.14	100–130	—	—
Shellac	900–2,000	—	—	—	1.1–2.7	66–90	—	—
Cold-molded composition	700–1,700	1.5–4.5	3,500–7,800	—	1.9–2.12	182–260	—	—
Hard rubber	1,500–10,000	0.5I	—	0.33×10^6	1.12–1.80	—	31B	80×10^{-6}

[a] C, Charpy.
[b] R, Rockwell.
[c] I, Izod.
[d] B, Brinell.

TABLE 12-5 ELECTRICAL PROPERTIES OF MOLDED COMPOSITIONS

Type of Material	Power Factor			Dielectric Constant		
	60 cycles/sec	1000 cycles/sec	10^6 cycles/sec	60 cycles/sec	1000 cycles/sec	10^6 cycles/sec
Phenolic laminated, paper base	0.02–0.15	0.02–0.10	0.02–0.06	4.5–6.5	4.5–6.0	4.0–5.5
Phenolic laminated, canvas base	0.05–0.30	0.05–0.20	0.04–0.10	5.0–9.0	4.5–8.0	4.0–6.0
Phenolic molded, woodflour filled	0.02–0.30	0.02–0.15	0.035–0.08	4.5–10.0	4.5–10.0	4.0–8.0
Phenolic molded, cellulose filled	0.05–0.30	0.04–0.15	0.04–0.10	4.5–10.0	4.5–10.0	4.0–8.0
Phenolic molded, fabric filled	0.05–0.30	0.04–0.15	0.04–0.10	4.5–15	4.5–15	4.0–10
Phenolic molded, mineral filled	0.020–0.50	0.01–0.20	0.005–0.10	4.5–20	4.5–15	4.0–10
Cast phenolics	0.070–0.50	0.030–0.30	0.04–0.13	7.0–30	6.0–25	5.5–15
Phenolic molded, transparent	0.06	0.04	0.019	5.5	5.0	4.5
Furfuryl-phenol molded, woodflour filled	—	0.04–0.15	0.01–0.06	—	4.0–8.0	6–7.5
Furfuryl-phenol molded, asbestos filled	—	0.1–0.15	0.06–0.15	—	4.5–20	5–18
Furfuryl-phenol molded, fabric filled	—	0.08–0.2	0.05–0.08	—	4.5–6	6
Furfuryl-phenol laminated, paper base	—	—	0.15–0.48	—	4.5	—
Furfury-phenol laminated, cloth base	—	—	0.41–0.64	—	4.5	—
Urea molded	0.050–0.13	0.035–0.07	0.035–0.040	8.0–10	8.0–9.0	6.9–7.5
Polystyrene molded	0.0001–0.0002	0.0001–0.0002	Under 0.0002	2.6	2.6	2.6
Acrylate molded	0.05–0.06	—	—	3.4–3.6	—	—
Methyl methacrylate, molded	0.06–0.08	—	0.02	4.0–4.4	—	2.8
Cellulose nitrate, molded	0.062–0.149	—	0.07–0.09	6.7–7.3	—	6.2
Cellulose acetate, molded	0.03–0.05	0.035–0.07	0.035–0.07	4.5–7.0	4.5–6.5	4.0–4.5
Ethyl cellulose, molded	0.03–0.06	0.025	—	2.6–2.9	3.9	—
Shellac, molded filled	0.004–0.018P[a]	—	—	3–4	—	—
Cold-molded composition	0.2	—	0.07	15.0	—	6.0
Hard rubber	—	—	0.003–0.008	2.8	2.9–3.0	3.0

Type of Material	Loss Factor			Resistivity (megohm-cm)	Dielectric Strength Step Test (V/mil)
	60 cycles/sec	1000 cycles/sec	10^6 cycles/sec		
Phenolic laminated, paper base	0.1–1.0	0.1–0.6	0.10–0.30	$2 \times 10^6 – 1 \times 10^8$	400–600
Phenolic laminated, canvas base	0.25–2.0	0.15–1.0	0.15–1.0	$3 \times 10^5 – 4 \times 10^7$	250–400
Phenolic molded, woodflour filled	0.10–3.0	0.20–1.5	0.15–0.80	$10^4 – 10^6$	200–300
Phenolic molded, cellulose filled	0.25–3.0	0.20–1.5	0.15–1.0	$10^4 – 10^5$	200–300
Phenolic molded, fabric filled	0.25–4.5	0.20–2.0	0.15–1.0	$10^3 – 10^5$	200–300
Phenolic molded, mineral filled	0.10–4.0	0.045–3.0	0.020–1.0	$10^3 – 10^8$	200–375
Cast phenolics	0.70–16.0	0.20–4.0	0.20–2.0	$10^4 – 10^7$	120–300
Phenolic molded, transparent	—	—	—	7.5×10^5	325
Furfuryl-phenol molded, woodflour filled	—	—	—	$10^{10} – 10^{12}$	400–600
Furfuryl-phenol molded, asbestos filled	—	—	—	$10^9 – 10^{11}$	200–500
Furfuryl-phenol molded, fabric filled	—	—	—	10^{10}	200–500
Furfuryl-phenol laminated, paper base	—	—	—	—	900–1800
Furfuryl-phenol laminated, cloth base	—	—	—	—	300–700
Urea molded	0.40–1.2	0.28–0.65	0.24–0.30	$1–10 \times 10^6$	275–325
Polystyrene molded	0.00026–0.00053	0.00026–0.00053	Under 0.0005	Over 10^{10}	500–525
Acrylate molded	—	—	—	1.0×10^{15}	500
Methyl methacrylate, molded	—	—	—	1.0×10^{15}	480
Cellulose nitrate, molded	—	—	—	$2–30 \times 10^{10}$	660–780
Cellulose acetate, molded	0.13–0.30	1.15–0.40	0.15–0.40	$10^6 – 10^8$	300–350
Ethyl cellulose, molded	—	—	—	—	1500
Shellac, molded filled	0.016–0.16P	—	—	$10^8 – 10^{16}$	200–600
Cold-molded composition	—	—	—	1.3×10^{12}	85–100
Hard rubber	—	—	—	$10^{12} – 10^{15}$	250–1000

[a]P, Paper filled.

TABLE 12-6   STRENGTH-WEIGHT RELATIONSHIPS FOR PLASTIC AND SOME COMMON
STRUCTURAL MATERIALS[a]

Material	Specific Gravity	Tensile Strength (psi)	Tensile Strength / Specific Gravity	Tensile Strength / (Specific Gravity)^2
Chrome-vanadium steel	7.85	164,000	20,800	2,700
Structural steel	7.83	65,000	8,300	1,060
Cast iron, gray	7.03	5,000	5,000	720
Titanium alloy	4.7	145,000	31,000	6,600
Aluminum alloy	2.85	6,000	20,000	7,200
Magnesium alloy	1.81	44,000	24,300	13,400
Glass fabric laminate	1.9	45,000	23,600	12,500
Asbestos cloth laminate	1.7	9,000	5,300	3,100
Paper laminate	1.33	20,000	15,000	11,300
Cellulose acetate	1.3	5,000	3,900	3,000
Methyl methacrylate	1.18	8,500	7,200	6,100
Polystrene	1.06	5,500	5,200	4,900
Polyethylene	0.92	1,300	1,400	1,500
Sitka spruce	0.40	17,000	42,500	106,000

[a] The values given here are average values. Different compositions, heat treatments, etc., of a given material may differ widely
from the values shown here.

ADVANCED FRP COMPOSITES IN CONSTRUCTION

Advanced Fiber Reinforced Plastic (FRP) Composites have the potential to become a very
useful and attractive construction material in the 21st century.

FRP composites have the potential to be used as stand-alone structural members, as
reinforcement to prestressed and non-prestressed concrete members, or in the repair or renovation of other structural materials.

FRP composites consist of a polymer matrix reinforced with fibers. The polymer matrix can be thermosetting plastic like polyester or thermoplastic like nylon. The polymer
matix plays a very important role in the behavior of the composite. The polymer matrix
transfers stresses between the fibers, protects the fibers, and prevents the buckling of
fibers. A polymer matrix made of thermosetting plastic provides good thermal stability and
chemcal resistance. However, thermosetting polymers generally have a short shelf-life
and a low strain-to-failure value. Thermoplastic polymers offer higher strain-to-failure
value and an unlimited storage life when adequately stored. Because of the lower viscosity of thermosetting polymers and the relative ease of combining them with fibers, the
production of FRP members has been more frequently completed with thermosetting
polymers. The most common types of fiber used in FRP applications are made of glass,
carbon, or Aramid. The major fiber parameters that affect the performance of the composite
are fiber orientation, fiber length, fiber shape, and the fiber composition. For the FRP
composites to perform adequately in civil engineering applications, the stresses must be
distributed internally within the composite. This is effectively accomplished with the
bonding mechanism. The bonding mechanism between the matrix and the fibers relies
on both chemical and mechanical processes. Coupling agents are used to enhance the

chemical bond between fibers and matrix while fiber shape is used to enhance the mechanical bond.

For reinforced concrete structures, FRP composites can be used as reinforcement to concrete members subjected to flexure, shear, and compression, replacing traditional steel rebars. The concrete members can include one-dimensional FRP elements like rods or bars. They can also include multi-dimensional elements such as grids.

For prestressed concrete structures, FRP tendons can be used to prestress bonded or unbonded members instead of high-strength prestress steel. For example, a precast post-tensioned bridge of 9m (30 ft) span and a 5.2m (17 ft) width was erected at a cement plant in 1991. It has a 180mm (7 in.) width thick deck slab supported by three longitudinal girders. Cables of three different materials were used to prestress the slab. Thirty Glass Fiber Reinforced Plastic (GFRP) cables were used to prestress one-third of the length of the slab; thirty Carbon Fiber Reinforced Plastic (CFRP) cables were used to prestress the next third of the length; and steel cable prestressing was adopted for the remaining length. Each GRFP and CRFP cable consists of 7 4-mm 0.156 in. diameter rods. The initial prestress and final prestress after all losses were set at 0.6 P and 0.5 P respectively. Plastic ducts housing the cables were grouted with high-strength epoxy and mortar for bonding purposes, and the temporary anchorages were removed. Monitoring of bridge deflections and stresses in cables and concrete was carried out to assess losses in the cables and the actual deflections under moving loads.

FRP composites are also manufactured as structural shapes that can be used to construct civil engineering and structural systems. For example, an antenna structure can be built using FRP members. "The Arch," is an 84 ft high, 160 ft clear span, fiberglass tripod used to develop antennae patterns for the U.S. Naval Command Control Ocean Systems Center at Point Loma, CA. FRP structural shapes and FRP studs and nuts were used for the structure because of their unique combination of RF transparency and strength. The corrosion resistance and low maintenance of fiberglass were additional benefits for "The Arch," which replaces an older wooden tower built in 1952 with steel bolts (which gave interference). Another application for FRP structures can be seen in Niagara Falls, which creates an aggressive and moist environment. This environment resulted in the deterioration of the steel observation deck. To avoid a repetition of this situation and expensive long-term repairs, the members of the steel deck were replaced with FRP components.

FRP composites have also been used for the repair and renovation of structural systems made of other construction materials. Repair and renovation using FRP materials have been conducted on structures made with masonry, concrete, timber, and steel. Bonding FRP plates to the surface has strengthened masonry structures. Concrete columns in seismic sensitive areas have been reinforced by wrapping FRP products around the column as a part of general seismic upgrading system.

The major applications of FRP in construction occur where special performance requirements are required. Such applications include structures subject to severe chemical attack, projects dealing with electrical and reactor equipment, or supports of magnetic resonance imaging (MRI) medical equipment.

The design of composites parts and structures with fiber reinforced composites is not covered by most current design codes. As such, these designs are performed using basic engineering principles rather than standard design equations. The engineering principles should be applied to the design after establishing the actual material properties of the composites and the factors that could affect these properties.

PROBLEMS

12.1 List seven outstanding characteristics of plastics.

12.2 Define organic plastic.

12.3 List three general classifications of organic plastics.

12.4 Describe a thermoplastic plastic and list several.

12.5 Describe a thermosetting plastic and list several.

12.6 Describe a chemically setting plastic and list several.

12.7 Explain the difference between polymerization and condensation.

12.8 Describe the procedure for manufacturing organic plastics.

12.9 List and describe the methods of forming and fabricating plastics.

12.10 List the mechanical properties of organic plastics and discuss how they compare to other materials of construction.

12.11 List the types of polymer matrix and of fiber used in the production of FRP components.

12.12 Discuss the main applications of FRP components in construction applications.

REFERENCES

American Concrete Institute Committee 440, "State of the Art Report for Concrete Structures," ACI 440R-96, *American Concrete Institute*, Farmington Hills, 68 pp. 1996.

American Society for Testing Materials, *Symposium on Plastics*, Philadelphia, 1938.

GILMORE, G. and SPENCER, R., "Injection Molding Process," *Modern Plastics*, No. 27 (April 1950), p. 143.

MACTAGGORT, E.F., and CHAMBERS, H.H., *Plastic and Building*, Sir Isaac Pitman & Sons, Ltd., London, 1951.

MILLS, A.P., HAYWARD, H.W., and RADER, L.F., *Materials of Construction*, John Wiley & Sons, Inc., New York, 1955.

MOORE, H.F., and MOORE, M.B., *Materials of Engineering*, McGraw-Hill Book Company, New York, 1953.

13 MISCELLANEOUS MATERIALS

GLASS

Glass is a hard, brittle, inorganic product, ordinarily transparent or translucent, made by the fusion of silica, flux, and a stabilizer, and cooled without crystallizing. Glass can be rolled, blown, cast, or pressed for a variety of uses. Glass is also referred to as a glazing material, and generally there are four types of glazing material:

1. **Annealed Glass:** This glass has been subjected to a slow, controlled cooling process during manufacture to control residual stresses so that it can be cut or subjected to other fabrication. Regular polished plated, float, sheet, rolled, and some patterned surface glasses are examples of annealed glass. When breakage occurs it does so in large pieces with jagged edges.
2. **Tempered Glass:** This type of glass is heated and then cooled in a special way that makes it many times stronger than ordinary glass. When broken, it crumbles into small pieces that minimizes the chance of serious personal injury. It cannot be cut, however, but rather must be purchased in the exact size.
3. **Laminated Glass:** This glass is made by heating layers of ordinary glass with a resilient plastic material between them to bind them together. If it does break, only a small hole is created.
4. **Rigid Plastic:** This glazing material is made from one or more layers of plastic bonded together.

In talking about glass in general terms the majority of glass is made up of three ingredients: silica or sand, potash, and lime. These three substances are melted at high temperature to form glass. Silica is the primary ingredient (96 percent). Potash serves to reduce the melting point of silica so that the furnaces can handle the manufacturing operation. Lime is added to the system to make the glass more stable. Borosilicates may be added to reduce the thermal expansion of the glass and this type of glass is known as "Pyrex." Other ingredients may also be added to change characteristics such as color and the like.

Glass is extremely stable and will not deteriorate. It will only break when an external force is applied. When glass breaks it is always in tension and at right angles to the tension producing it. Even if the force applied is in compression or shear the breakage will always occur in tension. Further, glass breaks from the outside to the inside. Like most objects, when glass is struck in compression it bends. The struck face is in compression and the back face is in tension. This back face in tension reaches the elastic limit and failure occurs.

Generally glass breakage results from either an external force or thermal shock. The external force is the sudden shock or blow to the glass or a steady pressure over time. Thermal shock occurs when one face is cooled or heated suddenly and the coolness or heat is not transmitted to the other side and a differential in the contraction or expansion occurs. Failure results, as the stress created is too much.

From an engineering point of view, glass is extremely weak and codes and standards have been established to deal with the utilization of glass in engineering projects. When glass is manufactured it must meet specific standards such as the American National Standards Institute (ANSI) Z97.1 Safety Glazing Material Used in Buildings–Performance Specifications and Methods of Test. ANSI indicates that the 100 foot-pound-force and 150 foot pound-force energy levels chosen in the standard were established as practically related to those situations in which limited acceleration path precluded, in most cases, the possibility of an individual developing his/her full kinetic energy potential. ANSI further states that the safe performance criteria is raised to 400 foot pound-force impact level for unlimited acceleration paths in which it might be reasonable to expect that an energetic person might develop something approaching his/her full impact velocity. As such, glass is controlled by these minimum safety values and these values have been adopted by the various building code officials throughout the country.

As one example of the codes, the various building codes state that glazing, operable or inoperable, in shower and bathtub doors and enclosures with a horizontal edge less than 6 ft above the room floor level or less than 70 in. above the compartment floor is a specific hazardous location. As such, and because of the hazardous location, any glazing material utilized in a shower or bathtub situation must meet the ANSI Standard described above.

Quickly, via the ANSI test results, the reader will find out that annealed glass cannot pass the ANSI test. Annealed glass simply does not meet the ANSI Standard. When annealed glass is utilized it has to be utilized in situations that are not in the traveled path of individuals. One example of the use of annealed glass would be in windows in which the bottom horizontal edge is at least 36 in. from the floor level. The reason for this is that when an individual walks or runs into annealed glass or strikes it the jagged edges that result from the breakage cause considerable damage to the individual. As such, today it is required to utilize safety glazing in situations where a potential injury to an individual may result.

CONCRETE BLOCK

Concrete blocks are made by mixing portland cement and water with aggregates such as sand, gravel, crushed stone, expanded slag, expanded shale or clay, cinders, or pumice. Concrete blocks are made in 4, 6, 8, 10 or 12 in. thick and are 8 in. high and 16 in. long. However, the most common size is $8 \times 8 \times 16$ and these numbers are nominal dimensions. Actual dimensions vary somewhat depending on the shrinkage but are normally approximately $\frac{3}{8}$ in. less than the actual.

In addition to the normal type of concrete block, special sizes are often made that serve a specific purpose. Some of these special blocks include solid block, half block, corner block, pier block, chimney block, jamb block, and decorative block.

Concrete block is normally referred to as Concrete Masonry Units (CMUs). Concrete block is laid in courses (horizontal rows) and in vertical planes called wythes. Concrete block, when constructed, can be reinforced or unreinforced. Reinforced concrete block can be reinforced in both the vertical and horizontal direction and provides, when the cavities are filled with concrete, the same effect as columns. This provides for an extremely strong wall.

For a number of years concrete block was hidden from the public in building terms. In other words, concrete block was the wall structure that was hidden behind the brick or veneer facing. Today, concrete block has come into its own, with several structures being built with the block exposed. It is not unusual to see a reinforced concrete structure 8 stories or more. Concrete block provides for a strong, long-lasting structure at a cost far less than structures utilizing other material. As such, it is extremely important that all codes and standards be adhered to and followed.

BRICK

Today brick is defined as a solid masonry unit of clay or shale, formed into a rectangular prism while plastic, and then burned or fired in a kiln.

Dealing with plastic conditions, clay worked with water can take the form of soft mud, stiff mud, or dry press green brick. Green brick is the condition of the material before the drying and burning process.

In the soft mud process brick is produced by molding relatively wet clay (20 to 30 percent moisture) in a mold, applying light pressure to smooth the surface, and completely filling the mold. When the inside of the mold is sanded to prevent sticking of the clay, the product is sand-struck brick. When the molds are wetted to prevent sticking, the product is water-struck brick.

In the stiff mud process the brick is produced by extruding a stiff but plastic clay (12 to 15 percent moisture) through a die. The clay rectangular column is then cut into separate bricks by wire cutters. The surface may be sanded or textured and glazes or enamels may be applied before the bricks move to the drying process.

In the dry press process, the powdered clay is deposited into a rectangular mold and a close fitting piston compresses the clay.

All brick, prior to entering the kiln, goes through a drying process in order to reduce the water content, thereby producing a stiffer green brick that is easier to handle. The drying process, in removing some of the moisture, also saves time and fuel while the green brick is in the kiln, for there is less moisture to deal with.

During the burning process (kiln) the green bricks shrink to various sizes, all dependent on the moisture content, material, and heat applied. In the end a normal finished brick is approximately $2\frac{1}{4}$ in. high by $3\frac{3}{4}$ in. wide by 8 in. long. However, the size may vary slightly as a result of the shrinking process.

There are many types of brick in use today. Building brick, which was originally referred to as common brick, is a brick manufactured for building purposes—not specifically treated for color or texture. Building brick in different parts of the country has different characteristics. Face brick is brick made especially for facing purposes, often treated to produce surface texture. This brick is made of selected clay, or treated to produce a desired color. This type of brick has a better durability and an excellent appearance. Glazed brick has one surface of each brick glazed for color. This type of brick is used for walls or partitions in hospitals, dairies, or laboratories where cleanliness is key to the success of the operation. Fire brick is brick made of refractory ceramic material which will resist high temperatures related to fireplaces, boilers, and the like. Fire brick is generally larger than rectangular brick. Cored brick are bricks made with 2 rows of 5 holes extending through their beds to reduce weight. Of importance here is the fact that there is no significant difference in strength of walls built with cored brick or solid brick. Floor brick is brick that is smooth and dense, highly resistant to abrasion, used as a finished floor surface. Paving brick is vitrified brick, especially suitable for use in pavements where resistance to abrasion is important. Finally, sewer brick is low absorption, abrasive-resistant brick intended for use in drainage structures.

When laid bricks take many positions depending on the design and specifications. When viewing a brick, if the brick is $2\frac{1}{4}$ in. high and 8 in. long the position is known as a Stretcher. If the brick is $2\frac{1}{4}$ in. high and $3\frac{3}{4}$ in. long, then we have a Header position. If the brick is $2\frac{1}{4}$ in. wide and 8 in. tall, this is known as the Soldier position. Finally, if the brick is $2\frac{1}{4}$ in. wide and $3\frac{3}{4}$ in. tall then the Rowlatch position is intended.

Specific names also apply when bricks are used in various combinations. These are known as bond patterns and three of the types are as follows. The running bond, which is utilized for most projects, is brick running in the Stretcher position for each layer separated by a mortar bed. In other words, the first layer consists of all brick in the Stretcher position and the second layer has the bricks in the Stretcher position, but offset by a half a brick. The third layer is the same as the first layer and the fourth layer is the same as the second layer, and so forth. The second method, the English cross bond, has the first layer in the Stretcher position and the second layer in the Header position with the third layer the same as the first layer and the fourth layer the same as the second layer and so forth. The third method, the Flemish bond, has the first layer in combination, Stretcher followed by Header followed by Stretcher, etc. The second layer is the Header followed by the Stretcher followed by the Header, etc. Layer three is a repeat of layer one and layer four is a repeat of layer two and so forth. In all these cases or combinations, the vertical bonds do not meet but overlap one another.

In addition to the above, it is also important to note that brick at times needs to be cut. As such when the brick is cut in half the short way making the brick $2\frac{1}{4}$ by $3\frac{3}{4}$ by 4 in. it is known as a Half or Bat. When the brick is cut the long way in half making it $2\frac{1}{4}$ by $1\frac{7}{8}$ by 8 in. it is known as a Queen Closure. When the brick is split in half to become $1\frac{1}{8}$ by $3\frac{3}{4}$ by 8 in. it is called a Split. When a corner is removed from a brick it is called a King Closure. When the brick is cut to be $2\frac{1}{4}$ by $3\frac{3}{4}$ by 2 in. deep it is referred to as a Quarter Closure and when it is cut to be $2\frac{1}{4}$ by $3\frac{3}{4}$ by 6 in. deep it is referred to as a Three-Quarter Closure.

In closing, bricks are commonly classified into three grades: Severe Weathering (SW), Moderate Weathering (MW), and No Weathering (NW). Further classifications can be made by compressive strength. Brick shall develop not less than 1500 psi for NW grade, 2500 psi for MW grade, and 3000 psi for SW grade on an average of 5 samples. Thus, a variety of classifications exist for brick.

MORTAR

Mortar in simple terms is defined as a plastic mixture used in masonry construction that can be troweled and hardens in place. The most common materials that mortar may contain are portland, hydraulic, or mortar cement; lime; fine aggregate; and water. The actual mix is determined by specifications and varies accordingly as to what strength or use is intended. As would be expected, the strength of the mortar can be increased by increasing the proportion of cement to sand and by properly controlling the other ingredients.

Each material that makes up the mortar batch contributes to the overall quality of the mix. As such, cement, lime, sand, and water each individually provide a specific property and collectively provide for the desired end product. Specifically, cement provides for the strength and durability; lime offers workability and elasticity; sand is a filler reducing the drying shrinkage, and enhances the strength; and finally water is the mixing agent which adds to the workability and allows the cement hydration process to take place.

As one can imagine, any change within the proportion of each of these ingredients can alter the final product. Changes in the cement content will alter the strength of the mix, its workability, and setting characteristics. A change in the lime content creates changes in the workability of the mix. Changes in the sand type also affect the strength of the mix. An increase or decrease in the water content will make for a dryer or wetter mix resulting in a change of compressive strength.

Therefore all materials utilized for mortar and their proportions are controlled via specifications as set forth by the American Society for Testing and Materials (ASTM). These specifications relate to all materials that might be utilized for the making of mortar and include portland cement, masonry cement, blended hydraulic cements, quicklime, hydrated lime, aggregates (both natural and manufactured) and mortar for unit masonry.

The Uniform Building Code (UBC) and the Building Officials and Code Administrators Building Code (BOCA), along with a number of other Building Code agencies, specify that mortar types must comply with the ASTM Standard C270 "Mortar for Unit Masonry." As such four types of mortar exist: M, S, N, and O. Table 13-1 shows the proportions by volume (cementitious material) for each mortar mix.

If for some reason the designer chooses not to utilize the mortar proportion table, the designer may utilize Table 13-2, Specified Compressive Strength of Masonry. This table allows one to mix the approved materials as long as the mortar meets the compressive strength stated.

Given the fact that four mortar types exist the question arises as to what application does each type apply to. As such and as one would expect the codes provide for that answer. Table 13-3, Masonry and Mortar Types provide a table that lists the types of masonry construction along with the types of mortar permitted for each application.

SHOTCRETE

Shotcrete is defined as mortar or concrete pneumatically projected at high velocity onto a surface. The force of the jet impacting on the surface compacts the material. There are two types of shotcrete; namely, the wet-mixed process and the dry-mixed process (which is also referred to as gunite).

TABLE 13-1   MORTAR PROPORTIONS FOR UNIT MASONRY

		PORTLAND CEMENT OR BLENDED CEMENT[a]	MASONRY CEMENT[b]			MORTAR CEMENT[c]			HYDRATED LIME OR LIME PUTTY[a]	AGGREGATE MEASURED IN A DAMP, LOOSE CONDITION
MORTAR	*TYPE*		*M*	*S*	*N*	*M*	*S*	*N*		
Cement-lime	M	1	—	—	—	—	—	—	$\frac{1}{4}$	
	S	1	—	—	—	—	—	—	over $\frac{1}{4}$ to $\frac{1}{2}$	
	N	1	—	—	—	—	—	—	over $\frac{1}{2}$ to $1\frac{1}{4}$	Not less than $2\frac{1}{4}$
	O	1	—	—	—	—	—	—	over $1\frac{1}{4}$ to $2\frac{1}{2}$	and not more than 3 times
Mortar cement	M	1	—	—	—	—	—	1	—	the sum of
	M	—	—	—	—	1	—	—	—	of the separate
	S	$\frac{1}{2}$	—	—	—	—	—	1	—	volumes of
	S	—	—	—	—	—	1	—	—	cementitious
	N	—	—	—	—	—	—	1	—	materials.
Masonry cement	M	1	—	—	1	—	—	—	—	
	M	—	1	—	—	—	—	—	—	
	S	$\frac{1}{2}$	—	—	1	—	—	—	—	
	S	—	—	1	—	—	—	—	—	
	N	—	—	—	1	—	—	—	—	
	O	—	—	—	1	—	—	—	—	

[a] When plastic cement is used in lieu of portland cement, hydrated lime or putty may be added, but not in excess of one tenth of the volume of cement.

[b] Masonry cement conforming to the requirements of U.B.C. Standard No. 24-16.

[c] Mortar cement conforming to the requirements of U.B.C. Standard No. 24-19.

TABLE 13-2   SPECIFIED COMPRESSIVE STRENGTH OF MASONRY[A]

Mortar	*Type*	*Average Compressive Strength at 28 Days Min. psi (MPa)*	*Water Retention, min.*	*Air Content max. %*	*Aggregate Ratio (Measured in Damp, Loose Condition)*
Cement-Lime	M	2500 (17.2)	75	12	
	S	1800 (12.4)	75	12	
	N	750 (5.2)	75	14[B]	
	O	350 (2.4)	75	14[B]	
Masonry Cement	M	2500 (17.2)	75	...[C]	Not less than $2\frac{1}{4}$ and not
	S	1800 (12.4)	75	...[C]	more than $3\frac{1}{2}$ times the
	N	750 (5.2)	75	...[C]	sum of the separate
	O	350 (2.4)	75	...[C]	volumes of cementitious
	K	75 (.52)			materials

[A] Laboratory prepared mortar only

[B] When structural reinforcement is incorporated in cement-lime mortar, the maximum air content shall be 12%.

[C] When structural reinforcement is incorporated in masonry cement mortar, the maximum air content shall be 18%.

In the wet-mixed process (also known as Pneumatically Applied Concrete) a quality control batching and mixing operation takes place in relation to the concrete (which consists of portland cement, fine aggregate, coarse aggregate, and water). This batching and mixing operation

TABLE 13-3 MASONRY AND MORTAR TYPES

Types of masonry	Types of mortar permitted
Fire brick	Refractory air-setting mortar
Glass-block masonry	S or N
Grouted and filled cell masonry	M or S
Linings of existing masonry above or below grade	M or S
Masonry, above grade or interior	
Cavity walls and masonry-bonded hollow walls	
Design wind pressure over 20 psf	M or S
Design wind pressure 20 psf or less	M, S or N
Piers of hollow units	M or S
Piers of solid units	M, S or N
Walls of hollow units	M, S or N
Walls of solid units	M, S, N or O
Masonry in contact with earth	M or S
Masonry other than above	M, S or N
Nonloadbearing partitions and fireproofing	M, S, N, O or gypsum

1 psf = 4.882 kg/m^2.

is usually performed at a concrete mixing plant. The concrete is then transported to the construction site via concrete trucks. It is then transferred to a pump which in turn passes the concrete through a hose under pressure to a nozzle. Compressed air is injected at the nozzle and the material is jetted at a high velocity onto the surface which was to receive the material.

In the dry-mixed process (also known as Pneumatically Applied Mortar) cement and sand are mixed on site and fed into a feeder and into a hose where it is conveyed by compressed air through the hose to the nozzle. Water is then introduced into the nozzle under pressure and the water mixes with the cement and sand. The mortar (cement, sand, water) is then jetted from the nozzle onto the surface that is to receive the material.

In viewing the two processes it is obvious that the wet-mixed process has several advantages over the dry-mixed process. In the wet-mixed process the water is controlled as to its amount at the batching and mixing plant. As such, one has a better control of the accuracy of the water content and thus can better ensure the proper water-cement ratio and thus the strength of the mix design. In the dry-mixed process the nozzle operator controls the amount of water in the mix and thus controls the mix design. The mix may be of a lesser strength or of a greater strength. It is not until later and many days later that one finds out if the water added at the nozzle was the proper amount. Another advantage that the wet-mixed process has over the dry-mixed process is that there is assurance that the water is properly mixed with the other ingredients and no sand pockets exist. In other words all ingredients are mixed together well in the wet-mixed process whereas in the dry-mixed process one runs a risk that not all ingredients are mixed thoroughly and as a result the material may not adhere to the surface properly when placed upon it. A third advantage of the wet-mixed process over the dry-mixed process is that dust control is much more manageable with the wet-mixed process as there is considerably less dust.

The appropriate advantages of the dry-mixed process over the wet-mixed process is that the dry-mixed process is better suited for placing mixes containing lightweight porous aggregates and that in the dry-mixed process the use of longer hose lengths is possible. However, these advantages do not outweigh the fact that in the wet-mixed process control exists with all ingredients and especially with the water. The amount of water placed in the wet-mixed process is without question as it is known whereas in the dry-mixed process it is the guesswork of the nozzleperson. Thus, it should be obvious that the wet-mixed process provides a safer method in determining the end results.

It should also be understood that the strength of pneumatically applied concrete after 28 days of curing ranges from 3000 to 7000 psi. In addition values greater than 7000 psi have been received. Usually, specifications related to shotcrete call for values in the 3500 to 4500 psi range.

Overall the main advantages of the use of shotcrete for bridge repair, dams, swimming pools, foundations, and the like is its strength, high density, bonding ability, and its low shrinkage capabilities.

PROBLEMS

13.1. List and discuss the four types of glazing material utilized today.

13.2. Explain the failure mechanism of glass and the ANSI standards associated with such.

13.3. What is concrete block and how is it utilized in today's building industry?

13.4. List and explain the various types of concrete block utilized.

13.5. How is brick manufactured?

13.6. What is mortar and how is it different than concrete?

13.7. What are the four types of mortar and how do they differ?

13.8. Discuss what mortar types are best suited for Fire Brick, Glass Block Masonry, Piers of hollow walls and Non loadbearing partitions.

13.9. What is shotcrete?

13.10. What is the difference between the wet-mixed process and the dry-mixed process? What are the advantages of each?

REFERENCES

ADAMS, J. T., *The Complete Concrete, Masonry, and Brick Handbook*, Van Nostrand Reinhold Company, New York, 1983.

CLYDE, JAMES E., *Construction Inspection: A Field Guide to Practice*, Wiley Interscience, New York, 1979.

Masonry Design Manual, Third Edition, Masonry Industry Advancement Committee, Masonry Institute of America, 1979.

National Concrete Masonry Association, *A Manual of Facts on Concrete Masonry*, Mortars for Concrete Masonry, 1989.

VAN ORMAN, HALSEY, *Handbook of Home Construction*, Van Nostrand Reinhold Company, New York, 1982.

14

DESIGN FOR THE ENVIRONMENT

INTRODUCTION

Design for the Environment (DfE) is a term used to describe the systematic integration of environmental principles and considerations into the design of products, structures, and processes. The main objective of this practice is to design products that are environmentally compatible without compromising safety, economic viability and overall quality. Engineers all over the world are becoming increasingly conscious of the environmental demands placed on their design and construction practices. These demands stem from the realization that earth has both a limited amount of resources that can be harvested to meet human needs, and a limited capacity to absorb wastes that are generated throughout the lifetime of structures and products. As the earth's population continues to increase and the standards of living improve worldwide, global environmental stresses are on a steep climb. No action will undoubtedly have devastating implications for the future of humankind.

Material selection is a key component of a designer's practice. Since all building materials are derived from earth and are eventually returned to earth as residues, the environmental burden associated with their life cycle is increasingly becoming an important design criterion. It is such a life cycle thinking that forms the basis of DfE. The main features of DfE are minimizing materials and energy and maximizing reuse and recycling while keeping other criteria such as functionality, performance, aesthetics and price within acceptable design requirements.

Figure 14-1 depicts a simplified life cycle of a product. The life cycle stages are generic and can be applicable to a consumer product as well as a component of a commercial building. One can observe that each stage of the productís life cycle involves raw materials and energy inputs and waste outputs. The quantification of these inputs and outputs and the as-

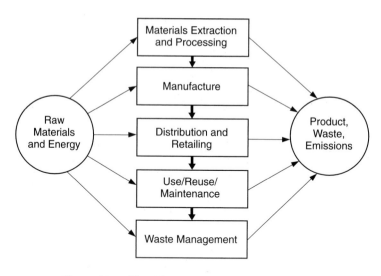

Figure 14-1   Simplified Stages of Product Life Cycle

sociated environmental impacts is achieved by a structured technique called Life Cycle Assessment (LCA). LCA provides a useful tool for decision making, and when combined with cost data and other design requirements, it provides the basis for environmentally compatible design.

This chapter addresses emerging topics on design for environment including material budgets, life cycle assessment, materials selection, concepts of environmental sustainability, use of recycled materials, environmental audits and design for disassembly and maintainability. It introduces students to environmental concepts that are on the cutting edge of engineering practice in the planning, design and construction of structures and processes.

THE PRINCIPLE OF SUSTAINABLE DEVELOPMENT

The United Nations World Commission on Environment and Development defined sustainable development as "development that meets the needs of the present without compromising the ability of future generations to meet their own needs". In order to better understand the complexity of this concept, one should look at the global conditions, which have led in the pursuit of sustainable development. They are as follows (Weston, 1993):

- Increasing population densities
- Concentration of population in urban areas
- Increasing rates of per capita resource use
- Overharvesting of renewable resources
- Exhaustion of nonrenewable resources
- Mismanagement of natural habitat
- Degradation of environmental quality

- Extinction of species
- Greater risks to individual human health, safety and security
- Increasing disparity in living standards
- Escalating terrorism, local warfare and threats to national security

Although the concept of sustainable development is an economic one, its implementation is a complex undertaking, which incorporates social, cultural, technological and educational issues in the local, regional and global scales. It involves both individual and collective efforts.

Some of the conditions driving sustainable development are closer to, and can be directly impacted by engineering practice. Here are some ways that engineers/designers can have a direct impact on achieving sustainable development:

1: **Use of ecologically benign materials:** The selection and use of materials and products that minimize the destruction of the global environment is one of the most important sustainable development concepts in engineering practice. Using materials derived from renewable resources, products made from recycled materials and materials that do not require toxic chemicals for manufacture, are some examples of sustainable practices.

2. **Energy efficient designs:** The great majority of energy used today comes from depletion of global natural resources. Energy valued more than $85 billion per year is used in commercial buildings in the United States. Implementation of energy saving designs contribute greatly to sustainable development.

3. **Decreasing risks to human health and the environment:** Minimizing use of toxic chemicals, reducing waste outputs to land, water and air, and designing structures that provide a healthy environment for the users are environmentally compatible practices.

Conceptualization of sustainable practices is a relatively simple task and one can contribute several more components of sustainable engineering practice. Quantification and decision making however, are hampered by the lack of well-established analysis tools and the availability of useable information. Significant advances have been made recently in developing methodologies, tools and information that help engineers incorporate sustainability principles in design and construction practices. These are summarized in the following sections.

LIFE CYCLE ASSESSMENT

Life Cycle Assessment (LCA) is a tool to evaluate the environmental impacts of a product or activity in a holistic manner across its entire life from raw material extraction to the return of all residues to earth. According to the US Environmental Protection Agency (USEPA, 1992), LCA includes four distinct but interrelated components:

1. *Scoping or Initiation:* This is the first step of LCA and entails the exact definition of the assessment. This is the step where the goals of the LCA are clearly defined, all the resources available to the LCA team are identified, and the boundary conditions that limit the extent of the LCA are well understood (Graendel and Allenby, 1995). LCA

is a very data-intensive process and one can easily understand that if it is not scoped adequately, it can easily become an endless, laborious exercise, or so limited that it may miss the systems aspect of the problem that needs to be solved.

2. *Inventory Analysis:* This is the most developed component of LCA. It entails quantification of energy, raw materials, air and water emissions, solid wastes, and other environmental releases throughout the life cycle of a product, from raw material extraction to disposal. A complete inventory analysis includes quantification of all inputs and outputs of the four major life cycle stages:

- Raw Materials Acquisition: Includes all activities associated with the extraction of all materials from earth including transportation. For metals, for example, this includes mining operations, ore processing and physical/chemical separation, and material refining up to the point where finished materials are produced.
- Manufacturing: Encompasses materials manufacture, product fabrication such as the fabrication of steel beams, and filling/packaging and distribution of components to the user.
- Use/Reuse/Maintenance: Entails all activities after distribution of the product to the user and includes all maintenance activities throughout its life in service.
- Recycle/Waste Management: Includes all activities associated with the product after its intended service life has been completed, such as recycling, waste processing, or disposal.

A systematic approach to inventory analysis is necessary. Every step in the life cycle of a product is carefully delineated and is classified within one of these stages. Boundary conditions are drawn around each process/activity component and data are gathered for each component pertaining to energy and raw material inputs and waste outputs, including transportation. Special considerations such as geographic distribution, alternative energy sources, and others must often be incorporated in the analysis. For example, if a product is to be incinerated at the end of its service life at an incinerator that produces useable energy, the appropriate energy value should be 'credited' to this product's energy budget.

3. *Impact Assessment:* This process entails the identification and quantification of the environmental impacts of a product or activity. The information obtained in the inventory analysis is used to quantify environmental burdens throughout the life cycle of the product. It is not difficult to understand that quantifying environmental impacts can be a very complicated process. There are two basic questions that are targeted in an impact assessment: first, identification of the impacts on specific environmental properties at each life cycle stage and second, the severity of each particular impact. Specific impacts on environmental properties include:

- Depletion of abiotic resources such as minerals, fossil fuels, and soil
- Depletion of biotic resources such as plant and animal species
- Toxicity to humans
- Toxicity to the environment
- Odor, noise, radiation and heat pollution
- Precipitation acidity

TABLE 14-1 RELATIVE RANKING OF ENVIRONMENTAL HAZARDS

High-Risk Problems	Medium-Risk Problems	Low-Risk Problems
Habitat alteration and destruction	Herbicides/pesticides	Oil spills
Species extinction and loss of biological diversity	Toxics, nutrients, biological oxygen demand and turbidity in surface waters	Groundwater pollution
Stratospheric ozone depletion	Acid deposition	Radionuclides
Global climate change	Airborne toxics including smog	Acid runoff to surface waters
		Thermal pollution

(adopted from Graedel and Allenby, 1995)

- Soil erosion
- Surface and ground water quality
- Indoor and outdoor air quality
- Ozone formation (ground level)
- Ozone depletion (stratosphere)

The Environmental Protection Agency's Science Advisory Committee has prioritized environmental hazards on the basis of several criteria and classified them on a relative scale as shown in Table 14-1.

4. *Improvement Assessment:* The goal of improvement assessment is to identify and evaluate various options that can lead to environmental improvements in processes and activities throughout the life cycle of the product. The results of the inventory analysis and impact assessment form the basis for executing an improvement assessment. Improvements can be made in all stages of a productís life cycle. For example:

- Raw material extraction: Use better ore mining techniques that minimize by-products, soil erosion, and water use, better tree farming management practices, use of recycled materials whenever possible
- Product design and manufacturing: Design products that can be easily disassembled into homogeneous, recyclable components; minimize use of hazardous chemicals in the manufacturing process by replacing with environmentally benign substitutes
- Packaging and distribution: Design lighter, recyclable, packaging; minimize transport distance
- Consumer use: Design energy efficient products, improve product maintenance
- Waste management: Design for disassembly and recyclability; identify components for better separation

It is clear that LCA is a data-intensive process that requires a systematic approach. Although the LCA concept is not new, its development as a design and management tool is still in its infancy. As such, large data gaps exist which make LCA application quite difficult. It is only very recently, that various governmental, academic and other organizations have undertaken the task of organizing electronic databases designed specifically for LCA applications. Government databases may include data collected as part of enforcing environmental compliance such as the US Environmental Protection Agency's Toxics Chemical Release Inventory (TRIS), the Integrated Risk Information System (IRIS) and the Aerometric Information Retrieval System (AIRS). Industrial data sources are less complete and are usually industry specific, such as: the American Iron and Steel Data Base, published by the American Iron and Steel Institute (AISI), the Coke and Coal Data Base, published by the American Coke and Coal Chemicals Institute (ACCI), and the Wood Data Base, published by the Western Wood Products Association. These databases are not LCA specific but contain information that can be valuable in conducting an LCA.

A limited number of LCA-specific databases are currently under development. Some of them are developed by private consultants and are not available for public use. An electronic database IDEMAT developed by the Delft University of Technology in the Netherlands, provides detailed information about materials, processes and components used in product manufacturing. The emphasis is on physical and environmental properties of construction materials.

In order to provide a systematic methodology for executing LCA, Graedel and Allenby (1995) introduced the matrix concept for materials and process audits. Matrices are used to condense information in order to increase the level of comprehension of large amounts of information over a wide spectrum of topics. Matrices are also very useful in comparing alternative designs or processes for the same product.

A matrix template has two axes, one consisting of the life cycle stages of a product, and the other consisting of specific issues within the evaluation category of interest. There are four categories that constitute the primary matrices: Manufacturing Matrix, Environmental Matrix, Toxicity/Exposure Matrix and The Social/Political Matrix. A system of symbols is used to denote the relative level of concern of each issue. An example of an environmental impact matrix for a specific product is shown in Figure 14-2.

A close examination of this matrix shows that the materials extraction, manufacturing and disposal life cycle stages of this product present the most significant environmental challenges and provide the greatest opportunity for improvement.

A summary matrix of all primary matrices can be used to compare the overall performance (toxicity/exposure, environmental, manufacturing and social/ political) impacts of alternative designs or products.

Although some scientists advocate that cost should not be considered in the LCA analysis, economic consideration is an essential criterion of the life cycle of a product or project. Life cycle costing includes, in addition to the conventional and liability costs (i.e. capital, labor, equipment, permits), environmental costs. Environmental costs can be divided in two broad categories, internal and external. Internal costs are those that are incurred by the various organizations directly involved in the product's life cycle such as, the raw material processor, or the manufacturer. Internal costs include environmental permits, pollution prevention/control equipment, insurance, remediation, hazardous waste management, et cetera. External costs are those that are not directly paid by an organization but are passed to the society as a whole. These costs include health care resulting from adverse environmental effects (smog, ozone depletion, waterborn diseases), resource depletion, acid deposition etc. External costs are very difficult to estimate because large-scale environmental impacts are not easy to quantify.

	Material Extraction	Manufact-uring	Packaging	Distribution	Consumer Use	Reuse/Recycle	Disposal
Resource Depletion	Moderate	None	None	None	None	None	Minor
Energy Consumption	Significant	Moderate	None	Minor	None	Minor	Moderate
Water Pollution	Moderate	Significant	None	None	None	Minor	None
Water Use	Moderate	Minor	Minor	None	None	None	None
Solid Waste	Minor	Moderate	Moderate	None	None	None	Significant
Air Emissions	Significant	Significant	Minor	Minor	None	None	Moderate

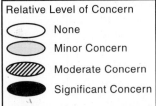

Relative Level of Concern

- None
- Minor Concern
- Moderate Concern
- Significant Concern

Figure 14-2 Example of an Environmental Impact Matrix

MATERIAL SELECTION

Material selection is one of the first design steps where engineers begin applying life cycle principles. Designers in the past were seldom concerned with issues of global sustainability, resource depletion or post-service fate of materials used in their designs. In order to take those issues into consideration, the designer must have a good understanding of the environmental impacts associated with the life cycle of the materials. Figure 14-3 shows a generic product life cycle. In selecting materials for design, engineers must consider environmental impacts imparted by each stage of the material's life cycle. Energy use varies from material to material and across the various life cycle stages. An indicator that is sometimes used for DfE purposes is the total embodied energy of a material at the point of use. In addition to energy use, there are several other factors that are considered in material selection including:

Resource Depletion

All building materials we use originate from earth. Raw materials are either stored in finite amounts in the earth's crust such as metals, minerals and fossil fuels, or are produced by biological activity such as wood. The resources stored close to the earth's surface, and are highly variable, both geographically and in terms of relative abundance. Both elements, and compounds that are formed from combinations of these elements, are of finite quantity in

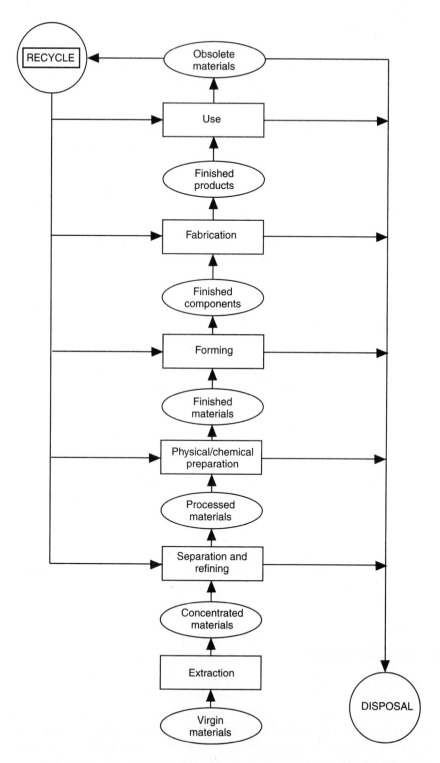

Figure 14-3   Generic Life Cycle of a Product (Graedel and Allenby, 1995)

the earth's crust. For all practical purposes they are considered to be non-renewable. One can look at the relative abundance of elements in the earth to make a first assessment of the sustained availability of a particular material. For example iron and aluminum make up about 6 and 8 percent of the earth's crust respectively and are much more abundant than copper, chromium and lead.

Since economically extractable elements are concentrated as deposits in specific geologic formations around the globe, they are subject to diffusion after they are used. For example, once a ton of metal has been extracted from earth, and although it may return to earth as residue, it has been depleted as a resource because it is not recoverable. The only recourse is to extend the useful life of the metal by reusing and/or recycling. Recycling requires that the metal is re-concentrated as scrap and re-processed to produce new products.

On the other hand, renewable natural resources are either self-sustained, such as solar energy, or can be reproduced at relatively short time scales such as wood. Wood is the most common renewable building material. The total wood consumption worldwide is about 3.2 billion metric tons annually. Industrial timber, used for lumber, particleboard, and plywood, makes up a little less than half of the worldwide consumption and is more than the consumption of steel and plastic combined. Developed countries are responsible for about 80 percent of the industrial timber consumption and only 60 percent of the production. Although wood is considered a renewable resource, poor management practices in the past have resulted in extensive deforestation in certain parts of the world. Today, only about 25 percent of the world's forests are managed for commercial wood production. Such management practices include replanting, soil conservation, and disease control.

Materials Extraction Impacts

The extraction of raw materials from the earth's crust is an energy intensive process with devastating environmental impacts. Due to low concentration of raw product in the deposit, mining operations require the removal of very large quantities of overburden soil and the processing of large volumes of ore. For example, more than 350 tons of overburden must be removed to recover 1 ton of copper. Ore processing requires substantial quantities of water and produces significant air pollution and groundwater pollution problems. Significant environmental impacts of mining include:

- Soil erosion and loss of topsoil
- Water quality impacts to both surface and groundwaters due to acid drainage
- Loss of habitat
- Destruction of flora and fauna
- Land subsidence
- Generation of solid waste (mine tailings)
- Socio-economic impacts (displacement of communities)
- Air pollution, especially particulate generation
- Noise and vibration

It is obvious that careful management practices must be implemented to mitigate the wide variety of environmental impacts of mining. One of the large environmental problems facing several communities today is the rehabilitation and reclamation of lands that have been devastated by strip mining operations in the past.

MATERIAL MINIMIZATION

Minimizing material use in the design of products and structures is a practice that leads to several environmental benefits such as decreasing natural resource depletion and lowering the amount of solid waste generated. In addition, project cost savings may be realized from such practices. Over the past two decades, industry has made significant improvements in making smaller, more compact products. The total consumption in carbon steel in the United States, for example, declined by more than 22 percent from 1970 to 1982. The construction industry, which accounted for over 21 percent of the carbon steel consumption in 1982, led other industries with about 36 percent decline. Such declines in material utilization come in part because of technological and design innovations. The use of computer simulation has led to better understanding of materials' structural properties and has contributed to reducing the mass of new structures. There are several ways that innovative engineering can contribute in materials minimization:

- Use of lightweight materials
- Use of thinner, composite sections
- Use of special reinforcements in structural members

Attention must be paid not to compromise other environmental attributes of a component while attempting to minimize the material mass. For example, composite, non-homogeneous materials are usually more difficult to recycle.

Reduction of Toxic Materials

Thousands of tons of asbestos have been used worldwide in the construction industry as insulation and flooring material since the early 1900's. In 1971, asbestos was listed as a hazardous air pollutant by the US EPA and since then several thousand pages of regulation and legal briefs have been written and millions of dollars have been spent to remove asbestos from buildings. Asbestos is only one of several materials that are used in construction which have been found to be toxic to humans and to impart adverse environmental impacts on the environment. Some common metals are included in this list. Lead was one of the most popular metals used in the paint industry. Lead pipes have been used in plumbing and construction since the time of the ancient Greek civilization. Lead and lead compounds have been used in lead-acid batteries, glass and ceramics, dyes, pigments and surface coatings, solder, plastics and rubber. It is estimated that over 250,000 tons of lead are discarded annually in municipal solid waste streams. Health effects of lead ingestion include retardation and brain damage, especially in children.

Mercury has long been recognized as a toxic substance. It is estimated that over 10,000 tons of mercury are released worldwide from both natural and manmade sources each year. Sources include smelters, power plants, thermostats, thermometers and light bulbs. Mercury has been found to accumulate in fish and is passed on the food chain. Health effects to humans include, poisoning and brain damage. Other toxic metals include cadmium, arsenic, chromium and nickel.

In addition to metals, hundreds of other compounds used in the processing and manufacture of building materials such as chlorinated solvents, polyaromatic hydrocarbons and

TABLE 14-2   SOLID WASTE BREAKDOWN BY CATEGORY IN THE US

Solid Waste Category	Approximate Percent by Weight
Agricultural	50%
Mining and Primary Raw Material Processing	37%
Industrial (non-hazardous)	7.4%
Industrial (hazardous)	0.6%
Municipal	5%

wood preserving agents (pentachlorophenol, creosote) have been identified by the US EPA as harmful to human health and the environment. The design engineer must consider the toxicity of materials in light of the hazard they pose during the service life of the structure and their potential to generate hazardous waste after the structure is demolished. Perhaps one of the most difficult issues is predicting the future hazard potential of materials that today are considered safe for use.

Recycling

More than 4 billion tons of solid waste is produced annually in the United States. Table 14-2 shows the solid waste distribution by category.

Table 14-3 shows a breakdown of municipal waste produced by category and the percentage of each fraction that was recovered in the US in 1995. Overall, 208 million tons of municipal waste was generated from which 56.2 million tons, or 27%, were recovered.

One can see that the amount of solid waste produced in the US is staggering. Most of this waste is landfilled, contributing to further degradation of the environment by polluting the land, water and atmosphere.

Recycling is the most important means by which we can extend the life of materials. It has two major environmental benefits: curbing the depletion of our natural resources and reducing waste disposal. Engineers must consider both maximizing the use of components made from recycled materials and incorporating materials in their design that are easily recyclable. The benefits derived from recycling can be substantial. According to the Institute of Scrap and Recycling Industries Inc., over two billion dollars in savings are realized in the USA annually in avoided solid waste costs by recycling scrap iron and steel. In addition to solid waste savings, using scrap iron to make new steel instead of virgin ore, results in 74 perecent savings in energy, 90 percent savings in virgin material use, 86 percent reduction in air pollution, 40 percent reduction in water use, 76 percent reduction in water pollution and 97 percent reduction in mining wastes.

It is estimated that over 4 million tons of construction related solid waste is generated annually in the US alone. This includes construction and demolition waste, and land clearing debris. At an average cost of $60 per ton for landfilling the total disposal cost climbs to $2.4 billion per year. It is also estimated that depending on local and market availability, over 80 percent of this waste can be recycled. Recycled materials in the construction industry include:

- Metal from heating, ventilating and air-conditioning systems
- Construction steel, beams, frames, trusses, reinforcing bars

TABLE 14-3 MUNICIPAL WASTE COMPOSITION AND RECOVERY IN THE USA IN 1995

Waste Type	Percent Produced by Weight	Percent Recycled by Weight
Paper	39.2%	40.0%
Food	6.7%	4.1%
Yard Waste	14.3%	30.3%
Glass	6.2%	27.3%
Metals	7.6%	39.7%
Plastics	9.1%	4.7%
Wood	7.1%	13.5%
Other	9.8%	10.9%

(source: US EPA530-R-97-015)

- Glass and aluminum
- Concrete and brick
- Lumber, particle board, plywood
- Roofing materials, ceiling tiles and drywall
- Land clearing debris

The degree of recycling is project specific and usually depends on local construction waste disposal fees, local recycled markets and prices. In a construction project recycling is the responsibility of many stakeholders including architects, design engineers, the owner, general contractor/construction manager and responsible local authorities. For this reason, in order to be properly implemented, it must be part of a project waste management plan. This plan should contain information and assign responsibilities including:

- Estimates of types and quantities of waste to be generated
- Identification of types of wastes that can be recycled
- A plan for collection, separation and storage of wastes and recyclables
- Identification of local recycling facilities and waste disposal facilities
- Cost-benefit analysis of recycling including how costs and savings will be handled contractually
- Assignment of recycling responsibilities throughout the project life

There are several opportunities in civil and transportation engineering projects to both recycle materials and use materials made from recyclables. It may take a little longer for the design professional to research various materials and implement waste management practices, but the effort is due to pay off both in reducing tangible project costs and helping the environment. Such opportunities include:

- Recycling of contaminated soils and re-use as fill material after treatment
- Use of recycled paints as primer or first coating in structural members and walls
- Use of particle board from recycled wood
- Use of recycled plastic lumber and other recycled building components
- Use of recycled ceiling acoustic tiles

- Use of recycled structural steel
- Use macadam that contains recycled fillers in paved areas
- Use mortar and lean concrete containing recycled materials as aggregate

Recycling can be dramatically improved if certain considerations are given in the design and manufacturing/construction of products and structures. Such considerations include:

- Design products to facilitate easy disassembly of components and materials at the end of service life
- Use homogeneous materials as much as possible instead of composites and blends of heterogeneous materials
- Increase the potential for refurbishing and replacing components instead of wasting entire systems or assemblies
- Fasten components together so that can be easily separated
- Avoid the use of hazardous materials
- Identify the composition of materials to facilitate easier separation and sorting

In order to ensure that environmental factors are adequately considered in the material selection process and more generally, to follow DfE principles, engineers may find the use of checklists useful. A checklist for material selection will include questions such as:

- Do any proposed materials come from limited natural resources?
- Has the use of hazardous materials been minimized?
- Are any proposed materials difficult to recycle?
- Has the use of recycled materials been maximized?
- Has material reduction been adequately considered?
- Has the number of different materials (or composites) been minimized?
- Have all material replacement/substitution options been adequately considered?

ENVIRONMENTAL QUALITY STANDARDS

Incorporating environmental quality in product development has become a goal of the modern corporate world. The need to develop environmental quality standards and mechanisms for uniform applications of such standards across the globe has been long recognized. The International Organization for Standardization (ISO) is a worldwide federation composed of members of more than 110 countries. ISO was founded in 1946 to promote the development and adoption of international standards. ISO has recently undertaken the task of developing generic standards that provide business with a structure to manage environmental quality. The ISO 14000 is a series of standards under development that can be voluntarily adopted by industry in order to guide organizations through the implementation of effective environmental management systems. The ISO offers standards, guidance documents for implementation and monitoring the compliance with these standards, and a certification program for organizations worldwide. The ISO 14000 series consists of six standards and the supporting documentation, some of which are still under development. They are as follows:

Environmental Management Systems: specifies requirements to enable an organization to develop policies and the management infrastructure for implementing environmental quality and evaluating progress.

Environmental Auditing: provides requirements and guidelines for performing environmental audits.

Environmental Performance Evaluation: provides a process for the organization to measure environmental performance on the basis of pre-determined uniform criteria.

Life Cycle Assessment: provides methodology for performing LCA including environmental impact and improvement analysis.

Environmental Labeling: Provides methodology and guidance to unify environmental labeling.

Environmental Aspects in Product Standards: provides DfE methodology and guidance.

Like the ISO 9000 standard series, which addresses product quality, the ISO 1400 standards are rapidly becoming a global industry requirement. By selecting products manufactured by ISO 14000 certified organizations, design professionals can take the first step toward applying DfE principles.

PROBLEMS

14.1 Develop an environmental impact matrix for cement production and assign relative weights to each matrix component.

14.2 Develop a list of recycled materials that could be used as aggregate substitutes in a) concrete and b) bituminous mixes.

14.3 Develop a list of environmental impacts of iron ore extraction.

14.4 Research and discuss the differences in recycling potential between thermoplastics and thermosetting plastics. Which type is preferable from the recycling point of view?

14.5 Develop a list of reasons that explain the substantial differences in recycling rates of paper, metals, plastics and glass.

REFERENCES

GRAEDEL, T. E. and ALLENBY, B. R., *Industrial Ecology*, Prentice Hall, 1995

US Environmental Protection Agency, Life-Cycle Assessment: Inventory Guidelines and Principles, EPA/600/R-92/036, 1996

US Environmental Protection Agency, Characterization of Municipal Solid Waste in the United States: 1996 Update, EPA530-R-97-015, 1997

WESTON, R. F., Toward Better Understanding the Concept of Sustainable Development, Roy F. Weston Inc., 1993

15 ENVIRONMENTAL CONSIDERATIONS IN CONSTRUCTION

Environmental issues have become of great importance in the construction industry. It is therefore necessary that engineers who are involved in the planning, design, and execution of construction projects become familiar with both the technical and regulatory issues associated with their projects. The basic premise is that every project must be performed in a manner that protects the environment and public health.

There are several environmental problems encountered in construction, but the most important a civil engineer may face can be divided as follows:

Handling of hazardous materials.

Designing and constructing waste-containment facilities.

Designing and executing environmental remediation projects.

This chapter presents a brief overview of these issues.

REGULATORY ISSUES

Environmental protection and restoration is driven by regulations. During the past 20 years many environmental laws has been adopted by various agencies in the United States. They range from national acts passed by Congress to laws adopted by local municipalities. These laws are enforced by the federal and local authorities. Failure to comply with environmental regulations may result in penalties ranging from fines to jail terms. As part of the environmental regulations, an environmental-permitting infrastructure has been developed. Environmental permits may be required to execute various project activities at a particular site. Engineers therefore must have an in-depth knowledge of the environmental regulatory requirements of projects of which they are in charge.

Information on applicable environmental laws, regulations, and environmental permits can be obtained through various regulatory agencies including the following:

The United States Environmental Protection Agency (USEPA).

The department of environmental protection or conservation of the particular state in which the project is executed.

County or municipal departments of health.

Various other state departments, such as the department of health and the department of transportation.

Hazardous Materials/Hazardous Waste

Since prehistoric times people have tried to improve their standard of living by improving the properties of materials to suit specific purposes. Since then, and especially during the technological revolution of the twentieth century, thousands of new materials have become part of everyday life. Along with the benefits that come from these materials, various hazards to human health are introduced in and away from the workplace. Such hazards come from routine or accidental exposure to harmful chemicals that adversely impact human health and the environment.

Hazardous materials are used in the construction industry either in the manufacturing of construction materials or in sites where construction activities occur. For example, a wide variety of solvents and other chemical agents are used in asphalt processing plants, cement production facilities, and steel plants. In addition different paints may contain a variety of harmful substances. It is therefore necessary that engineers understand the nature of various hazards that may arise in the workplace.

In 1983 the U.S. Occupational Safety and Health Administration (OSHA) published a regulation in the Federal Register (48 FR 53280) that mandates the communication of hazards to workers. This regulation is called the Hazard Communication standard, and makes provisions for hazard evaluation, container labeling, Material Safety Data Sheets, and worker education and training. The standard requires employers to develop a comprehensive program to communicate hazards to their employees that contains at minimum the preceding provisions.

Labels must be used in chemical containers to identify the contents. The labels according to OSHA must contain the following:

Chemical identification.

Hazard warning.

Name and address of manufacturer.

The OSHA standard mandates that all chemical manufacturers develop a Material Safety Data Sheet (MSDS) for each hazardous chemical they produce or import. The MSDS is intended to be a miniencyclopedia containing key information about the specific chemical. It includes the following information:

Identification of the chemical as used in the container label.

Manufacturer's name and address.

Emergency telephone number.

Telephone number for further information about the chemical.

List and identification of hazardous components.

Key physical and chemical characteristics of the substance, such as boiling point, vapor pressure, melting point and water solubility.

Fire and explosion hazard data such as flash point and special fire-fighting procedures.

Reactivity data such as hazardous decomposition and by-product formation.

Acute and chronic health-related hazards such as carcinogenicity, signs and symptoms of exposure, and emergency procedures.

Precautions for safe handling and use including waste disposal methods.

Control measures such as respiratory protection and ventilation requirements.

Monitoring of the project or facility environment may also be required if various hazardous materials are used.

Hazardous wastes often result from the use, improper disposal, or accidental releases of hazardous materials. Large volumes of hazardous wastes may be encountered or produced during the course of various construction activities. For example, it is not uncommon during foundation excavation for a project to encounter soils that have been contaminated by accidental releases of hazardous materials or routine industrial operations in the past. In cases in which old industrial or storage facilities are demolished the debris may contain contaminants at high levels. These wastes must be first classified, and then treated and disposed in a safe manner to protect the environment and public health. To achieve this, both physical and chemical properties of the waste must be determined.

Waste sampling.   Any time waste is encountered a comprehensive sampling plan is required to define the physical and chemical characteristics of the waste. Usually, the sampling is performed in a manner that the collected data are most representative of the bulk waste. In absence of a regulatory directive or specific sampling protocol, the following standard tests could be used (Bagchi, 1990):

1. ASTM Standard D140 for extremely viscous liquids.
2. ASTM Standard D346 for crushed or powdered wastes.
3. ASTM Standard D420 for soil or rocklike waste.
4. ASTM Standard D1452 for soil-type waste.
5. ASTM Standard D2234 for fly ash-type waste.

Chemical tests.   Two common types of chemical tests are usually performed.

Bulk Chemical Analysis.   The goal of this test is to identify the chemical composition of the waste. Currently there is no standard procedure available for all wastes. However, there are several test guidelines, provided by the Water Pollution Control Federation, ASTM, USEPA, and other organizations.

Leaching Test.   The most-used leach test is the USEPA Toxicity Characteristic Leaching Procedure (TCLP). It requires grinding the waste to ensure maximum surface contact area with an acidic extraction fluid. Depending on the environmental conditions at the disposal site that the waste will be exposed to, a leaching media should be determined (i.e., water leaching, acid leaching). The ratio of waste to extraction fluid is set to achieve a saturated solution. The concentration of various contaminants present in the leachate is obtained and compared against the regulatory levels set by USEPA. If the regulatory limits are exceeded, the waste is characterized as hazardous.

In addition to the EPA TCLP test a waste can be characterized as hazardous on the basis of the following criteria:

The waste is listed by EPA as hazardous.
The waste is corrosive.
The waste is ignitable.
The waste is reactive.

MATERIALS FOR CONSTRUCTION OF WASTE IMPOUNDMENTS

Waste impoundments (engineered landfills) are used for ultimate disposal of wastes. Both hazardous and nonhazardous wastes can be landfilled. Some hazardous wastes, however, cannot be disposed unless they undergo some kind of treatment. Landfills must be designed such that they do not allow contaminants to escape to the environment. Of particular concern is the potential of landfills to pollute the subsurface soils and groundwaters. This threat arises from the fact that as rainwater percolates through the landfill, it dissolves various pollutants and caries them downward to the underlying aquifers. The resulting mixture of pollutants is called leachate.

A major component in the design of engineered landfills is the liner and leachate collection systems. Liners are intended to prevent contaminants emanating from landfills from polluting the subsurface environment. Before the placement of any waste, the bottom and sides of the impoundment are lined with materials of very low permeability (bottom liners). After the landfill is filled to capacity, its surface is then sealed with another liner (surface liner or surface cap).

Various materials are used to construct liners for hazardous and nonhazardous waste disposal facilities. These materials can be broadly classified into three categories.

Compacted clay.   Naturally occurring clay soils are routinely used for lining landfills. The clay is brought to the site and is carefully compacted in layers ranging in thickness from 6 to 18 in. The following factors, parameters, and properties affect the construction and performance of clay liners:

Mineral composition of clay.
Grain size distribution and grain shape.
Soil fabric.
Water content at compaction.
Clay permeability.

Mechanical strength.

Clay compressibility.

Clay chemistry and organic content.

Chemical composition of leachate.

Admixtures.   Soils mixed with bentonite, asphalt, or cement have proved to be useful landfill liner materials.

Bentonite Amended Soil Liners.   Bentonite is a clay mineral of the smectite group, similar to montmorillonite. Bentonite absorbs water and swells heavily, creating a low-permeability material. This property makes it desirable as an additive to coarser soils to form a good liner material. Typically soil-amended liners are composed of well-graded sand mixed with bentonite. The percentage of bentonite may vary between 3 to 15 percent depending on the type of soil used. The desirable mixture is one that has low permeability and is stable. Mixing can occur either on site or in a mixing plant. Trial mixes must be tested to obtain the one yielding the lowest permeability.

Asphalt and Cement Admixture Liners.   Asphalt and cement have been used to produce soil-asphalt or soil-cement mixes with low permeability and acceptable performance. The USEPA has sponsored tests to evaluate the compatibility of several lining materials exposed to various wastes. These tests indicated that soil-cement mixes became stronger and less permeable with time. The soil-cement liners were also exposed to various wastes including toxic pesticide formulations, oil refinery sludge, and plastic wastes. Results indicated no seepage had occurred through the soil-cement liners following $2\frac{1}{2}$ years of exposure.

Synthetic membranes.   Synthetic membranes (or geomembrane) are manufactured by the plastics and rubber industries using polymeric materials. The properties of the polymers are controlled by the size of the molecules, the molecular size distribution, and their shape and structure. Additives are sometimes used to improve the manufacturing process, durability, and the performance of the final membrane. Several synthetic membranes with different chemical compositions are available in the market. The selection should be based on the cost of installation, the workability, the chemical compatibility with the waste, and the mechanical properties of the membrane. Membranes are manufactured in sheets of thicknesses ranging from 20 to 150 mil (1 mil=0.001 in.) and varying dimensions. One of the most important concerns in the use of geomembranes is obtaining good splices of the sheets in the field. Other important concerns include the puncture resistance of the polymer durability and resistance to ultraviolet (UV) light. Table 15-1 lists the most frequently used types of synthetic membranes and some of their characteristics.

Membranes undergo several tests including the following (Koerner, 1990):

Tensile strength (ASTM D412, D638, D882).

Seam strength (ASTM D413, D3083).

Tear resistance (ASTM D2263).

Impact resistance (ASTM D1709, D3029, D3998).

Puncture resistance (ASTM D2582, D3787).

Swelling resistance (ASTM D570).

TABLE 15-1 COMMONLY USED SYNTHETIC MEMBRANES AND THEIR CHARACTERISTICS

Membrane Type	Resistance to Ultraviolet Light and Weathering	Resistance to Temperature Variations	Strength Characteristics	Chemical Resistance	Seam Properties
Polyvinyl chloride	Poor	Poor	Good	Good	Good
Butyl rubber	Good	Good	Poor	Good (poor for hydrocarbons)	Poor
High-density polyethylene	Good	Good	Good	Good	Good
Ethylene propylene diene monomer	Good	Good	Good	Poor for hydrocarbons and chlorinated solvents	Poor
Chlorinated polyethelene	Good	Good	Good	Poor for petroleum products and chlorinated solvents	Poor

Ozone resistance (ASTM D518).

UV light resistance (ASTM D3334).

For hazardous waste landfills, the USEPA requires that composite double liners are used. Double liners are a combination of compacted clay and synthetic membranes. Sandwiched between them are sand layers that function as leachate collection and drainage conduits. Leachate is then pumped to the surface for treatment.

When the landfill has reached its capacity, a similar liner system is installed at the surface to prevent rainwater from percolating through the waste. The surface cap is then seeded with grass.

CONTAMINATED SOILS AND REMEDIAL OPTIONS

The type of hazardous waste most frequently encountered in construction projects is contaminated soil. As contaminants are released in the soil surface they migrate downward under the influence of gravity polluting large areas. Millions of tons of soil have been contaminated by unsafe industry practices, municipal waste disposal, and accidental releases of contaminants. Engineers therefore must be familiar with the latest techniques for remediating contaminated sites.

As a result of more restrictive regulations and shrinking availability of landfills, remediation of sites contaminated with industrial nonhazardous wastes has become a priority issue. The USEPA and the environmental industry have launched an enormous effort to develop innovative, cost-effective site remedial technologies. These technologies are in different stages of development, and the most promising ones are summarized subsequently.

Stabilization/solidification (S/S) technologies.   Stabilization and solidification refer to treatment processes that are designed to accomplish one or more of the following:

1. Improve the handling and physical characteristics of the waste.
2. Limit the mobility of toxic constituents of the waste.
3. Render the waste a usable material in construction or elsewhere

Contaminated soils in-situ, without disturbing the site, or ex-situ following excavation.

Stabilization and solidification of contaminated soils encompass a wide array of treatment processes. According to the USEPA, stabilization refers to those techniques that reduce the hazard potential of a waste by converting the contaminants into their least soluble, mobile, or toxic form. The physical nature and handling characteristics of the waste are not necessarily modified by stabilization. Solidification, conversely, refers to the techniques that encapsulate the waste in a monolithic solid of high structural integrity. The encapsulation may be of fine waste particles (microencapsulation), or coarser waste agglomerated particles or even large waste blocks (macroencapsulation).

The following are the most frequently used S/S techniques.

Cement-based S/S.   Cement-based S/S is a process in which contaminated soils are mixed with Portland cement and water to create a monolithic matrix. The contaminants are incorporated into the cement matrix and, in some cases, undergo physical-chemical changes that reduce their mobility. Cement-based S/S has been applied to wastes containing various metals such as arsenic, cadmium, chromium, copper, lead, nickel, and zinc. Cement has also been used with complex wastes, oils, and oil sludges, wastes containing vinyl chloride, resins, S/S plastics, asbestos, sulfides, and other materials.

Lime-based S/S.   Lime-based S/S involves similar mechanisms as the cement techniques, the difference being the generally higher pH resulting from lime treatment. *Lime* is a rather broad term for a variety of calcium-bearing compounds. Various forms of lime have been successfully used as a soil-stabilizing agent. The most commonly used products are hydrated high-calcium lime, monohydrated dolomitic lime, and calcitic quicklime. Lime has been known to be effective in stabilizing soils contaminated with both organic and inorganic substances. It is generally considered to be more effective than cement. However, interestingly enough, lime is not so widely used as cement-based S/S techniques are.

Pozzolanic S/S.   Pozzolanic S/S involves siliceous and aluminosilicate materials, which do not display cementing action alone, but form cementitious substances when combined with lime or cement and water at ambient temperatures. The primary contaminant mechanism is the physical entrapment of the contaminant in the pozzolanic matrix. Examples of common pozzolans are clay, fly ash, pumice, lime kiln dusts, and blast furnace slag. Pozzolanic S/S is generally much slower than cement or lime S/S, and involves much larger volume increases compared with the original waste product volume.

Thermoplastic Techniques.   Thermoplastic S/S is a process in which the waste materials do not react chemically with the encapsulating material. In this technology, a thermoplastic material, such as polyethylene or paraffin, is used to bind the waste constituents into a S/S mass. The process requires trained operators and specialized equipment, and is energy intensive.

Vitrification.   This is a thermal encapsulation method suited for soils containing extremely dangerous or low-level radioactive contaminants. It may be performed ex-situ, typically using an electric furnace, or in-situ using a pair of electrodes embedded in the soil. The waste is fused with silica into a glasslike material with a very low leach rate that is safe for disposal without further treatment. The in-situ process employs an array of electrodes placed into the soil to the desired depth. A large electrical current is used as the driving force establishing a thermal gradient in the soil. As the temperature rises above 1600°C, organics present in the soil are vaporized and then pyrolyzed. Inorganics thermally decompose or enter into melt reaction. The resulting product is monolithic glass like microcrystalline structure that is inert and stable. The process is extremely energy intensive and requires specialized labor and equipment.

Soil vapor extraction (SVE).   Soil vapor extraction is an in-situ technology that has been used extensively during the past 10 years to remove volatile organic compounds (VOCs) from contaminated soils in the vadose (unsaturated) zone. The principle of SVE operation is the transfer of VOCs from the undisolved contaminant (nonaqueous phase), contaminated moisture (aqueous phase), and the soil grains (solid phase) to the air contained in the soil pores (gaseous phase). In SVE applications air flow is induced through the soil matrix by applying vacuum in an extraction well with or without the assistance of air injection wells. Contaminated vapors are then pumped out of the soil and are treated above ground usually by activated carbon adsorption columns. Key factors influencing SVE are contaminant distribution at the site, the site hydrogeologic properties, and contaminant properties.

Soil washing/soil flushing technologies.   Soil washing refers to a process in which excavated contaminated soils are scrubbed in an aqueous stream to remove undesired contaminants. Soil flushing, conversely, refers to a similar process applied in-situ. In soil washing the soil is prepared for the washing process by screening to remove large objects like pieces of wood, concrete, and other debris. The screened soil is mixed thoroughly with water and sometimes other cleaning agents to strip the contaminants from the soil. The process occurs in devices that resemble washing machines. During the washing process the soils are segregated by size to separate the coarser components from the fine particles. The coarser grains can most of the time be washed free of contaminants and can ultimately be returned to fill the excavation. Finer soils, however, silt and clay sizes, require additional treatment. Soil flushing is an in-situ treatment process for removing organic or inorganic contaminants. In this process mixtures of water and extraction agents are pumped or sprayed into the contaminated soil zones. The wash fluid is then pumped out by recovery wells and is treated above ground.

Bioremediation.   Bioremediation refers to the application of natural microbial metabolic processes to remediate contaminated soils and other wastes. The technology has been applied both in-situ and ex-situ. The principle of this technology lies in the ability of certain types of bacteria to degrade chemical compounds. The bacteria degrative mechanism is the metabolic transformation of organic compounds into energy, biomass, and carbon dioxide. The conversion of organic carbon to inorganic carbon is called mineralization and occurs by enzymatic oxidation the presence of oxygen (aerobic degradation), or nitrite or sulfite (anaerobic degradation). Bacteria in the subsurface environment are either attached to the soil grains or held in suspension in the soil moisture or groundwater. Bacterial growth in such matrices is sensitive to various factors including toxicity of the contaminant, pH, temperature, and moisture conditions. To sustain bacterial growth it is often necessary to enhance the system by adding nutrients. The most common nutrient amendment is the addition of either phosphate or a reduced nitrogen source such as ammonia. Bioremediation has been found to treat soils contaminated with a variety of organics effectively including petroleum hydrocarbons and chlorinated solvents.

Incineration/thermal destruction.   Incineration refers to the process of combustion and ultimate destruction of hazardous wastes. In the combustion of contaminated soils, one of the major considerations is the complete destruction of chemical molecules. The process occurs in incinerators. Incinerators vary widely in size and method of operation. Some types of incinerators have been constructed to be mobile and can be moved from site to site. Others are large, centralized facilities. Important parameters to be considered in the incineration processes include (1) type of chemical constituents in the waste; (2) working temperatures; (3) residence time; and (4) turbulence. The soil must be exposed to high enough temperatures to allow complete oxidation of organic materials. The waste must remain in this temperature regime for the appropriate amount of time to ensure complete oxidation. Important incinerator design considerations include the physical and chemical properties of waste, auxiliary fuel needs, air pollution control system, and construction materials and cost. One of the most common types of incinerators is the rotary kiln—this type of incinerator is composed of a rotating primary combustion chamber and a secondary combustion chamber. Soil enters the rotating primary combustion chamber, which is heated by fossil fuel burners to temperatures ranging from 1200 to 1800°F. Volatilization and pyrolysis occur, and the combustion gases are burned on an auxiliary after the burner has been operating at temperatures in the range of 1400 to 2400°F. Residence times range from minutes to hours depending on the composition of the soil and type of the waste.

R E F E R E N C E S

BAGCHI, A., *Design, Construction, and Monitoring of Sanitary Landfill*, John Wiley & Sons, Inc., New York, 1990, 244 pp.

KOERNER, R. M., *Designing with Geosynthetics*, 2nd ed., Prentice Hall, Englewood Cliffs, N.J., 1990.

APPENDICES

SOIL

ASTM D422	319
ASTM D4318	326
ASTM D1556	337
ASTM D3282	343
ASTM D698	349
ASTM D1557	357

AGGREGATE AND CONCRETE

ASTM C29	365
ASTM C31	369
ASTM C78	374
ASTM C128	377
ASTM C136	382
ASTM C618	387
ASTM C666	390
ASTM C995	396

WOOD

ASTM D198	398

ASPHALT

ASTM D1559	417

METAL

ASTM E8	423
ASTM E23	444

Designation: D 422 – 63 (Reapproved 1990)$^{\epsilon1}$

Standard Test Method for
Particle-Size Analysis of Soils[1]

This standard is issued under the fixed designation D 422; the number immediately following the designation indicates the year of original adoption or, in the case of revision, the year of last revision. A number in parentheses indicates the year of last reapproval. A superscript epsilon (ϵ) indicates an editorial change since the last revision or reapproval.

$^{\epsilon1}$ NOTE—Section 19 was added editorially in September 1990.

1. Scope

1.1 This test method covers the quantitative determination of the distribution of particle sizes in soils. The distribution of particle sizes larger than 75 µm (retained on the No. 200 sieve) is determined by sieving, while the distribution of particle sizes smaller than 75 µm is determined by a sedimentation process, using a hydrometer to secure the necessary data (Notes 1 and 2).

NOTE 1—Separation may be made on the No. 4 (4.75-mm), No. 40 (425-µm), or No. 200 (75-µm) sieve instead of the No. 10. For whatever sieve used, the size shall be indicated in the report.

NOTE 2—Two types of dispersion devices are provided: (1) a high-speed mechanical stirrer, and (2) air dispersion. Extensive investigations indicate that air-dispersion devices produce a more positive dispersion of plastic soils below the 20-µm size and appreciably less degradation on all sizes when used with sandy soils. Because of the definite advantages favoring air dispersion, its use is recommended. The results from the two types of devices differ in magnitude, depending upon soil type, leading to marked differences in particle size distribution, especially for sizes finer than 20 µm.

2. Referenced Documents

2.1 *ASTM Standards:*

D 421 Practice for Dry Preparation of Soil Samples for Particle-Size Analysis and Determination of Soil Constants[2]

E 11 Specification for Wire-Cloth Sieves for Testing Purposes[3]

E 100 Specification for ASTM Hydrometers[4]

3. Apparatus

3.1 *Balances*—A balance sensitive to 0.01 g for weighing the material passing a No. 10 (2.00-mm) sieve, and a balance sensitive to 0.1 % of the mass of the sample to be weighed for weighing the material retained on a No. 10 sieve.

3.2 *Stirring Apparatus*—Either apparatus A or B may be used.

3.2.1 Apparatus A shall consist of a mechanically oper-

ated stirring device in which a suitably mounted electric motor turns a vertical shaft at a speed of not less than 10 000 rpm without load. The shaft shall be equipped with a replaceable stirring paddle made of metal, plastic, or hard rubber, as shown in Fig. 1. The shaft shall be of such length that the stirring paddle will operate not less than ¾ in. (19.0 mm) nor more than 1½ in. (38.1 mm) above the bottom of the dispersion cup. A special dispersion cup conforming to either of the designs shown in Fig. 2 shall be provided to hold the sample while it is being dispersed.

3.2.2 Apparatus B shall consist of an air-jet dispersion cup[5] (Note 3) conforming to the general details shown in Fig. 3 (Notes 4 and 5).

NOTE 3—The amount of air required by an air-jet dispersion cup is of the order of 2 ft³/min; some small air compressors are not capable of supplying sufficient air to operate a cup.

NOTE 4—Another air-type dispersion device, known as a dispersion tube, developed by Chu and Davidson at Iowa State College, has been shown to give results equivalent to those secured by the air-jet dispersion cups. When it is used, soaking of the sample can be done in the sedimentation cylinder, thus eliminating the need for transferring the slurry. When the air-dispersion tube is used, it shall be so indicated in the report.

NOTE 5—Water may condense in air lines when not in use. This water must be removed, either by using a water trap on the air line, or by blowing the water out of the line before using any of the air for dispersion purposes.

3.3 *Hydrometer*—An ASTM hydrometer, graduated to read in either specific gravity of the suspension or grams per litre of suspension, and conforming to the requirements for hydrometers 151H or 152H in Specifications E 100. Dimensions of both hydrometers are the same, the scale being the only item of difference.

3.4 *Sedimentation Cylinder*—A glass cylinder essentially 18 in. (457 mm) in height and 2½ in. (63.5 mm) in diameter, and marked for a volume of 1000 mL. The inside diameter shall be such that the 1000-mL mark is 36 ± 2 cm from the bottom on the inside.

3.5 *Thermometer*—A thermometer accurate to 1°F (0.5°C).

3.6 *Sieves*—A series of sieves, of square-mesh woven-wire cloth, conforming to the requirements of Specification E 11. A full set of sieves includes the following (Note 6):

---

[1] This test method is under the jurisdiction of ASTM Committee D-18 on Soil and Rock and is the direct responsibility of Subcommittee D18.03 on Texture, Plasticity, and Density Characteristics of Soils.

Current edition approved Nov. 21, 1963. Originally published 1935. Replaces D 422 – 62.

[2] *Annual Book of ASTM Standards*, Vol 04.08.

[3] *Annual Book of ASTM Standards*, Vol 14.02.

[4] *Annual Book of ASTM Standards*, Vol 14.03.

[5] Detailed working drawings for this cup are available at a nominal cost from the American Society for Testing and Materials, 1916 Race St., Philadelphia, PA 19103. Order Adjunct No. 12-404220-00.

ASTM D 422

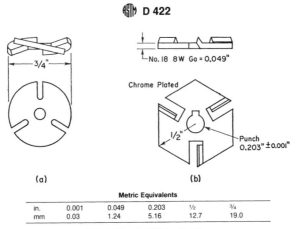

Metric Equivalents

in.	0.001	0.049	0.203	½	¾
mm	0.03	1.24	5.16	12.7	19.0

FIG. 1   Detail of Stirring Paddles

3-in. (75-mm)	No. 10 (2.00-mm)
2-in. (50-mm)	No. 20 (850-μm)
1½-in. (37.5-mm)	No. 40 (425-μm)
1-in. (25.0-mm)	No. 60 (250-μm)
¾-in. (19.0-mm)	No. 140 (106-μm)
⅜-in. (9.5-mm)	No. 200 (75-μm)
No. 4 (4.75-mm)	

NOTE 6—A set of sieves giving uniform spacing of points for the graph, as required in Section 17, may be used if desired. This set consists of the following sieves:

3-in. (75-mm)	No. 16 (1.18-mm)
1½-in. (37.5-mm)	No. 30 (600-μm)
¾-in. (19.0-mm)	No. 50 (300-μm)
⅜-in. (9.5-mm)	No. 100 (150-μm)
No. 4 (4.75-mm)	No. 200 (75-μm)
No. 8 (2.36-mm)	

3.7 *Water Bath or Constant-Temperature Room*—A water bath or constant-temperature room for maintaining the soil suspension at a constant temperature during the hydrometer analysis. A satisfactory water tank is an insulated tank that maintains the temperature of the suspension at a convenient constant temperature at or near 68°F (20°C). Such a device is illustrated in Fig. 4. In cases where the work is performed in a room at an automatically controlled constant temperature, the water bath is not necessary.

3.8 *Beaker*—A beaker of 250-mL capacity.

3.9 *Timing Device*—A watch or clock with a second hand.

4. Dispersing Agent

4.1 A solution of sodium hexametaphosphate (sometimes called sodium metaphosphate) shall be used in distilled or demineralized water, at the rate of 40 g of sodium hexametaphosphate/litre of solution (Note 7).

NOTE 7—Solutions of this salt, if acidic, slowly revert or hydrolyze back to the orthophosphate form with a resultant decrease in dispersive action. Solutions should be prepared frequently (at least once a month) or adjusted to pH of 8 or 9 by means of sodium carbonate. Bottles containing solutions should have the date of preparation marked on them.

4.2 All water used shall be either distilled or demineralized water. The water for a hydrometer test shall

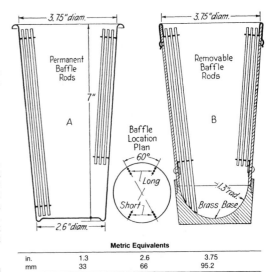

Metric Equivalents

in.	1.3	2.6	3.75
mm	33	66	95.2

FIG. 2   Dispersion Cups of Apparatus

be brought to the temperature that is expected to prevail during the hydrometer test. For example, if the sedimentation cylinder is to be placed in the water bath, the distilled or demineralized water to be used shall be brought to the temperature of the controlled water bath; or, if the sedimentation cylinder is used in a room with controlled temperature, the water for the test shall be at the temperature of the room. The basic temperature for the hydrometer test is 68°F (20°C). Small variations of temperature do not introduce differences that are of practical significance and do not prevent the use of corrections derived as prescribed.

⚡ D 422

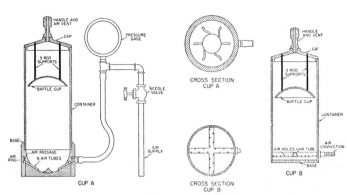

FIG. 3   Air-Jet Dispersion Cups of Apparatus B

5. Test Sample

5.1 Prepare the test sample for mechanical analysis as outlined in Practice D 421. During the preparation procedure the sample is divided into two portions. One portion contains only particles retained on the No. 10 (2.00-mm) sieve while the other portion contains only particles passing the No. 10 sieve. The mass of air-dried soil selected for purpose of tests, as prescribed in Practice D 421, shall be sufficient to yield quantities for mechanical analysis as follows:

5.1.1 The size of the portion retained on the No. 10 sieve shall depend on the maximum size of particle, according to the following schedule:

Nominal Diameter of Largest Particles, in. (mm)	Approximate Minimum Mass of Portion, g
⅜ (9.5)	500
¾ (19.0)	1000
1 (25.4)	2000
1½ (38.1)	3000
2 (50.8)	4000
3 (76.2)	5000

5.1.2 The size of the portion passing the No. 10 sieve shall be approximately 115 g for sandy soils and approximately 65 g for silt and clay soils.

5.2 Provision is made in Section 5 of Practice D 421 for weighing of the air-dry soil selected for purpose of tests, the separation of the soil on the No. 10 sieve by dry-sieving and washing, and the weighing of the washed and dried fraction retained on the No. 10 sieve. From these two masses the percentages retained and passing the No. 10 sieve can be calculated in accordance with 12.1.

Note 8—A check on the mass values and the thoroughness of pulverization of the clods may be secured by weighing the portion passing the No. 10 sieve and adding this value to the mass of the washed and oven-dried portion retained on the No. 10 sieve.

SIEVE ANALYSIS OF PORTION RETAINED ON NO. 10 (2.00-mm) SIEVE

6. Procedure

6.1 Separate the portion retained on the No. 10 (2.00-mm) sieve into a series of fractions using the 3-in. (75-mm),

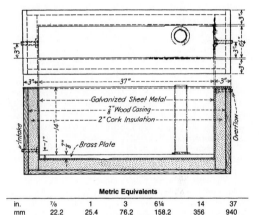

		Metric Equivalents				
in.	⅞	1	3	6¼	14	37
mm	22.2	25.4	76.2	158.2	356	940

FIG. 4   Insulated Water Bath

2-in. (50-mm), 1½-in. (37.5-mm), 1-in. (25.0-mm), ¾-in. (19.0-mm), ⅜-in. (9.5-mm), No. 4 (4.75-mm), and No. 10 sieves, or as many as may be needed depending on the sample, or upon the specifications for the material under test.

6.2 Conduct the sieving operation by means of a lateral and vertical motion of the sieve, accompanied by a jarring action in order to keep the sample moving continuously over the surface of the sieve. In no case turn or manipulate fragments in the sample through the sieve by hand. Continue sieving until not more than 1 mass % of the residue on a sieve passes that sieve during 1 min of sieving. When mechanical sieving is used, test the thoroughness of sieving by using the hand method of sieving as described above.

6.3 Determine the mass of each fraction on a balance conforming to the requirements of 3.1. At the end of weighing, the sum of the masses retained on all the sieves used should equal closely the original mass of the quantity sieved.

D 422

7. Determination of Composite Correction for Hydrometer Reading

7.1 Equations for percentages of soil remaining in suspension, as given in 14.3, are based on the use of distilled or demineralized water. A dispersing agent is used in the water, however, and the specific gravity of the resulting liquid is appreciably greater than that of distilled or demineralized water.

7.1.1 Both soil hydrometers are calibrated at 68°F (20°C), and variations in temperature from this standard temperature produce inaccuracies in the actual hydrometer readings. The amount of the inaccuracy increases as the variation from the standard temperature increases.

7.1.2 Hydrometers are graduated by the manufacturer to be read at the bottom of the meniscus formed by the liquid on the stem. Since it is not possible to secure readings of soil suspensions at the bottom of the meniscus, readings must be taken at the top and a correction applied.

7.1.3 The net amount of the corrections for the three items enumerated is designated as the composite correction, and may be determined experimentally.

7.2 For convenience, a graph or table of composite corrections for a series of 1° temperature differences for the range of expected test temperatures may be prepared and used as needed. Measurement of the composite corrections may be made at two temperatures spanning the range of expected test temperatures, and corrections for the intermediate temperatures calculated assuming a straight-line relationship between the two observed values.

7.3 Prepare 1000 mL of liquid composed of distilled or demineralized water and dispersing agent in the same proportion as will prevail in the sedimentation (hydrometer) test. Place the liquid in a sedimentation cyclinder and the cylinder in the constant-temperature water bath, set for one of the two temperatures to be used. When the temperature of the liquid becomes constant, insert the hydrometer, and, after a short interval to permit the hydrometer to come to the temperature of the liquid, read the hydrometer at the top of the meniscus formed on the stem. For hydrometer 151H the composite correction is the difference between this reading and one; for hydrometer 152H it is the difference between the reading and zero. Bring the liquid and the hydrometer to the other temperature to be used, and secure the composite correction as before.

8. Hygroscopic Moisture

8.1 When the sample is weighed for the hydrometer test, weigh out an auxiliary portion of from 10 to 15 g in a small metal or glass container, dry the sample to a constant mass in an oven at 230 ± 9°F (110 ± 5°C), and weigh again. Record the masses.

9. Dispersion of Soil Sample

9.1 When the soil is mostly of the clay and silt sizes, weigh out a sample of air-dry soil of approximately 50 g. When the soil is mostly sand the sample should be approximately 100 g.

9.2 Place the sample in the 250-mL beaker and cover with 125 mL of sodium hexametaphosphate solution (40 g/L). Stir until the soil is thoroughly wetted. Allow to soak for at least 16 h.

9.3 At the end of the soaking period, disperse the sample further, using either stirring apparatus A or B. If stirring apparatus A is used, transfer the soil - water slurry from the beaker into the special dispersion cup shown in Fig. 2, washing any residue from the beaker into the cup with distilled or demineralized water (Note 9). Add distilled or demineralized water, if necessary, so that the cup is more than half full. Stir for a period of 1 min.

NOTE 9—A large size syringe is a convenient device for handling the water in the washing operation. Other devices include the wash-water bottle and a hose with nozzle connected to a pressurized distilled water tank.

9.4 If stirring apparatus B (Fig. 3) is used, remove the cover cap and connect the cup to a compressed air supply by means of a rubber hose. A air gage must be on the line between the cup and the control valve. Open the control valve so that the gage indicates 1 psi (7 kPa) pressure (Note 10). Transfer the soil - water slurry from the beaker to the air-jet dispersion cup by washing with distilled or demineralized water. Add distilled or demineralized water, if necessary, so that the total volume in the cup is 250 mL, but no more.

NOTE 10—The initial air pressure of 1 psi is required to prevent the soil - water mixture from entering the air-jet chamber when the mixture is transferred to the dispersion cup.

9.5 Place the cover cap on the cup and open the air control valve until the gage pressure is 20 psi (140 kPa). Disperse the soil according to the following schedule:

Plasticity Index	Dispersion Period, min
Under 5	5
6 to 20	10
Over 20	15

Soils containing large percentages of mica need be dispersed for only 1 min. After the dispersion period, reduce the gage pressure to 1 psi preparatory to transfer of soil - water slurry to the sedimentation cylinder.

10. Hydrometer Test

10.1 Immediately after dispersion, transfer the soil - water slurry to the glass sedimentation cylinder, and add distilled or demineralized water until the total volume is 1000 mL.

10.2 Using the palm of the hand over the open end of the cylinder (or a rubber stopper in the open end), turn the cylinder upside down and back for a period of 1 min to complete the agitation of the slurry (Note 11). At the end of 1 min set the cylinder in a convenient location and take hydrometer readings at the following intervals of time (measured from the beginning of sedimentation), or as many as may be needed, depending on the sample or the specification for the material under test: 2, 5, 15, 30, 60, 250, and 1440 min. If the controlled water bath is used, the sedimentation cylinder should be placed in the bath between the 2- and 5-min readings.

NOTE 11—The number of turns during this minute should be approximately 60, counting the turn upside down and back as two turns.

⏚ D 422

Any soil remaining in the bottom of the cylinder during the first few turns should be loosened by vigorous shaking of the cylinder while it is in the inverted position.

10.3 When it is desired to take a hydrometer reading, carefully insert the hydrometer about 20 to 25 s before the reading is due to approximately the depth it will have when the reading is taken. As soon as the reading is taken, carefully remove the hydrometer and place it with a spinning motion in a graduate of clean distilled or demineralized water.

NOTE 12—It is important to remove the hydrometer immediately after each reading. Readings shall be taken at the top of the meniscus formed by the suspension around the stem, since it is not possible to secure readings at the bottom of the meniscus.

10.4 After each reading, take the temperature of the suspension by inserting the thermometer into the suspension.

11. Sieve Analysis

11.1 After taking the final hydrometer reading, transfer the suspension to a No. 200 (75-µm) sieve and wash with tap water until the wash water is clear. Transfer the material on the No. 200 sieve to a suitable container, dry in an oven at 230 ± 9°F (110 ± 5°C) and make a sieve analysis of the portion retained, using as many sieves as desired, or required for the material, or upon the specification of the material under test.

CALCULATIONS AND REPORT

12. Sieve Analysis Values for the Portion Coarser than the No. 10 (2.00-mm) Sieve

12.1 Calculate the percentage passing the No. 10 sieve by dividing the mass passing the No. 10 sieve by the mass of soil originally split on the No. 10 sieve, and multiplying the result by 100. To obtain the mass passing the No. 10 sieve, subtract the mass retained on the No. 10 sieve from the original mass.

12.2 To secure the total mass of soil passing the No. 4 (4.75-mm) sieve, add to the mass of the material passing the No. 10 sieve the mass of the fraction passing the No. 4 sieve and retained on the No. 10 sieve. To secure the total mass of soil passing the ⅜-in. (9.5-mm) sieve, add to the total mass of soil passing the No. 4 sieve, the mass of the fraction passing the ⅜-in. sieve and retained on the No. 4 sieve. For the remaining sieves, continue the calculations in the same manner.

12.3 To determine the total percentage passing for each sieve, divide the total mass passing (see 12.2) by the total mass of sample and multiply the result by 100.

13. Hygroscopic Moisture Correction Factor

13.1 The hydroscopic moisture correction factor is the ratio between the mass of the oven-dried sample and the air-dry mass before drying. It is a number less than one, except when there is no hygroscopic moisture.

14. Percentages of Soil in Suspension

14.1 Calculate the oven-dry mass of soil used in the hydrometer analysis by multiplying the air-dry mass by the hygroscopic moisture correction factor.

14.2 Calculate the mass of a total sample represented by the mass of soil used in the hydrometer test, by dividing the oven-dry mass used by the percentage passing the No. 10

TABLE 1   Values of Correction Factor, α, for Different Specific Gravities of Soil Particles[A]

Specific Gravity	Correction Factor[A]
2.95	0.94
2.90	0.95
2.85	0.96
2.80	0.97
2.75	0.98
2.70	0.99
2.65	1.00
2.60	1.01
2.55	1.02
2.50	1.03
2.45	1.05

[A] For use in equation for percentage of soil remaining in suspension when using Hydrometer 152H.

(2.00-mm) sieve, and multiplying the result by 100. This value is the weight W in the equation for percentage remaining in suspension.

14.3 The percentage of soil remaining in suspension at the level at which the hydrometer is measuring the density of the suspension may be calculated as follows (Note 13): For hydrometer 151H:

$$P = [(100\,000/W) \times G/(G - G_1)](R - G_1)$$

NOTE 13—The bracketed portion of the equation for hydrometer 151H is constant for a series of readings and may be calculated first and then multiplied by the portion in the parentheses.

For hydrometer 152H:

$$P = (Ra/W) \times 100$$

where:
a = correction faction to be applied to the reading of hydrometer 152H. (Values shown on the scale are computed using a specific gravity of 2.65. Correction factors are given in Table 1),
P = percentage of soil remaining in suspension at the level at which the hydrometer measures the density of the suspension,
R = hydrometer reading with composite correction applied (Section 7),
W = oven-dry mass of soil in a total test sample represented by mass of soil dispersed (see 14.2), g,
G = specific gravity of the soil particles, and
G_1 = specific gravity of the liquid in which soil particles are suspended. Use numerical value of one in both instances in the equation. In the first instance any possible variation produces no significant effect, and in the second instance, the composite correction for R is based on a value of one for G_1.

15. Diameter of Soil Particles

15.1 The diameter of a particle corresponding to the percentage indicated by a given hydrometer reading shall be calculated according to Stokes' law (Note 14), on the basis that a particle of this diameter was at the surface of the suspension at the beginning of sedimentation and had settled to the level at which the hydrometer is measuring the density of the suspension. According to Stokes' law:

$$D = \sqrt{[30n/980(G - G_1)]} \times L/T$$

where:
D = diameter of particle, mm,

✪ D 422

n = coefficient of viscosity of the suspending medium (in this case water) in poises (varies with changes in temperature of the suspending medium),

L = distance from the surface of the suspension to the level at which the density of the suspension is being measured, cm. (For a given hydrometer and sedimentation cylinder, values vary according to the hydrometer readings. This distance is known as effective depth (Table 2)),

T = interval of time from beginning of sedimentation to the taking of the reading, min,

G = specific gravity of soil particles, and

G_I = specific gravity (relative density) of suspending medium (value may be used as 1.000 for all practical purposes).

NOTE 14—Since Stokes' law considers the terminal velocity of a single sphere falling in an infinity of liquid, the sizes calculated represent the diameter of spheres that would fall at the same rate as the soil particles.

15.2 For convenience in calculations the above equation may be written as follows:

$$D = K\sqrt{L/T}$$

where:

K = constant depending on the temperature of the suspension and the specific gravity of the soil particles. Values of K for a range of temperatures and specific gravities are given in Table 3. The value of K does not change for a series of readings constituting a test, while values of L and T do vary.

15.3 Values of D may be computed with sufficient accuracy, using an ordinary 10-in. slide rule.

NOTE 15—The value of L is divided by T using the A- and B-scales, the square root being indicated on the D-scale. Without ascertaining the value of the square root it may be multiplied by K, using either the C- or CI-scale.

16. Sieve Analysis Values for Portion Finer than No. 10 (2.00-mm) Sieve

16.1 Calculation of percentages passing the various sieves used in sieving the portion of the sample from the hydrometer test involves several steps. The first step is to calculate the mass of the fraction that would have been retained on the No. 10 sieve had it not been removed. This mass is equal to the total percentage retained on the No. 10 sieve (100 minus total percentage passing) times the mass of the total sample represented by the mass of soil used (as calculated in 14.2), and the result divided by 100.

16.2 Calculate next the total mass passing the No. 200 sieve. Add together the fractional masses retained on all the sieves, including the No. 10 sieve, and subtract this sum from the mass of the total sample (as calculated in 14.2).

16.3 Calculate next the total masses passing each of the other sieves, in a manner similar to that given in 12.2.

16.4 Calculate last the total percentages passing by dividing the total mass passing (as calculated in 16.3) by the total mass of sample (as calculated in 14.2), and multiply the result by 100.

17. Graph

17.1 When the hydrometer analysis is performed, a graph

TABLE 2 Values of Effective Depth Based on Hydrometer and Sedimentation Cylinder of Specified Sizes[A]

Hydrometer 151H		Hydrometer 152H			
Actual Hydrometer Reading	Effective Depth, L, cm	Actual Hydrometer Reading	Effective Depth, L, cm	Actual Hydrometer Reading	Effective Depth, L, cm
1.000	16.3	0	16.3	31	11.2
1.001	16.0	1	16.1	32	11.1
1.002	15.8	2	16.0	33	10.9
1.003	15.5	3	15.8	34	10.7
1.004	15.2	4	15.6	35	10.6
1.005	15.0	5	15.5		
1.006	14.7	6	15.3	36	10.4
1.007	14.4	7	15.2	37	10.2
1.008	14.2	8	15.0	38	10.1
1.009	13.9	9	14.8	39	9.9
1.010	13.7	10	14.7	40	9.7
1.011	13.4	11	14.5	41	9.6
1.012	13.1	12	14.3	42	9.4
1.013	12.9	13	14.2	43	9.2
1.014	12.6	14	14.0	44	9.1
1.015	12.3	15	13.8	45	8.9
1.016	12.1	16	13.7	46	8.8
1.017	11.8	17	13.5	47	8.6
1.018	11.5	18	13.3	48	8.4
1.019	11.3	19	13.2	49	8.3
1.020	11.0	20	13.0	50	8.1
1.021	10.7	21	12.9	51	7.9
1.022	10.5	22	12.7	52	7.8
1.023	10.2	23	12.5	53	7.6
1.024	10.0	24	12.4	54	7.4
1.025	9.7	25	12.2	55	7.3
1.026	9.4	26	12.0	56	7.1
1.027	9.2	27	11.9	57	7.0
1.028	8.9	28	11.7	58	6.8
1.029	8.6	29	11.5	59	6.6
1.030	8.4	30	11.4	60	6.5
1.031	8.1				
1.032	7.8				
1.033	7.6				
1.034	7.3				
1.035	7.0				
1.036	6.8				
1.037	6.5				
1.038	6.2				

[A] Values of effective depth are calculated from the equation:

$$L = L_1 + \frac{1}{2}[L_2 - (V_B/A)]$$

where:
L = effective depth, cm,
L_1 = distance along the stem of the hydrometer from the top of the bulb to the mark for a hydrometer reading, cm,
L_2 = overall length of the hydrometer bulb, cm,
V_B = volume of hydrometer bulb, cm^3, and
A = cross-sectional area of sedimentation cylinder, cm^2
Values used in calculating the values in Table 2 are as follows:
For both hydrometers, 151H and 152H:
L_2 = 14.0 cm
V_B = 67.0 cm^3
A = 27.8 cm^2
For hydrometer 151H:
L_1 = 10.5 cm for a reading of 1.000
    = 2.3 cm for a reading of 1.031
For hydrometer 152H:
L_1 = 10.5 cm for a reading of 0 g/litre
    = 2.3 cm for a reading of 50 g/litre

of the test results shall be made, plotting the diameters of the particles on a logarithmic scale as the abscissa and the percentages smaller than the corresponding diameters to an

⬙ D 422

TABLE 3 Values of K for Use in Equation for Computing Diameter of Particle in Hydrometer Analysis

Temperature, °C	Specific Gravity of Soil Particles								
	2.45	2.50	2.55	2.60	2.65	2.70	2.75	2.80	2.85
16	0.01510	0.01505	0.01481	0.01457	0.01435	0.01414	0.01394	0.01374	0.01356
17	0.01511	0.01486	0.01462	0.01439	0.01417	0.01396	0.01376	0.01356	0.01338
18	0.01492	0.01467	0.01443	0.01421	0.01399	0.01378	0.01359	0.01339	0.01321
19	0.01474	0.01449	0.01425	0.01403	0.01382	0.01361	0.01342	0.1323	0.01305
20	0.01456	0.01431	0.01408	0.01386	0.01365	0.01344	0.01325	0.01307	0.01289
21	0.01438	0.01414	0.01391	0.01369	0.01348	0.01328	0.01309	0.01291	0.01273
22	0.01421	0.01397	0.01374	0.01353	0.01332	0.01312	0.01294	0.01276	0.01258
23	0.01404	0.01381	0.01358	0.01337	0.01317	0.01297	0.01279	0.01261	0.01243
24	0.01388	0.01365	0.01342	0.01321	0.01301	0.01282	0.01264	0.01246	0.01229
25	0.01372	0.01349	0.01327	0.01306	0.01286	0.01267	0.01249	0.01232	0.01215
26	0.01357	0.01334	0.01312	0.01291	0.01272	0.01253	0.01235	0.01218	0.01201
27	0.01342	0.01319	0.01297	0.01277	0.01258	0.01239	0.01221	0.01204	0.01188
28	0.01327	0.01304	0.01283	0.01264	0.01244	0.01225	0.01208	0.01191	0.01175
29	0.01312	0.01290	0.01269	0.01249	0.01230	0.01212	0.01195	0.01178	0.01162
30	0.01298	0.01276	0.01256	0.01236	0.01217	0.01199	0.01182	0.01165	0.01149

arithmetic scale as the ordinate. When the hydrometer analysis is not made on a portion of the soil, the preparation of the graph is optional, since values may be secured directly from tabulated data.

18. Report

18.1 The report shall include the following:

18.1.1 Maximum size of particles,

18.1.2 Percentage passing (or retained on) each sieve, which may be tabulated or presented by plotting on a graph (Note 16),

18.1.3 Description of sand and gravel particles:

18.1.3.1 Shape—rounded or angular,

18.1.3.2 Hardness—hard and durable, soft, or weathered and friable,

18.1.4 Specific gravity, if unusually high or low,

18.1.5 Any difficulty in dispersing the fraction passing the No. 10 (2.00-mm) sieve, indicating any change in type and amount of dispersing agent, and

18.1.6 The dispersion device used and the length of the dispersion period.

NOTE 16—This tabulation of graph represents the gradation of the sample tested. If particles larger than those contained in the sample were removed before testing, the report shall so state giving the amount and maximum size.

18.2 For materials tested for compliance with definite specifications, the fractions called for in such specifications shall be reported. The fractions smaller than the No. 10 sieve shall be read from the graph.

18.3 For materials for which compliance with definite specifications is not indicated and when the soil is composed almost entirely of particles passing the No. 4 (4.75-mm) sieve, the results read from the graph may be reported as follows:

(1) Gravel, passing 3-in. and retained on No. 4 sieve . . . . %
(2) Sand, passing No. 4 sieve and retained on No. 200 sieve . . . . %
   (a) Coarse sand, passing No. 4 sieve and retained on No. 10 sieve . . . . %
   (b) Medium sand, passing No. 10 sieve and retained on No. 40 sieve . . . . %
   (c) Fine sand, passing No. 40 sieve and retained on No. 200 sieve . . . . %
(3) Silt size, 0.074 to 0.005 mm . . . . %
(4) Clay size, smaller than 0.005 mm . . . . %
   Colloids, smaller than 0.001 mm . . . . %

18.4 For materials for which compliance with definite specifications is not indicated and when the soil contains material retained on the No. 4 sieve sufficient to require a sieve analysis on that portion, the results may be reported as follows (Note 17):

SIEVE ANALYSIS

Sieve Size	Percentage Passing
3-in.	. . . .
2-in.	. . . .
1½-in.	. . . .
1-in.	. . . .
¾-in.	. . . .
⅜-in.	. . . .
No. 4 (4.75-mm)	. . . .
No. 10 (2.00-mm)	. . . .
No. 40 (425-μm)	. . . .
No. 200 (75-μm)	. . . .

HYDROMETER ANALYSIS

0.074 mm	. . . .
0.005 mm	. . . .
0.001 mm	. . . .

NOTE 17—No. 8 (2.36-mm) and No. 50 (300-μm) sieves may be substituted for No. 10 and No. 40 sieves.

19. Keywords

19.1 grain-size; hydrometer analysis; hygroscopic moisture; particle-size; sieve analysis

 **Designation: D 4318 – 95a**

Standard Test Method for
Liquid Limit, Plastic Limit, and Plasticity Index of Soils[1]

This standard is issued under the fixed designation D 4318; the number immediately following the designation indicates the year of original adoption or, in the case of revision, the year of last revision. A number in parentheses indicates the year of last reapproval. A superscript epsilon (ϵ) indicates an editorial change since the last revision or reapproval.

1. Scope*

1.1 This test method covers the determination of the liquid limit, plastic limit, and the plasticity index of soils as defined in Section 3.

1.1.1 Two procedures for preparing test specimens are provided as follows: *Wet preparation procedure*, as described in 10.1. *Dry preparation procedure*, as described in 10.2. The procedure to be used shall be specified by the requesting authority. If no procedure is specified, use the wet preparation procedure.

1.1.2 Two methods for determining the liquid limit are provided as follows: *Method A*, Multipoint test as described in Sections 11 and 12. *Method B*, One-point test as described in Sections 14 and 15. The method to be used shall be specified by the requesting authority. If no method is specified, use Method A.

1.1.3 The plastic limit test procedure is described in Sections 16, 17, and 18. The plastic limit test is performed on material prepared for the liquid limit test.

1.1.4 The procedure for calculating the plasticity index is given in Section 19.

1.2 The liquid limit and plastic limit of soils (along with the shrinkage limit) are often collectively referred to as the Atterberg limits. These limits distinguished the boundaries of the several consistency states of plastic soils.

1.3 The multipoint liquid limit method is generally more precise than the one-point method. It is recommended that the multipoint method be used in cases where results may be subject to dispute, or where greater precision is required.

1.4 Because the one-point method requires the operator to judge when the test specimen is approximately at its liquid limit, it is particularly not recommended for use by inexperienced operators.

1.5 The correlations on which the calculations of the one-point method are based may not be valid for certain soils, such as organic soils or soils from a marine environment. It is strongly recommended that the liquid limit of these soils be determined by the multipoint method.

1.6 The liquid and plastic limits of many soils that have been allowed to dry before testing may be considerably different from values obtained on undried samples. If the liquid and plastic limits of soils are used to correlate or estimate the engineering behavior of soils in their natural

moist state, samples should not be permitted to dry before testing unless data on dried samples are specifically desired.

1.7 The composition and concentration of soluble salts in a soil affect the values of the liquid and plastic limits as well as the water content values of soils (see Method D 2216). Special consideration should therefore be given to soils from a marine environment or other sources where high soluble salt concentrations may be present. The degree to which the salts present in these soils are diluted or concentrated must be given careful consideration.

1.8 Since the tests described herein are performed only on that portion of a soil which passes the 425-μm (No. 40) sieve, the relative contribution of this portion of the soil to the properties of the sample as a whole must be considered when using these tests to evaluate properties of a soil.

1.9 The values stated in acceptable metric units are to be regarded as the standard. The values given in parentheses are for information only.

1.10 *This standard does not purport to address all of the safety concerns, if any, associated with its use. It is the responsibility of the user of this standard to establish appropriate safety and health practices and determine the applicability of regulatory limitations prior to use.*

2. Referenced Documents

2.1 *ASTM Standards:*

C 670 Practice for Preparing Precision and Bias Statements for Test Methods for Construction Materials[2]

C 702 Methods for Reducing Field Samples of Aggregate to Testing Size[3]

D 75 Practice for Sampling Aggregates[2]

D 420 Practice for Investigating and Sampling Soil and Rock for Engineering Purposes[2]

D 653 Terminology Relating to Soil, Rock, and Contained Fluids[2]

D 1241 Specification for Materials for Soil-Aggregate Subbase, Base, and Surface Courses[2]

D 2216 Test Method for Laboratory Determination of Water (Moisture) Content of Soil and Rock[2]

D 2487 Classification of Soils for Engineering Purposes (Unified Soil Classification System)[2]

D 2488 Practice for Description and Identification of Soils (Visual-Manual Procedure)[2]

D 3282 Practice for Classification of Soils and Soil-Aggregate Mixtures for Highway Construction Purposes[3]

---

[1] This test method is under the jurisdiction of ASTM Committee D-18 on Soil and Rock and is the direct responsibility of Subcommittee D18.03 on Texture, Plasticity and Density Characteristics of Soils.

Current edition approved Dec. 10, 1995. Published April 1996. Originally published as D 4318 – 83. Last previous edition D 4318 – 95.

---

[2] *Annual Book of ASTM Standards*, Vol 04.02.
[3] *Annual Book of ASTM Standards*, Vol 04.08.

* A Summary of Changes section appears at the end of standard.

D 4318

D 4753 Specification for Evaluating, Selecting, and Specifying Balances and Scales for Use in Soil and Rock Testing[3]

E 11 Specification for Wire-Cloth Sieves for Testing Purposes[2]

3. Terminology

3.1 *Definitions:*

3.1.1 The definitions of terms in this test method are in accordance with Terminology D 653.

3.2 *Description of Terms Specific to This Standard:*

3.2.1 *Atterberg limits*—Originally, six "limits of consistency" of fine-grained soils were defined by Albert Atterberg: the upper limit of viscous flow, the liquid limit, the sticky limit, the cohesion limit, the plastic limit, and the shrinkage limit. In current engineering usage, the term usually refers only to the liquid limit, plastic limit, and in some references, the shrinkage limit.

3.2.2 *consistency*—the relative ease with which a soil can be deformed.

3.2.3 *liquid limit (LL, w_L)*—the water content, in percent, of a soil at the arbitrarily defined boundary between the semi-liquid and plastic states.

3.2.3.1 *Discussion*—The undrained shear strength of soil at the liquid limit is considered to be approximately 2 kPa (0.28 psi).

3.2.4 *plastic limit (PL, w_p)*—the water content, in percent, of a soil at the boundary between the plastic and semi-solid states.

3.2.5 *plastic soil*—a soil which has a range of water content over which it exhibits plasticity and which will retain its shape on drying.

3.2.6 *plasticity index (PI)*—the range of water content over which a soil behaves plastically. Numerically, it is the difference between the liquid limit and the plastic limit.

3.2.7 *liquidity index*—the ratio, expressed as a percentage of (1) the natural water content of a soil minus its plastic limit, to (2) its plasticity index.

3.2.8 *activity number (A)*—the ratio of (1) the plasticity index of a soil to (2) the percent by weight of particles having an equivalent diameter smaller than 0.002 mm.

4. Summary of Test Method

4.1 The sample is processed to remove any material retained on a 425-μm (No. 40) sieve. The liquid limit is determined by performing trials in which a portion of the sample is spread in a brass cup, divided in two by a grooving tool, and then allowed to flow together from the shocks caused by repeatedly dropping the cup in a standard mechanical device. The multipoint liquid limit, Method A, requires three or more trials over a range of water contents to be performed and the data from the trials plotted or calculated to make a relationship from which the liquid limit is determined. The one-point liquid limit, Method B, uses the data from two trials at one water content multiplied by a correction factor to determine the liquid limit.

4.2 The plastic limit is determined by alternately pressing together and rolling into a 3.2-mm (⅛-in.) diameter thread a small portion of plastic soil until its water content is reduced to a point at which the thread crumbles and can no longer be pressed together and rerolled. The water content of the soil at

this point is reported as the plastic limit.

4.3 The plasticity index is calculated as the difference between the liquid limit and the plastic limit.

5. Significance and Use

5.1 This test method is used as an integral part of several engineering classification systems to characterize the fine-grained fractions of soils (see Test Method D 2487 and Practice D 3282) and to specify the fine-grained fraction of construction materials (see Specification D 1241). The liquid limit, plastic limit, and plasticity index of soils are also used extensively, either individually or together, with other soil properties to correlate with engineering behavior such as compressibility, permeability, compactibility, shrink-swell, and shear strength.

5.2 The liquid and plastic limits of a soil can be used with the natural water content of the soil to express its relative consistency or liquidity index and can be used with the percentage finer than 2-μm size to determine its activity number.

5.3 These methods are sometimes used to evaluate the weathering characteristics of clay-shale materials. When subjected to repeated wetting and drying cycles, the liquid limits of these materials tend to increase. The amount of increase is considered to be a measure of a shale's susceptibility to weathering.

5.4 The liquid limit of a soil containing substantial amounts of organic matter decreases dramatically when the soil is oven-dried before testing. Comparison of the liquid limit of a sample before and after oven-drying can therefore be used as a qualitative measure of organic matter content of a soil.

6. Apparatus

6.1 *Liquid Limit Device*—A mechanical device consisting of a brass cup suspended from a carriage designed to control its drop onto a hard rubber base. Figure 1 shows the essential features and critical dimensions of the device. The device may be operated by either a hand crank or electric motor.

6.1.1 *Base*—A hard rubber base having a D Durometer hardness of 80 to 90, and a resilience such that an 8-mm (⁵⁄₁₆-in.) diameter polished steel ball, when dropped from a height of 25 cm (9.84 in.) will have an average rebound of at least 77 % but no more than 90 %. Conduct resilience tests on the finished base with the feet attached. Details for measuring the resilience of the base are given in Appendix A.

6.1.2 *Rubber Feet*, supporting the base, designed to provide isolation of the base from the work surface, and having an A Durometer hardness no greater than 60 as measured on the finished feet attached to the base.

6.1.3 *Cup*, brass, with a weight, including cup hanger, of 185 to 215 g.

6.1.4 *Cam*—designed to raise the cup smoothly and continuously to its maximum height, over a distance of at least 180° of cam rotation, without developing an upward or downward velocity of the cup when the cam follower leaves the cam. (The preferred cam motion is a uniformly accelerated lift curve.)

NOTE 1—The cam and follower design in Fig. 1 is for uniformly accelerated (parabolic) motion after contact and assures that the cup has no velocity at drop off. Other cam designs also provide this feature and

D 4318

DIMENSIONS

LETTER	A△	B△	C△	E△	F	G	H	J△	K△	L△	M△
MM	54 ± 0.5	2 ± 0.1	27 ± 0.5	56 ± 2.0	32	10	16	60 ± 1.0	50 ± 2.0	150 ± 2.0	125 ± 2.0

LETTER	N	P	R	T	U△	V	W	Z
MM	24	28	24	45	47 ± 1.0	3.8	13	6.5

△ ESSENTIAL DIMENSIONS

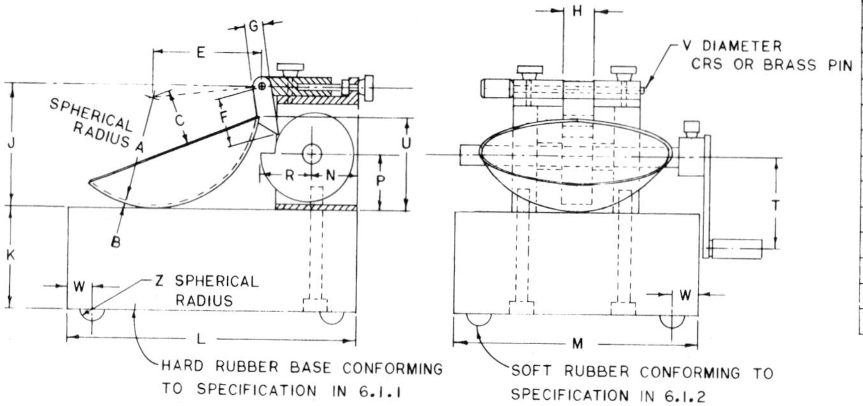

CAM ANGLE DEGREES	CAM RADIUS
0	0.742 R
30	0.753 R
60	0.764 R
90	0.773 R
120	0.784 R
150	0.796 R
180	0.818 R
210	0.854 R
240	0.901 R
270	0.945 R
300	0.974 R
330	0.995 R
360	1.000 R

FIG. 1 Hand-Operated Liquid Limit Device

may be used. However, if the cam-follower lift pattern is not known, zero velocity at drop off can be assured by carefully filing or machining the cam and follower so that the cup height remains constant over the last 20 to 45° of cam rotation.

6.1.5 *Carriage*, constructed in a way that allows convenient but secure adjustment of the height of drop of the cup to 10 mm (0.394 in.), and designed such that the cup and cup hanger assembly is only attached to the carriage by means of a removable pin.

6.1.6 *Motor Drive (Optional)*—As an alternative to the hand crank shown in Fig. 1, the device may be equipped with a motor to turn the cam. Such a motor must turn the cam at 2 ± 0.1 revolutions per second and must be isolated from the rest of the device by rubber mounts or in some other way that prevents vibration from the motor being transmitted to the rest of the apparatus. It must be equipped with an ON-OFF switch and a means of conveniently positioning the cam for height of drop adjustments. The results obtained using a motor-driven device must not differ from those obtained using a manually operated device.

6.2 *Flat Grooving Tool*—A tool made of plastic or noncorroding-metal having the dimensions shown in Fig. 2. The design of the tool may vary as long as the essential dimensions are maintained. The tool may, but need not, incorporate the gage for adjusting the height of drop of the liquid limit device.

NOTE 2—Prior to the adoption of this test method, a curved grooving tool was specified as part of the apparatus for performing the liquid limit test. The curved tool is not considered to be as accurate as the flat tool

described in 6.2 since it does not control the depth of the soil in the liquid limit cup. However, there are some data which indicate that typically the liquid limit is slightly increased when the flat tool is used instead of the curved tool.

6.3 *Gage*—A metal gage block for adjusting the height of drop of the cup, having the dimensions shown in Fig. 3. The design of the tool may vary provided the gage will rest securely on the base without being susceptible to rocking, and the edge which contacts the cup during adjustment is straight, at least 10 mm (⅜ in.) wide, and without bevel or radius.

6.4 *Containers*—Small corrosion-resistant containers with snug-fitting lids for water content specimens. Aluminum or stainless steel cans 2.5 cm (1 in.) high by 5 cm (2 in.) in diameter are appropriate.

6.5 *Balance*, conforming to Specification D 4753, Class GP1.

6.6 *Storage Container*—A container in which to store the prepared soil specimen that will not contaminate the specimen in any way, and will prevent moisture loss. A porcelain, glass, or plastic dish about 11.4 cm (4½ in.) in diameter and a plastic bag large enough to enclose the dish and be folded over is adequate.

6.7 *Ground Glass Plate*—A ground glass plate at least 30 cm (12 in.) square by 1 cm (⅜ in.) thick for rolling plastic limit threads.

6.8 *Spatula*—A spatula or pill knife having a blade about 2 cm (¾ in.) wide, and about 10 to 13 cm (3 to 4 in.) long.

6.9 *Sieve*—A 20.3-cm (8-in.) diameter, 425-μm (No. 40)

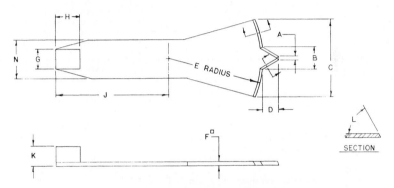

D 4318

DIMENSIONS

LETTER	A△	B△	C△	D△	E△	F△
MM	2	11	40	8	50	2
	± 0.1	± 0.2	± 0.5	± 0.1	± 0.5	± 0.1
LETTER	G	H	J	K△	L△	N
MM	10	13	60	10	60 DEG	20
	MINIMUM			± 0.05	± 1 DEG	

△ ESSENTIAL DIMENSIONS

□ BACK AT LEAST 15 MM FROM TIP

NOTE : DIMENSION A SHOULD BE 1.9–2.0 AND DIMENSION D
SHOULD BE 8.0–8.1 WHEN NEW TO ALLOW FOR
ADEQUATE SERVICE LIFE

FIG. 2   Grooving Tool (Optional Height-of-Drop Gage Attached)

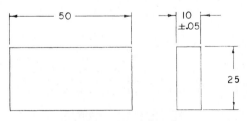

DIMENSIONS IN MILLIMETRES

FIG. 3   Height of Drop Gage

sieve conforming to the requirements of Specification E 11 and having a rim at least 5 cm (2 in.) above the mesh. A 2-mm (No. 10) sieve meeting the same requirements may also be needed.

6.10 *Wash Bottle*, or similar container for adding controlled amounts of water to soil and washing fines from coarse particles.

6.11 *Drying Oven*, thermostatically controlled, preferably of the forced-draft type, capable of continuously maintaining a temperature of 110 ± 5°C (230 ± 9°F) throughout the drying chamber.

6.12 *Washing Pan*, round, flat-bottomed, at least 7.6 cm

(3 in.) deep, and slightly larger at the bottom than a 20.3-cm (8-in.) diameter sieve.

7. Reagents and Materials

7.1 *Purity of Water*—Where distilled water is referred to in this test method, either distilled or demineralized water may be used.

8. Sampling

8.1 Samples may be taken from any location that satisfies testing needs. However, Methods C 702, Practice D 75, and Practice D 420 should be used as guides for selecting and preserving samples from various types of sampling operations. Samples which will be prepared using the wet preparation procedure (10.1) must be kept at their natural water content prior to preparation.

8.2 Where sampling operations have preserved the natural stratification of a sample, the various strata must be kept separated and tests performed on the particular stratum of interest with as little contamination as possible from other strata. Where a mixture of materials will be used in construction, combine the various components in such proportions that the resultant sample represents the actual construction case.

8.3 Where data from this test method are to be used for correlation with other laboratory or field test data, use the

🏛 **D 4318**

same material as used for these tests where possible.

8.4 Obtain a representative portion from the total sample sufficient to provide 150 to 200 g of material passing the 425-μm (No. 40) sieve. Free flowing samples may be reduced by the methods of quartering or splitting. Cohesive samples shall be mixed thoroughly in a pan with a spatula, or scoop and a representative portion scooped from the total mass by making one or more sweeps with a scoop through the mixed mass.

9. Calibration of Apparatus

9.1 *Inspection of Wear:*

9.1.1 *Liquid Limit Device*—Determine that the liquid limit device is clean and in good working order. Check the following specific points.

9.1.1.1 *Wear of Base*—The spot on the base where the cup makes contact should be worn no greater than 10 mm (³⁄₈ in.) in diameter. If the wear spot is greater than this, the base can be machined to remove the worn spot provided the resurfacing does not make the base thinner than specified in 6.1 and the other dimensional relationships are maintained.

9.1.1.2 *Wear of Cup*—Replace the cup when the grooving tool has worn a depression in the cup 0.1 mm (0.004 in.) deep or when the rim of the cup has been reduced to half its original thickness. Verify that the cup is firmly attached to the cup hanger.

9.1.1.3 *Wear of Cup Hanger*—Verify that the cup hanger pivot does not bind and is not worn to an extent that allows more than 3 mm (¹⁄₈ in.) side-to-side movement of the lowest point on the rim.

9.1.1.4 *Wear of Cam*—The cam shall not be worn to an extent that the cup drops before the cup hanger (cam follower) loses contact with the cam.

9.1.2 *Grooving Tools*—Inspect grooving tools for wear on a frequent and regular basis. The rapidity of wear depends on the material from which the tool is made and the types of soils being tested. Soils containing a large proportion of sand particles may cause rapid wear of grooving tools; therefore, when testing these materials, tools should be inspected more frequently than for other soils.

NOTE 3—The width of the tip of grooving tools is conveniently checked using a pocket-sized measuring magnifier equipped with a millimetre scale. Magnifiers of this type are available from most laboratory supply companies. The depth of the tip of grooving tools can be checked using the depth measuring feature of vernier calipers.

9.2 *Adjustment of Height of Drop*—Adjust the height of drop of the cup so that the point on the cup that comes in contact with the base rises to a height of 10 ± 0.2 mm. See Fig. 4 for proper location of the gage relative to the cup during adjustment.

NOTE 4—A convenient procedure for adjusting the height of drop is as follows: place a piece of masking tape across the outside bottom of the cup parallel with the axis of the cup hanger pivot. The edge of the tape away from the cup hanger should bisect the spot on the cup that contacts the base. For new cups, placing a piece of carbon paper on the base and allowing the cup to drop several times will mark the contact spot. Attach the cup to the device and turn the crank until the cup is raised to its maximum height. Slide the height gage under the cup from the front, and observe whether the gage contacts the cup or the tape. (See Fig. 4.) If the tape and cup are both contacted, the height of drop is approximately correct. If not, adjust the cup until simultaneous contact is made. Check adjustment by turning the crank at 2 revolutions per second while holding the gage in position against the tape and cup. If a faint ringing or clicking sound is heard without the cup rising from the gage, the adjustment is correct. If no ringing is heard or if the cup rises from the gage, readjust the height of drop. If the cup rocks on the gage during this checking operation, the cam follower pivot is excessively worn and the worn parts should be replaced. Always remove tape after completion of adjustment operation.

10. Preparation of Test Specimens

10.1 *Wet Preparation*—Except where the dry method of specimen preparation is specified (10.2), prepare specimens for test as described in the following sections.

10.1.1 *Samples Passing the 425-μm (No. 40) Sieve:*

10.1.1.1 When by visual and manual procedures it is determined that the sample has little or no material retained on a 425-μm (No. 40) sieve, prepare a specimen of 150 to 200 g by mixing thoroughly with distilled or demineralized water on the glass plate using the spatula. If desired, soak the soil in a storage dish with a small amount of water to soften the soil before the start of mixing. Adjust the water content of the soil to bring it to a consistency that would require 25 to 35 blows of the liquid limit device to close the groove (Note 5).

10.1.1.2 If, during mixing, a small percentage of material is encountered that would be retained on a 425-μm (No. 40) sieve, remove these particles by hand (if possible). If it is impractical to remove the coarser material by hand, remove small percentages (less than about 15 %) of coarser material by working the specimen through a 425-μm sieve using a piece of rubber sheeting, rubber stopper, or other convenient

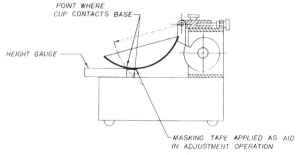

FIG. 4   Calibration for Height of Drop

D 4318

device provided the operation does not distort the sieve or degrade material that would be retained if the washing method described in 10.1.2 were used. If larger percentages of coarse material are encountered during mixing, or it is considered impractical to remove the coarser material by the methods just described, wash the sample as described in 10.1.2. When the coarse particles found during mixing are concretions, shells, or other fragile particles, do not crush these particles to make them pass a 425-μm sieve, but remove by hand or by washing.

10.1.1.3 Place the mixed soil in the storage dish, cover to prevent loss of moisture, and allow to stand for at least 16 h (overnight). After the standing period and immediately before starting the test, thoroughly remix the soil.

NOTE 5—The time taken to adequately mix a soil will vary greatly, depending on the plasticity and initial water content. Initial mixing times of more than 30 min may be needed for stiff, fat clays.

10.1.2 *Samples Containing Material Retained on a 425-μm (No. 40) Sieve:*

10.1.2.1 Select a sufficient quantity of soil at natural water content to provide 150 to 200 g of material passing the 425-μm (No. 40) sieve. Place in a pan or dish and add sufficient water to cover the soil. Allow to soak until all lumps have softened and the fines no longer adhere to the surfaces of the coarse particles (Note 6).

NOTE 6—In some cases, the cations of salts present in tap water will exchange with the natural cations in the soil and significantly alter the test results if tap water is used in the soaking and washing operations. Unless it is known that such cations are not present in the tap water, distilled or demineralized water should be used. As a general rule, water containing more than 100 mg/L of dissolved solids should not be used for washing operations.

10.1.2.2 When the sample contains a large percentage of material retained on the 425-μm (No. 40) sieve, perform the following washing operation in increments, washing no more than 0.5 kg (1 lb) of material at one time. Place the 425-μm sieve in the bottom of the clean pan. Pour the soil-water mixture onto the sieve. If gravel or coarse sand particles are present, rinse as many of these as possible with small quantities of water from a wash bottle, and discard. Alternatively, pour the soil water mixture over a 2.00-mm (No. 10) sieve nested atop the 425-μm sieve, rinse the fine material through and remove the 2.00-mm sieve. After washing and removing as much of the coarser material as possible, add sufficient water to the pan to bring the level to about 13 mm (½ in.) above the surface of the 425- μm sieve. Agitate the slurry by stirring with the fingers while raising and lowering the sieve in the pan and swirling the suspension so that fine material is washed from the coarse particles. Disaggregate fine soil lumps that have not slaked by gently rubbing them over the sieve with the fingertips. Complete the washing operation by raising the sieve above the water surface and rinsing the material retained with a small amount of clean water. Discard material retained on the 425-μm sieve.

10.1.2.3 Reduce the water content of the material passing the 425 μm (No. 40) sieve until it approaches the liquid limit. Reduction of water content may be accomplished by one or a combination of the following methods: (*a*) exposing to air currents at room temperature, (*b*) exposing to warm air currents from a source such as an electric hair dryer, (*c*) decanting clear water from surface of the suspension, (*d*)

filtering in a buchner funnel or using filter candles, or (*e*) draining in a colander or plaster of paris dish lined with high retentivity,[4] high wet-strength filter paper. If a plaster of paris dish is used, take care that the dish never becomes sufficiently saturated that it fails to absorb water into its surface. Thoroughly dry dishes between uses. During evaporation and cooling, stir the sample often enough to prevent overdrying of the fringes and soil pinnacles on the surface of the mixture. For soil samples containing soluble salts, use a method of water reduction (*a* or *b*) that will not eliminate the soluble salts from the test specimen.

10.1.2.4 Thoroughly mix the material passing the 425-μm sieve on the glass plate using the spatula. Adjust the water content of the mixture, if necessary, by adding small increments of distilled or demineralized water or by allowing the mixture to dry at room temperature while mixing on the glass plate. The soil should be at a water content that will result in closure of the groove in 25 to 35 blows. Put the mixed soil in the storage dish, cover to prevent loss of moisture, and allow to stand for at least 16 h. After the standing period and immediately before starting the test, thoroughly remix the soil.

10.2 *Dry Preparation:*

10.2.1 Select sufficient soil to provide 150 to 200 g of material passing the 425-μm (No. 40) sieve after processing. Dry the sample at room temperature or in an oven at a temperature not exceeding 60°C until the soil clods will pulverize readily. Disaggregation is expedited if the sample is not allowed to completely dry. However, the soil should have a dry appearance when pulverized.

10.2.2 Pulverize the sample in a mortar with a rubber-tipped pestle or in some other way that does not cause breakdown of individual grains. When the coarse particles found during pulverization are concretions, shells, or other fragile particles, do not crush these particles to make them pass a 425-μm (No. 40) sieve, but remove by hand or other suitable means, such as washing.

10.2.3 Separate the sample on a 425-μm (No. 40) sieve, shaking the sieve by hand to assure thorough separation of the finer fraction. Return the material retained on the 425-μm sieve to the pulverizing apparatus and repeat the pulverizing and sieving operations as many times as necessary to assure that all fine material has been disaggregated and material retained on the 425-μm sieve consists only of individual sand or gravel grains.

10.2.4 Place material remaining on the 425-μm (No. 40) sieve after the final pulverizing operations in a dish and soak in a small amount of water. Stir the soil water mixture and pour over a 425-μm sieve, catching the water and any suspended fines in the washing pan. Pour this suspension into a dish containing the dry soil previously sieved through the 425-μm sieve. Discard material retained on the 425-μm sieve.

10.2.5 Proceed as described in 10.1.2.3 and 10.1.2.4.

MULTIPOINT LIQUID LIMIT—METHOD A

11. Procedure

11.1 Place a portion of the prepared soil in the cup of the

---

[4] S and S 595 filter paper available in 320-mm circles has proven satisfactory.

liquid limit device at the point where the cup rests on the base, squeeze it down, and spread it into the cup to a depth of about 10 mm at its deepest point, tapering to form an approximately horizontal surface. Take care to eliminate air bubbles from the soil pat, but form the pat with as few strokes as possible. Keep the unused soil in the storage dish. Cover the storage dish with a wet towel (or use other means) to retain the moisture in the sample.

11.2 Form a groove in the soil pat by drawing the tool, beveled edge forward, through the soil on a line joining the highest point to the lowest point on the rim of the cup. When cutting the groove, hold the grooving tool against the surface of the cup and draw in an arc, maintaining the tool perpendicular to the surface of the cup throughout its movement. See Fig. 5. In soils where a groove cannot be made in one stroke without tearing the soil, cut the groove with several strokes of the grooving tool. Alternatively, cut the groove to slightly less than required dimensions with a spatula and use the grooving tool to bring the groove to final dimensions. Exercise extreme care to prevent sliding the soil pat relative to the surface of the cup.

11.3 Verify that no crumbs of soil are present on the base or the underside of the cup. Lift and drop the cup by turning the crank at a rate of 1.9 to 2.1 drops per second until the two halves of the soil pat come in contact at the bottom of the groove along a distance of 13 mm (½ in.). See Fig. 6.

NOTE 7—Use of a scale is recommended to verify that the groove has closed 13 mm (½ in.)

11.4 Verify that an air bubble has not caused premature closing of the groove by observing that both sides of the groove have flowed together with approximately the same shape. If a bubble has caused premature closing of the groove, reform the soil in the cup, adding a small amount of soil to make up for that lost in the grooving operation and repeat 11.1 to 11.3. If the soil slides on the surface of the cup,

repeat 11.1 through 11.3 at a higher water content. If, after several trials at successively higher water contents, the soil pat continues to slide in the cup or if the number of blows required to close the groove is always less than 25, record that the liquid limit could not be determined, and report the soil as nonplastic without performing the plastic limit test.

11.5 Record the number of drops, N, required to close the groove. Remove a slice of soil approximately the width of the spatula, extending from edge to edge of the soil cake at right angles to the groove and including that portion of the groove in which the soil flowed together, place in a weighed container, and cover.

11.6 Return the soil remaining in the cup to the storage dish. Wash and dry the cup and grooving tool and reattach the cup to the carriage in preparation for the next trial.

11.7 Remix the entire soil specimen in the storage dish adding distilled water to increase the water content of the soil and decrease the number of blows required to close the groove. Repeat 11.1 through 11.6 for at least two additional trials producing successively lower numbers of blows to close the groove. One of the trials shall be for a closure requiring 25 to 35 blows, one for closure between 20 and 30 blows, and one trial for a closure requiring 15 to 25 blows.

11.8 Determine the water content, W^n, of the soil specimen from each trial in accordance with Test Method D 2216. Initial weighings should be performed immediately after completion of the test. If the test is to be interrupted for more than about 15 minutes, the specimens already obtained should be weighed at the time of the interruption.

12. Calculation

12.1 Plot the relationship between the water content, W^n, and the corresponding number of drops, N, of the cup on a semilogarithmic graph with the water content as ordinates on the arithmetical scale, and the number of drops as abscissas

FIG. 5 Grooved Soil Pat in Liquid Limit Device

D 4318

FIG. 6 Soil Pat After Groove Has Closed

on a logarithmic scale. Draw the best straight line through the three or more plotted points.

12.2 Take the water content corresponding to the intersection of the line with the 25-drop abscissa as the liquid limit of the soil. Computational methods may be substituted for the graphical method for fitting a straight line to the data and determining the liquid limit.

ONE-POINT LIQUID LIMIT—METHOD B

13. Preparation of Test Specimens

13.1 Prepare the specimen in the same manner in accordance with Section 10, except that at mixing, adjust the water content to a consistency requiring 20 to 30 drops of the liquid limit cup to close the groove.

14. Procedure

14.1 Proceed as described in 11.1 through 11.5 except that the number of blows required to close the groove shall be 20 to 30. If less than 20 or more than 30 blows are required, adjust the water content of the soil and repeat the procedure.

14.2 Immediately after removing a water content specimen as described in 11.5, reform the soil in the cup, adding a small amount of soil to make up for that lost in the grooving and water content sampling orientations. Repeat 11.2 through 11.5, and, if the second closing of the groove requires the same number of drops or no more than two drops difference, secure another water content specimen. Otherwise, remix the entire specimen and repeat.

NOTE 8—Excessive drying or inadequate mixing will cause the number of blows to vary.

14.3 Determine water contents of specimens in accordance with 11.8.

15. Calculation

15.1 Determine the liquid limit for each water content specimen using one of the following equations:

$$LL = W^n \left(\frac{N}{25}\right)^{0.121}$$

or

$$LL = kW^n$$

where:

N = number of blows causing closure of the groove at water content,
W^n = water content, and
k = factor given in Table 1.

The liquid limit is the average of the two trial liquid limit values.

15.2 If the difference between the two trial liquid limit values is greater than one percentage point, repeat the test.

TABLE 1 Factors for Obtaining Liquid Limit from Water Content and Number of Drops Causing Closure of Groove

N (Number of Drops)	k (Factor for Liquid Limit)
20	0.974
21	0.979
22	0.985
23	0.990
24	0.995
25	1.000
26	1.005
27	1.009
28	1.014
29	1.018
30	1.022

D 4318

PLASTIC LIMIT

16. Preparation of Test Specimen

16.1 Select a 20-g portion of soil from the material prepared for the liquid limit test, either after the second mixing before the test, or from the soil remaining after completion of the test. Reduce the water content of the soil to a consistency at which it can be rolled without sticking to the hands by spreading or mixing continuously on the glass plate or in the storage dish. The drying process may be accelerated by exposing the soil to the air current from an electric fan, or by blotting with paper that does not add any fiber to the soil, such as hard surface paper toweling or high wet-strength filter paper.

17. Procedure

17.1 From the 20-g mass, select a portion of 1.5 to 2.0 g. Form the test specimen into an ellipsoidal mass. Roll this mass between the palm or fingers and the ground-glass plate with just sufficient pressure to roll the mass into a thread of uniform diameter throughout its length (Note 9). The thread shall be further deformed on each stroke so that its diameter reaches 3.2 mm (⅛ in.), taking no more than 2 min (Note 10). The amount of hand or finger pressure required will vary greatly, according to the soil. Fragile soils of low plasticity are best rolled under the outer edge of the palm or at the base of the thumb.

NOTE 9—A normal rate of rolling for most soils should be 80 to 90 strokes per minute, counting a stroke as one complete motion of the hand forward and back to the starting position. This rate of rolling may have to be decreased for very fragile soils.

NOTE 10—A 3.2-mm (⅛-in.) diameter rod or tube is useful for frequent comparison with the soil thread to ascertain when the thread has reached the proper diameter.

17.1.1 When the diameter of the thread becomes 3.2 mm, break the thread into several pieces. Squeeze the pieces together, knead between the thumb and first finger of each hand, reform into an ellipsoidal mass, and reroll. Continue this alternate rolling to a thread 3.2 mm in diameter, gathering together, kneading and rerolling, until the thread crumbles under the pressure required for rolling and the soil can no longer be rolled into a 3.2-mm diameter thread (see Fig. 7). It has no significance if the thread breaks into threads of shorter length. Roll each of these shorter threads to 3.2 mm in diameter. The only requirement for continuing the test is that they are able to be reformed into an ellipsoidal mass and rolled out again. The operator shall at no time attempt to produce failure at exactly 3.2 mm diameter by allowing the thread to reach 3.2 mm, then reducing the rate of rolling or the hand pressure, or both, while continuing the rolling without further deformation until the thread falls apart. It is permissible, however, to reduce the total amount of deformation for feebly plastic soils by making the initial diameter of the ellipsoidal mass nearer to the required 3.2-mm final diameter. If crumbling occurs when the thread has a diameter greater than 3.2 mm, this shall be considered a satisfactory end point, provided the soil has been previously rolled into a thread 3.2 mm in diameter. Crumbling of the thread will manifest itself differently with the various types of soil. Some soils fall apart in numerous small aggregations of particles, others may form an outside tubular layer that starts splitting at both ends. The splitting progresses toward the middle, and finally, the thread falls apart in many small platy particles. Fat clay soils require much pressure to deform the thread, particularly as they approach the plastic limit. With these soils, the thread breaks into a series of barrel-shaped segments about 3.2 to 9.5 mm (⅛ to ⅜ in.) in length.

17.2 Gather the portions of the crumbled thread together and place in a weighed container. Immediately cover the container.

17.3 Select another 1.5 to 2.0-g portion of soil from the

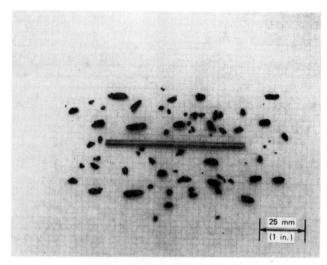

FIG. 7   Lean Clay Soil at the Plastic Limit

D 4318

TABLE 2 Table of Precision Estimates[A]

Material and Type Index	Standard Deviation[B]	Acceptable Range of Two Results[B]
Single-operator precision:		
Liquid Limit	0.8	2.4
Plastic Limit	0.9	2.6
Multilaboratory precision:		
Liquid Limit	3.5	9.9
Plastic Limit	3.7	10.6

[A] The figures given in Column 2 are the standard deviations that have been found to be appropriate for the test results described in Column 1. The figures given in Column 3 are the limits that should not be exceeded by the difference between the two properly conducted tests.

[B] These numbers represent, respectively, the (1S) and (D2S) limits as described in Practice C 670.

original 20-g specimen and repeat the operations described in 17.1 and 17.2 until the container has at least 6 g of soil.

17.4 Repeat 17.1 through 17.3 to make another container holding at least 6 g of soil. Determine the water content of the soil contained in the containers in accordance with Test Method D 2216.

18. Calculation

18.1 Compute the average of the two water contents. Repeat the test if the difference between the two water contents is greater than the acceptable range for two results listed in Table 2 for single-operator precision. The plastic limit is the average of the two water contents.

PLASTICITY INDEX

19. Calculation

19.1 Calculate the plasticity index as follows:

$$PI = LL - PL$$

where:
LL = liquid limit,
PL = plastic limit.

19.1.1 Both LL and PL are whole numbers. If either the liquid limit or plastic limit could not be determined, or if the plastic limit is equal to or greater than the liquid limit, report the soil as nonplastic, NP.

20. Report

20.1 Report the following information:

20.1.1 Sample identifying information,

20.1.2 Any special specimen selection process used, such as removal of sand lenses from undisturbed sample,

20.1.3 Report sample as air-dried if the sample was air-dried before or during preparation,

20.1.4 Liquid limit, plastic limit, and plasticity index to the nearest whole number, omitting the percent designation. If the liquid limit or plastic limit tests could not be performed, or if the plastic limit is equal to or greater than the liquid limit, report the soil as nonplastic, NP,

20.1.5 Estimate of the percentage of sample retained on the 425-μm (No. 40) sieve, and

20.1.6 Procedure by which liquid limit was performed, if it differs from the multipoint method.

21. Precision and Bias

21.1 *Precision*—Criteria for judging the acceptability of liquid limit and plastic limit test results obtained by this test method on material are given in Table 2. The estimates of precision are based on the results of an interlaboratory study that included eleven laboratories performing the multipoint test (Method A) on three replicate samples of soil having a liquid limit of 64 and a plastic limit of 22.

21.2 *Bias*—There is no acceptable reference value for this test method; therefore, bias cannot be determined.

22. Keywords

22.1 activity; Atterberg limits; liquid limit; plasticity index; plastic limit

SUMMARY OF CHANGES

This section identifies the location of changes to this test method that have been incorporated since the last issue. Committee D-18 has highlighted those changes that affect the technical interpretation or use of this specification.

(1) In Section 6.1.1, the lower limit of the rebound requirement for liquid limit bases was changed from 80 % to 77 %.

(2) Section 10.1.2.3 was changed to restore two options for reducing the water content (*c* and *d*) that were deleted from the method during a previous revision.

APPENDIX

(Nonmandatory Information)

X1.1 A device for measuring the resilience of liquid limit device bases is shown in Fig. X1.1. The device consists of a clear acrylic plastic tube and cap, a 5/16-in. diameter steel ball, and a small bar magnet. The cylinder may be cemented to the cap or threaded as shown. The small bar magnet is held in the recess of the cap and the steel ball is fixed into the recess in the underside of the cap with the bar magnet. The cylinder is then turned upright and placed on the top surface of the base to be tested. Holding the tube lightly against the liquid limit device base with one hand, release the ball by

D 4318

CLEAR PLASTIC (SUCH AS ACRYLIC)
CAP AND TUBE

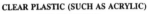

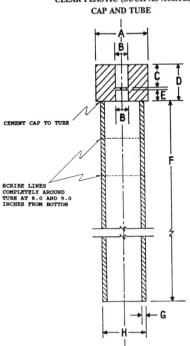

CEMENT CAP TO TUBE

SCRIBE LINES
COMPLETELY AROUND
TUBE AT 8.0 AND 9.0
INCHES FROM BOTTOM

TABLE OF MEASUREMENTS

DIMENSION	DESCRIPTION	ENGLISH, in.	METRIC, mm
A	DIAM. OF CAP	1 1/2	38.10
B	DIAM. OF HOLE	3/8	9.52
C	DEPTH OF HOLE	10/16	15.88
D	HEIGHT OF CAP	1	25.40
E	DEPTH OF HOLE	5/16	7.94
F	LENGTH OF TUBE	10	254.00
G	WALL THICKNESS	1/8	3.18
H	O.D. OF TUBE	1 1/4	31.75

FIG. X1.1   Resilience Tester

pulling the magnet out of the cap. Use the scale markings on the outside of the cylinder to determine the highest point reached by the bottom of the ball. Repeat the drop at least three times, placing the tester in a different location for each drop. Tests should be conducted at room temperature.

Designation: D 1556 – 90 (Reapproved 1996)[ε1]

Standard Test Method for
Density and Unit Weight of Soil in Place by the Sand-Cone Method[1]

This standard is issued under the fixed designation D 1556; the number immediately following the designation indicates the year of original adoption or, in the case of revision, the year of last revision. A number in parentheses indicates the year of last reapproval. A superscript epsilon (ε) indicates an editorial change since the last revision or reapproval.

[ε1] NOTE—Editorial changes were made in July 1996.

1. Scope

1.1 This test method may be used to determine the in-place density and unit weight of soils using a sand cone apparatus.

1.2 This test method is applicable for soils without appreciable amounts of rock or coarse materials in excess of 1½ in. (38 mm) in diameter.

1.3 This test method may also be used for the determination of the in-place density and unit weight of undisturbed or *in situ* soils, provided the natural void or pore openings in the soil are small enough to prevent the sand used in the test from entering the voids. The soil or other material being tested should have sufficient cohesion or particle attraction to maintain stable sides on a small hole or excavation, and be firm enough to withstand the minor pressures exerted in digging the hole and placing the apparatus over it, without deforming or sloughing.

1.4 This test method is not suitable for organic, saturated, or highly plastic soils that would deform or compress during the excavation of the test hole. This test method may not be suitable for soils consisting of unbound granular materials that will not maintain stable sides in the test hole, soils containing appreciable amounts of coarse material larger than 1½ in. (38 mm), and granular soils having high void ratios.

1.5 When materials to be tested contain appreciable amounts of particles larger than 1½ in. (38 mm), or when test hole volumes larger than 0.1 ft³ (2830 cm³) are required, Test Method D 4914 or D 5030 are applicable.

1.6 It is common practice in the engineering profession to concurrently use pounds to represent both a unit of mass (lbm) and a unit of force (lbf). This implicitly combines two separate systems of units, that is, the absolute system and the gravitational system. It is scientifically undesirable to combine the use of two separate sets of inch-pound units within a single standard. This test method has been written using the gravitational system of units when dealing with the inch-pound system. In this system the pound (lbf) represents a unit of force (weight). However, the use of balances or scales recording pounds of mass (lbm), or the recording of density

in lbm/ft³ should not be regarded as non-conformance with this test method.

1.7 *This standard does not purport to address all of the safety concerns, if any, associated with its use. It is the responsibility of the user of this standard to establish appropriate safety and health practices and determine the applicability of regulatory limitations prior to use.*

2. Referenced Documents

2.1 *ASTM Standards:*

D 653 Terms and Symbols Relating to Soil and Rock[2]

D 698 Test Methods for Moisture-Density Relations of Soils and Soil-Aggregate Mixtures Using 5.5-lb (2.49-kg) Rammer and 12-in. (304.8-mm) Drop[2]

D 1557 Test Methods for Moisture-Density Relations of Soils and Soil-Aggregate Mixtures Using 10-lb (4.54-kg) Rammer and 18-in. (457-mm) Drop[2]

D 2216 Method for Laboratory Determination of Water (Moisture) Content of Soil, Rock, and Soil-Aggregate Mixtures[2]

D 3584 Practice for Indexing Papers and Reports on Soil and Rock for Engineering Purposes[2]

D 4253 Test Method for Maximum Index Density of Soils Using a Vibratory Table[2]

D 4254 Test Method for Minimum Index Density of Soils and Calculation of Relative Density[2]

D 4643 Method for Determination of Water (Moisture) Content of Soil by the Microwave Oven Method[2]

D 4718 Practice for Correction of Unit Weight and Water Content for Soils Containing Oversize Particles[2]

D 4753 Specification for Evaluating, Selecting, and Specifying Balances and Scales for Use in Soil and Rock Testing[2]

D 4914 Test Method for Density and Unit Weight of Soil and Rock in Place by the Sand Replacement Method[2]

D 4944 Test Method for Field Determination of Water (Moisture) Content of Soil by the Calcium Carbide Gas Pressure Tester Method[3]

D 4959 Test Method for Determination of Water (Moisture) Content of Soil by Direct Heating Method[3]

D 5030 Test Methods for Density and Unit Weight of Soil and Rock in Place by the Water Replacement Method[3]

---

[1] This test method is under the jurisdiction of ASTM Committee D-18 on Soil and Rock and is the direct responsibility of Subcommittee D18.08 on Special and Construction Control Tests.

Current edition approved June 29, 1990. Published November 1990. Originally published as D 1556 – 58 T. Last previous edition D 1556 – 82[ε1]

[2] *Annual Book of ASTM Standards*, Vol 04.08.
[3] *Annual Book of ASTM Standards*, Vol 04.09.

⊕ D 1556

3. Terminology

3.1 *Definitions*—All definitions are in accordance with Terminology D 653.

4. Summary of Test Method

4.1 A test hole is hand excavated in the soil to be tested and all the material from the hole is saved in a container. The hole is filled with free flowing sand of a known density, and the volume is determined. The in-place wet density of the soil is determined by dividing the wet mass of the removed material by the volume of the hole. The water content of the material from the hole is determined and the dry mass of the material and the in-place dry density are calculated using the wet mass of the soil, the water content, and the volume of the hole.

5. Significance and Use

5.1 This test method is used to determine the density of compacted soils placed during the construction of earth embankments, road fill, and structural backfill. It often is used as a basis of acceptance for soils compacted to a specified density or percentage of a maximum density determined by a test method, such as Test Methods D 698 or D 1557.

5.2 This test method can be used to determine the in-place density of natural soil deposits, aggregates, soil mixtures, or other similar material.

5.3 The use of this test method is generally limited to soil in an unsaturated condition. This test method is not recommended for soils that are soft or friable (crumble easily) or in a moisture condition such that water seeps into the hand excavated hole. The accuracy of the test may be affected for soils that deform easily or that may undergo a volume change in the excavated hole from vibration, or from standing or walking near the hole during the test (see Note 1).

NOTE 1—When testing in soft conditions or in soils near saturation, volume changes may occur in the excavated hole as a result of surface loading, personnel performing the test, and the like. This can sometimes be avoided by the use of a platform that is supported some distance from the hole. As it is not always possible to detect when a volume change has taken place, test results should always be compared to the theoretical saturation density, or the zero air voids line on the dry density versus water content plot. Any in-place density test on compacted soils that calculates to be more than 95 % saturation is suspect and an error has probably occurred, or the volume of the hole has changed during testing.

6. Apparatus

6.1 *Sand-Cone Density Apparatus*, consisting of the following:

6.1.1 An attachable jar or other sand container having a volume capacity in excess of that required to fill the test hole and apparatus during the test.

6.1.2 A detachable appliance consisting of a cylindrical valve with an orifice approximately ½ in. (13 mm) in diameter, attached to a metal funnel and sand container on one end, and a large metal funnel (sand-cone) on the other end. The valve will have stops to prevent rotating past the completely open or completely closed positions. The appliance will be constructed of metal sufficiently rigid to prevent distortion or volume changes in the cone. The walls of the cone will form an angle of approximately 60 degrees with the

base to allow uniform filling with sand.

6.1.3 A metal base plate or template with a flanged center hole cast or machined to receive the large funnel (cone) of the appliance described in 6.1.2. The base plate may be round or square and will be a minimum of 3 in. (75 mm) larger than the funnel (sand-cone). The plate will be flat on the bottom and have sufficient thickness or stiffness to be rigid. Plates with raised edges, ridges, ribs, or other stiffners of approximately ⅜ to ½ in. (10 to 13 mm) high may be used.

6.1.4 The mass of the sand required to fill the apparatus and base plate will be determined in accordance with the instructions in Annex A1 prior to use.

6.1.5 The details for the apparatus shown in Fig. 1 represents the minimum acceptable dimensions suitable for testing soils having maximum particle sizes of approximately 1½ in. (38 mm) and test hole volumes of approximately 0.1 ft³ (2830 cm³). When the material being tested contains a small amount of oversize and isolated larger particles are encountered, the test should be moved to a new location. Larger apparatus and test hole volumes are needed when particles larger than 1½ in. (38 mm) are prevalent. The apparatus described here represents a design that has proven satisfactory. Larger apparatus, or other designs of similar proportions may be used as long as the basic principles of the sand volume determination are observed. When test hole volumes larger than 0.1 ft³ (5660 cm³) are required Test

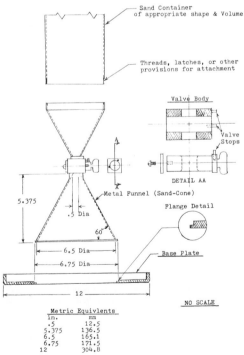

FIG. 1   **Density Apparatus**

ASTM D 1556

Method D 4914 should be utilized.

6.2 *Sand*—Sand must be clean, dry, uniform in density and grading, uncemented, durable, and free-flowing. Any gradation may be used that has a uniformity coefficient ($C_u = D_{60}/D_{10}$) less than 2.0, a maximum particle size smaller than 2.0 mm (No. 10 sieve), and less than 3 % by weight passing 250 µm (No. 60 sieve). Uniformly graded sand is needed to prevent segregation during handling, storage, and use. Sand free of fines and fine sand particles is required to prevent significant bulk-density changes with normal daily changes in atmospheric humidity. Sand comprised of durable, natural subrounded, or rounded particles is desirable. Crushed sand or sand having angular particles may not be free-flowing, a condition that can cause bridging resulting in inaccurate density determinations (see Note 2). In selecting a sand from a potential source, a gradation and bulk-density determinations in accordance with the procedure in Annex A2 should be made on each container or bag of sand. To be an acceptable sand, the bulk-density variation between any one determination shall not be greater than 1 % of the average. Before using sand in density determinations, it shall be dried, then allowed to reach an air-dried state in the general location where it is to be used (see Note 3). Sand shall not be re-used without removing any contaminating soil, checking the gradation, drying and redetermining the bulk-density (see Note 4). Bulk-density tests of the sand will be made at time intervals not exceeding 14 days, always after any significant changes in atmospheric humidity, before reusing, and before use of a new batch from a previously approved supplier (see Note 5).

NOTE 2—Some manufactured (crushed) sands such as blasting sand have been successfully used with good reproducibility. The reproducibility of test results using angular sand should be checked under laboratory controlled testing situations before selecting an angular sand for use.

NOTE 3—Many organizations have found it beneficial to store sands in moisture resistant containers. Sand should be stored in dry areas protected from weather. The use of a lighted bulb or other heat source in, or adjacent to the storage containers has also been found to be beneficial in areas of high humidity.

NOTE 4—As a general rule, reclaiming sand after testing is not desirable.

NOTE 5—Most sands have a tendency to absorb moisture from the atmosphere. A very small amount of absorbed moisture can make a substantial change in bulk-density. In areas of high humidity, or where the humidity changes frequently, the bulk-density may need to be determined more often than the 14 day maximum interval indicated. The need for more frequent checks can be determined by comparing the results of different bulk-density tests on the same sand made in the same conditions of use over a period of time.

6.3 *Balances or Scales*—Meeting Specification D 4753, with 5.0 g readability, or better, to determine the mass of sand and excavated soils. A balance or scale having a minimum capacity of 20 kg and 5.0 g readability is suitable for determining the mass of the sand and the excavated soil when apparatus with the dimensions shown in Fig. 1 is used.

6.4 *Drying Equipment*—Equipment corresponding to the method used for determining water content as specified in Test Methods D 2216, D 4643, D 4959, or D 4944.

6.5 *Miscellaneous Equipment*—Knife, small pick, chisel, small trowel, screwdriver, or spoons for digging test holes, large nails or spikes for securing the base plate; buckets with lids, plastic-lined cloth sacks, or other suitable containers for retaining the density samples, moisture sample, and density sand respectively; small paint brush, calculator, notebook or test forms, etc.

7. Procedure

7.1 Select a location/elevation that is representative of the area to be tested, and determine the density of the soil in-place as follows:

7.1.1 Inspect the cone apparatus for damage, free rotation of the valve, and properly matched baseplate. Fill the cone container with conditioned sand for which the bulk-density has been determined in accordance with Annex A2, and determine the total mass.

7.1.2 Prepare the surface of the location to be tested so that it is a level plane. The base plate may be used as a tool for striking off the surface to a smooth level plane.

7.1.3 Seat the base plate on the plane surface, making sure there is contact with the ground surface around the edge of the flanged center hole. Mark the outline of the base plate to check for movement during the test, and if needed, secure the plate against movement using nails pushed into the soil adjacent to the edge of the plate, or by other means, without disturbing the soil to be tested.

7.1.4 In soils where leveling is not successful, or surface voids remain, the volume horizontally bounded by the funnel, plate and ground surface must be determined by a preliminary test. Fill the space with sand from the apparatus, determine the mass of sand used to fill the space, refill the apparatus, and determine a new initial mass of apparatus and sand before proceeding with the test. After this measurement is completed, carefully brush the sand from the prepared surface (see Note 6).

NOTE 6—A second calibrated apparatus may be taken to the field when this condition is anticipated (instead of refilling and making a second determination). The procedure in 7.1.4 may be used for each test when the best possible accuracy is desired, however, it is usually not needed for most production testing where a relatively smooth surface is obtainable.

7.1.5 Dig the test hole through the center hole in the base plate, being careful to avoid disturbing or deforming the soil that will bound the hole. Test hole volumes are to be as large as practical to minimize errors and will in no case be smaller than the volumes indicated in Table 1 for the maximum size of soil particle removed from the test hole. The sides of the hole should slope slightly inward and the bottom should be reasonably flat or concave. The hole should be kept as free as possible of pockets, overhangs, and sharp obtrusions since these affect the accuracy of the test. Soils that are essentially granular require extreme care and may require digging a conical-shaped test hole. Place all excavated soil, and any soil loosened during digging, in a moisture tight container that is marked to identify the test number. Take care to avoid losing any materials. Protect this material from any loss of moisture

TABLE 1   Minimum Test Hole Volumes Based on Maximum Size of Included Particle

Maximum Particle Size		Minimum Test Hole Volumes	
in.	(mm)	cm³	ft³
½	(12.7)	1415	0.05
1	(25.4)	2125	0.075
1½	(38)	2830	0.1

D 1556

until the mass has been determined and a specimen has been obtained for a water content determination.

7.1.6 Clean the flange of the base plate hole, invert the sand-cone apparatus and seat the sand-cone funnel into the flanged hole at the same position as marked during calibration (see Annex A1). Eliminate or minimize vibrations in the test area due to personnel or equipment. Open the valve and allow the sand to fill the hole, funnel, and base plate. Take care to avoid jarring or vibrating the apparatus while the sand is running. When the sand stops flowing, close the valve.

7.1.7 Determine the mass of the apparatus with the remaining sand, record, and calculate the mass of sand used.

7.1.8 Determine and record the mass of the moist material that was removed from the test hole. When oversized material corrections are required, determine the mass of the oversized material on the appropriate sieve and record, taking care to avoid moisture losses. When required, make appropriate corrections for the oversized material using Practice D 4718.

7.1.9 Mix the material thoroughly, and either obtain a representative specimen for water content determination, or use the entire sample.

7.1.10 Determine the water content in accordance with Test Method D 2216, D 4643, D 4944, or D 4959. Correlations to Method D 2216 will be performed when required by other test methods.

7.2 Water content specimens must be large enough and selected in such a way that they represent all the material obtained from the test hole. The minimum mass of the water content specimens is that required to provide water content values accurate to 1.0 %.

8. Calculation

8.1 Calculations shown are for mass in grams and volumes in cubic centimetres. Other units are permissible provided the appropriate conversion factors are used to maintain consistency of units throughout the calculations. See 1.6 for additional comments on the usage of inch-pound units.

8.2 Calculate the volume of the test hole as follows:

$$V = (M_1 - M_2)/\rho_1$$

where:

V = volume of the test hole, cm^3,
M_1 = mass of the sand used to fill the test hole, funnel and base plate, g (from 7.1.7),
M_2 = mass of the sand used to fill the funnel and base plate (from Annex A1.2.2.6), g, and
ρ_1 = bulk density of the sand (from Annex A2.3.5), g/cm^3.

8.3 Calculate the dry mass of material removed from the test hole as follows:

$$M_4 = 100 \, M_3/(w + 100)$$

where:

w = water content of the material removed from test hole, %, (from 7.1.10),
M_3 = moist mass of the material from test hole, g, (from 7.1.8), and
M_4 = dry mass of material from test hole, g, or multiply by 0.002205 for lb.

8.4 Calculate the in-place wet and dry density of the material tested as follows:

$$\rho_m = M_3/V$$
$$\rho_d = M_4/V$$

where:

V = volume of the test hole, cm^3 (from 8.2),
M_3 = moist mass of the material from the test hole, g, (from 7.1.8),
M_4 = dry mass of the material from the test hole, g, (from 8.3),
ρ_m = wet density of the tested material g/cm^3 or its wet unit weight, γ_m in lb/ft^3 where $\gamma_m = \rho_m \times 62.43$, and
ρ_d = dry density of the tested material, g/cm^3 or its dry unit weight, γ_d in lb/ft^3 where $\gamma_d = \rho_d \times 62.43$.

8.5 It may be desired to express the in-place density as a percentage of some other density, for example, the laboratory densities determined in accordance with Test Method D 698, D 1557, D 4253 or D 4254. This relation can be determined by dividing the in-place density by the laboratory density and multiplying by 100. Calculations for determining relative density are provided in Test Method D 4254. Corrections for oversize material, if required, should be performed in accordance with Practice D 4718.

9. Report

9.1 Report, as a minimum, the following information:

9.1.1 Test location, elevation, thickness of layer tested, or other pertinent data to locate or identify the test.

9.1.2 Test hole volume, cm^3 or ft^3.

9.1.3 In-place wet density, g/cm^3 or lb/ft^3.

9.1.4 In-place dry density, ρ_d, g/cm^3.

9.1.5 In-place dry unit weight, KN/m^3 ($\rho_d \times 9.807$), or lb/ft^3 ($\rho_d \times 62.43$), expressed to the nearest .1 KN/m^3, or 1.0 for lb/ft^3.

9.1.6 In-place water content of the soil expressed as a percentage of dry mass, and the test method used.

9.1.7 Test apparatus identity and calibrated volume.

9.1.8 Bulk density of the sand used, g/cm^3, or lb/ft^3.

9.1.9 Visual description of the soil or material designation.

9.1.10 Mass and percentage of oversized particles and the size sieve used, if performed.

9.1.11 Comments on the test, as applicable.

9.1.12 If the in-place dry density or unit weight is expressed as a percentage of another value, include the following:

9.1.12.1 The laboratory test method used.

9.1.12.2 The comparative dry density or unit weight value and water content used.

9.1.12.3 Correction for oversized material and details, if applicable.

9.1.12.4 The comparative percentage of the in-place material to the comparison value.

9.1.13 If the in-place density, unit weight, or water content are to be used for acceptance, include the acceptance criteria applicable to the test.

10. Precision and Bias

10.1 *Statement of Precision*—Due to the nature of the soil or rock materials tested by the method it is either not feasible or too costly at this time to produce multiple specimens that

D 1556

have uniform physical properties. Any variation observed in the data is just as likely to be due to specimen variation as to operator or laboratory testing variation. Subcommittee D18.08 welcomes proposals that would allow for development of a valid precision statement.

10.2 *Statement of Bias*—There is no accepted reference value for this test method, therefore, bias cannot be determined.

10.3 While no formal round-robin testing has been completed, it is estimated by Subcommittee D18.08 from available data that the results of two properly conducted tests performed by a skilled operator on the same material at a given time and location should not differ by more than approximately 2 lb/ft^3 (3.2 Kg/m^3). Tests performed by unskilled operators on the same material would be expected to yield substantially greater differences.

11. Keywords

11.1 The following keywords are applicable to this test method in accordance with Practice D 3584: acceptance tests; compaction tests; degree of compaction; density tests; earthfill; embankments; field control density; field tests; inplace density; inplace dry density; *in situ* density; relative density; sand cone; soil compaction; soil tests; unit weight.

ANNEXES

(Mandatory Information)

A1. CALIBRATION OF SAND CONE APPARATUS

A1.1 Scope

A1.1.1 This annex describes the procedure for determining the mass of sand contained in the funnel and base plate of the sand-cone apparatus.

A1.1.2 The mass of sand contained in the apparatus and base plate is dependent on the bulk-density of the sand. Consequently, this procedure must be performed for each apparatus whenever there are changes in the sand bulk-densities.

A1.2 Calibration Procedure

A1.2.1 Calibration of the apparatus can be accomplished by either of two methods:

A1.2.1.1 *Method A*—By determining the mass of calibrated sand that can be contained in each funnel and base plate set, or

A1.2.1.2 *Method B*—By determining the volume of sand needed to fill each funnel and base plate set and applying this volume constant whenever new sand bulk-densities are calculated.

A1.2.1.3 Since the mass of sand contained in the apparatus funnel and base plate is dependent on the bulk density of the sand, if Method A is used, it must be repeated whenever the bulk-density of the sand changes.

A1.2.2 All determinations of mass are to be made to the nearest 5 g.

A1.2.3 *Method A:*

A1.2.3.1 Fill the apparatus with sand that is dried and conditioned to the same state anticipated during use in testing.

A1.2.3.2 Determine the mass of the apparatus filled with sand, g.

A1.2.3.3 Place the base plate on a clean, level, plane surface. Invert the container/apparatus and seat the funnel in the flanged center hole in the base plate. Mark and identify the apparatus and base plate so that the same apparatus and plate can be matched and reseated in the same position during testing.

A1.2.3.4 Open the valve fully until the sand flow stops, making sure the apparatus, base plate, or plane surface are not jarred or vibrated before the valve is closed.

A1.2.3.5 Close the valve sharply, remove the apparatus and determine the mass of the apparatus and remaining sand. Calculate the mass of sand used to fill the funnel and base plate as the difference between the initial and final mass.

A1.2.3.6 Repeat the procedure a minimum of three times. The maximum variation between any one determination and the average will not exceed 1 %. Use the average of the three determinations for this value in the test calculations.

A1.2.4 *Method B (Optional):*

A1.2.4.1 When large numbers of tests and batches of sand are anticipated, it may be advantageous to determine the volume of each apparatus and base plate. Baring damage to the apparatus or mismatching of the base plates, this volume will remain constant, and will eliminate the need to repeat Method A when the sand bulk-density changes (see Note A1.1). If this alternative is chosen, the calculations in the field test must be altered to determine the total volume of the sand in the field test hole and apparatus. The volume of the apparatus is then subtracted to determine the volume of the test hole.

A1.2.4.2 Determine the mass of sand required to fill the apparatus funnel and base plate in accordance with A1.2.3, following steps A1.2.3.1 through A1.2.3.6 for each batch of sand.

A1.2.4.3 Calculate the volume of the funnel and base plate by dividing the bulk-density of the sand (as determined in Annex A2) by the mass of sand found in A1.2.3.6. Perform a minimum of three determinations and calculate an average value. The maximum volume variation between any one determination and the average will not exceed 1 %. Use the average of the values when performing test calculations.

Note A1.1—The sand-cone apparatus should be routinely inspected for damage that may affect the volume of the cone. Dings, out-of-round, or other damage will affect the volume and will necessitate a redetermination of the volume (if repairable).

⏚ D 1556

A2. CALIBRATION OF DENSITY SAND

A2.1 Scope

A2.1.1 This annex is used for determining the bulk-density (calibration) of the sand for use in this test method.

A2.1.2 The calibration determines an average density of the sand for use in calculating the volume of the test hole.

A2.2 Equipment Required

A2.2.1 *Container*—Select a container of known volume that is approximately the same size and allows the sand to fall approximately the same distance as the hole excavated during a field test. The $1/30$ ft^3 (944 cm^3) and $1/13.33$ ft^3 (2124 cm^3) molds specified in Test Methods D 698, or the 0.1 ft^3 (2830 cm^3) mold specified in Test Method D 4253 are recommended. Alternatively, cast duplicates of actual test holes may be used. This is accomplished by forming plaster of paris negatives in actual test holes over a range of test volumes, and using these as forms for portland cement concrete castings. These should be cast against a flat plane surface, and after the removal of the negative, sealed water tight and the volume determined in accordance with the procedure in Test Method D 4253 (10.2.2).

A2.2.1.1 Determine the container volume to 1 % using water in accordance with the procedures described in Test Method D 4253.

A2.2.2 *Sand-Cone Apparatus*—Use a sand cone apparatus of the same size and design as will be used during field testing.

A2.2.2.1 Flow characteristics through different valve assemblies have been shown to cause different bulk-density values. Bulk-density determinations will be required for each apparatus set unless other assemblies are determined to provide the same results.

A2.2.3 *Balance or Scale*—A balance or scale having a sufficient capacity to determine the mass of the calibration container filled with sand. For 0.500 ft^3 (14 200 cm^3) containers, a balance having a minimum capacity of 50 lb (20 kg) and meeting the requirements of Specification D 4753 for 0.01 lb (5 g) readability is required.

A2.2.4 *Metal Straightedge*, about 2 in. (50 cm) wide, at least $\frac{1}{8}$ in. (3 mm) thick, and length approximately 1.5 times the diameter of the calibration container.

A2.3 Bulk-Density Determination

A2.3.1 Fill the assembled apparatus with sand. The sand is to be dried and conditioned to the same state anticipated during use.

A2.3.2 Determine and record the mass of the calibration container when empty.

A2.3.3 *Method A (Preferred):*

A2.3.3.1 When the calibration container has the same diameter as the flanged center hole in the base plate, invert and center the sand filled apparatus and base plate on the calibration container.

A2.3.3.2 Fully open the valve and allow the sand to fill the container. When the sand flow stops, close the valve.

A2.3.3.3 Determine the mass of the apparatus and remaining sand. Calculate the net mass of sand in the calibration container by subtracting the mass of sand contained in the cone and base plate (as determined in Annex A1) and record.

A2.3.4 *Method B (Alternative):*

A2.3.4.1 Invert and support the apparatus over the calibration container so that the sand falls approximately the same distance and location as in a field test, and fully open the valve.

A2.3.4.2 Fill the container until it just overflows and close the valve. Using a minimum number of strokes and taking care not to jar or densify the sand, carefully strike off excess sand to a smooth level surface. Any vibration or jarring during the bulk-density determination will result in settling and densifying the sand, leading to erroneous results.

A2.3.4.3 Clean any sand from the outside of the calibration container. Determine the mass of the container and sand. Record the net mass of the sand by subtracting the mass of the empty container.

A2.3.5 Perform at least three bulk-density determinations and calculate the average. The maximum variation between any one determination and the average will not exceed 1 %. Repeated determinations not meeting these requirements indicates nonuniform sand density, and the sand source should be re-evaluated for suitability. The average value obtained is to be used in the test calculations.

A2.4 Calculation

A2.4.1 Calculate the bulk-density of the sand as follows:

$$\rho_1 = M_5/V_1$$

where:

ρ_1 = bulk-density of the sand, g/cm^3, (multiply by 9.807 for KN/m^3, or 62.43 for lb/ft^3),

M_5 = mass of the sand to fill the calibration container, g, (from A2.3.4.3), and

V_1 = volume of the calibration container, cm^3 (from A2.2.1.1).

Designation: D 3282 – 93

Standard
Classification of Soils and Soil-Aggregate Mixtures for Highway Construction Purposes[1]

This standard is issued under the fixed designation D 3282; the number immediately following the designation indicates the year of original adoption or, in the case of revision, the year of last revision. A number in parentheses indicates the year of last reapproval. A superscript epsilon (ε) indicates an editorial change since the last revision or reapproval.

This standard has been approved for use by agencies of the Department of Defense. Consult the DoD Index of Specifications and Standards for the specific year of issue which has been adopted by the Department of Defense.

1. Scope

1.1 This standard describes a procedure for classifying mineral and organomineral soils into seven groups based on laboratory determination of particle-size distribution, liquid limit, and plasticity index. It may be used when a precise engineering classification is required, especially for highway construction purposes. Evaluation of soils within each group is made by means of a *group index*, which is a value calculated from an empirical formula.

NOTE 1—The group classification, including the group index, should be useful in determining the relative quality of the soil material for use in earthwork structures, particularly embankments, subgrades, subbases, and bases. However, for the detailed design of important structures, additional data concerning strength or performance characteristics of the soil under field conditions will usually be required.

1.2 *This standard does not purport to address all of the safety problems, if any, associated with its use. It is the responsibility of the user of this standard to establish appropriate safety and health practices and determine the applicability of regulatory limitations prior to use.*

2. Referenced Documents

2.1 *ASTM Standards:*
D 420 Guide for Investigating and Sampling Soil and Rock[2]
D 421 Practice for Dry Preparation of Soil Samples for Particle-Size Analysis and Determination of Soil Constants[2]
D 422 Test Method for Particle-Size Analysis of Soils[2]
D 653 Terminology Relating to Soil, Rock, and Contained Fluids[2]
D 1140 Test Method for Amount of Material in Soils Finer Than the No. 200 (75-μm) Sieve[2]
D 1452 Practice for Soil Investigation and Sampling by Auger Borings[2]
D 1586 Test Method for Penetration Test and Split-Barrel Sampling of Soils[2]
D 1587 Practice for Thin-Walled Tube Sampling of Soils[2]
D 2217 Practice for Wet Preparation of Soil Samples for Particle-Size Analysis and Determination of Soil Constants[2]

D 4318 Test Method for Liquid Limit, Plastic Limit, and Plasticity Index of Soils[2]
2.2 *AASHTO Document:*
M 145 The Classification of Soils and Soil Aggregate Mixtures for Highway Construction Purposes[3]

3. Terminology

3.1 *Descriptions of Terms Specific to This Standard:*
3.1.1 The following terms are frequently used in this standard. These terms differ slightly from those given in Terminology D 653, but are used here to maintain consistency with common highway usage.
3.1.2 *boulders*—rock fragments, usually rounded by weathering or abrasion, that will be retained on a 3-in. (75-mm) sieve.
3.1.3 *coarse sand*—particles of rock or soil that will pass a No. 10 (2-mm) sieve and be retained on a No. 40 (425-μm) sieve.
3.1.4 *fine sand*—particles of rock or soil that will pass a No. 40 (425-μm) sieve and be retained on a No. 200 (75-μm) sieve.
3.1.5 *gravel*—particles of rock that will pass a 3-in. (75-mm) sieve and be retained on a No. 10 (2-mm) sieve.
3.1.6 *silt-clay (combined silt and clay)*—fine soil and rock particles that will pass a No. 200 (75-μm) sieve.
3.1.6.1 *silty*—fine-grained material that has a plasticity index of 10 or less.
3.1.6.2 *clayey*—fine-grained material that has a plasticity index of 11 or more.

4. Significance and Use

4.1 The standard described classifies soils from any geographic location into groups (including group indexes) based on the results of prescribed laboratory tests to determine the particle-size characteristics, liquid limit, and plasticity index.

4.2 The assigning of a group symbol and group index can be used to aid in the evaluation of the significant properties of the soil for highway and airfield purposes.

4.3 The various groupings of this classification system correlate in a general way with the engineering behavior of soils. Also, in a general way, the engineering behavior of a soil varies inversely with its group index. Therefore, this standard provides a useful first step in any field or laboratory investigation for geotechnical engineering purposes.

---

[1] This standard is under the jurisdiction of ASTM Committee D-18 on Soil and Rock and is the direct responsibility of Subcommittee D18.07 on Identification and Classification of Soils.
Current edition approved Sept. 15, 1993. Published November 1993. Originally published as D 3282 – 73. Last previous edition D 3282 – 92.
[2] *Annual Book of ASTM Standards*, Vol 04.08.

[3] Available from American Asssociation of State Highway and Transportation Officials, 444 N Capitol St., NW, Suite 225, Washington, DC 20001.

D 3282

5. Apparatus

5.1 *Apparatus for Preparation of Samples*—See Practices D 421 or D 2217.

5.2 *Apparatus for Particle-Size Analysis*—See Test Methods D 1140 and D 422.

5.3 *Apparatus for Liquid Limit and Plastic Limit Tests*—See Test Method D 4318.

6. Sampling

6.1 Conduct field investigations and sampling in accordance with one or more of the following procedures:

6.1.1 Guide D 420,

6.1.2 Practice D 1452,

6.1.3 Method D 1586,

6.1.4 Practice D 1587.

7. Test Sample

7.1 Test samples shall represent that portion of the field sample finer than the 3-in. (75-mm) sieve and shall be obtained as follows:

7.1.1 Air-dry the field sample,

7.1.2 Weigh the field sample,

7.1.3 Separate the field sample into two fractions on a 3-in. (75-mm) sieve,

7.1.4 Weigh the fraction retained on the 3-in. (75-mm) sieve. Compute the percentage of plus 3-in. material in the field sample, and note this percentage as auxiliary information, and

7.1.5 Thoroughly mix the fraction passing the 3-in. (75-mm) sieve and select the test samples.

NOTE 2—If visual examination indicates that no boulder size material is present, omit 7.1.3 and 7.1.4.

7.2 Prepare the test sample in accordance with Practices D 421 or D 2217. Determine the percentage of the sample finer than a No. 10 (2-mm) sieve.

NOTE 3—It is recommended that the method for wet preparation be used for soils containing organic matter or irreversible mineral colloids.

8. Testing Procedure

8.1 Determine the percentage of the test sample finer than a No. 200 (75-µm) sieve in accordance with Test Methods D 1140 or D 422.

NOTE 4—For granular materials the percentage of the sample finer than a No. 40 (425-µm) sieve must also be determined.

8.2 Determine the liquid limit and the plasticity index of a portion of the test sample passing a No. 40 (425-µm) sieve in accordance with Test Method D 4318.

9. Classification Procedure

9.1 Using the test data determined in Section 8, classify the soil into the appropriate group or subgroup, or both, in accordance with Tables 1 or 2. Use Fig. 1 to classify silt-clay materials on the basis of liquid limit and plasticity index values.

NOTE 5—All limiting values are shown as whole numbers. If fractional numbers appear on test reports, convert to the nearest whole numbers for the purpose of classification.

9.1.1 With the required test data available, proceed from left to right in Tables 1 or 2 and the correct classification will be found by the process of elimination. The first group from the left into which the test data will fit is the correct classification.

NOTE 6—Classification of materials in the various groups applies only to the fraction passing the 3-in. (75-mm) sieve. Therefore, any specification regarding the use of A-1, A-2, or A-3 materials in construction should state whether boulders (retained on 3-in. sieve) are permitted.

10. Description of Classification Groups

10.1 *Granular Materials,* containing 35 % or less passing the No. 200 (75-µm) sieve:

10.1.1 *Group A-1*—The typical material of this group is a well-graded mixture of stone fragments or gravel, coarse sand, fine sand, and a nonplastic or feebly-plastic soil binder. However, this group also includes stone fragments, gravel, coarse sand, volcanic cinders, etc., without a soil binder.

10.1.1.1 Subgroup A-1-a includes those materials consisting predominantly of stone fragments or gravel, either with or without a well-graded binder of fine material.

10.1.1.2 Subgroup A-1-b includes those materials consisting predominantly of coarse sand, either with or without a well-graded soil binder.

10.1.2 *Group A-3*—The typical material of this group is fine beach sand or fine desert-blow sand without silty or clay fines, or with a very small amount of nonplastic silt. This group also includes stream-deposited mixtures of poorly-graded fine sand and limited amounts of coarse sand and gravel.

10.1.3 *Group A-2*—This group includes a wide variety of "granular" materials which are borderline between the materials falling in Groups A-1 and A-3, and the silt-clay materials of Groups A-4, A-5, A-6, and A-7. It includes all materials containing 35 % or less passing a No. 200 (75-µm) sieve which cannot be classified in Groups A-1 or A-3, due to the fines content or the plasticity indexes, or both, in excess of the limitations for those groups.

10.1.3.1 Subgroups A-2-4 and A-2-5 include various granular materials containing 35 % or less passing a No. 200 (75-µm) sieve and with a minus No. 40 (425-µm) portion having the characteristics of Groups A-4 and A-5, respectively. These groups include such materials as gravel and coarse sand with silt contents or plasticity indexes in excess of the limitations of Group A-1 and fine sand with nonplastic-silt content in excess of the limitations of Group A-3.

10.1.3.2 Subgroups A-2-6 and A-2-7 include materials similar to those described under Subgroups A-2-4 and A-2-5, except that the fine portion contains plastic clay having the characteristics of the A-6 or A-7 group, respectively.

10.2 *Silt-Clay Materials,* containing more than 35 % passing a No. 200 (75-µm) sieve:

10.2.1 *Group A-4*—The typical material of this group is a nonplastic or moderately plastic silty soil usually having 75 % or more passing a No. 200 (75-µm) sieve. This group also includes mixtures of fine silty soil and up to 64 % of sand and gravel retained on a No. 200 sieve.

10.2.2 *Group A-5*—The typical material of this group is similar to that described under Group A-4, except that it is usually of diatomaceous or micaceous character and may be highly elastic as indicated by the high liquid limit.

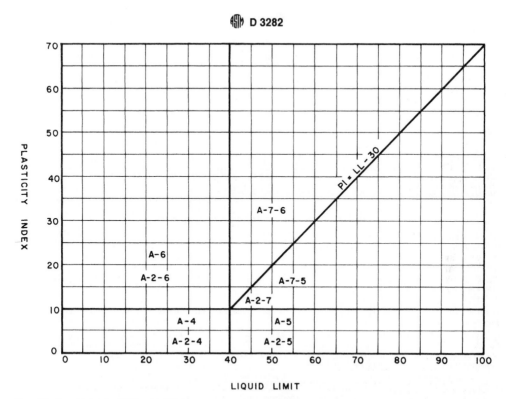

⏚⏚ **D 3282**

NOTE—A-2 soils contain less than 35 % finer than 200 sieve.

FIG. 1   Liquid Limit and Plasticity Index Ranges for Silt-Clay Materials

10.2.3 *Group A-6*—The typical material of this group is a plastic clay soil usually having 75 % or more passing a No. 200 (75-μm) sieve. This group also includes mixtures of fine clayey soil and up to 64 % of sand and gravel retained on a No. 200 sieve. Materials of this group usually have a high volume change between wet and dry states.

10.2.4 *Group A-7*—The typical material of this group is similar to that described under Group A-6, except that it has the high liquid limits characteristic of Group A-5 and may be elastic as well as subject to high-volume change.

10.2.4.1 Subgroup A-7-5 includes those materials with moderate plasticity indexes in relation to the liquid limit and which may be highly elastic as well as subject to considerable volume change.

10.2.4.2 Subgroup A-7-6 includes those materials with high plasticity indexes in relation to liquid limit and which are subject to extremely high volume change.

NOTE 7—Highly organic soils (peat or muck) may be classified in Group A-8. Classification of these materials is based on visual inspection and is not dependent on the percentage passing the No. 200 (75-μm) sieve, liquid limit, or plasticity index. The material is composed primarily of partially-decayed organic matter, generally has a fibrous texture, a dark brown or black color, and an odor of decay. These organic materials are unsuitable for use in embankments and subgrades. They are highly compressible and have low strength.

11. Group Index Computation

11.1 The classifications obtained from Tables 1 or 2 may be modified by the addition of a group-index value. Group-index values should always be shown in parentheses after the group symbol as A-2-6(3), A-4(5), A-6(12), A-7-5(17), etc.

11.1.1 Calculate the group index from the following empirical formula:

$$\text{Group index} = (F - 35)[0.2 + 0.005(LL - 40)] + 0.01(F - 15)(PI - 10)$$

where:

F = percentage passing No. 200 (75-μm) sieve, expressed as a whole number (this percentage is based only on the material passing the 3-in. (75-mm) sieve),

LL = liquid limit, and

PI = plasticity index.

11.1.2 If the calculated group index is negative, report the group index as zero (0).

11.1.3 If the soil is nonplastic and when the liquid limit cannot be determined, report the group index as zero (0).

11.1.4 Report the group index to the nearest whole number.

11.1.5 The group index value may be estimated using Fig. 2 by determining the partial group index due to the liquid limit and that due to the plasticity index, then obtaining the total of the two partial group indexes.

◆ D 3282

TABLE 1 Classification of Soils and Soil-Aggregate Mixtures

General Classification	Granular Materials (35 % or less passing No. 200)			Silt-Clay Materials (More than 35 % passing No. 200)			
Group Classification	A-1	A-3[A]	A-2	A-4	A-5	A-6	A-7
Sieve analysis, % passing:							
No. 10 (2.00 mm)	. . .	. . .	. . .	. . .	. . .	. . .	. . .
No. 40 (425 μm)	50 max	51 min	. . .	. . .	. . .	. . .	. . .
No. 200 (75 μm)	25 max	10 max	35 max	36 min	36 min	36 min	36 min
Characteristics of fraction passing No. 40 (425 μm):							
Liquid limit	. . .	. . .	B	40 max	41 min	40 max	41 min
Plasticity index	6 max		B	10 max	10 max	11 min	11 min
General rating as subgrade	Excellent to Good			Fair to Poor			

[A] The placing of A-3 before A-2 is necessary in the "left to right elimination process" and does not indicate superiority of A-3 over A-2.
[B] See Table 2 for values.
Reprinted with permission of American Association of State Highway and Transportation Officials

TABLE 2 Classification of Soils and Soil-Aggregate Mixtures

General Classification	Granular Materials (35 % or less passing No. 200)							Silt-Clay Materials (More than 35 % passing No. 200)			
	A-1		A-3	A-2				A-4	A-5	A-6	A-7
Group classification	A-1-a	A-1-b		A-2-4	A-2-5	A-2-6	A-2-7				A-7-5, A-7-6
Sieve analysis, % passing:											
No. 10 (2.00 mm)	50 max	. . .	. . .	. . .	. . .	. . .	. . .	. . .	. . .	. . .	. . .
No. 40 (425 μm)	30 max	50 max	51 min	. . .	. . .	. . .	. . .	. . .	. . .	. . .	. . .
No. 200 (75 μm)	15 max	25 max	10 max	35 max	35 max	35 max	35 max	36 min	36 min	36 min	36 min
Characteristics of fraction passing No. 40 (425 μm):											
Liquid limit	. . .		. . .	40 max	41 min	40 max	41 min	40 max	41 min	40 max	41 min
Plasticity index	6 max		N.P.	10 max	10 max	11 min	11 min	10 max	10 max	11 min	11 min[A]
Usual types of significant constituent materials	Stone Fragments, Gravel and Sand		Fine Sand	Silty or Clayey Gravel and Sand				Silty Soils		Clayey Soils	
General rating as subgrade	Excellent to Good							Fair to Poor			

[A] Plasticity index of A-7-5 subgroup is equal to or less than *LL* minus 30. Plasticity index of A-7-6 subgroup is greater than *LL* minus 30 (see Fig. 1).
Reprinted with permission of American Association of State Highway and Transportation Officials.

11.1.6 The group index of soils in the A-2-6 and A-2-7 subgroups shall be calculated using only the *PI* portion of the formula (or Fig. 2).

11.2 The following examples illustrate the calculations for the group index:

11.2.1 Assume that an A-6 material has 55 % passing a No. 200 (75-μm) sieve, a liquid limit of 40, and a plasticity index of 25, then:

Group index = $(55 - 35)[0.2 + 0.005(40 - 40)]$
$\qquad + [0.01(55 - 15)(25 - 10)] = 4.0 + 6.0 = 10$

11.2.2 Assume that an A-7 material has 80 % passing a No. 200 (75-μm) sieve, a liquid limit of 90, and a plasticity index of 50, then:

Group index = $(80 - 35)[0.2 \pm 0.005(90 - 40)]$
$\qquad + [0.01(80 - 15)(50 - 10)] = 20.3 + 26.0 = 46.3$ (report as 46)

11.2.3 Assume that an A-4 material has 60 % passing a No. 200 (75-μm) sieve, a liquid limit of 25, and a plasticity index of 1, then

Group index = $(60 - 35)[0.2 + 0.005(25 - 40)]$
$\qquad + [0.01(60 - 15)(1 - 10)] = 25 \times (0.2 - 0.075)$
$\qquad\qquad + 0.01(45)(-9) = 3.1 - 4.1 = -1.0$ (report as 0)

11.2.4 Assume that an A-2-7 material has 30 % passing a No. 200 (75-μm) sieve, a liquid limit of 50, and a plasticity index of 30, then

Group index = $0.01(30 - 15)(30 - 10)$
$\qquad = 3.0$ or 3 (note that only the *PI* portion of the formula was used)

12. Discussion of Group Index

12.1 The empirical group index formula devised for approximate within-group evaluation of the "clayey-granular materials" and the "silt-clay materials" is based on the following assumptions:

NOTE 8—Group index values should only be used to compare soils within the same group and not between groups.

12.1.1 Materials falling within Groups A-1-a, A-1-b, A-2-4, A-2-5, and A-3 are satisfactory as subgrade when properly drained and compacted under moderate thickness of pavement (base or surface course, or both) of a type suitable for traffic to be carried or can be made satisfactory by additions of small amounts of natural or artificial binders.

12.1.2 Materials falling within the "clayey granular" Groups A-2-6 and A-2-7 and the "silt-clay" Groups A-4, A-5, A-6, and A-7 will range in quality as subgrade from the approximate equivalent of the good A-2-4 and A-2-5 subgrades to fair and poor subgrades requiring a layer of subbase material or an increased thickness of base course over that required in 12.1.1, in order to furnish adequate support for traffic loads.

12.1.3 A minimum of 35 % passing a No. 200 (75-μm)

D 3282

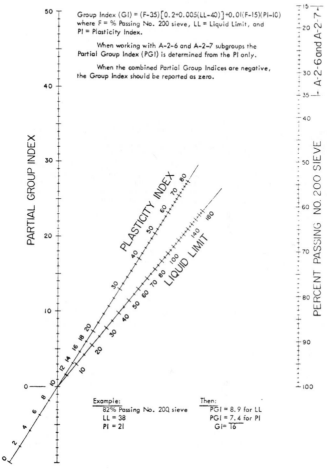

Reprinted with permission of American Association of State Highway and Transportation Officials

FIG. 2 Group Index Chart

sieve is assumed to be critical if plasticity is neglected, but the critical minimum is only 15 % when affected by plasticity indexes greater than 10.

12.1.4 Liquid limits of 40 and above are assumed to be critical.

12.1.5 Plasticity indexes of 10 and above are assumed to be critical.

12.2 There is no upper limit of group index value obtained by use of the formula: The adopted critical values of percentage passing the No. 200 (75-µm) sieve, liquid limit, and plasticity index, are based on an evaluation of subgrade, subbase, and base-course materials by several highway organizations that use the tests involved in this classification system.

12.3 Under average conditions of good drainage and thorough compaction, the supporting value of a material as subgrade may be assumed as an inverse ratio to its group index; that is, a group index of 0 indicates a "good" subgrade material and a group index of 20 or greater indicates a "very poor" subgrade material.

13. Keywords

13.1 Atterberg limits; classification; clay; gradation; highways; sand gravel; silt; soil classification; soil tests

D 3282

APPENDIX

(Nonmandatory Information)

X1. RATIONALE

X1.1 The 1992 edition differs from the previous edition in that the title was changed to better indicate the use of the standard.

 **Designation: D 698 – 91ᵉ¹**

Test Method for
Laboratory Compaction Characteristics of Soil Using Standard Effort (12,400 ft-lbf/ft³ (600 kN-m/m³))[1]

This standard is issued under the fixed designation D 698; the number immediately following the designation indicates the year of original adoption or, in the case of revision, the year of last revision. A number in parentheses indicates the year of last reapproval. A superscript epsilon (ε) indicates an editorial change since the last revision or reapproval.

ε¹ NOTE—Equation 1 was corrected editorially in January 1997.

1. Scope

1.1 This test method covers laboratory compaction procedures used to determine the relationship between water content and dry unit weight of soils (compaction curve) compacted in a 4 or 6-in. (101.6 or 152.4-mm) diameter mold with a 5.5-lbf (24.4-N) rammer dropped from a height of 12 in. (305 mm) producing a compactive effort of 12,400 ft-lbf/ft³ (600 kN-m/m³).

NOTE 1—The equipment and procedures are similar as those proposed by R. R. Proctor (*Engineering News Record*—September 7, 1933) with this one major exception: his rammer blows were applied as "12 inch firm strokes" instead of free fall, producing variable compactive effort depending on the operator, but probably in the range 15,000 to 25,000 ft-lbf/ft³ (700 to 1,200 kN-m/m³). The standard effort test (see 3.2.2) is sometimes referred to as the Proctor Test.

NOTE 2—Soils and soil-aggregate mixtures should be regarded as natural occurring fine- or coarse-grained soils or composites or mixtures of natural soils, or mixtures of natural and processed soils or aggregates such as silt, gravel, or crushed rock.

1.2 This test method applies only to soils that have 30 % or less by weight of particles retained on the ¾-inch (19.0-mm) sieve.

NOTE 3—For relationships between unit weights and water contents of soils with 30 % or less by weight of material retained on the ¾-in. (19.0-mm) sieve to unit weights and water contents of the fraction passing ¾-in. (19.0-mm) sieve, see Practice D 4718.

1.3 Three alternative procedures are provided. The procedure used shall be as indicated in the specification for the material being tested. If no procedure is specified, the choice should be based on the material gradation.

1.3.1 *Procedure A:*
1.3.1.1 *Mold*—4-in. (101.6-mm) diameter.
1.3.1.2 *Material*—Passing No. 4 (4.75-mm) sieve.
1.3.1.3 *Layers*—Three.
1.3.1.4 *Blows per layer*—25.
1.3.1.5 *Use*—May be used if 20 % or less by weight of material is retained on the No. 4 (4.75-mm) sieve.
1.3.1.6 *Other Use*—If this procedure is not specified, materials that meet these gradation requirements may be tested using Procedures B or C.
1.3.2 *Procedure B:*
1.3.2.1 *Mold*—4-in. (101.6-mm) diameter.

1.3.2.2 *Material*—Passing ⅜-in. (9.5-mm) sieve.
1.3.2.3 *Layers*—Three.
1.3.2.4 *Blows per layer*—25.
1.3.2.5 *Use*—Shall be used if more than 20 % by weight of the material is retained on the No. 4 (4.75-mm) sieve and 20 % or less by weight of the material is retained on the ⅜-in. (9.5-mm) sieve.
1.3.2.6 *Other Use*—If this procedure is not specified, materials that meet these gradation requirements may be tested using Procedure C.
1.3.3 *Procedure C:*
1.3.3.1 *Mold*—6-in. (152.4-mm) diameter.
1.3.3.2 *Material*—Passing ¾-inch (19.0-mm) sieve.
1.3.3.3 *Layers*—Three.
1.3.3.4 *Blows per layer*—56.
1.3.3.5 *Use*—Shall be used if more than 20 % by weight of the material is retained on the ⅜-in. (9.5-mm) sieve and less than 30 % by weight of the material is retained on the ¾-in. (19.0-mm) sieve.
1.3.4 The 6-in. (152.4-mm) diameter mold shall not be used with Procedure A or B.

NOTE 4—Results have been found to vary slightly when a material is tested at the same compactive effort in different size molds.

1.4 If the test specimen contains more than 5 % by weight oversize fraction (coarse fraction) and the material will not be included in the test, corrections must be made to the unit weight and water content of the specimen or to the appropriate field in place density test specimen using Practice D 4718.

1.5 This test method will generally produce a well defined maximum dry unit weight for non-free draining soils. If this test method is used for free draining soils the maximum unit weight may not be well defined, and can be less than obtained using Test Methods D 4253.

1.6 The values in inch-pound units are to be regarded as the standard. The values stated in SI units are provided for information only.

1.6.1 In the engineering profession it is customary practice to use, interchangeably, units representing both mass and force, unless dynamic calculations ($F = Ma$) are involved. This implicitly combines two separate systems of units, that is, the absolute system and the gravimetric system. It is scientifically undesirable to combine the use of two separate systems within a single standard. This test method has been written using inch-pound units (gravimetric system) where the pound (lbf) represents a unit of force. The use of mass (lbm) is for convenience of units and is not intended to

---

[1] This test method is under the jurisdiction of ASTM Committee D-18 on Soil and Rock and is the direct responsibility of Subcommittee D18.03 on Texture, Plasticity and Density Characteristics of Soils.
Current edition approved Nov. 19, 1991. Published January 1992.

◆◐ D 698

convey the use is scientifically correct. Conversions are given in the SI system in accordance with Practice E 380. The use of balances or scales recording pounds of mass (lbm), or the recording of density in lbm/ft³ should not be regarded as nonconformance with this standard.

1.7 *This standard does not purport to address all of the safety concerns, if any, associated with its use. It is the responsibility of the user of this standard to establish appropriate safety and health practices and determine the applicability of regulatory limitations prior to use.*

2. Referenced Documents

2.1 *ASTM Standards:*
C 127 Test Method for Specific Gravity and Absorption of Coarse Aggregate[2]
C 136 Method for Sieve Analysis of Fine and Coarse Aggregate[2]
D 422 Test Method for Particle Size Analysis of Soils[3]
D 653 Terminology Relating to Soil, Rock, and Contained Fluids[3]
D 854 Test Method for Specific Gravity of Soils[3]
D 1557 Test Methods for Moisture-Density Relations of Soils and Soil Aggregate Mixtures Using 10-lb (4.54-kg.) Rammer and 18-in. (457 mm) Drop[3]
D 2168 Test Methods for Calibration of Laboratory Mechanical-Rammer Soil Compactors[3]
D 2216 Test Method for Laboratory Determination of Water (Moisture) Content of Soil, Rock and Soil-Aggregate Mixtures[3]
D 2487 Test Method for Classification of Soils for Engineering Purposes[3]
D 2488 Practice for Description of Soils (Visual-Manual Procedure)[3]
D 4220 Practices for Preserving and Transporting Soil Samples[3]
D 4253 Test Methods for Maximum Index Density of Soils Using a Vibratory Table[3]
D 4718 Practice for Correction of Unit Weight and Water Content for Soils Containing Oversize Particles[3]
D 4753 Specification for Evaluating, Selecting and Specifying Balances and Scales For Use in Soil and Rock Testing[3]
E 1 Specification for ASTM Thermometers[4]
E 11 Specification for Wire-Cloth Sieves for Testing Purposes[5]
E 319 Practice for the Evaluation of Single-Pan Mechanical Balances[5]
E 380 Practice for Use of the International System of Units (SI) (the Modernized Metric System)[5]

3. Terminology

3.1 *Definitions*—See Terminology D 653 for general definitions.

3.2 *Description of Terms Specific to This Standard:*

3.2.1 *oversize fraction (coarse fraction), P_c in %*—the portion of total sample not used in performing the compac-

tion test; it may be the portion of total sample retained on the No. 4 (4.75-mm), ⅜-in. (9.5-mm), or ¾-in. (19.0-mm) sieve.

3.2.2 *standard effort*—the term for the 12,400 ft-lbf/ft³ (600 kN-m/m³) compactive effort applied by the equipment and procedures of this test.

3.2.3 *standard maximum dry unit weight, γ_{dmax} in lbf/ft³ (kN/m³)*—the maximum value defined by the compaction curve for a compaction test using standard effort.

3.2.4 *standard optimum water content, w_o in %*—the water content at which a soil can be compacted to the maximum dry unit weight using standard compactive effort.

3.2.5 *test fraction (finer fraction), P_F in %*—the portion of the total sample used in performing the compaction test; it is the fraction passing the No. 4 (4.75-mm) sieve in Procedure A, minus ⅜-in. (9.5-mm) sieve in Procedure B, or minus ¾-in. (19.0-mm) sieve in Procedure C.

4. Summary of Test Method

4.1 A soil at a selected water content is placed in three layers into a mold of given dimensions, with each layer compacted by 25 or 56 blows of a 5.5-lbf (24.4-N) rammer dropped from a distance of 12-in. (305-mm), subjecting the soil to a total compactive effort of about 12,400 ft-lbf/ft³ (600 kN-m/m³). The resulting dry unit weight is determined. The procedure is repeated for a sufficient number of water contents to establish a relationship between the dry unit weight and the water content for the soil. This data, when plotted, represents a curvilinear relationship known as the compaction curve. The values of optimum water content and standard maximum dry unit weight are determined from the compaction curve.

5. Significance and Use

5.1 Soil placed as engineering fill (embankments, foundation pads, road bases) is compacted to a dense state to obtain satisfactory engineering properties such as, shear strength, compressibility, or permeability. Also, foundation soils are often compacted to improve their engineering properties. Laboratory compaction tests provide the basis for determining the percent compaction and water content needed to achieve the required engineering properties, and for controlling construction to assure that the required compaction and water contents are achieved.

5.2 During design of an engineered fill, shear, consolidation, permeability, or other tests require preparation of test specimens by compacting at some water content to some unit weight. It is common practice to first determine the optimum water content (w_o) and maximum dry unit weight (γ_{dmax}) by means of a compaction test. Test specimens are compacted at a selected water content (w), either wet or dry of optimum (w_o) or at optimum (w_o), and at a selected dry unit weight related to a percentage of maximum dry unit weight (γ_{dmax}). The selection of water content (w), either wet or dry of optimum (w_o) or at optimum (w_o) and the dry unit weight (γ_{dmax}) may be based on past experience, or a range of values may be investigated to determine the necessary percent of compaction.

6. Apparatus

6.1 *Mold Assembly*—The molds shall be cylindrical in

D 698

shape, made of rigid metal and be within the capacity and dimensions indicated in 6.1.1 or 6.1.2 and Figs. 1 and 2. The walls of the mold may be solid, split, or tapered. The "split" type may consist of two half-round sections, or a section of pipe split along one element, which can be securely locked together to form a cylinder meeting the requirements of this section. The "tapered" type shall an internal diameter taper that is uniform and not more than 0.200 in./ft (16.7- mm/m) of mold height. Each mold shall have a base plate and an extension collar assembly, both made of rigid metal and constructed so they can be securely attached and easily detached from the mold. The extension collar assembly shall have a height extending above the top of the mold of at least 2.0 in. (50.8-mm) which may include an upper section that flares out to form a funnel provided there is at least a 0.75 in. (19.0-mm) straight cylindrical section beneath it. The extension collar shall align with the inside of the mold. The bottom of the base plate and bottom of the centrally recessed area that accepts the cylindrical mold shall be planar.

6.1.1 *Mold, 4 in.*—A mold having a 4.000 ± 0.016-in. (101.6 ± 0.4-mm) average inside diameter, a height of 4.584 ± 0.018-in. (116.4 ± 0.5-mm) and a volume of 0.0333 ± 0.0005 ft³ (944 ± 14 cm³). A mold assembly having the minimum required features is shown in Fig. 1.

6.1.2 *Mold, 6 in.*—A mold having a 6.000 ± 0.026-in. (152.4 ± 0.7-mm) average inside diameter, a height of 4.584 ± 0.018-in. (116.4 ± 0.5-mm), and a volume of 0.075 ± 0.0009 ft³ (2124 ± 25 cm³). A mold assembly having the minimum required features is shown in Fig. 2.

6.2 *Rammer*—A rammer, either manually operated as described further in 6.2.1 or mechanically operated as described in 6.2.2. The rammer shall fall freely through a distance of 12 ± 0.05-in. (304.8 ± 1.3-mm) from the surface of the specimen. The mass of the rammer shall be 5.5 ± 0.02-lbm (2.5 ± 0.01-kg), except that the mass of the mechanical rammers may be adjusted as described in Test Methods D 2168, see Note 5. The striking face of the rammer shall be planar and circular, except as noted in 6.2.2.3, with a diameter when new of 2.000 ± 0.005-in. (50.80 ± 0.13-mm). The rammer shall be replaced if the striking face becomes worn or bellied to the extent that diameter exceeds 2.000 ± 0.01-in. (50.80 ± 0.25-mm).

NOTE 5—It is a common and acceptable practice in the inch-pound system to assume that the mass of the rammer is equal to its mass determined using either a kilogram or pound balance and 1 lbf is equal to 1 lbm or 0.4536 kg. or 1 N is equal to 0.2248 lbm or 0.1020 kg.

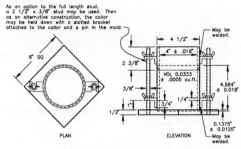

FIG. 1 4.0-in. Cylindrical Mold

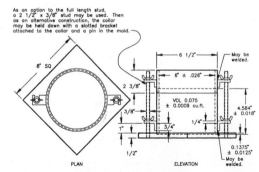

FIG. 2 6.0-in. Cylindrical Mold

6.2.1 *Manual Rammer*—The rammer shall be equipped with a guide sleeve that has sufficient clearance that the free fall of the rammer shaft and head is not restricted. The guide sleeve shall have at least four vent holes at each end (eight holes total) located with centers ¾ ± ¹/₁₆-in. (19.0 ± 1.6-mm) from each end and spaced 90 degrees apart. The minimum diameter of the vent holes shall be ³/₈-in. (9.5-mm). Additional holes or slots may be incorporated in the guide sleeve.

6.2.2 *Mechanical Rammer-Circular Face*—The rammer shall operate mechanically in such a manner as to provide uniform and complete coverage of the specimen surface. There shall be 0.10 ± 0.03-in. (2.5 ± 0.8-mm) clearance between the rammer and the inside surface of the mold at its smallest diameter. The mechanical rammer shall meet the calibration requirements of Test Methods D 2168. The mechanical rammer shall be equipped with a positive mechanical means to support the rammer when not in operation.

6.2.2.3 *Mechanical Rammer-Sector Face*—When used with the 6-in. (152.4-mm) mold, a sector face rammer may be used in place of the circular face rammer. The specimen contact face shall have the shape of a sector of a circle of radius equal to 2.90 ± 0.02-in. (73.7 ± 0.5-mm). The rammer shall operate in such a manner that the vertex of the sector is positioned at the center of the specimen.

6.3 *Sample Extruder (optional)*—A jack, frame or other device adapted for the purpose of extruding compacted specimens from the mold.

6.4 *Balance*—A class GP5 balance meeting the requirements of Specification D 4753 for a balance of 1-g readability.

6.5 *Drying Oven*—Thermostatically controlled, preferably of a forced-draft type and capable of maintaining a uniform temperature of 230 ± 9°F (110 ± 5°C) throughout the drying chamber.

6.6 *Straightedge*—A stiff metal straightedge of any convenient length but not less than 10-in. (254-mm). The total length of the straightedge shall be machined straight to a tolerance of ±0.005-in. (±0.1-mm). The scraping edge shall be beveled if it is thicker than ⅛-in. (3-mm).

6.7 *Sieves*—¾-in. (19.0-mm), ³/₈-in. (9.5-mm), and No. 4 (4.75-mm), conforming to the requirements of Specification E 11.

6.8 *Mixing Tools*—Miscellaneous tools such as mixing

pan, spoon, trowel, spatula, etc., or a suitable mechanical device for thoroughly mixing the sample of soil with increments of water.

7. Calibration

7.1 Perform calibrations before initial use, after repairs or other occurrences that might affect the test results, at intervals not exceeding 1,000 test specimens, or annually, whichever occurs first, for the following apparatus:

7.1.2 *Balance*—Evaluate in accordance with Specification D 4753.

7.1.3 *Molds*—Determine the volume as described in Annex 1.

7.1.4 *Manual Rammer*—Verify the free fall distance, rammer mass, and rammer face in accordance with Section 6.2. Verify the guide sleeve requirements in accordance with Section 6.2.1.

7.1.5 *Mechanical Rammer*—Calibrate and adjust the mechanical rammer in accordance with Test Methods D 2168. In addition, the clearance between the rammer and the inside surface of the mold shall be verified in accordance with 6.2.2.

8. Test Sample

8.1 The required sample mass for Procedures A and B is approximately 35-lbm (16-kg), and for Procedure C is approximately 65-lbm (29-kg) of dry soil. Therefore, the field sample should have a moist mass of at least 50-lbm (23-kg) and 100-lbm (45-kg), respectively.

8.2 Determine the percentage of material retained on the No. 4 (4.75-mm), ³⁄₈-in. (9.5-mm), or ³⁄₄-in. (19.0-mm) sieve as appropriate for choosing Procedure A, B, or C. Make this determination by separating out a representative portion from the total sample and determining the percentages passing the sieves of interest by Test Methods D 422 or Method C 136. It is only necessary to calculate percentages for the sieve or sieves for which information is desired.

9. Preparation of Apparatus

9.1 Select the proper compaction mold in accordance with the procedure (A, B, or C) being used. Determine and record its mass to the nearest gram. Assemble the mold, base and extension collar. Check the alignment of the inner wall of the mold and mold extension collar. Adjust if necessary.

9.2 Check that the rammer assembly is in good working condition and that parts are not loose or worn. Make any necessary adjustments or repairs. If adjustments or repairs are made, the rammer must be recalibrated.

10. Procedure

10.1 *Soils:*

10.1.1 Do not reuse soil that has been previously laboratory compacted.

10.1.2 When using this test method for soils containing hydrated halloysite, or where past experience with a particular soil indicates that results will be altered by air drying, use the moist preparation method (see 10.2).

10.1.3 Prepare the soil specimens for testing in accordance with 10.2 (preferred) or with 10.3.

10.2 *Moist Preparation Method (preferred)*—Without previously drying the sample, pass it through a No. 4 (4.75-

mm), ³⁄₈-in. (9.5-mm), or ³⁄₄-in. (19.0-mm) sieve, depending on the procedure (A, B, or C) being used. Determine the water content of the processed soil.

10.2.1 Prepare at least four (preferably five) specimens having water contents such that they bracket the estimated optimum water content. A specimen having a water content close to optimum should be prepared first by trial additions of water and mixing (see Note 6). Select water contents for the rest of the specimens to provide at least two specimens wet and two specimens dry of optimum, and water contents varying by about 2 %. At least two water contents are necessary on the wet and dry side of optimum to accurately define the dry unit weight compaction curve (see 10.5). Some soils with very high optimum water content or a relatively flat compaction curve may require larger water content increments to obtain a well defined maximum dry unit weight. Water content increments should not exceed 4 %.

NOTE 6—With practice it is usually possible to visually judge a point near optimum water content. Typically, soil at optimum water content can be squeezed into a lump that sticks together when hand pressure is released, but will break cleanly into two sections when "bent". At water contents dry of optimum soils tend to crumble; wet of optimum soils tend to stick together in a sticky cohesive mass. Optimum water content is typically slightly less than the plastic limit.

10.2.2 Use approximately 5-lbm (2.3-kg) of the sieved soil for each specimen to be compacted using Procedure A or B, or 13-lbm (5.9-kg) using Procedure C. To obtain the specimen water contents selected in 10.2.1, add or remove the required amounts of water as follows: to add water, spray it into the soil during mixing; to remove water, allow the soil to dry in air at ambient temperature or in a drying apparatus such that the temperature of the sample does not exceed 140°F (60°C). Mix the soil frequently during drying to maintain an even water content distribution. Thoroughly mix each specimen to ensure even distribution of water throughout and then place in a separate covered container and allow to stand in accordance with Table 1 prior to compaction. For the purpose of selecting a standing time, the soil may be classified using Test Method D 2487, Practice D 2488 or data on other samples from the same material source. For referee testing, classification shall be by Test Method D 2487.

10.3 *Dry Preparation Method*—If the sample is too damp to be friable, reduce the water content by air drying until the material is friable. Drying may be in air or by the use of drying apparatus such that the temperature of the sample does not exceed 140°F (60°C). Thoroughly break up the aggregations in such a manner as to avoid breaking individual particles. Pass the material through the appropriate sieve: No. 4 (4.75-mm), ³⁄₈-in. (9.5-mm), or ³⁄₄-in. (19.0-mm). When preparing the material by passing over the ³⁄₄-in. sieve for compaction in the 6-in. mold, break up aggregations sufficiently to at least pass the ³⁄₈-in. sieve in order to facilitate the distribution of water throughout the soil in later mixing.

TABLE 1   Required Standing Times of Moisturized Specimens

Classification	Minimum Standing Time, h
GW, GP, SW, SP	No Requirement
GM, SM	3
All other soils	16

◀️ D 698

TABLE 2 Metric Equivalents for Figs. 1 and 2

in.	mm
0.016	0.41
0.026	0.66
0.032	0.81
0.028	0.71
1/2	12.70
2 1/2	63.50
2 5/8	66.70
4	101.60
4 1/2	114.30
4.584	116.43
4 3/4	120.60
6	152.40
6 1/2	165.10
6 5/8	168.30
6 3/4	171.40
8 1/4	209.60

ft³	cm³
1/30 (0.0333)	943
0.0005	14
1/13.333 (0.0750)	2,124
0.0011	31

10.3.1 Prepare at least four (preferably five) specimens in accordance with 10.2.1.

10.3.2 Use approximately 5-lbm (2.3-kg) of the sieved soil for each specimen to be compacted using Procedure A or B, or 13-lbm (5.9-kg) using Procedure C. Add the required amounts of water to bring the water contents of the specimens to the values selected in 10.3.1. Follow the specimen preparation procedure specified in 10.2.2 for drying the soil or adding water into the soil and curing each test specimen.

10.4 *Compaction*—After curing, if required, each specimen shall be compacted as follows:

10.4.1 Determine and record the mass of the mold or mold and base plate.

10.4.2 Assemble and secure the mold and collar to the base plate. The mold shall rest on a uniform rigid foundation, such as provided by a cylinder or cube of concrete with a mass of not less than 200-lbm (91-kg). Secure the base plate to the rigid foundation. The method of attachment to the rigid foundation shall allow easy removal of the assembled mold, collar and base plate after compaction is completed.

10.4.3 Compact the specimen in three layers. After compaction, each layer should be approximately equal in thickness. Prior to compaction, place the loose soil into the mold and spread into a layer of uniform thickness. Lightly tamp the soil prior to compaction until it is not in a fluffy or loose state, using either the manual compaction rammer or a 2-in. (5-mm) diameter cylinder. Following compaction of each of the first two layers, any soil adjacent to the mold walls that has not been compacted or extends above the compacted surface shall be trimmed. The trimmed soil may be included with the additional soil for the next layer. A knife or other suitable device may be used. The total amount of soil used shall be such that the third compacted layer slighly extends into the collar, but does not exceed 1/4-in. (6-mm) above the top of the mold. If the third layer does extend above the top of the mold by more than 1/4-in. (6-mm), the specimen shall be discarded. The specimen shall be discarded when the last blow on the rammer for the third layer results in the bottom

of the rammer extending below the top of the compaction mold.

10.4.4 Compact each layer with 25 blows for the 4-in. (101.6-mm) mold or with 56 blows for the 6-in. (152.4-mm) mold.

NOTE 7—When compacting specimens wetter than optimum water content, uneven compacted surfaces can occur and operator judgement is required as to the average height of the specimen.

10.4.5 In operating the manual rammer, take care to avoid lifting the guide sleeve during the rammer upstroke. Hold the guide sleeve steady and within 5° of vertical. Apply the blows at a uniform rate of approximately 25 blows/min and in such a manner as to provide complete, uniform coverage of the specimen surface.

10.4.6 Following compaction of the last layer, remove the collar and base plate from the mold, except as noted in 10.4.7. A knife may be used to trim the soil adjacent to the collar to loosen the soil from the collar before removal to avoid disrupting the soil below the top of the mold.

10.4.7 Carefully trim the compacted specimen even with the top of the mold by means of the straightedge scraped across the top of the mold to form a plane surface even with the top of the mold. Initial trimming of the specimen above the top of the mold with a knife may prevent the soil from tearing below the top of the mold. Fill any holes in the top surface with unused or trimmed soil from the specimen, press in with the fingers, and again scrape the straightedge across the top of the mold. Repeat the appropriate preceding operations on the bottom of the specimen when the mold volume was determined without the base plate. For very wet or dry soils, soil or water may be lost if the base plate is removed. For these situations, leave the base plate attached to the mold. When the base plate is left attached, the volume of the mold must be calibrated with the base plate attached to the mold rather than a plastic or glass plate as noted in Annex 1, A1.4.

10.4.8 Determine and record the mass of the specimen and mold to the nearest gram. When the base plate is left attached, determine and record the mass of the specimen, mold and base plate to the nearest gram.

10.4.9 Remove the material from the mold. Obtain a specimen for water content by using either the whole specimen (preferred method) or a representative portion. When the entire specimen is used, break it up to facilitate drying. Otherwise, obtain a portion by slicing the compacted specimen axially through the center and removing about 500-g of material from the cut faces. Obtain the water content in accordance with Test Method D 2216.

10.5 Following compaction of the last specimen, compare the wet unit weights to ensure that a desired pattern of obtaining data on each side of the optimum water content will be attained for the dry unit weight compaction curve. Plotting the wet unit weight and water content of each compacted specimen can be an aid in making the above evaluation. If the desired pattern is not obtained, additional compacted specimens will be required. Generally, one water content value wet of the water content defining the maximum wet unit weight is sufficient to ensure data on the wet side of optimum water content for the maximum dry unit weight.

⚛ D 698

11. Calculation

11.1 Calculate the dry unit weight and water content of each compacted specimen as explained in 11.3 and 11.4. Plot the values and draw the compaction curve as a smooth curve through the points (see example, Fig. 3). Plot dry unit weight to the nearest 0.1 lbf/ft³ (0.2 kN/m³) and water content to the nearest 0.1 %. From the compaction curve, determine the optimum water content and maximum dry unit weight. If more than 5 % by weight of oversize material was removed from the sample, calculate the corrected optimum water content and maximum dry unit weight of the total material using Practice D 4718. This correction may be made to the appropriate field in place density test specimen rather than to the laboratory test specimen.

11.2 Plot the 100 % saturation curve. Values of water content for the condition of 100 % saturation can be calculated as explained in 11.5 (see example, Fig. 3).

NOTE 8—The 100 % saturation curve is an aid in drawing the compaction curve. For soils containing more than approximately 10 % fines at water contents well above optimum, the two curves generally become roughly parallel with the wet side of the compaction curve between 92 % to 95 % saturation. Theoretically, the compaction curve cannot plot to the right of the 100 % saturation curve. If it does, there is an error in specific gravity, in measurements, in calculations, in test procedures, or in plotting.

NOTE 9—The 100 % saturation curve is sometimes referred to as the zero air voids curve or the complete saturation curve.

11.3 *Water Content, w*—Calculate in accordance with Test Method D 2216.

11.4 *Dry Unit Weights*—Calculate the moist density (Eq. 1), the dry density (Eq. 2), and then the dry unit weight (Eq. 3) as follows:

$$\rho_m = \frac{(M_t - M_{md})}{1000\ V} \qquad (1)$$

where:
ρ_m = moist density of compacted specimen, Mg/m³,
M_t = mass of moist specimen and mold, kg,
M_{md} = mass of compaction mold, kg, and
V = volume of compaction mold, m³ (see Annex 1)

$$\rho_d = \rho_m / (1 + w/100) \qquad (2)$$

where:
ρ_d = dry density of compacted specimen, Mg/m³, and
w = water content, %.

$$\gamma_d = 62.43\ \rho_d \text{ in lbf/ft}^3 \qquad (3)$$

or

$$\gamma_d = 9.807\ \rho_d \text{ in kN/m}^3$$

where:
γ_d = dry unit weight of compacted specimen.

11.5 To calculate points for plotting the 100 % saturation curve or zero air voids curve select values of dry unit weight, calculate corresponding values of water content corresponding to the condition of 100 % saturation as follows:

$$w_{sat}\ \frac{(\gamma_w)(G_s) - \gamma_d}{(\gamma_d)(G_s)} \times 100 \qquad (4)$$

where:
w_{sat} = water content for complete saturation, %,
γ_w = unit weight of water, 62.43 lbf/ft³ (9.807 kn/m³),
γ_d = dry unit weight of soil, and
G_s = specific gravity of soil.

NOTE 10—Specific gravity may be estimated for the test specimen on the basis of test data from other samples of the same soil classification and source. Otherwise, a specific gravity test (Test Method C 127, Test Method D 854, or both) is necessary.

12. Report

12.1 The report shall contain the following information:

12.1.1 Procedure used (A, B, or C).

12.1.2 Preparation method used (moist or dry).

12.1.3 As received water content if determined.

12.1.4 Standard optimum water content, to the nearest 0.5 %.

12.1.5 Standard maximum dry unit weight, to the nearest 0.5 lbf/ft³.

12.1.6 Description of rammer (manual or mechanical).

12.1.7 Soil sieve data when applicable for determination of procedure (A, B, or C) used.

12.1.8 Description of material used in test, by Practice D 2488, or classification by Test Method D 2487.

12.1.9 Specific gravity and method of determination.

12.1.10 Origin of material used in test, for example, project, location, depth, and the like.

12.1.11 Compaction curve plot showing compaction points used to establish compaction curve, and 100 % saturation curve, point of maximum dry unit weight and optimum water content.

12.1.12 Oversize correction data if used, including the oversize fraction (coarse fraction), P_c in %.

13. Precision and Bias

13.1 *Precision*—Data are being evaluated to determine the precision of this test method. In addition, pertinent data

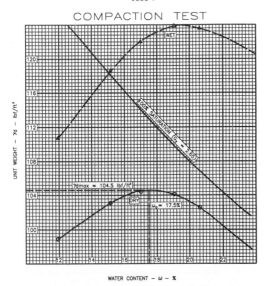

COMPACTION TEST

WATER CONTENT — ω – %

FIG. 3 Example Compaction Curve Plotting

D 698

is being solicited from users of the test method.

13.2 *Bias*—It is not possible to obtain information on bias because there is no other method of determining the values of standard maximum dry unit weight and optimum water content.

14. Keywords

14.1 NT—impact compaction using standard effort; RT—density; RT—moisture-density curves; RT—proctor test; UF—compaction characteristics; UF—soil compaction; USE—laboratory tests

ANNEX

(Mandatory Information)

A1. VOLUME OF COMPACTION MOLD

A1.1 Scope

A1.1.1 This annex describes the procedure for determining the volume of a compaction mold.

A1.1.2 The volume is determined by a water-filled method and checked by a linear-measurement method.

A1.2 Apparatus

A1.2.1 In addition to the apparatus listed in Section 6 the following items are required:

A1.2.1.1 *Vernier or Dial Caliper*—having a measuring range of at least 0 to 6 in. (0 to 150 mm) and readable to at least 0.001 in. (0.02 mm).

A1.2.1.2 *Inside Micrometer*—having a measuring range of at least 2 to 12 in. (50 to 300 mm) and readable to at least 0.001 in. (0.02 mm).

A1.2.1.3 *Plastic or Glass Plates*—Two plastic or glass plates approximately 8 in. square by ¼ in. thick (200 by 200 mm by 6 mm).

A1.2.1.4 *Thermometer*—0 to 50°C range, 0.5°C graduations, conforming to the requirements of Specification E 1.

A1.2.1.5 *Stopcock grease* or similar sealant.

A1.2.1.6 *Miscellaneous equipment*—Bulb syringe, towels, etc.

A1.3 Precautions

A1.3.1 Perform this procedure in an area isolated from drafts or extreme temperature fluctuations.

A1.4 Procedure

A1.4.1 *Water-Filling Method:*

A1.4.1.1 Lightly grease the bottom of the compaction mold and place it on one of the plastic or glass plates. Lightly grease the top of the mold. Be careful not to get grease on the inside of the mold. If it is necessary to use the base plate, as noted in 10.4.7, place the greased mold onto the base plate and secure with the locking studs.

A1.4.1.2 Determine the mass of the greased mold and both plastic or glass plates to the nearest 0.01-lbm (1-g) and record. When the base plate is being used in lieu of the bottom plastic or glass plate determine the mass of the mold, base plate and a single plastic or glass plate to be used on top of the mold to the nearest 0.01-lbm (1-g) and record.

A1.4.1.3 Place the mold and the bottom plastic or glass plate on a firm, level surface and fill the mold with water to slightly above its rim.

A1.4.1.4 Slide the second plate over the top surface of the mold so that the mold remains completely filled with water and air bubbles are not entrapped. Add or remove water as necessary with a bulb syringe.

A1.4.1.5 Completely dry any excess water from the outside of the mold and plates.

A1.4.1.6 Determine the mass of the mold, plates and water and record to the nearest 0.01-lbm (1-g).

A1.4.1.7 Determine the temperature of the water in the mold to the nearest 1°C and record. Determine and record the absolute density of water from Table A1.1.

A1.4.1.8 Calculate the mass of water in the mold by subtracting the mass determined in A1.4.1.2 from the mass determined in A1.4.1.6.

A1.4.1.9 Calculate the volume of water by dividing the mass of water by the density of water and record to the nearest 0.0001 ft³ (1 cm³).

A1.4.1.10 When the base plate is used for the calibration of the mold volume repeat A1.4.1.3 through A1.4.1.9.

A1.4.2 *Linear Measurement Method:*

A1.4.2.1 Using either the vernier caliper or the inside micrometer, measure the diameter of the mold 6 times at the top of the mold and 6 times at the bottom of the mold, spacing each of the six top and bottom measurements equally around the circumference of the mold. Record the values to the nearest 0.001-in. (0.02-mm).

A1.4.2.2 Using the vernier caliper, measure the inside height of the mold by making three measurements equally spaced around the circumference of the mold. Record values to the nearest 0.001-in. (0.02-mm).

A1.4.2.3 Calculate the average top diameter, average bottom diameter and average height.

A1.4.2.4 Calculate the volume of the mold and record to the nearest 0.0001 ft³ (1 cm³) as follows:

TABLE A1 Density of Water[A]

Temperature, °C (°F)	Density of Water, g/ml
18 (64.4)	0.99862
19 (66.2)	0.99843
20 (68.0)	0.99823
21 (69.8)	0.99802
22 (71.6)	0.99779
23 (73.4)	0.99756
24 (75.2)	0.99733
25 (77.0)	0.99707
26 (78.8)	0.99681

[A] Values other than shown may be obtained by referring to the Handbook of Chemistry and Physics, Chemical Rubber Publishing Co., Cleveland, Ohio.

⏚ D 698

$$V = \frac{(\pi)(h)(d_t + d_b)^2}{(16)(1728)} \text{ (inch-pound)}$$

$$V = \frac{(\pi)(h)(d_t + d_b)^2}{(16)(10^3)} \text{ (SI)}$$

where:
V = volume of mold, ft³ (cm³),
h = average height, in. (mm),
d_t = average top diameter, in. (mm),
d_b = average bottom diameter, in. (mm),
$1/1728$ = constant to convert in³ to ft³, and
$1/10^3$ = constant to convert mm³ to cm³.

A1.5 Comparison of Results

A1.5.1 The volume obtaine
within the volume tolerance re

A1.5.2 The difference betv
not exceed 0.5 % of the nomi

A1.5.3 Repeat the determi
teria are not met.

A1.5.4 Failure to obtain s:
the two methods, even after
that the mold is badly deform

A1.5.5 Use the volume of t
water-filling method as the as:
lating the moist and dry densi

Designation: D 1557 – 91ᵉ¹

Test Method for
Laboratory Compaction Characteristics of Soil Using Modified Effort (56,000 ft-lbf/ft³ (2,700 kN-m/m³))[1]

This standard is issued under the fixed designation D 1557; the number immediately following the designation indicates the year of original adoption or, in the case of revision, the year of last revision. A number in parentheses indicates the year of last reapproval. A superscript epsilon (ε) indicates an editorial change since the last revision or reapproval.

ᵉ¹ NOTE—Equation 1 was corrected editorially in January 1997.

1. Scope

1.1 This test method covers laboratory compaction procedures used to determine the relationship between water content and dry unit weight of soils (compaction curve) compacted in a 4- or 6-in. (101.6 or 152.4 mm) diameter mold with a 10-lbf. (44.5-N) rammer dropped from a height of 18 in. (457 mm) producing a compactive effort of 56,000 ft-lbf/ft³ (2,700 kN-m/m³).

NOTE 1—Soils and soil-aggregate mixtures should be regarded as natural occurring fine- or coarse-grained soils or composites or mixtures of natural soils, or mixtures of natural and processed soils or aggregates such as silt, gravel, or crushed rock.

NOTE 2—The equipment and procedures are the same as proposed by the U.S. Corps of Engineers in 1945. The modified effort test (see 3.2.2) is sometimes referred to as the Modified Proctor Compaction Test.

1.2 This test method applies only to soils that have 30 % or less by weight of their particles retained on the ¾-in. (19.0-mm) sieve.

NOTE 3—For relationships between unit weights and water contents of soils with 30 % or less by weight of material retained on the ¾-in. (19.0-mm) sieve to unit weights and water contents of the fraction passing the ¾-in. (19.0-mm) sieve, see Practice D 4718.

1.3 Three alternative procedures are provided. The procedure used shall be as indicated in the specification for the material being tested. If no procedure is specified, the choice should be based on the material gradation.

1.3.1 *Procedure A:*

1.3.1.1 *Mold*—4-in. (101.6-mm) diameter.

1.3.1.2 *Material*—Passing No. 4 (4.75-mm) sieve.

1.3.1.3 *Layers*—Five.

1.3.1.4 *Blows per layer*—25.

1.3.1.5 *Use*—May be used if 20 % or less by weight of the material is retained on the No. 4 (4.75-mm) sieve.

1.3.1.6 *Other Use*—If this procedure is not specified, materials that meet these gradation requirements may be tested using Procedures B or C.

1.3.2 *Procedure B:*

1.3.2.1 *Mold*—4-in. (101.6-mm) diameter.

1.3.2.2 *Material*—Passing ⅜-in. (9.5-mm) sieve.

1.3.2.3 *Layers*—Five.

1.3.2.4 *Blows per layer*—25.

1.3.2.5 *Use*—Shall be used if more than 20 % by weight of the material is retained on the No. 4 (4.75-mm) sieve and 20 % or less by weight of the material is retained on the ⅜-in. (9.5-mm) sieve.

1.3.2.6 *Other Use*—If this procedure is not specified, materials that meet these gradation requirements may be tested using Procedure C.

1.3.3 *Procedure C:*

1.3.3.1 *Mold*—6-in. (152.4-mm) diameter.

1.3.3.2 *Material*—Passing ¾-inch (19.0-mm) sieve.

1.3.3.3 *Layers*—Five.

1.3.3.4 *Blows per layer*—56.

1.3.3.5 *Use*—Shall be used if more than 20 % by weight of the material is retained on the ⅜-in. (9.53-mm) sieve and less than 30 % by weight of the material is retained on the ¾-in. (19.0-mm) sieve.

1.3.4 The 6-in. (152.4-mm) diameter mold shall not be used with Procedure A or B.

NOTE 4—Results have been found to vary slightly when a material is tested at the same compactive effort in different size molds.

1.4 If the test specimen contains more than 5 % by weight oversize fraction (coarse fraction) and the material will not be included in the test, corrections must be made to the unit weight and water content of the test specimen or to the appropriate field in place density test specimen using Practice D 4718.

1.5 This test method will generally produce well defined maximum dry unit weight for non-free draining soils. If this test method is used for free draining soils the maximum unit weight may not be well defined, and can be less than obtained using Test Methods D 4253.

1.6 The values in inch-pound units are to be regarded as the standard. The values stated in SI units are provided for information only.

1.6.1 In the engineering profession it is customary practice to use, interchangeably, units representing both mass and force, unless dynamic calculations ($F = Ma$) are involved. This implicitly combines two separate systems of units, that is, the absolute system and the gravimetric system. It is scientifically undesirable to combine the use of two separate systems within a single standard. This test method has been written using inch-pound units (gravimetric system) where the pound (lbf) represents a unit of force. The use of mass (lbm) is for convenience of units and is not intended to convey the use is scientifically correct. Conversions are given in the SI system in accordance with Practice E 380. The use

[1] This test method is under the jurisdiction of ASTM Committee D-18 on Soil and Rock and is the direct responsibility of Subcommittee D18.03 on Texture, Plasticity and Density Characteristics of Soils.

Current edition approved Nov. 18, 1991. Published January 1992. Originally published as D 1557 – 58. Last previous edition D 1557 – 78 (1990).ᵉ¹

⚶ D 1557

of balances or scales recording pounds of mass (lbm), or the recording of density in lbm/ft³ should not be regarded as nonconformance with this standard.

1.7 *This standard does not purport to address all of the safety concerns, if any, associated with its use. It is the responsibility of the user of this standard to establish appropriate safety and health practices and determine the applicability of regulatory limitations prior to use.*

2. Referenced Documents

2.1 *ASTM Standards:*

C 127 Test Method for Specific Gravity and Absorption of Coarse Aggregate[2]

C 136 Method for Sieve Analysis of Fine and Coarse Aggregates[2]

D 422 Method for Particle Size Analysis of Soils[3]

D 653 Terminology Relating to Soil, Rock, and Contained Fluids[3]

D 698 Test Method for Laboratory Compaction Characteristics of Soil Using Standard Effort [12,400 ft-lbf/ft³ (600 kN-mJ/m³)][3]

D 854 Test Method for Specific Gravity of Soils[3]

D 2168 Test Methods for Calibration of Laboratory Mechanical-Rammer Soil Compactors[3]

D 2216 Test Method for Laboratory Determination of Water (Moisture) Content of Soil, Rock and Soil Aggregate Mixtures[3]

D 2487 Test Method for Classification of Soils for Engineering Purposes[3]

D 2488 Practice for Description and Identification of Soils (Visual-Manual Procedure)[3]

D 4220 Practices for Preserving and Transporting Soil Samples[3]

D 4253 Test Methods for Maximum Index Density of Soils Using a Vibratory Table[3]

D 4718 Practice for Correction of Unit Weight and Water Content for Soils Containing Oversize Particles[3]

D 4753 Specification for Evaluating, Selecting and Specifying Balances and Scales For Use in Soil and Rock Testing[3]

E 1 Specification for ASTM Thermometers[4]

E 11 Specification for Wire-Cloth Sieves for Testing Purposes[4]

E 319 Practice for the Evaluation of Single-Pan Mechanical Balances[4]

E 380 Practice for Use of the SI International System of Units (SI)[4]

3. Terminology

3.1 *Definitions*—See Terminology D 653 for general definitions.

3.2 *Description of Terms Specific to This Standard:*

3.2.1 *modified effort*—the term for the 56 000 ft-lbf/ft³ (2700 kN-m/m³) compactive effort applied by the equipment and procedures of this test.

3.2.2 *modified maximum dry unit weight,* γ_{dmax} *(lbf/ft³*

(kN/m³))—the maximum value defined by the compaction curve for a compaction test using modified effort.

3.2.3 *modified optimum water content,* w_o *(%)*—the water content at which the soil can be compacted to the maximum dry unit weight using modified compactive effort.

3.2.4 *oversize fraction (coarse fraction),* P_c *(%)*—the portion of total sample not used in performing the compaction test; it may be the portion of total sample retained on the No. 4 (3.74-mm), ³/₈-in. (9.5-mm), or ³/₄-in. (19.0-mm) sieve.

3.2.5 *test fraction (finer fraction),* P_F *(%)*—the portion of the total sample used in performing the compaction test; it may be fraction passing the No. 4 (4.75-mm) sieve in Procedure A, minus ³/₈-in. (9.5-mm) sieve in Procedure B, or minus ³/₄-in. (19.0-mm) sieve in Procedure C.

4. Summary of Test Method

4.1 A soil at a selected water content is placed in five layers into a mold of given dimensions, with each layer compacted by 25 or 56 blows of a 10-lbf (44.5-N) rammer dropped from a distance of 18-in. (457-mm), subjecting the soil to a total compactive effort of about 56 000 ft—lbf/ft³ (2700 kN-m/m³). The resulting dry unit weight is determined. The procedure is repeated for a sufficient number of water contents to establish a relationship between the dry unit weight and the water content for the soil. This data, when plotted, represents a curvilinear relationship known as the compaction curve. The values of optimum water content and modified maximum dry unit weight are determined from the compaction curve.

5. Significance and Use

5.1 Soil placed as engineering fill (embankments, foundation pads, road bases) is compacted to a dense state to obtain satisfactory engineering properties such as, shear strength, compressibility, or permeability. Also, foundation soils are often compacted to improve their engineering properties. Laboratory compaction tests provide the basis for determining the percent compaction and water content needed to achieve the required engineering properties, and for controlling construction to assure that the required compaction and water contents are achieved.

5.2 During design of an engineered fill, shear, consolidation, permeability, or other tests require preparation of test specimens by compacting at some water content to some unit weight. It is common practice to first determine the optimum water content (w_o) and maximum dry unit weight (γ_{dmax}) by means of a compaction test. Test specimens are compacted at a selected water content (w), either wet or dry of optimum (w_o) or at optimum (w_o), and at a selected dry unit weight related to a percentage of maximum dry unit weight (γ_{dmax}). The selection of water content (w), either wet or dry of optimum (w_o) or at optimum (w_o) and the dry unit weight (γ_{dmax}) may be based on past experience, or a range of values may be investigated to determine the necessary percent of compaction.

6. Apparatus

6.1 *Mold Assembly*—The molds shall be cylindrical in shape, made of rigid metal and be within the capacity and dimensions indicated in 6.1.1 or 6.1.2 and Figs. 1 and 2. The walls of the mold may be solid, split, or tapered. The "split"

---

[2] *Annual Book of ASTM Standards*, Vol 04.02.
[3] *Annual Book of ASTM Standards*, Vol 04.08.
[4] *Annual Book of ASTM Standards*, Vol 14.02.

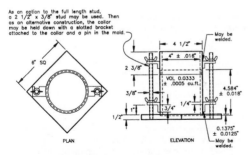

As an option to the full length stud,
a 2 1/2" x 3/8" stud may be used. Then
as an alternative construction, the collar
may be held down with a slotted bracket
attached to the collar and a pin in the mold.

PLAN   ELEVATION

see Table 2 for metric equivalents

FIG. 1   Cylindrical Mold, 4.0-in.

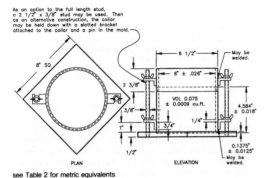

As an option to the full length stud,
a 2 1/2" x 3/8" stud may be used. Then
as an alternative construction, the collar
may be held down with a slotted bracket
attached to the collar and a pin in the mold.

PLAN   ELEVATION

see Table 2 for metric equivalents

FIG. 2   Cylindrical Mold, 6.0-in.

type, may consist of two half-round sections, or a section of pipe split along one element, which can be securely locked together to form a cylinder meeting the requirements of this section. The "tapered" type shall an internal diameter taper that is uniform and not more than 0.200 in./ft (16.7 mm/m) of mold height. Each mold shall have a base plate and an extension collar assembly, both made of rigid metal and constructed so they can be securely attached and easily detached from the mold. The extension collar assembly shall have a height extending above the top of the mold of at least 2.0 in. (50.8 mm) which may include an upper section that flares out to form a funnel provided there is at least a 0.75-in. (19.0-mm) straight cylindrical section beneath it. The extension collar shall align with the inside of the mold. The bottom of the base plate and bottom of the centrally recessed area that accepts the cylindrical mold shall be planar.

6.1.1 *Mold, 4 in.*—A mold having a 4.000 ± 0.016-in. (101.6 ± 0.4-mm) average inside diameter, a height of 4.584 ± 0.018 in. (116.4 ± 0.5 mm) and a volume of 0.0333 ± 0.0005 ft³ (944 ± 14 cm³). A mold assembly having the minimum required features is shown in Fig. 1.

6.1.2 *Mold, 6 in.*—A mold having a 6.000 ± 0.026-in. (152.4 ± 0.7-mm) average inside diameter, a height of 4.584 ± 0.018 in. (116.4 ± 0.5 mm), and a volume of 0.075 ± 0.0009 ft³ (2124 ± 25 cm³). A mold assembly having the

minimum required features is shown in Fig. 2.

6.2 *Rammer*—A rammer, either manually operated as described further in 6.2.1 or mechanically operated as described in 6.2.2. The rammer shall fall freely through a distance of 18 ± 0.05 in. (457.2 ± 1.6 mm) from the surface of the specimen. The mass of the rammer shall be 10 ± 0.02 lbm (4.54 ± 0.01 kg), except that the mass of the mechanical rammers may be adjusted as described in Test Methods D 2168 (see Note 5). The striking face of the rammer shall be planar and circular, except as noted in 6.2.2.3, with a diameter when new of 2.000 ± 0.005 in. (50.80 ± 0.13 mm). The rammer shall be replaced if the striking face becomes worn or bellied to the extent that the diameter exceeds 2.000 ± 0.01 in. (50.80 ± 0.25 mm).

NOTE 5—It is a common and acceptable practice in the inch-pound system to assume that the mass of the rammer is equal to its mass determined using either a kilogram or pound balance and 1 lbf is equal to 1 lbm or 0.4536 kg or 1 N is equal to 0.2248 lbm or 0.1020 kg.

6.2.1 *Manual Rammer*—The rammer shall be equipped with a guide sleeve that has sufficient clearance that the free fall of the rammer shaft and head is not restricted. The guide sleeve shall have at least four vent holes at each end (eight holes total) located with centers ¾ ± ¹/₁₆ in. (19.0 ± 1.6 mm) from each end and spaced 90° apart. The minimum diameter of the vent holes shall be ⅜ in. (9.5 mm). Additional holes or slots may be incorporated in the guide sleeve.

6.2.2 *Mechanical Rammer-Circular Face*—The rammer shall operate mechanically in such a manner as to provide uniform and complete coverage of the specimen surface. There shall be 0.10 ± 0.03-in. (2.5 ± 0.8-mm) clearance between the rammer and the inside surface of the mold at its smallest diameter. The mechanical rammer shall meet the calibration requirements of Test Methods D 2168. The mechanical rammer shall be equipped with a positive mechanical means to support the rammer when not in operation.

6.2.2.3 *Mechanical Rammer-Sector Face*—When used with the 6.0-in. (152.4-mm) mold, a sector face rammer may be used in place of the circular face rammer. The specimen contact face shall have the shape of a sector of a circle of radius equal to 2.90 ± 0.02 in. (73.7 ± 0.5 mm). The rammer shall operate in such a manner that the vertex of the sector is positioned at the center of the specimen.

6.3 *Sample Extruder (optional)*—A jack, frame or other device adapted for the purpose of extruding compacted specimens from the mold.

6.4 *Balance*—A class GP5 balance meeting the requirements of Specification D 4753 for a balance of 1-g readability.

6.5 *Drying Oven*—Thermostatically controlled, preferably of a forced-draft type and capable of maintaining a uniform temperature of 230 ± 9°F (110 ± 5°C) throughout the drying chamber.

6.6 *Straightedge*—A stiff metal straightedge of any convenient length but not less than 10 in. (254 mm). The total length of the straightedge shall be machined straight to a tolerance of ±0.005 in. (±0.1 mm). The scraping edge shall be beveled if it is thicker than ⅛ in. (3 mm).

6.7 *Sieves*—¾-in. (19.0-mm), ⅜-in. (9.5-mm), and No. 4 (4.75-mm), conforming to the requirements of Specification E 11.

6.8 *Mixing Tools*—Miscellaneous tools such as mixing pan, spoon, trowel, spatula, spray bottle, etc., or a suitable mechanical device for thoroughly mixing the sample of soil with increments of water.

7. Test Sample

7.1 The required sample mass for Procedures A and B is approximately 35 lbm (16 kg), and for Procedure C is approximately 65 lbm (29 kg) of dry soil. Therefore, the field sample should have a moist mass of at least 50 lbm (23 kg) and 100 lbm (45 kg), respectively.

7.2 Determine the percentage of material retained on the No. 4 (4.75-mm), ⅜-in. (9.5-mm), or ¾-in. (19.0-mm) sieve as appropriate for choosing Procedure A, B, or C. Make this determination by separating out a representative portion from the total sample and determining the percentages passing the sieves of interest by Test Methods D 422 or C 136. It is only necessary to calculate percentages for the sieve or sieves for which information is desired.

8. Preparation of Apparatus

8.1 Select the proper compaction mold in accordance with the procedure (A, B, or C) being used. Determine and record its mass to the nearest gram. Assemble the mold, base and extension collar. Check the alignment of the inner wall of the mold and mold extension collar. Adjust if necessary.

8.2 Check that the rammer assembly is in good working condition and that parts are not loose or worn. Make any necessary adjustments or repairs. If adjustments or repairs are made, the rammer must be recalibrated.

9. Calibration

9.1 Perform calibrations before initial use, after repairs or other occurrences that might affect the test results, at intervals not exceeding 1000 test specimens, or annually, whichever occurs first, for the following apparatus:

9.1.2 *Balance*—Evaluate in accordance with Specification D 4753.

9.1.3 *Molds*—Determine the volume as described in Annex A1.

9.1.4 *Manual Rammer*—Verify the free fall distance, rammer mass, and rammer face in accordance with 6.2. Verify the guide sleeve requirements in accordance with 6.2.1.

9.1.5 *Mechanical Rammer*—Calibrate and adjust the mechanical rammer in accordance with Test Methods D 2168. In addition, the clearance between the rammer and the inside surface of the mold shall be verified in accordance with 6.2.2.

10. Procedure

10.1 *Soils:*

10.1.1 Do not reuse soil that has been previously laboratory compacted.

10.1.2 When using this test method for soils containing hydrated halloysite, or where past experience with a particular soil indicates that results will be altered by air drying, use the moist preparation method (see 10.2).

10.1.3 Prepare the soil specimens for testing in accordance with 10.2 (preferred) or with 10.3.

10.2 *Moist Preparation Method (preferred)*—Without previously drying the sample, pass it through a No. 4 (4.75-mm), ⅜-in. (9.5-mm), or ¾-in. (19.0-mm) sieve, depending on the procedure (A, B, or C) being used. Determine the water content of the processed soil.

10.2.1 Prepare at least four (preferably five) specimens having water contents such that they bracket the estimated optimum water content. A specimen having a water content close to optimum should be prepared first by trial additions of water and mixing (see Note 6). Select water contents for the rest of the specimens to provide at least two specimens wet and two specimens dry of optimum, and water contents varying by about 2 %. At least two water contents are necessary on the wet and dry side of optimum to accurately define the dry unit weight compaction curve (see 10.5). Some soils with very high optimum water content or a relatively flat compaction curve may require larger water content increments to obtain a well defined maximum dry unit weight. Water content increments should not exceed 4 %.

NOTE 6—With practice it is usually possible to visually judge a point near optimum water content. Typically, soil at optimum water content can be squeezed into a lump that sticks together when hand pressure is released, but will break cleanly into two sections when "bent". At water contents dry of optimum soils tend to crumble; wet of optimum soils tend to stick together in a sticky cohesive mass. Optimum water content is typically slightly less than the plastic limit.

10.2.2 Use approximately 5 lbm (2.3 kg) of the sieved soil for each specimen to be compacted using Procedure A or B, or 13 lbm (5.9 kg) using Procedure C. To obtain the specimen water contents selected in 10.2.1, add or remove the required amounts of water as follows: to add water, spray it into the soil during mixing; to remove water, allow the soil to dry in air at ambient temperature or in a drying apparatus such that the temperature of the sample does not exceed 140°F (60°C). Mix the soil frequently during drying to maintain even water content distribution. Thoroughly mix each specimen to ensure even distribution of water throughout and then place in a separate covered container and allow to stand in accordance with Table 1 prior to compaction. For the purpose of selecting a standing time, the soil may be classified by Test Method D 2487, Practice D 2488 or data on other samples from the same material source. For referee testing, classification shall be by Test Method D 2487.

10.3 *Dry Preparation Method*—If the sample is too damp to be friable, reduce the water content by air drying until the material is friable. Drying may be in air or by the use of drying apparatus such that the temperature of the sample does not exceed 140°F (60°C). Thoroughly break up the aggregations in such a manner as to avoid breaking individual particles. Pass the material through the appropriate sieve: No. 4 (4.75 mm), ⅜ in. (9.5 mm), or ¾ in. (19.0 mm). When preparing the material by passing over the ¾ in. sieve for compaction in the 6 in. mold, break up aggregations sufficiently to at least pass the ⅜ in. sieve in order to

TABLE 1 Required Standing Times of Moisturized Specimens

Classification	Minimum Standing Time, h
GW, GP, SW, SP	no requirement
GM, SM	3
All other soils	16

D 1557

TABLE 2   Metric Equivalents for Figs. 1 and 2

in.	mm
0.016	0.41
0.026	0.66
0.032	0.81
0.028	0.71
1/2	12.70
2½	63.50
2⅝	66.70
4	101.60
4½	114.30
4.584	116.43
4¾	120.60
6	152.40
6½	165.10
6⅝	168.30
6¾	171.40
8¼	208.60

ft³	cm³
1/30 (0.0333)	943
0.0005	14
1/13.333 (0.0750)	2,124
0.0011	31

facilitate the distribution of water throughout the soil in later mixing.

10.3.1 Prepare at least four (preferably five) specimens in accordance with 10.2.1.

10.3.2 Use approximately 5 lbm (2.3 kg) of the sieved soil for each specimen to be compacted using Procedure A or B, or 13 lbm (5.9 kg) using Procedure C. Add the required amounts of water to bring the water contents of the specimens to the values selected in 10.3.1. Follow the specimen preparation procedure specified in 10.2.2 for drying the soil or adding water into the soil and curing each test specimen.

10.4 *Compaction*—After curing, if required, each specimen shall be compacted as follows:

10.4.1 Determine and record the mass of the mold or mold and base plate.

10.4.2 Assemble and secure the mold and collar to the base plate. The mold shall rest on a uniform rigid foundation, such as provided by a cylinder or cube of concrete with a mass of not less than 200 lbm (91 kg). Secure the base plate to the rigid foundation. The method of attachment to the rigid foundation shall allow easy removal of the assembled mold, collar and base plate after compaction is completed.

10.4.3 Compact the specimen in five layers. After compaction, each layer should be approximately equal in thickness. Prior to compaction, place the loose soil into the mold and spread into a layer of uniform thickness. Lightly tamp the soil prior to compaction until it is not in a fluffy or loose state, using either the manual compaction rammer or a 2 in. (5 mm) diameter cylinder. Following compaction of each of the first four layers, any soil adjacent to the mold walls that has not been compacted or extends above the compacted surface shall be trimmed. The trimmed soil may be included with the additional soil for the next layer. A knife or other suitable device may be used. The total amount of soil used shall be such that the fifth compacted layer slightly extends into the collar, but does not exceed ¼ in. (6 mm) above the top of the mold. If the fifth layer does extend above the top of the mold by more than ¼ in. (6 mm), the specimen shall be discarded. The specimen shall be discarded when the last

blow on the rammer for the fifth layer results in the bottom of the rammer extending below the top of the compaction mold.

10.4.4 Compact each layer with 25 blows for the 4 in. (101.6 mm) mold or with 56 blows for the 6 in. (152.4 mm) mold.

NOTE 7—When compacting specimens wetter than optimum water content, uneven compacted surfaces can occur and operator judgment is required as to the average height of the specimen.

10.4.5 In operating the manual rammer, take care to avoid lifting the guide sleeve during the rammer upstroke. Hold the guide sleeve steady and within 5° of vertical. Apply the blows at a uniform rate of approximately 25 blows/min and in such a manner as to provide complete, uniform coverage of the specimen surface.

10.4.6 Following compaction of the last layer, remove the collar and base plate from the mold, except as noted in 10.4.7. A knife may be used to trim the soil adjacent to the collar to loosen the soil from the collar before removal to avoid disrupting the soil below the top of the mold.

10.4.7 Carefully trim the compacted specimen even with the top and bottom of the mold by means of the straightedge scraped across the top and bottom of the mold to form a plane surface even with the top and bottom of the mold. Initial trimming of the specimen above the top of the mold with a knife may prevent tearing out soil below the top of the mold. Fill any holes in either surface with unused or trimmed soil from the specimen, press in with the fingers, and again scrape the straightedge across the top and bottom of the mold. Repeat the appropriate preceding operations on the bottom of the specimen when the mold volume was determined without the base plate. For very wet or dry soils, soil or water may be lost if the base plate is removed. For these situations, leave the base plate attached to the mold. When the base plate is left attached, the volume of the mold must be calibrated with the base plate attached to the mold rather than a plastic or glass plate as noted in Annex A1 (A1.4.1).

10.4.8 Determine and record the mass of the specimen and mold to the nearest gram. When the base plate is left attached, determine and record the mass of the specimen, mold and base plate to the nearest gram.

10.4.9 Remove the material from the mold. Obtain a specimen for water content by using either the whole specimen (preferred method) or a representative portion. When the entire specimen is used, break it up to facilitate drying. Otherwise, obtain a portion by slicing the compacted specimen axially through the center and removing about 500 g of material from the cut faces. Obtain the water content in accordance with Test Method D 2216.

10.5 Following compaction of the last specimen, compare the wet unit weights to ensure that a desired pattern of obtaining data on each side of the optimum water content will be attained for the dry unit weight compaction curve. Plotting the wet unit weight and water content of each compacted specimen can be an aid in making the above evaluation. If the desired pattern is not obtained, additional compacted specimens will be required. Generally, one water content value wet of the water content defining the maximum wet unit weight is sufficient to ensure data on the wet

⟨STM⟩ **D 1557**

side of optimum water content for the maximum dry unit weight.

11. Calculation

11.1 Calculate the dry unit weight and water content of each compacted specimen as explained in 11.3 and 11.4. Plot the values and draw the compaction curve as a smooth curve through the points (see example, Fig. 3). Plot dry unit weight to the nearest 0.1 lbf/ft³ (0.2 kN/m³) and water content to the nearest 0.1 %. From the compaction curve, determine the optimum water content and maximum dry unit weight. If more than 5 % by weight of oversize material was removed from the sample, calculate the corrected optimum water content and corrected maximum dry unit weight of the total material using Practice D 4718. This correction may be made to the appropriate field in place density test specimen rather than to the laboratory test specimen.

11.2 Plot the 100 % saturation curve. Values of water content for the condition of 100 % saturation can be calculated as explained in 11.5 (see example, Fig. 3).

NOTE 8—The 100 % saturation curve is an aid in drawing the compaction curve. For soils containing more than approximately 10 % fines at water contents well above optimum, the two curves generally become roughly parallel with the wet side of the compaction curve between 92 % to 95 % saturation. Theoretically, the compaction curve cannot plot to the right of the 100 % saturation curve. If it does, there is an error in specific gravity, in measurements, in calculations, in test procedures, or in plotting.

NOTE 9—The 100 % saturation curve is sometimes referred to as the zero air voids curve or the complete saturation curve.

11.3 *Water Content, w*—Calculate in accordance with Test Method D 2216.

11.4 *Dry Unit Weights*—Calculate the moist density (Eq 1), the dry density (Eq 2), and then the dry unit weight (Eq 3) as follows:

$$\rho_m = \frac{(M_t - M_{md})}{1000 \, V} \quad (1)$$

where:
ρ_m = moist density of compacted specimen, Mg/m³,
M_t = mass of moist specimen and mold, kg,
M_{md} = mass of compaction mold, kg, and
V = volume of compaction mold, m³ (see Annex A1).

$$\rho_d = \rho_m / (1 + w/100) \quad (2)$$

where:
ρ_d = dry density of compacted specimen, Mg/m³, and
w = water content, %.

$$\gamma_d = 62.43 \, \rho_d \text{ in lbf/ft}^3 \quad (3)$$
or
$$\gamma_d = 9.807 \, \rho_d \text{ in kN/m}^3$$

where:
γ_d = dry unit weight of compacted specimen.

11.5 To calculate points for plotting the 100 % saturation curve or zero air voids curve select values of dry unit weight, calculate corresponding values of water content corresponding to the condition of 100 % saturation as follows:

$$w_{sat} = \frac{(\gamma_w) G_s - \gamma d}{(\gamma d)(G_s)} \times 100 \quad (4)$$

where:
w_{sat} = water content for complete saturation, %,
γ_w = unit weight of water, 62.43 lbf/ft³ (9.807 kN/m³),
γ_d = dry unit weight of soil, and
G_s = specific gravity of soil.

NOTE 10—Specific gravity may be estimated for the test specimen on the basis of test data from other samples of the same soil classification and source. Otherwise, a specific gravity test (Test Method D 854) is necessary.

12. Report

12.1 Report the following information:
12.1.1 Procedure used (A, B, or C).
12.1.2 Preparation method used (moist or dry).
12.1.3 As-received water content, if determined.
12.1.4 Modified optimum water content, to the nearest 0.5 %.
12.1.5 Modified maximum (optimum) dry unit weight, to the nearest 0.5 lbf/ft³.
12.1.6 Description of rammer (manual or mechanical).
12.1.7 Soil sieve data when applicable for determination of procedure (A, B, or C) used.
12.1.8 Description of material used in test, by Practice D 2488, or classification by Test Method D 2487.
12.1.9 Specific gravity and method of determination.
12.1.10 Origin of material used in test, for example, project, location, depth, and the like.
12.1.11 Compaction curve plot showing compaction points used to establish compaction curve, and 100 % saturation curve, point of maximum dry unit weight and optimum water content.
12.1.12 Oversize correction data if used, including the oversize fraction (coarse fraction), P_c in %.

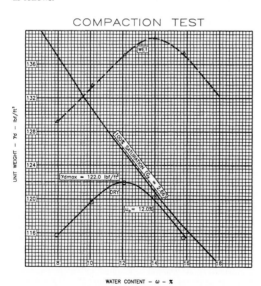

COMPACTION TEST

WATER CONTENT – ω – %

FIG. 3 Example Compaction Curve Plotting

⟨STM⟩ D 1557

13. Precision and Bias

13.1 *Precision*—Data are being evaluated to determine the precision of this test method. In addition, pertinent data is being solicited from users of the test method.

13.2 *Bias*—It is not possible to obtain information on bias because there is no other method of determining the values of modified maximum dry unit weight and optimum water content.

14. Keywords

14.1 compaction characteristics; density; impact compaction using modified effort; laboratory tests; modified proctor test; moisture-density curves; soil compaction

ANNEX

(Mandatory Information)

A1. VOLUME OF COMPACTION MOLD

A1.1 Scope

A1.1.1 This annex describes the procedure for determining the volume of a compaction mold.

A1.1.2 The volume is determined by a water-filled method and checked by a linear-measurement method.

A1.2 Apparatus

A1.2.1 In addition to the apparatus listed in Section 6, the following items are required:

A1.2.1.1 *Vernier or Dial Caliper*, having a measuring range of at least 0 to 6 in. (0 to 150 mm) and readable to at least 0.001 in. (0.02 mm).

A1.2.1.2 *Inside Micrometer*, having a measuring range of at least 2 to 12 in. (50 to 300 mm) and readable to at least 0.001 in. (0.02 mm).

A1.2.1.3 *Plastic or Glass Plates*—Two plastic or glass plates about 8 in.^{2} by $\frac{1}{4}$ in. thick (200 mm^2 by 6 mm).

A1.2.1.4 *Thermometer*—0 to 50°C range, 0.5°C graduations, conforming to the requirements of Specification E 1.

A1.2.1.5 *Stopcock Grease* or similar sealant.

A1.2.1.6 *Miscellaneous equipment*—Bulb syringe, towels, etc.

A1.3 Precautions

A1.3.1 Perform this procedure in an area isolated from drafts or extreme temperature fluctuations.

A1.4 Procedure

A1.4.1 *Water-Filling Method:*

A1.4.1.1 Lightly grease the bottom of the compaction mold and place it on one of the plastic or glass plates. Lightly grease the top of the mold. Be careful not to get grease on the inside of the mold. If it is necessary to use the base plate, as noted in 10.4.7, place the greased mold onto the base plate and secure with the locking studs.

A1.4.1.2 Determine the mass of the greased mold and both plastic or glass plates to the nearest 0.01 lbm (1 g) and record. When the base plate is being used in lieu of the bottom plastic or glass plate determine the mass of the mold, base plate and a single plastic or glass plate to be used on top of the mold to the nearest 0.01 lbm (1 g) and record.

A1.4.1.3 Place the mold and the bottom plate on a firm, level surface and fill the mold with water to slightly above its rim.

A1.4.1.4 Slide the second plate over the top surface of the mold so that the mold remains completely filled with water and air bubbles are not entrapped. Add or remove water as necessary with a bulb syringe.

A1.4.1.5 Completely dry any excess water from the outside of the mold and plates.

A1.4.1.6 Determine the mass of the mold, plates and water and record to the nearest 0.01 lbm (1 g).

A1.4.1.7 Determine the temperature of the water in the mold to the nearest 1°C and record. Determine and record the absolute density of water from Table A1.1.

A1.4.1.8 Calculate the mass of water in the mold by subtracting the mass determined in A1.4.1.2 from the mass determined in A1.4.1.6.

A1.4.1.9 Calculate the volume of water by dividing the mass of water by the density of water and record to the nearest 0.0001 ft^3 (1 cm^3).

A1.4.1.10 When the base plate is used for the calibration of the mold volume repeat steps A1.4.1.3 through A1.4.1.9.

A1.4.2 *Linear Measurement Method:*

A1.4.2.1 Using either the vernier caliper or the inside micrometer, measure the diameter of the mold six times at the top of the mold and six times at the bottom of the mold spacing each of the six top and bottom measurements equally around the circumference of the mold. Record the values to the nearest 0.001 in. (0.02 mm).

A1.4.2.2 Using the vernier caliper, measure the inside height of the mold by making three measurements equally spaced around the circumference of the mold. Record values to the nearest 0.001 in. (0.02 mm).

A1.4.2.3 Calculate the average top diameter, average bottom diameter and average height.

A1.4.2.4 Calculate the volume of the mold and record to

TABLE A1.1 Density of Water[A]

Temperature, °C (°F)	Density of Water, γ/ml
18 (64.4)	0.99862
19 (66.2)	0.99843
20 (68.0)	0.99823
21 (69.8)	0.99802
22 (71.6)	0.99779
23 (73.4)	0.99756
24 (75.2)	0.99733
25 (77.0)	0.99707
26 (78.8)	0.99681

[A] Values other than shown may be obtained by referring to the *Handbook of Chemistry and Physics*, Chemical Rubber Publishing Co., Cleveland, OH.

the nearest 0.0001 ft³ (1 cm³) using Eq A1a (for inch-pound) or A1b (for SI):

$$V = \frac{(\pi)(h)(d_t + d_b)^2}{(16)(1728)}$$ (A1a)

$$V = \frac{(\pi)(h)(d_t + d_b)^2}{(16)(10^3)}$$ (A1b)

where:
V = volume of mold, ft³ (cm³),
h = average height, in. (mm),
d_t = average top diameter, in. (mm),
d_b = average bottom diameter, in. (mm),
$1/1728$ = constant to convert in³ to ft³, and
$1/10^3$ = constant to convert mm³ to cm³.

A1.5 Comparison of Results

A1.5.1 The volume obtained by either method should be within the volume tolerance requirements of 6.1.1 and 6.1.2.

A1.5.2 The difference between the two methods should not exceed 0.5 % of the nominal volume of the mold.

A1.5.3 Repeat the determination of volume if these criteria are not met.

A1.5.4 Failure to obtain satisfactory agreement between the two methods, even after several trials, is an indication that the mold is badly deformed and should be replaced.

A1.5.5 Use the volume of the mold determined using the water-filling method as the assigned volume value for calculating the moist and dry density (see 11.4).

Designation: C 29/C 29M – 97

AMERICAN SOCIETY FOR TESTING AND MATERIALS
100 Barr Harbor Dr., West Conshohocken, PA 19428
Reprinted from the Annual Book of ASTM Standards. Copyright ASTM
If not listed in the current combined index, will appear in the next edition.

American Association of State
Highway and Transportation Officials Standard
AASHTO No.: T 19/T19M

Standard Test Method for
Bulk Density ("Unit Weight") and Voids in Aggregate[1]

This standard is issued under the fixed designation C 29/C 29M; the number immediately following the designation indicates the year of original adoption or, in the case of revision, the year of last revision. A number in parentheses indicates the year of last reapproval. A superscript epsilon (ε) indicates an editorial change since the last revision or reapproval.

This test method has been approved for use by agencies of the Department of Defense. Consult the DoD Index of Specifications and Standards for the specific year of issue which has been adopted by the Department of Defense.

1. Scope

1.1 This test method covers the determination of bulk density ("unit weight") of aggregate in a compacted or loose condition, and calculated voids between particles in fine, coarse, or mixed aggregates based on the same determination. This test method is applicable to aggregates not exceeding 5 in. [125 mm] in nominal maximum size.

NOTE 1—Unit weight is the traditional terminology used to describe the property determined by this test method, which is weight per unit volume (more correctly, mass per unit volume or density).

1.2 The values stated in either inch-pound units or SI units are to be regarded separately as standard, as appropriate for a specification with which this test method is used. An exception is with regard to sieve sizes and nominal size of aggregate, in which the SI values are the standard as stated in Specification E 11. Within the text, SI units are shown in brackets. The values stated in each system may not be exact equivalents; therefore each system must be used independently of the other, without combining values in any way.

1.3 *This standard does not purport to address all of the safety concerns, if any, associated with its use. It is the responsibility of the user of this standard to establish appropriate safety and health practices and determine the applicability of regulatory limitations prior to use.*

2. Referenced Documents

2.1 *ASTM Standards:*
C 125 Terminology Relating to Concrete and Concrete Aggregates[2]
C 127 Test Method for Specific Gravity and Absorption of Coarse Aggregate[2]
C 128 Test Method for Specific Gravity and Absorption of Fine Aggregate[2]
C 138 Test Method for Unit Weight, Yield, and Air Content (Gravimetric) of Concrete[2]
C 670 Practice for Preparing Precision and Bias Statements for Test Methods for Construction Materials[2]
C 702 Practice for Reducing Samples of Aggregate to Testing Size[2]
D 75 Practice for Sampling Aggregates[3]

D 123 Terminology Relating to Textiles[4]
E 11 Specification for Wire Cloth and Sieves for Testing Purposes[5]
2.2 *AASHTO Standard:*
T19/T19M Method of Test for Unit Weight and Voids in Aggregate[6]

3. Terminology

3.1 *Definitions*—Definitions are in accordance with Terminology C 125 unless otherwise indicated.
3.1.1 *bulk density, n—of aggregate,* the mass of a unit volume of bulk aggregate material, in which the volume includes the volume of the individual particles and the volume of the voids between the particles. Expressed in lb/ft^3 [kg/m^3].
3.1.2 *unit weight, n*—weight (mass) per unit volume. (Deprecated term–used preferred term **bulk density.**)

DISCUSSION—Weight is equal to the mass of the body multiplied by the acceleration due to gravity. Weight may be expressed in absolute units (newtons, poundals) or in gravitational units (kgf, lbf), for example: on the surface of the earth, a body with a mass of 1 kg has a weight of 1 kgf (approximately 9.81 N), or a body with a mass of 1 lb has a weight of 1 lbf (approximately 4.45 N or 32.2 poundals). Since weight is equal to mass times the acceleration due to gravity, the weight of a body will vary with the location where the weight is determined, while the mass of the body remains constant. On the surface of the earth, the force of gravity imparts to a body that is free to fall an acceleration of approximately 9.81 m/s^2 (32.2 ft/s^2). 　　　D 123.

3.2 *Description of Term Specific to this Standard:*
3.2.1 *voids, n—in unit volume of aggregate,* the space between particles in an aggregate mass not occupied by solid mineral matter.

DISCUSSION—Voids within particles, either permeable or impermeable, are not included in voids as determined by this test method.

4. Significance and Use

4.1 This test method is often used to determine bulk density values that are necessary for use for many methods of selecting proportions for concrete mixtures.

4.2 The bulk density also may be used for determining mass/volume relationships for conversions in purchase agreements. However, the relationship between degree of compaction of aggregates in a hauling unit or stockpile and that achieved in this test method is unknown. Further,

---

[1] This test method is under the jurisdiction of ASTM Committee C-9 on Concrete and Concrete Aggregates and is the direct responsibility of Subcommittee C09.20 on Normal Weight Aggregates.
Current edition approved July 10, 1997. Published September 1997. Originally published as C 29 – 20 T. Last previous edition C 29/C 29M – 91a.
[2] *Annual Book of ASTM Standards,* Vol 04.02.
[3] *Annual Book of ASTM Standards,* Vol 04.03.

[4] *Annual Book of ASTM Standards,* Vol 07.01.
[5] *Annual Book of ASTM Standards,* Vol 14.02.
[6] Available from American Association of State Highway and Transportation Officials, 444 N. Capitol St. NW, Suite 225, Washington, DC 20001.

⬦ C 29/C 29M

aggregates in hauling units and stockpiles usually contain absorbed and surface moisture (the latter affecting bulking), while this test method determines the bulk density on a dry basis.

4.3 A procedure is included for computing the percentage of voids between the aggregate particles based on the bulk density determined by this test method.

5. Apparatus

5.1 *Balance*—A balance or scale accurate within 0.1 % of the test load at any point within the range of use, graduated to at least 0.1 lb [0.05 kg]. The range of use shall be considered to extend from the mass of the measure empty to the mass of the measure plus its contents at 120 lb/ft³ [1920 kg/m³].

5.2 *Tamping Rod*—A round, straight steel rod, ⅝ in. [16 mm] in diameter and approximately 24 in. [600 mm] in length, having the tamping end or both ends rounded to a hemispherical tip, the diameter of which is ⅝ in. (16 mm).

5.3 *Measure*—A cylindrical metal measure, preferably provided with handles. It shall be watertight, with the top and bottom true and even, and sufficiently rigid to retain its form under rough usage. The measure shall have a height approximately equal to the diameter, but in no case shall the height be less than 80 % nor more than 150 % of the diameter. The capacity of the measure shall conform to the limits in Table 1 for the aggregate size to be tested. The thickness of metal in the measure shall be as described in Table 2. The top rim shall be smooth and plane within 0.01 in. [0.25 mm] and shall be parallel to the bottom within 0.5° (Note 2). The interior wall of the measure shall be a smooth and continuous surface.

NOTE 2—The top rim is satisfactorily plane if a 0.01-in. [0.25-mm] feeler gage cannot be inserted between the rim and a piece of ¼-in. [6-mm] or thicker plate glass laid over the measure. The top and bottom are satisfactorily parallel if the slope between pieces of plate glass in contact with the top and bottom does not exceed 0.87 % in any direction.

5.3.1 If the measure also is to be used for testing for bulk density of freshly-mixed concrete according to Test Method C 138, the measure shall be made of steel or other suitable metal not readily subject to attack by cement paste. Reactive materials, such as aluminum alloys are permitted, where as a consequence of an initial reaction, a surface film is formed which protects the metal against further corrosion.

5.3.2 Measures larger than nominal 1 ft³ (28 L) capacity shall be made of steel for rigidity, or the minimum thicknesses of metal listed in Table 2 shall be suitably increased.

5.4 *Shovel or Scoop*—A shovel or scoop of convenient size for filling the measure with aggregate.

5.5 *Calibration Equipment*—A piece of plate glass, preferably at least ¼ in. [6 mm] thick and at least 1 in. [25 mm] larger than the diameter of the measure to be calibrated. A supply of water-pump or chassis grease that can be placed on the rim of the container to prevent leakage.

6. Sampling

6.1 Obtain the sample in accordance with Practice D 75, and reduce to test sample size in accordance with Practice C 702.

TABLE 1 Capacity of Measures

Nominal Maximum Size of Aggregate		Capacity of Measure[A]	
in.	mm	ft³	L (m³)
½	12.5	¹⁄₁₀	2.8 (0.0028)
1	25.0	⅓	9.3 (0.0093)
1½	37.5	½	14 (0.014)
3	75	1	28 (0.028)
4	100	2½	70 (0.070)
5	125	3½	100 (0.100)

[A] The indicated size of measure shall be used to test aggregates of a nominal maximum size equal to or smaller than that listed. The actual volume of the measure shall be at least 95 % of the nominal volume listed.

TABLE 2 Requirements for Measures

Capacity of Measure	Thickness of Metal, min		
	Bottom	Upper 1½ in. or 38 mm of wall[A]	Remainder of wall
Less than 0.4 ft³	0.20 in.	0.10 in.	0.10 in.
0.4 ft³ to 1.5 ft³, incl	0.20 in.	0.20 in.	0.12 in.
over 1.5 to 2.8 ft³, incl	0.40 in.	0.25 in.	0.15 in.
over 2.8 to 4.0 ft³, incl	0.50 in.	0.30 in.	0.20 in.
Less than 11 L	5.0 mm	2.5 mm	2.5 mm
11 to 42 L, incl	5.0 mm	5.0 mm	3.0 mm
over 42 to 80 L, incl	10.0 mm	6.4 mm	3.8 mm
over 80 to 133 L, incl	13.0 mm	7.6 mm	5.0 mm

[A] The added thickness in the upper portion of the wall may be obtained by placing a reinforcing band around the top of the measure.

7. Test Sample

7.1 The size of the sample shall be approximately 125 to 200 % of the quantity required to fill the measure, and shall be handled in a manner to avoid segregation. Dry the aggregate sample to essentially constant mass, preferably in an oven at 230 ± 9°F [110 ± 5°C].

8. Calibration of Measure

8.1 Fill the measure with water at room temperature and cover with a piece of plate glass in such a way as to eliminate bubbles and excess water.

8.2 Determine the mass of the water in the measure using the balance described in 5.1.

8.3 Measure the temperature of the water and determine its density from Table 3, interpolating if necessary.

8.4 Calculate the volume, V, of the measure by dividing the mass of the water required to fill the measure by its density. Alternatively, calculate the factor for the measure $(1/V)$ by dividing the density of the water by the mass required to fill the measure.

NOTE 3—For the calculation of bulk density, the volume of the measure in SI units should be expressed in cubic metres, or the factor as $1/m^3$. However, for convenience the size of the measure may be expressed in litres.

8.5 Measures shall be recalibrated at least once a year or whenever there is reason to question the accuracy of the calibration.

9. Selection of Procedure

9.1 The shoveling procedure for loose bulk density shall be used only when specifically stipulated. Otherwise, the compact bulk density shall be determined by the rodding procedure for aggregates having a nominal maximum size of

◊ C 29/C 29M

TABLE 3 Density of Water

Temperature		lb/ft³	kg/m³
°F	°C		
60	15.6	62.366	999.01
65	18.3	62.336	998.54
70	21.1	62.301	997.97
73.4	23.0	62.274	997.54
75	23.9	62.261	997.32
80	26.7	62.216	996.59
85	29.4	62.166	995.83

1½ in. [37.5 mm] or less, or by the jigging procedure for aggregates having a nominal maximum size greater than 1½ in. [37.5 mm] and not exceeding 5 in. [125 mm].

10. Rodding Procedure

10.1 Fill the measure one-third full and level the surface with the fingers. Rod the layer of aggregate with 25 strokes of the tamping rod evenly distributed over the surface. Fill the measure two-thirds full and again level and rod as above. Finally, fill the measure to overflowing and rod again in the manner previously mentioned. Level the surface of the aggregate with the fingers or a straightedge in such a way that any slight projections of the larger pieces of the coarse aggregate approximately balance the larger voids in the surface below the top of the measure.

10.2 In rodding the first layer, do not allow the rod to strike the bottom of the measure forcibly. In rodding the second and third layers, use vigorous effort, but not more force than to cause the tamping rod to penetrate to the previous layer of aggregate.

NOTE 4—In rodding the larger sizes of coarse aggregate, it may not be possible to penetrate the layer being consolidated, especially with angular aggregates. The intent of the procedure will be accomplished if vigorous effort is used.

10.3 Determine the mass of the measure plus its contents, and the mass of the measure alone, and record the values to the nearest 0.1 lb [0.05 kg].

11. Jigging Procedure

11.1 Fill the measure in three approximately equal layers as described in 10.1, compacting each layer by placing the measure on a firm base, such as a cement-concrete floor, raising the opposite sides alternately about 2 in. [50 mm], and allowing the measure to drop in such a manner as to hit with a sharp, slapping blow. The aggregate particles, by this procedure, will arrange themselves in a densely compacted condition. Compact each layer by dropping the measure 50 times in the manner described, 25 times on each side. Level the surface of the aggregate with the fingers or a straightedge in such a way that any slight projections of the larger pieces of the coarse aggregate approximately balance the larger voids in the surface below the top of the measure.

11.2 Determine the mass of the measure plus its contents, and the mass of the measure alone, and record the values to the nearest 0.1 lb [0.05 kg].

12. Shoveling Procedure

12.1 Fill the measure to overflowing by means of a shovel or scoop, discharging the aggregate from a height not to exceed 2 in. [50 mm] above the top of the measure. Exercise

care to prevent, so far as possible, segregation of the particle sizes of which the sample is composed. Level the surface of the aggregate with the fingers or a straightedge in such a way that any slight projections of the larger pieces of the coarse aggregate approximately balance the larger voids in the surface below the top of the measure.

12.2 Determine the mass of the measure plus its contents, and the mass of the measure alone, and record the values to the nearest 0.1 lb [0.05 kg].

13. Calculation

13.1 *Bulk Density*—Calculate the bulk density for the rodding, jigging, or shoveling procedure as follows:

$$M = (G - T)/V \tag{1}$$

or

$$M = (G - T) \times F \tag{2}$$

where:
M = bulk density of the aggregate, lb/ft³ [kg/m³],
G = mass of the aggregate plus the measure, lb [kg],
T = mass of the measure, lb [kg],
V = volume of the measure, ft³ [m³], and
F = factor for measure, ft⁻³ [m⁻³].

13.1.1 The bulk density determined by this test method is for aggregate in an oven-dry condition. If the bulk density in terms of saturated-surface-dry (SSD) condition is desired, use the exact procedure in this test method, and then calculate the SSD bulk density using the following formula:

$$M_{SSD} = M[1 + (A/100)] \tag{3}$$

where:
M_{SSD} = bulk density in SSD condition, lb/ft³ [kg/m³], and
A = % absorption, determined in accordance with Test Method C 127 or Test Method C 128.

13.2 *Void Content*—Calculate the void content in the aggregate using the bulk density determined by either the rodding, jigging, or shoveling procedure, as follows:

$$\% \text{ Voids} = 100[(S \times W) - M]/(S \times W) \tag{4}$$

where:
M = bulk density of the aggregate, lb/ft³ [kg/m³],
S = bulk specific gravity (dry basis) as determined in accordance with Test Method C 127 or Test Method C 128, and
W = density of water, 62.3 lb/ft³ [998 kg/m³].

14. Report

14.1 Report the results for the bulk density to the nearest 1 lb/ft³ [10 kg/m³] as follows:

14.1.1 Bulk density by rodding, or
14.1.2 Bulk density by jigging, or
14.1.3 Loose bulk density.

14.2 Report the results for the void content to the nearest 1 % as follows:

14.2.1 Voids in aggregate compacted by rodding, %, or
14.2.2 Voids in aggregate compacted by jigging, %, or
14.2.3 Voids in loose aggregate, %.

15. Precision and Bias

15.1 The following estimates of precision for this test method are based on results from the AASHTO Materials Reference Laboratory (AMRL) Proficiency Sample Pro-

C 29/C 29M

gram, with testing conducted using this test method and AASHTO Method T19/T19M. There are no significant differences between the two test methods. The data are based on the analyses of more than 100 paired test results from 40 to 100 laboratories.

15.2 *Coarse Aggregate* (*bulk density*):

15.2.1 *Single-Operator Precision*—The single-operator standard deviation has been found to be 0.88 lb/ft³ [14 kg/m³] (1s). Therefore, results of two properly conducted tests by the same operator on similar material should not differ by more than 2.5 lb/ft³ [40 kg/m³] (d2s).

15.2.2 *Multilaboratory Precision*—The multilaboratory standard deviation has been found to be 1.87 lb/ft³ [30 kg/m³] (1s). Therefore, results of two properly conducted tests from two different laboratories on similar material should not differ by more than 5.3 lb/ft³ [85 kg/m³] (d2s).

15.2.3 These numbers represent, respectively, the (1s) and (d2s) limits as described in Practice C 670. The precision estimates were obtained from the analysis of AMRL proficiency sample data for bulk density by rodding of normal weight aggregates having a nominal maximum aggregate size of 1 in. [25.0 mm], and using a ½-ft³ [14-L] measure.

15.3 *Fine Aggregate* (*bulk density*):

15.3.1 *Single-Operator Precision*—The single-operator standard deviation has been found to be 0.88 lb/ft³ [14 kg/m³] (1s). Therefore, results of two properly conducted

tests by the same operator on similar material should not differ by more than 2.5 lb/ft³ [40 kg/m³] (d2s).

15.3.2 *Multilaboratory Precision*—The multilaboratory standard deviation has been found to be 2.76 lb/ft³ [44 kg/m³] (1s). Therefore, results of two properly conducted tests from two different laboratories on similar material should not differ by more than 7.8 lb/ft³ [125 kg/m³] (d2s).

15.3.3 These numbers represent, respectively, the (1s) and (d2s) limits as described in Practice C 670. The precision estimates were obtained from the analysis of AMRL proficiency sample data for loose bulk density from laboratories using a ¹⁄₁₀-ft³ [2.8-L] measure.

15.4 No precision data on void content are available. However, as the void content in aggregate is calculated from bulk density and bulk specific gravity, the precision of the voids content reflects the precision of these measured parameters given in 15.2 and 15.3 of this test method and in Test Methods C 127 and C 128.

15.5 *Bias*—The procedure in this test method for measuring bulk density and void content has no bias because the values for bulk density and void content can be defined only in terms of a test method.

16. Keywords

16.1 aggregates; bulk density; coarse aggregate; density; fine aggregate; unit weight; voids in aggregates

Designation: C 31/C 31M – 96

AMERICAN SOCIETY FOR TESTING AND MATERIALS
100 Barr Harbor Dr., West Conshohocken, PA 19428
Reprinted from the Annual Book of ASTM Standards. Copyright ASTM
If not listed in the current combined index, will appear in the next edition.

Standard Practice for
Making and Curing Concrete Test Specimens in the Field[1]

This standard is issued under the fixed designation C 31/C 31M; the number immediately following the designation indicates the year of original adoption or, in the case of revision, the year of last revision. A number in parentheses indicates the year of last reapproval. A superscript epsilon (ϵ) indicates an editorial change since the last revision or reapproval.

This practice has been approved for use by agencies of the Department of Defense. Consult the DoD Index of Specifications and Standards for the specific year of issue which has been adopted by the Department of Defense.

1. Scope

1.1 This practice covers procedures for making and curing cylindrical and prismatic beam specimens from representative samples of fresh concrete for a construction project.

1.2 The concrete used to make the molded specimens shall have the same levels of slump, air content, and percentage of coarse aggregate as the concrete it represents. This practice is not satisfactory for making specimens from concretes not having a measurable slump or requiring other sizes and shapes of specimens to represent a product or structure.

1.3 The values stated in either inch-pound units or SI units shall be regarded separately as standard. The SI units are shown in brackets. The values stated may not be exact equivalents; therefore each system must be used independently of the other. Combining values from the two units may result in nonconformance.

1.4 *This standard does not purport to address all of the safety concerns, if any, associated with its use. It is the responsibility of the user of this standard to establish appropriate safety and health practices and determine the applicability of regulatory limitations prior to use.*

1.5 The text of this standard references notes which provide explanatory material. These notes shall not be considered as requirements of the standard.

2. Referenced Documents

2.1 *ASTM Standards:*
C 138 Test Method for Unit Weight, Yield, and Air Content (Gravimetric) of Concrete[2]
C 143 Test Method for Slump of Hydraulic Cement Concrete[2]
C 172 Practice for Sampling Freshly Mixed Concrete[2]
C 173 Test Method for Air Content of Freshly Mixed Concrete by the Volumetric Method[2]
C 192 Practice for Making and Curing Concrete Test Specimens in the Laboratory[2]
C 231 Test Method for Air Content of Freshly Mixed Concrete by the Pressure Method[2]
C 470 Specification for Molds for Forming Concrete Test Cylinders Vertically[2]

C 511 Specification for Moist Cabinets, Moist Rooms, and Water Storage Tanks Used in the Testing of Hydraulic Cements and Concretes[3]
C 617 Practice for Capping Cylindrical Concrete Specimens[2]
C 1064 Test Method for Temperature of Freshly Mixed Portland-Cement Concrete[2]
2.2 *American Concrete Institute Publication:*[4]
CP-1 Concrete Field Testing Technician, Grade I

3. Significance and Use

3.1 This practice provides standardized requirements for making, curing, protecting, and transporting concrete test specimens under field conditions.

3.2 If the specimens are made and standard cured, as stipulated herein, the resulting test data are able to be used for the following purposes:

3.2.1 Acceptance testing for specified strength,

3.2.2 Checking adequacy of mixture proportions for strength, and

3.2.3 Quality control.

3.3 If the specimens are made and field cured, as stipulated herein, the resulting test data are able to be used for the following purposes:

3.3.1 Determination of the time the structure is permitted to be put in service,

3.3.2 Comparison with test results of standard cured specimens or with test results from various in-place test methods,

3.3.3 Adequacy of curing and protection of concrete in the structure, or

3.3.4 Form or shoring removal time requirements.

4. Apparatus

4.1 *Molds, General*—Molds for specimens or fastenings thereto in contact with the concrete shall be made of steel, cast iron, or other nonabsorbent material, nonreactive with concrete containing portland or other hydraulic cements. Molds shall hold their dimensions and shape under all conditions of use. Molds shall be watertight during use as judged by their ability to hold water poured into them. Provisions for tests of water leakage are given in the Test Methods for Elongation, Absorption, and Water Leakage

---

[1] This practice is under the jurisdiction of ASTM Committee C-9 on Concrete and Concrete Aggregates and is the direct responsibility of Subcommittee C09.61 on Testing Concrete for Strength.
Current edition approved Aug. 10, 1996. Published October 1996. Originally published as C 31 – 20. Last previous edition C 31 – 95.
[2] *Annual Book of ASTM Standards*, Vol 04.02.

[3] *Annual Book of ASTM Standards*, Vol 04.01.
[4] Available from American Concrete Institute, P.O. Box 19150, Detroit, MI 48219-0150.

C 31/C 31M

section of Specification C 470. A suitable sealant, such as heavy grease, modeling clay, or microcrystalline wax shall be used where necessary to prevent leakage through the joints. Positive means shall be provided to hold base plates firmly to the molds. Reusable molds shall be lightly coated with mineral oil or a suitable nonreactive form release material before use.

4.2 *Cylinder Molds:*

4.2.1 *Molds for Casting Specimens Vertically*—Molds for casting concrete test specimens shall conform to the requirements of Specification C 470.

4.3 *Beam Molds*—Beam molds shall be rectangular in shape and of the dimensions required to produce the specimens stipulated in 5.2. The inside surfaces of the molds shall be smooth. The sides, bottom, and ends shall be at right angles to each other and shall be straight and true and free of warpage. Maximum variation from the nominal cross section shall not exceed 1/8 in. [3 mm] for molds with depth or breadth of 6 in. [150 mm] or more. Molds shall produce specimens at least as long but not more than 1/16 in. [2 mm] shorter than the required length in 5.2.

4.4 *Tamping Rod*—A round, straight steel rod with the dimensions conforming to those in Table 1, having the tamping end or both ends rounded to a hemispherical tip of the same diameter as the rod.

4.5 *Vibrators*—Internal vibrators shall be electric motor driven flexible shaft type. The frequency or vibration shall be 7000 vibrations per minute or greater while in use. The outside diameter or side dimension of the vibrating element shall be at least 0.75 in. [20 mm] and not greater than 1.5 in. [40 mm]. The combined length of the shaft and vibrating element shall exceed the maximum depth of the section being vibrated by at least 3 in. [75 mm]. For cylinders, the diameter of the vibrating element must be no more than one fourth the diameter of the cylinder. For beams, the diameter of the vibrating element must be no more than one third the width of the mold. A vibrating-reed tachometer shall be used to check the frequency of vibration.

4.6 *Mallet*—A mallet with a rubber or rawhide head weighing 1.25 ± 0.50 lb [0.6 ± 0.2 kg] shall be used.

4.7 *Small Tools*—Shovels, hand-held floats, scoops, and a vibrating-reed tachometer shall be provided.

4.8 *Slump Apparatus*—The apparatus for measurement of slump shall conform to the requirements of Test Method C 143.

4.9 *Sampling Receptacle*—The receptacle shall be a suitable heavy gage metal pan, wheelbarrow, or flat, clean nonabsorbent board of sufficient capacity to allow easy remixing of the entire sample with a shovel or trowel.

4.10 *Air Content Apparatus*—The apparatus for measuring air content shall conform to the requirements of Test Methods C 173 or C 231.

5. Testing Requirements

5.1 *Cylindrical Specimens*—Compressive or splitting tensile strength specimens shall be cylinders cast and hardened in an upright position, with a length equal to twice the diameter. The standard specimen shall be the 6 by 12-in. [150 by 300-mm] cylinder when the nominal maximum size of the coarse aggregate does not exceed 2 in. [50 mm] (Notes 1 and 2). When the nominal maximum size of the coarse aggregate does exceed 2 in. [50 mm], either the concrete sample shall be treated by wet sieving as described in Practice C 172 or the diameter of the cylinder shall be at least three times the nominal maximum size of coarse aggregate in the concrete. For acceptance testing for specified strength, cylinders smaller than 6 by 12 in. [150 by 300 mm] shall not be used, unless another size is specified (Note 3).

NOTE 1—The nominal maximum size is the smallest sieve opening through which the entire amount of aggregate is required to pass.

NOTE 2—When molds in SI units are required and not available, equivalent inch-pound unit size mold should be permitted.

NOTE 3—For uses other than acceptance testing for specified strength, a 4 by 8 in. [100 by 200 mm] or 5 by 10 in. [125 by 250 mm] cylinder may be suitable. However, the diameter of any cylinder shall be at least three times the nominal maximum size of the coarse aggregate in the concrete (Note 1). When cylinders smaller than the standard size are used, within-test variability has been shown to be higher but not to a statistically significant degree. The compressive strength results are affected by a number of factors including cylinder size.

5.2 *Rectangular Beam Specimens*—Flexural strength specimens shall be rectangular beams of concrete cast and hardened in the horizontal position. The length shall be at least 2 in. [50 mm] greater than three times the depth as tested. The ratio of width to depth as molded shall not exceed 1.5. The standard beam shall be 6 by 6 in. [150 by 150 mm] in cross section, and shall be used for concrete with nominal maximum size coarse aggregate up to 2 in. [50 mm] (Note 2). When the nominal maximum size of the coarse aggregate exceeds 2 in. [50 mm], the smaller cross sectional dimension of the beam shall be at least three times the nominal maximum size of the coarse aggregate. Unless required by project specifications, beams made in the field shall not have a width or depth of less than 6 in. [150 mm].

5.3 *Field Technicians*—The field technicians making and curing specimens for acceptance testing shall be certified ACI Field Testing Technicians, Grade I or equivalent. Equivalent personnel certification programs shall include both written and performance examinations, as outlined in ACI CP-1.

6. Sampling Concrete

6.1 The samples used to fabricate test specimens under this standard shall be obtained in accordance with Practice C 172 unless an alternative procedure has been approved.

6.2 Record the identity of the sample with respect to the location of the concrete represented and the time of casting.

7. Slump, Air Content, and Temperature

7.1 *Slump*—Measure and record the slump of each batch of concrete from which specimens are made immediately after remixing in the receptacle, as required in Test Method C 143.

TABLE 1   Tamping Rod and Rodding Requirements

Diameter of Cylinder, in. (mm)	Rod Dimensions		Number of Roddings/Layer
	Diameter of Rod, in. (mm)	Length of Rod, in. (mm)	
< 6 (150)	⅜ (10)	12 (300)	25
6 (150)	⅝ (16)	24 (600)	25
8 (200)	⅝ (16)	24 (600)	50
10 (250) (or greater)	⅝ (16)	24 (600)	75

∰ C 31/C 31M

TABLE 2 Specimen Size, Type, and Molding Requirements

Specimen Type and Size, as Depth, in. (mm)	Mode of Consolidation	Number of Layers	Approximate Depth of Layer, in. (mm)
Cylinders:			
12 (300) or less	rodding	3 equal	4 (100) or less
Over 12 (300)	rodding	as required	4 (100) or less
12 (300) or less	vibration	2 equal	6 (150) or less
12 (300) to 18 (450)	vibration	2 equal	half depth of specimen
Over 18 (450)	vibration	3 or more	8 (200) or as near as practicable
Beams:			
6 (150) to 8 (200)	rodding	2 equal	half depth of specimen
Over 8 (200)	rodding	3 or more	4 (100)
6 (150) to 8 (200)	vibration	1	depth of specimen
Over 8 (200)	vibration	2 or more	8 (200) as near as practicable

7.2 *Air Content*—Determine and record the air content in accordance with either Test Method C 173 or Test Method C 231. The concrete used in performing the air content test shall not be used in fabricating test specimens.

7.3 *Temperature*—Determine and record the temperature in accordance with Test Method C 1064.

NOTE 4—Some specifications may require the measurement of the unit weight of concrete. The volume of concrete produced per batch may be desired on some projects. Also, additional information on the air content measurements may be desired. Test Method C 138 is used to measure the unit weight, yield, and gravimetric air content of freshly mixed concrete.

8. Molding Specimens

8.1 *Place of Molding*—Mold specimens promptly on a level, rigid surface, free of vibration and other disturbances, at a place as near as practicable to the location where they are to be stored.

8.2 *Casting Cylinders*—From Tables 1, 2, and 3, determine method of consolidation, the number and approximate depth of the layers and number of roddings per layer. If consolidation is by rodding, select the size of the tamping rod from Table 2. If consolidation is by internal vibration, select the proper vibrator to meet the requirements in 4.5. Select a small tool, such as a scoop, blunted trowel, or shovel, of a size and shape large enough so each amount of concrete obtained from the sampling receptacle will be representative and small enough so concrete is not lost when being placed in the mold. While placing the concrete in the mold, move the small tool around the perimeter of the mold opening to ensure an even distribution of the concrete and minimize segregation. Each layer of concrete shall be consolidated as required. In placing the final layer, add an amount of concrete that will fill the mold after consolidation. Underfilled or overfilled molds shall be adjusted with representative concrete during consolidation of the top layer.

8.3 *Casting Beams*—From Tables 2 and 3 determine the method of consolidation, the approximate depth, and number of layers. If consolidation is by rodding use the ⅝-in. [16-mm] tamping rod. Determine the number of roddings

TABLE 3 Method of Consolidation Requirements

NOTE—Use method of consolidation in Table 3 unless another is specified.

Slump in. (mm)	Method of Consolidation
>3 (75)	rodding
1 to 3 (25 to 75)	rodding or vibration
<1 (25)	vibration

per layer, one for each 2 in.² [14 cm²] of the top surface area of the beam. If consolidation is by internal vibration, select the proper vibrator to meet the requirements in 4.5. Select a small tool, such as a scoop, blunted trowel, or shovel, of the size and shape large enough so each amount of concrete obtained from the sampling receptacle is representative and small enough so concrete is not lost when placed in the mold. While placing the concrete in the mold, move the small tool around the opening to ensure even distribution of the concrete and minimize segregation. Each layer shall be consolidated as required. In placing the final layer, add an amount of concrete that will fill the mold after consolidation. Underfilled or overfilled molds shall be adjusted with representative concrete during consolidation of the top layer.

8.4 *Consolidation*—The methods of consolidation for this practice are rodding or internal vibration.

8.4.1 *Rodding*—Place the concrete in the mold, in the required number of layers of approximately equal volume. Rod each layer with the rounded end of the rod using the required number of roddings. Rod the bottom layer throughout its depth. Distribute the roddings uniformly over the cross section of the mold. For each upper layer allow the rod to penetrate about ½ in. [12 mm] into the underlying layer when the depth of the layer is less than 4 in. [100 mm], and about 1 in. [25 mm] when the depth is 4 in. [100 mm] or more. After each layer is rodded, tap the outsides of the mold lightly 10 to 15 times with the mallet, to close any holes left by rodding and to release any large air bubbles that may have been trapped. Use an open hand to tap light-gage single-use cylinder molds which are susceptible to damage if tapped with a mallet. After tapping, spade the concrete along the sides and ends of beam molds with a trowel or other suitable tool.

8.4.2 *Vibration*—Maintain a uniform time period for duration of vibration for the particular kind of concrete, vibrator, and specimen mold involved. The duration of vibration required will depend upon the workability of the concrete and the effectiveness of the vibrator. Usually sufficient vibration has been applied as soon as the surface of the concrete has become relatively smooth. Continue vibration only long enough to achieve proper consolidation of the concrete. Overvibration may cause segregation. Fill the molds and vibrate in the required number of approximately equal layers. Place all the concrete for each layer in the mold before starting vibration of that layer. In compacting the specimen, the vibrator shall not be allowed to rest on the bottom or sides of the mold. Carefully withdraw the vibrator in such

⬡ C 31/C 31M

a manner that no air pockets are left in the specimen. When placing the final layer, avoid overfilling by more than ¼ in. [6 mm].

8.4.2.1 *Cylinders*—Use three insertions of the vibrator at different points for each layer. Allow the vibrator to penetrate through the layer being vibrated, and into the layer below, approximately 1 in. [25 mm]. After each layer is vibrated, tap the outsides of the mold lightly 10 to 15 times with the mallet, to close any holes that remain and to release any large air bubbles that may have been trapped. Use an open hand to tap light-gage single-use molds which are susceptible to damage if tapped with a mallet.

8.4.2.2 *Beams*—Insert the vibrator at intervals not exceeding 6 in. [150 mm] along the center line of the long dimension of the specimen. For specimens wider than 6 in., use alternating insertions along two lines. Allow the shaft of the vibrator to penetrate into the bottom layer approximately 1 in. (25 mm). After each layer is vibrated, tap the outsides of the mold lightly 10 to 15 times with the mallet to close any holes left by vibrating and to release any large air bubbles that may have been trapped.

8.5 *Finishing*—After consolidation, strike off excess concrete from the surface and float or trowel as required. Perform all finishing with the minimum manipulation necessary to produce a flat even surface that is level with the rim or edge of the mold and that has no depressions or projections larger than ⅛ in. [3.3 mm].

8.5.1 *Cylinders*—After consolidation, finish the top surfaces by striking them off with the tamping rod where the consistency of the concrete permits or with a wood float or trowel. If desired, cap the top surface of freshly made cylinders with a thin layer of stiff portland cement paste which is permitted to harden and cure with the specimen. See section on Capping Materials of Practice C 617.

8.5.2 *Beams*—After consolidation of the concrete, use a hand-held float to strike off the top surface to the required tolerance to produce a flat, even surface.

8.6 *Identification*—Mark the specimens to positively identify them and the concrete they represent. Use a method that will not alter the top surface of the concrete. Do not mark the removable caps. Upon removal of the molds, mark the test specimens to retain their identities.

9. Curing

9.1 *Protection*—Immediately after finishing, precautions shall be taken to prevent evaporation and loss of water from the specimens. Protect the outside surfaces of cardboard molds from contact with wet burlap or other sources of water. Cover specimens with a nonabsorbent, nonreactive plate or sheet of impervious plastic. When wetted burlap is used over the plate or plastic sheet to help retard evaporation, the burlap must not be in contact with the surface of the concrete.

9.2 *Standard Curing*—Standard curing is the curing method used when the specimens are made and cured for the purposes stated in 3.2.

9.2.1 *Storage*—Immediately after finishing move the specimens to an initial curing place for storage (Note 5). If cylinders in the single use molds are moved, lift and support the cylinders from the bottom of the molds with a large trowel or similar device. If the top surface is marred during

movement to place of initial storage, immediately refinish.

9.2.2 *Initial Curing*—After molding, the specimens shall be stored in a temperature range between 60 to 80°F [16 to 27°C] and in a moist environment preventing any loss of moisture up to 48 h (Note 5). At all times the temperature in and between specimens shall be controlled by shielding from direct rays of the sun and radiant heating devices. Specimens that are to be transported to the laboratory for final curing of Section 9.2.3 before 48 h shall remain in the molds in a moist environment, until they are received in the laboratory, demolded and placed in final curing. If specimens are not transported within 48 h, the molds shall be removed within 24 ± 8 h and final curing used until transported (see 10.1).

NOTE 5—It may be necessary to create an environment during initial curing to provide satisfactory moisture and to control the temperature. The specimens may be immersed immediately in saturated limewater, and/or stored in tightly constructed wooden boxes, damp sand pits, temporary buildings at construction sites, under wet burlap, or in heavyweight closed plastic bags. Immersing in saturated limewater is not acceptable for specimens in cardboard or other molds that expand when immersed in water. Other suitable methods may be used provided the foregoing requirements limiting specimen temperature and moisture loss are met. The temperature may be controlled by ventilation, or thermostatically controlled cooling devices, or by heating devices such as stoves, light bulbs, or thermostatically controlled heating elements. Temperature record of the specimens may be established by means of maximum-minimum thermometers. Early age results may be lower when stored near 60°F [16°C] and higher when stored near 80°F [27°C].

9.2.3 *Final Curing:*

9.2.3.1 *Cylinders*—Upon completion of initial curing and within 30 min after removing the molds, store specimens in a moist condition with free water maintained on their surfaces at all times at a temperature of 73 ± 3°F [23 ± 2°C]. Temperatures between 68 and 86°F [20 and 30°C] are permitted for a period not to exceed 3 h immediately prior to test if free moisture is maintained on the surfaces of the specimen at all times, except when capping with sulfur mortar capping compound. When capping with this material, the ends of the cylinder will be dried as described in Practice C 617. Specimens shall not be exposed to dripping or running water. The curing requirements for water storage tanks, moist rooms and cabinets are in Specification C 511.

9.2.3.2 *Beams*—Beams are to be cured the same as cylinders (see 9.2.3.1) except for a minimum of 20 h prior to testing, they shall be stored in water saturated with calcium hydroxide at 73 ± 3°F [23 ± 2°C]. Drying of the surfaces of the beam shall be prevented between removal from limewater and completion of testing.

NOTE 6—Relatively small amounts of surface drying of flexural specimens can induce tensile stresses in the extreme fibers that will markedly reduce the indicated flexural strength.

9.3 *Field Curing*—Field curing is the curing method used for the specimens made and cured as stated in 3.3.

9.3.1 *Cylinders*—Store cylinders in or on the structure as near to the point of deposit of the concrete represented as possible. Protect all surfaces of the cylinders from the elements in as near as possible the same way as the formed work. Provide the cylinders with the same temperature and

C 31/C 31M

moisture environment as the structural work. Test the specimens in the moisture condition resulting from the specified curing treatment. To meet these conditions, specimens made for the purpose of determining when a structure is permitted to be put in service shall be removed from the molds at the time of removal of form work.

9.3.2 *Beams*—As nearly as practicable, cure beams in the same manner as the concrete in the structure. At the end of 48 ± 4 h after molding, take the molded specimens to the storage location and remove from the molds. Store specimens representing pavements of slabs on grade by placing them on the ground as molded, with their top surfaces up. Bank the sides and ends of the specimens with earth or sand that shall be kept damp, leaving the top surfaces exposed to the specified curing treatment. Store specimens representing structure concrete as near the point in the structure they represent as possible, and afford them the same temperature protection and moisture environment as the structure. At the end of the curing period leave the specimens in place exposed to the weather in the same manner as the structure. Remove all beam specimens from field storage and store in limewater at 73 ± 3°F [23 ± 2°C] for 24 ± 4 h immediately before time of testing to ensure uniform moisture condition from specimen to specimen. Observe the precautions given in 9.2.3.2 to guard against drying between time of removal from curing to testing.

10. Transportation of Specimens to Laboratory

10.1 Prior to transporting, specimens shall be cured and protected as required in Section 9. During transportation, the specimens must be protected with suitable cushioning material to prevent damage from jarring and from freezing temperatures, or moisture loss. Excessive moisture loss is prevented by securely wrapping the specimens in plastic or surrounding them with wet sand or wet saw dust. The time for transportation shall not exceed 4 h.

11. Report

11.1 Report the following information to the laboratory that will test the specimens:

11.1.1 Identification number,

11.1.2 Location of concrete represented by the samples,

11.1.3 Date, time and name of individual molding specimens,

11.1.4 Slump, air content, and concrete temperature, test results and results of any other tests on the fresh concrete and any deviations from referenced standard test methods, and

11.1.5 Curing method.

12. Keywords

12.1 beams; casting samples; concrete; curing; cylinders; testing

Designation: C 78 – 94

Standard Test Method for
Flexural Strength of Concrete (Using Simple Beam with Third-Point Loading)[1]

This standard is issued under the fixed designation C 78; the number immediately following the designation indicates the year of original adoption or, in the case of revision, the year of last revision. A number in parentheses indicates the year of last reapproval. A superscript epsilon (ε) indicates an editorial change since the last revision or reapproval.

This test method has been approved for use by agencies of the Department of Defense. Consult the DoD Index of Specifications and Standards for the specific year of issue which has been adopted by the Department of Defense.

1. Scope

1.1 This test method covers the determination of the flexural strength of concrete by the use of a simple beam with third-point loading.

1.2 The values stated in inch-pound units are to be regarded as the standard. The SI equivalent of inch-pound units has been rounded where necessary for practical application.

1.3 *This standard does not purport to address all of the safety concerns, if any, associated with its use. It is the responsibility of the user of this standard to establish appropriate safety and health practices and determine the applicability of regulatory limitations prior to use.*

2. Referenced Documents

2.1 *ASTM Standards:*
C 31 Practice for Making and Curing Concrete Test Specimens in the Field[2]
C 42 Test Method for Obtaining and Testing Drilled Cores and Sawed Beams of Concrete[2]
C 192 Practice for Making and Curing Concrete Test Specimens in the Laboratory[2]
C 617 Practice for Capping Cylindrical Concrete Specimens[2]
C 1077 Practice for Laboratories Testing Concrete and Concrete Aggregates for Use in Construction and Criteria for Laboratory Evaluation[2]
E 4 Practices for Force Verification of Testing Machines[3]

3. Significance and Use

3.1 This test method is used to determine the flexural strength of specimens prepared and cured in accordance with Test Methods C 42 or Practices C 31 or C 192. Results are calculated and reported as the modulus of rupture. The strength determined will vary where there are differences in specimen size, preparation, moisture condition, curing, or where the beam has been molded or sawed to size.

3.2 The results of this test method may be used to determine compliance with specifications or as a basis for

proportioning, mixing and placement operations. It is used in testing concrete for the construction of slabs and pavements (Note 1).

4. Apparatus

4.1 The testing machine shall conform to the requirements of the sections on Basis of Verification, Corrections, and Time Interval Between Verifications of Practices E 4. Hand operated testing machines having pumps that do not provide a continuous loading in one stroke are not permitted. Motorized pumps or hand operated positive displacement pumps having sufficient volume in one continuous stroke to complete a test without requiring replenishment are permitted and shall be capable of applying loads at a uniform rate without shock or interruption.

4.2 *Loading Apparatus*—The third point loading method shall be used in making flexure tests of concrete employing bearing blocks which will ensure that forces applied to the beam will be perpendicular to the face of the specimen and applied without eccentricity. A diagram of an apparatus that accomplishes this purpose is shown in Fig. 1.

4.2.1 All apparatus for making flexure tests of concrete shall be capable of maintaining the specified span length and distances between load-applying blocks and support blocks constant within ±0.05 in. (±1.3 mm).

4.2.2 Reactions shall be parallel to the direction of the applied forces at all times during the test and the ratio of distance between the point of load application and nearest reaction to the depth of the beam shall not be less than one.

4.2.3 If an apparatus similar to that illustrated in Fig. 1 is used: the load-applying and support blocks should not be more than 2½ in. (64 mm) high, measured from the center or the axis of pivot, and should extend entirely across or beyond the full width of the specimen. Each case-hardened bearing surface in contact with the specimen shall not depart from a plane by more than 0.002 in. (0.05 mm) and shall be a portion of a cylinder, the axis of which is coincidental with either the axis of the rod or center of the ball, whichever the block is pivoted upon. The angle subtended by the curved surface of each block should be at least 45° (0.79 rad). The load-applying and support blocks shall be maintained in a vertical position and in contact with the rod or ball by means of spring-loaded screws that hold them in contact with the pivot rod or ball. The uppermost bearing plate and center point ball in Fig. 1 may be omitted when a spherically seated bearing block is used, provided one rod and one ball are used as pivots for the upper load-applying blocks.

[1] This test method is under the jurisdiction of ASTM Committee C-9 on Concrete and Concrete Aggregates and is the direct responsibility of Subcommittee C09.61 on Testing Concrete for Strength.
Current edition approved April 15, 1994. Published June 1994. Originally published as C 78 – 30T. Last previous edition C 78 – 84.
[2] *Annual Book of ASTM Standards*, Vol 04.02.
[3] *Annual Book of ASTM Standards*, Vol 03.01.

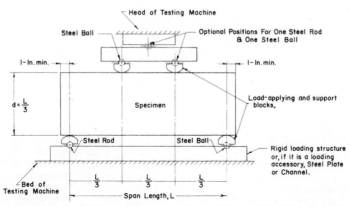

NOTE 1—This apparatus may be used inverted. If the testing machine applies force through a spherically seated head, the center pivot may be omitted, provided one load-applying block pivots on a rod and the other on a ball.

NOTE 2—1 in. = 25.4 mm.

FIG. 1 Diagrammatic View of a Suitable Apparatus for Flexure Test of Concrete by Third-Point Loading Method

5. Testing

5.1 The test specimen shall conform to all requirements of Test Method C 42 or Practices C 31 or C 192 applicable to beam and prism specimens and shall have a test span within 2 % of being three times its depth as tested. The sides of the specimen shall be at right angles with the top and bottom. All surfaces shall be smooth and free of scars, indentations, holes, or inscribed identification marks.

5.2 The technician performing the flexural strength test should be certified as an ACI Technician—Grade II, or by an equivalent written and performance test program.

NOTE 1—The testing laboratory performing this test method may be evaluated in accordance with Practice C 1077.

6. Procedure

6.1 Flexural tests of moist-cured specimens shall be made as soon as practical after removal from moist storage. Surface drying of the specimen results in a reduction in the measured flexural strength.

6.2 When using molded specimens, turn the test specimen on its side with respect to its position as molded and center it on the support blocks. When using sawed specimens, position the specimen so that the tension face corresponds to the top or bottom of the specimen as cut from the parent material. Center the loading system in relation to the applied force. Bring the load-applying blocks in contact with the surface of the specimen at the third points and apply a load of between 3 and 6 % of the estimated ultimate load. Using 0.004 in. (0.10 mm) and 0.015 in. (0.38 mm) leaf-type feeler gages, determine whether any gap between the specimen and the load-applying or support blocks is greater or less than each of the gages over a length of 1 in. (25 mm) or more. Grind, cap, or use leather shims on the specimen contact surface to eliminate any gap in excess of 0.004 in. (0.10 mm) in width. Leather shims shall be of uniform ¼ in. (6.4 mm) thickness, 1 to 2 in. (25 to 50 mm) width, and shall extend across the full width of the specimen. Gaps in excess of 0.015 in. (0.38 mm) shall be eliminated only by capping or

grinding. Grinding of lateral surfaces should be minimized inasmuch as grinding may change the physical characteristics of the specimens. Capping shall be in accordance with the applicable sections of Practice C 617.

6.3 Load the specimen continuously and without shock. The load shall be applied at a constant rate to the breaking point. Apply the load at a rate that constantly increases the extreme fiber stress between 125 and 175 psi/min (0.86 and 1.21 MPa/min), when calculated in accordance with 8.1, until rupture occurs.

7. Measurement of Specimens After Test

7.1 Take three measurements across each dimension (one at each edge and at the center) to the nearest 0.05 in. (1 mm) to determine the average width, average depth, and line of fracture location of the specimen at the section of failure. If fracture occurs at a capped section, include the cap thickness in measurement.

8. Calculation

8.1 If the fracture initiates in the tension surface within the middle third of the span length, calculate the modulus of rupture as follows:

$$R = PL/bd^2$$

where:

R = modulus of rupture, psi, or MPa,

P = maximum applied load indicated by the testing machine, lbf, or N,

L = span length, in., or mm,

b = average width of specimen, in., or mm, at the fracture, and

d = average depth of specimen, in., or mm, at the fracture.

NOTE 2—The weight of the beam is not included in the above calculation.

8.2 If the fracture occurs in the tension surface outside of the middle third of the span length by not more than 5 % of the span length, calculate the modulus of rupture as follows:

⚙ C 78

$$R = 3Pa/bd^2$$

where:

a = average distance between line of fracture and the nearest support measured on the tension surface of the beam, in., (or mm).

NOTE 3—The weight of the beam is not included in the above calculation.

8.3 If the fracture occurs in the tension surface outside of the middle third of the span length by more than 5 % of the span length, discard the results of the test.

9. Report

9.1 Report the following information:

9.1.1 Identification number,

9.1.2 Average width to the nearest 0.05 in. (1 mm),

9.1.3 Average depth to the nearest 0.05 in. (1 mm),

9.1.4 Span length in inches (or millimeters),

9.1.5 Maximum applied load in pound-force (or newtons),

9.1.6 Modulus of rupture calculated to the nearest 5 psi (0.05 MPa),

9.1.7 Curing history and apparent moisture condition of the specimens at the time of test,

9.1.8 If specimens were capped, ground, or if leather shims were used,

9.1.9 Whether sawed or molded and defects in specimens, and

9.1.10 Age of specimens.

10. Precision and Bias

10.1 *Precision*—The coefficient of variation of test results has been observed to be dependent on the strength level of the beams.[4] The single operator coefficient of variation has been found to be 5.7 %. Therefore, results of two properly conducted tests by the same operator on beams made from the same batch sample should not differ from each other by more than 16 %. The multilaboratory coefficient of variation has been found to be 7.0 %. Therefore, results of two different laboratories on beams made from the same batch sample should not differ from each other by more than 19 %.

10.2 *Bias*—Since there is no accepted standard for determining bias in this test method, no statement on bias is made.

11. Keywords

11.1 beams; concrete; flexural strength testing; modulus of rupture

---

[4] See "Improved Concrete Quality Control Procedures Using Third Point Loading" by P. M. Carrasquillo and R. L. Carrasquillo, Research Report 119-1F, Project 3-9-87-1119, Center For Transportation Research, The University of Texas at Austin, November 1987, for possible guidance as to the relationship of strength and variability.

 **Designation: C 128 – 93**

AMERICAN SOCIETY FOR TESTING AND MATERIALS
100 Barr Harbor Dr., West Conshohocken, PA 19428
Reprinted from the Annual Book of ASTM Standards. Copyright ASTM
If not listed in the current combined index, will appear in the next edition.

Standard Test Method for
Specific Gravity and Absorption of Fine Aggregate[1]

This standard is issued under the fixed designation C 128; the number immediately following the designation indicates the year of original adoption or, in the case of revision, the year of last revision. A number in parentheses indicates the year of last reapproval. A superscript epsilon (ϵ) indicates an editorial change since the last revision or reapproval.

This standard has been approved for use by agencies of the Department of Defense. Consult the DoD Index of Specifications and Standards for the specific year of issue which has been adopted by the Department of Defense.

1. Scope

1.1 This test method covers the determination of bulk and apparent specific gravity, 23/23°C (73.4/73.4°F), and absorption of fine aggregate.

1.2 This test method determines (after 24 h in water) the bulk specific gravity and the apparent specific gravity as defined in Terminology E 12, the bulk specific gravity on the basis of weight of saturated surface-dry aggregate, and absorption as defined in Definitions C 125.

NOTE 1—The subcommittee is considering revising Test Methods C 127 and C 128 to use the term "density" instead of "specific gravity" for coarse and fine aggregate, respectively.

1.3 The values stated in SI units are to be regarded as the standard.

1.4 *This standard does not purport to address all of the safety problems, if any, associated with its use. It is the responsibility of the user of this standard to establish appropriate safety and health practices and determine the applicability of regulatory limitations prior to use.*

2. Referenced Documents

2.1 *ASTM Standards:*
C 29/C 29M Test Method for Unit Weight and Voids in Aggregate[2,3]
C 70 Test Method for Surface Moisture in Fine Aggregate[2]
C 125 Terminology Relating to Concrete and Concrete Aggregates[2,3]
C 127 Test Method for Specific Gravity and Absorption of Coarse Aggregate[2,3]
C 188 Test Method for Density of Hydraulic Cement[4]
C 566 Test Method for Total Moisture Content of Aggregate by Drying[2]
C 670 Practice for Preparing Precision and Bias Statements for Test Methods for Construction Materials[2,3,4]
C 702 Practice for Reducing Field Samples of Aggregate to Testing Size[2]
D 75 Practices for Sampling Aggregates[2,3]
E 12 Terminology Relating to Density and Specific Gravity of Solids, Liquids, and Gases[2,5]

E 380 Practice for Use of the International System of Units (SI) (the Modernized Metric System)[6]
2.2 *AASHTO Standard:*
AASHTO No. T 84 Specific Gravity and Absorption of Fine Aggregates[7]

3. Significance and Use

3.1 Bulk specific gravity is the characteristic generally used for calculation of the volume occupied by the aggregate in various mixtures containing aggregate including portland cement concrete, bituminous concrete, and other mixtures that are proportioned or analyzed on an absolute volume basis. Bulk specific gravity is also used in the computation of voids in aggregate in Test Method C 29 and the determination of moisture in aggregate by displacement in water in Test Method C 70. Bulk specific gravity determined on the saturated surface-dry basis is used if the aggregate is wet, that is, if its absorption has been satisfied. Conversely, the bulk specific gravity determined on the oven-dry basis is used for computations when the aggregate is dry or assumed to be dry.

3.2 Apparent specific gravity pertains to the relative density of the solid material making up the constituent particles not including the pore space within the particles that is accessible to water. This value is not widely used in construction aggregate technology.

3.3 Absorption values are used to calculate the change in the weight of an aggregate due to water absorbed in the pore spaces within the constituent particles, compared to the dry condition, when it is deemed that the aggregate has been in contact with water long enough to satisfy most of the absorption potential. The laboratory standard for absorption is that obtained after submerging dry aggregate for approximately 24 h in water. Aggregates mined from below the water table may have a higher absorption when used, if not allowed to dry. Conversely, some aggregates when used may contain an amount of absorbed moisture less than the 24 h-soaked condition. For an aggregate that has been in contact with water and that has free moisture on the particle surfaces, the percentage of free moisture can be determined by deducting the absorption from the total moisture content determined by Test Method C 566 by drying.

---

[1] This test method is under the jurisdiction of ASTM Committee C-9 on Concrete and Concrete Aggregates and is the direct responsibility of Subcommittee C09.20 on Normal Weight Aggregates.
Current edition approved April 15, 1993. Published June 1993. Originally published as C 128 – 36. Last previous edition C 128 – 88.
[2] *Annual Book of ASTM Standards*, Vol 04.02.
[3] *Annual Book of ASTM Standards*, Vol 04.03.
[4] *Annual Book of ASTM Standards*, Vol 04.01.
[5] *Annual Book of ASTM Standards*, Vol 15.05.

[6] *Annual Book of ASTM Standards*, Vol 14.02. Excerpts in all volumes.
[7] Available from American Association of State Highway and Transportation Officials, 444 North Capitol St. N.W., Suite 225, Washington, DC 20001.

4. Apparatus

4.1 *Balance*—A balance or scale having a capacity of 1 kg or more, sensitive to 0.1 g or less, and accurate within 0.1 % of the test load at any point within the range of use for this test. Within any 100-g range of test load, a difference between readings shall be accurate within 0.1 g.

4.2 *Pycnometer*—A flask or other suitable container into which the fine aggregate test sample can be readily introduced and in which the volume content can be reproduced within ±0.1 cm³. The volume of the container filled to mark shall be at least 50 % greater than the space required to accommodate the test sample. A volumetric flask of 500 cm³ capacity or a fruit jar fitted with a pycnometer top is satisfactory for a 500-g test sample of most fine aggregates. A Le Chatelier flask as described in Test Method C 188 is satisfactory for an approximately 55-g test sample.

4.3 *Mold*—A metal mold in the form of a frustum of a cone with dimensions as follows: 40 ± 3 mm inside diameter at the top, 90 ± 3 mm inside diameter at the bottom, and 75 ± 3 mm in height, with the metal having a minimum thickness of 0.8 mm.

4.4 *Tamper*—A metal tamper weighing 340 ± 15 g and having a flat circular tamping face 25 ± 3 mm in diameter.

5. Sampling

5.1 Sampling shall be accomplished in general accordance with Practice D 75.

6. Preparation of Test Specimen

6.1 Obtain approximately 1 kg of the fine aggregate from the sample using the applicable procedures described in Practice C 702.

6.1.1 Dry it in a suitable pan or vessel to constant weight at a temperature of 110 ± 5°C (230 ± 9°F). Allow it to cool to comfortable handling temperature, cover with water, either by immersion or by the addition of at least 6 % moisture to the fine aggregate, and permit to stand for 24 ± 4 h.

6.1.2 As an alternative to 6.1.1, where the absorption and specific gravity values are to be used in proportioning concrete mixtures with aggregates used in their naturally moist condition, the requirement for initial drying to constant weight may be eliminated and, if the surfaces of the particles have been kept wet, the 24-h soaking may also be eliminated.

NOTE 2—Values for absorption and for specific gravity in the saturated surface-dry condition may be significantly higher for aggregate not oven dried before soaking than for the same aggregate treated in accordance with 6.1.1.

6.2 Decant excess water with care to avoid loss of fines, spread the sample on a flat nonabsorbent surface exposed to a gently moving current of warm air, and stir frequently to secure homogeneous drying. If desired, mechanical aids such as tumbling or stirring may be employed to assist in achieving the saturated surface-dry condition. Continue this operation until the test specimen approaches a free-flowing condition. Follow the procedure in 6.2.1 to determine whether or not surface moisture is present on the constituent fine aggregate particles. It is intended that the first trial of the cone test will be made with some surface water in the specimen. Continue drying with constant stirring and test at frequent intervals until the test indicates that the specimen

has reached a surface-dry condition. If the first trial of the surface moisture test indicates that moisture is not present on the surface, it has been dried past the saturated surface-dry condition. In this case thoroughly mix a few millilitres of water with the fine aggregate and permit the specimen to stand in a covered container for 30 min. Then resume the process of drying and testing at frequent intervals for the onset of the surface-dry condition.

6.2.1 *Cone Test for Surface Moisture*—Hold the mold firmly on a smooth nonabsorbent surface with the large diameter down. Place a portion of the partially dried fine aggregate loosely in the mold by filling it to overflowing and heaping additional material above the top of the mold by holding it with the cupped fingers of the hand holding the mold. Lightly tamp the fine aggregate into the mold with 25 light drops of the tamper. Each drop should start about 5 mm (0.2 in.) above the top surface of the fine aggregate. Permit the tamper to fall freely under gravitational attraction on each drop. Adjust the starting height to the new surface elevation after each drop and distribute the drops over the surface. Remove loose sand from the base and lift the mold vertically. If surface moisture is still present, the fine aggregate will retain the molded shape. When the fine aggregate slumps slightly it indicates that it has reached a surface-dry condition. Some angular fine aggregate or material with a high proportion of fines may not slump in the cone test upon reaching a surface-dry condition. This may be the case if fines become airborne upon dropping a handful of the sand from the cone test 100 to 150 mm onto a surface. For these materials the saturated surface-dry condition should be considered as the point that one side of the fine aggregate slumps slightly upon removing the mold.

NOTE 3—The following criteria have also been used on materials that do not readily slump:

(1) *Provisional Cone Test*—Fill the cone mold as described in 6.2.1 except only use 10 drops of the tamper. Add more fine aggregate and use 10 drops of the tamper again. Then add material two more times using 3 and 2 drops of the tamper, respectively. Level off the material even with the top of the mold, remove loose material from the base; and lift the mold vertically.

(2) *Provisional Surface Test*—If airborne fines are noted when the fine aggregate is such that it will not slump when it is at a moisture condition, add more moisture to the sand, and at the onset of the surface-dry condition, with the hand lightly pat approximately 100 g of the material on a flat, dry, clean, dark or dull nonabsorbent surface such as a sheet of rubber, a worn oxidized, galvanized, or steel surface, or a black-painted metal surface. After 1 to 3 s remove the fine aggregate. If noticeable moisture shows on the test surface for more than 1 to 2 s then surface moisture is considered to be present on the fine aggregate.

(3) Colorimetric procedures described by Kandhal and Lee, Highway Research Record No. 307, p. 44.

(4) For reaching the saturated surface-dry condition on a single size material that slumps when wet, hard-finish paper towels can be used to surface dry the material until the point is just reached where the paper towel does not appear to be picking up moisture from the surfaces of the fine aggregate particles.

7. Procedure

7.1 Make and record all weight determinations to 0.1 g.

7.2 Partially fill the pycnometer with water. Immediately introduce into the pycnometer 500 ± 10 g of saturated surface-dry fine aggregate prepared as described in Section 6, and fill with additional water to approximately 90 % of capacity. Roll, invert, and agitate the pycnometer to elimi-

C 128

nate all air bubbles (Note 4). Adjust its temperature to 23 ± 1.7°C (73.4 ± 3°F), if necessary by immersion in circulating water, and bring the water level in the pycnometer to its calibrated capacity. Determine the total weight of the pycnometer, specimen, and water.

NOTE 4—It normally takes about 15 to 20 min to eliminate air bubbles. Dipping the tip of a paper towel into the pycnometer has been found to be useful in dispersing the foam that sometimes builds up when eliminating the air bubbles. Optionally, a small amount of isopropyl alcohol may be used to disperse the foam. Do *not* use either of these procedures when using the alternate method described in 7.2.1.

7.2.1 *Alternative to Weighing in 7.2*—The quantity of added water necessary to fill the pycnometer at the required temperature may be determined volumetrically using a buret accurate to 0.15 mL. Compute the total weight of the pycnometer, specimen, and water as follows:

$$C = 0.9975\, V_a + S + W \qquad (1)$$

where:
C = weight of pycnometer with specimen and water to calibration mark, g,
V_a = volume of water added to pycnometer, mL,
S = weight of the saturated surface-dry specimen, and
W = weight of the empty pycnometer, g.

7.2.2 *Alternative to the Procedure in 7.2*—Use a Le Chatelier flask initially filled with water to a point on the stem between the 0 and the 1-mL mark. Record this initial reading with the flask and contents within the temperature range of 23 ± 1.7°C (73.4 ± 3°F). Add 55 ± 5 g of fine aggregate in the saturated surface-dry condition (or other weight as necessary to result in raising the water level to some point on the upper series of gradation). After all fine aggregate has been introduced, place the stopper in the flask and roll the flask in an inclined position, or gently whirl it in a horizontal circle so as to dislodge all entrapped air, continuing until no further bubbles rise to the surface (Note 5). Take a final reading with the flask and contents within 1°C (1.8°F) of the original temperature.

NOTE 5—When using the Le Chatelier flask method, the operator may use a small measured amount (not to exceed 1 mL) of isopropyl alcohol to eliminate foam appearing on the water surface. The volume of alcohol used must be subtracted from the final reading (R_2).

7.3 Remove the fine aggregate from the pycnometer, dry to constant weight at a temperature of 110 ± 5°C (230 ± 9°F), cool in air at room temperature for 1 ± ½ h, and weigh.

7.3.1 If the Le Chatelier flask method is used, a separate sample portion is needed for the determination of absorption. Weigh a separate 500 ± 10-g portion of the saturated surface-dry fine aggregate, dry to constant weight, and reweigh.

7.4 Determine the weight of the pycnometer filled to its calibration capacity with water at 23 ± 1.7°C (73.4 ± 3°F).

7.4.1 *Alternative to Weighing in 7.4*—The quantity of water necessary to fill the empty pycnometer at the required temperature may be determined volumetrically using a buret accurate to 0.15 mL. Calculate the weight of the pycnometer filled with water as follows:

$$B = 0.9975\, V + W \qquad (2)$$

where:
B = weight of flask filled with water, g,

V = volume of flask, mL, and
W = weight of the flask empty, g.

8. Bulk Specific Gravity

8.1 Calculate the bulk specific gravity, 23/23°C (73.4/73.4°F), as defined in Terminology E 12, as follows:

$$\text{Bulk sp gr} = A/(B + S - C) \qquad (3)$$

where:
A = weight of oven-dry specimen in air, g,
B = weight of pycnometer filled with water, g,
S = weight of the saturated surface-dry specimen, and
C = weight of pycnometer with specimen and water to calibration mark, g.

8.1.1 If the Le Chatelier flask method was used, calculate the bulk specific gravity, 23/23°C, as follows:

$$\text{Bulk sp gr} = \frac{S_1 (A/S)}{0.9975\,(R_2 - R_1)} \qquad (4)$$

where:
S_1 = weight of saturated surface-dry specimen used in Le Chatelier flask, g,
R_1 = initial reading of water level in Le Chatelier flask, and
R_2 = final reading of water level in Le Chatelier flask.

9. Bulk Specific Gravity (Saturated Surface-Dry Basis)

9.1 Calculate the bulk specific gravity, 23/23°C (73.4/73.4°F), on the basis of weight of saturated surface-dry aggregate as follows:

$$\text{Bulk sp gr (saturated surface-dry basis)} = S/(B + S - C) \qquad (5)$$

9.1.1 If the Le Chatelier flask method was used, calculate the bulk specific gravity, 23/23°C, on the basis of saturated surface-dry aggregate as follows:

$$\text{Bulk sp gr (saturated surface-dry basis)} = \frac{S_1}{0.9975\,(R_2 - R_1)} \qquad (6)$$

10. Apparent Specific Gravity

10.1 Calculate the apparent specific gravity, 23/23°C (73.4/73.4°F), as defined in Terminology E 12, as follows:

$$\text{Apparent sp gr} = A/(B + A - C) \qquad (7)$$

11. Absorption

11.1 Calculate the percentage of absorption, as defined in Terminology C 125, as follows:

$$\text{Absorption, \%} = [(S - A)/A] \times 100 \qquad (8)$$

12. Report

12.1 Report specific gravity results to the nearest 0.01 and absorption to the nearest 0.1 %. The Appendix gives mathematical interrelationships among the three types of specific gravities and absorption. These may be useful in checking the consistency of reported data or calculating a value that was not reported by using other reported data.

12.2 If the fine aggregate was tested in a naturally moist condition other than the oven dried and 24 h-soaked condition, report the source of the sample and the procedures used to prevent drying prior to testing.

⬡ C 128

TABLE 1   Precision

	Standard Deviation (1S) [A]	Acceptable Range of Two Results (D2S) [A]
Single-Operator Precision:		
Bulk specific gravity (dry)	0.011	0.032
Bulk specific gravity (SSD)	0.0095	0.027
Apparent specific gravity	0.0095	0.027
Absorption [B], %	0.11	0.31
Multilaboratory Precision:		
Bulk specific gravity (dry)	0.023	0.066
Bulk specific gravity (SSD)	0.020	0.056
Apparent specific gravity	0.020	0.056
Absorption [B], %	0.23	0.66

[A] These numbers represent, respectively, the (1S) and (D2S) limits as described in Practice C 670. The precision estimates were obtained from the analysis of combined AASHTO Materials Research Laboratory reference sample data from laboratories using 15 to 19 h saturation times and other laboratories using 24 ± 4 h saturation time. Testing was performed on normal weight aggregates, and started with aggregates in the oven-dry condition.

[B] Precision estimates are based on aggregates with absorptions of less than 1 % and may differ for manufactured fine aggregates and fine aggregates having absorption values greater than 1 %.

13. Precision and Bias

13.1 *Precision*—The estimates of precision of this test method (listed in Table 1) are based on results from the AASHTO Materials Reference Laboratory Reference Sample Program, with testing conducted by this test method and AASHTO Method T 84. The significant difference between the methods is that Test Method C 128 requires a saturation period of 24 ± 4 h, and Method T 84 requires a saturation period of 15 to 19 h. This difference has been found to have an insignificant effect on the precision indices. The data are based on the analyses of more than 100 paired test results from 40 to 100 laboratories.

13.2 *Bias*—Since there is no accepted reference material suitable for determining the bias for this test method, no statement on bias is being made.

14. Keywords

14.1 absorption; aggregate; fine aggregate; specific gravity

APPENDIX

(Nonmandatory Information)

X1. INTERRELATIONSHIPS BETWEEN SPECIFIC GRAVITIES AND ABSORPTION AS DEFINED IN TEST METHODS C 127 AND C 128

X1.1 Let:
S_d = bulk specific gravity (dry-basis),
S_s = bulk specific gravity (SSD-basis),
S_a = apparent specific gravity, and
A = absorption in %.

Then:

(1)
$$S_s = (1 + A/100)S_d$$

(2)
$$S_a = \frac{1}{\dfrac{1}{S_d} - \dfrac{A}{100}} = \frac{S_d}{1 - \dfrac{AS_d}{100}}$$

(2a)
$$\text{or } S_a = \frac{1}{\dfrac{1 + A/100}{S_s} - \dfrac{A}{100}}$$
$$= \frac{S_s}{1 - \dfrac{A}{100}(S_s - 1)}$$

(3)
$$A = \left(\frac{S_s}{S_d} - 1\right)100$$

(4)
$$A = \left(\frac{S_a - S_s}{S_a(S_s - 1)}\right)100$$

⟨ASTM⟩ C 128

Appendices

 **Designation: C 136 – 96a**

AMERICAN SOCIETY FOR TESTING AND MATERIALS
100 Barr Harbor Dr., West Conshohocken, PA 19428
Reprinted from the Annual Book of ASTM Standards. Copyright ASTM
If not listed in the current combined index, will appear in the next edition.

Standard Test Method for
Sieve Analysis of Fine and Coarse Aggregates[1]

This standard is issued under the fixed designation C 136; the number immediately following the designation indicates the year of original adoption or, in the case of revision, the year of last revision. A number in parentheses indicates the year of last reapproval. A superscript epsilon (ϵ) indicates an editorial change since the last revision or reapproval.

This standard has been approved for use by agencies of the Department of Defense. Consult the DoD Index of Specifications and Standards for the specific year of issue which has been adopted by the Department of Defense.

1. Scope

1.1 This test method covers the determination of the particle size distribution of fine and coarse aggregates by sieving.

1.2 Some specifications for aggregates which reference this method contain grading requirements including both coarse and fine fractions. Instructions are included for sieve analysis of such aggregates.

1.3 The values stated in SI units are to be regarded as the standard. The values in parentheses are provided for information purposes only. Specification E 11 designates the size of sieve frames with inch units as standard, but in this test method the frame size is designated in SI units exactly equivalent to the inch units.

1.4 *This standard does not purport to address all of the safety concerns, if any, associated with its use. It is the responsibility of the user of this standard to establish appropriate safety and health practices and determine the applicability of regulatory limitations prior to use.*

2. Referenced Documents

2.1 *ASTM Standards:*
C 117 Test Method for Materials Finer Than 75-μm (No. 200) Sieve in Mineral Aggregates by Washing[2]
C 125 Terminology Relating to Concrete and Concrete Aggregates[2]
C 670 Practice for Preparing Precision and Bias Statements for Test Methods for Construction Materials[2]
C 702 Practice for Reducing Field Samples of Aggregate to Testing Size[2]
D 75 Practice for Sampling Aggregates[3]
E 11 Specification for Wire-Cloth Sieves for Testing Purposes[4]
2.2 *AASHTO Standard:*
AASHTO No. T 27 Sieve Analysis of Fine and Coarse Aggregates[5]

3. Terminology

3.1 *Definitions*—For definitions of terms used in this standard, refer to Terminology C 125.

4. Summary of Test Method

4.1 A sample of dry aggregate of known mass is separated through a series of sieves of progressively smaller openings for determination of particle size distribution.

5. Significance and Use

5.1 This test method is used primarily to determine the grading of materials proposed for use as aggregates or being used as aggregates. The results are used to determine compliance of the particle size distribution with applicable specification requirements and to provide necessary data for control of the production of various aggregate products and mixtures containing aggregates. The data may also be useful in developing relationships concerning porosity and packing.

5.2 Accurate determination of material finer than the 75-μm (No. 200) sieve cannot be achieved by use of this method alone. Test Method C 117 for material finer than 75-μm sieve by washing should be employed.

6. Apparatus

6.1 *Balances*—Balances or scales used in testing fine and coarse aggregate shall have readability and accuracy as follows:

6.1.1 For fine aggregate, readable to 0.1 g and accurate to 0.1 g or 0.1 % of the test load, whichever is greater, at any point within the range of use.

6.1.2 For coarse aggregate, or mixtures of fine and coarse aggregate, readable and accurate to 0.5 g or 0.1 % of the test load, whichever is greater, at any point within the range of use.

6.2 *Sieves*—The sieve cloth shall be mounted on substantial frames constructed in a manner that will prevent loss of material during sieving. The sieve cloth and standard sieve frames shall conform to the requirements of Specification E 11. Nonstandard sieve frames shall conform to the requirements of Specification E 11 as applicable.

NOTE 1—It is recommended that sieves mounted in frames larger than standard 203.2-mm (8 in.) diameter be used for testing coarse aggregate to reduce the possibility of overloading the sieves. See 8.3.

6.3 *Mechanical Sieve Shaker*—A mechanical sieving device, if used, shall create motion of the sieves to cause the particles to bounce, tumble, or otherwise turn so as to present different orientations to the sieving surface. The sieving action shall be such that the criterion for adequacy of sieving described in 8.4 is met in a reasonable time period.

[1] This test method is under the jurisdiction of ASTM Committee C-9 on Concrete and Concrete Aggregates and is the direct responsibility of Subcommittee C09.20 on Normal Weight Aggregates.
Current edition approved Nov. 10, 1996. Published January 1997. Originally published as C 136 – 38 T. Last previous edition C 136 – 96.
[2] *Annual Book of ASTM Standards,* Vol 04.02.
[3] *Annual Book of ASTM Standards,* Vol 04.03.
[4] *Annual Book of ASTM Standards,* Vol 14.02.
[5] Available from American Association of State Highway and Transportation Officials, 444 North Capitol St. N.W., Suite 225, Washington, DC 20001.

C 136

NOTE 2—Use of a mechanical sieve shaker is recommended when the size of the sample is 20 kg or greater, and may be used for smaller samples, including fine aggregate. Excessive time (more than approximately 10 min) to achieve adequate sieving may result in degradation of the sample. The same mechanical sieve shaker may not be practical for all sizes of samples, since the large sieving area needed for practical sieving of a large nominal size coarse aggregate very likely could result in loss of a portion of the sample if used for a small sample of coarse aggregate or fine aggregate.

6.4 *Oven*—An oven of appropriate size capable of maintaining a uniform temperature of 110 ± 5°C (230 ± 9°F).

7. Sampling

7.1 Sample the aggregate in accordance with Practice D 75. The size of the field sample shall be the quantity shown in Practice D 75 or four times the quantity required in 7.4 and 7.5 (except as modified in 7.6), whichever is greater.

7.2 Thoroughly mix the sample and reduce it to an amount suitable for testing using the applicable procedures described in Practice C 702. The sample for test shall be approximately the quantity desired when dry and shall be the end result of the reduction. Reduction to an exact predetermined quantity shall not be permitted.

NOTE 3—Where sieve analysis, including determination of material finer than the 75-μm sieve, is the only testing proposed, the size of the sample may be reduced in the field to avoid shipping excessive quantities of extra material to the laboratory.

7.3 *Fine Aggregate*—The size of the test sample, after drying, shall be 300 g minimum.

7.4 *Coarse Aggregate*—The size of the test sample of coarse aggregate shall conform with the following:

Nominal Maximum Size, Square Openings, mm (in.)	Test Sample Size, min, kg (lb)
9.5 (⅜)	1 (2)
12.5 (½)	2 (4)
19.0 (¾)	5 (11)
25.0 (1)	10 (22)
37.5 (1½)	15 (33)
50 (2)	20 (44)
63 (2½)	35 (77)
75 (3)	60 (130)
90 (3½)	100 (220)
100 (4)	150 (330)
125 (5)	300 (660)

7.5 *Coarse and Fine Aggregate Mixtures*—The size of the test sample of coarse and fine aggregate mixtures shall be the same as for coarse aggregate in 7.4.

7.6 *Samples of Large Size Coarse Aggregate*—The size of sample required for aggregate with 50-mm nominal maximum size or larger is such as to preclude convenient sample reduction and testing as a unit except with large mechanical splitters and sieve shakers. As an option when such equipment is not available, instead of combining and mixing sample increments and then reducing the field sample to testing size, conduct the sieve analysis on a number of approximately equal sample increments such that the total mass tested conforms to the requirement of 7.4.

7.7 In the event that the amount of material finer than the 75-μm (No. 200) sieve is to be determined by Test Method C 117, proceed as follows:

7.7.1 For aggregates with a nominal maximum size of 12.5 mm (1/2 in.) or less, use the same test sample for testing by Test Method C 117 and this method. First test the sample in accordance with Test Method C 117 through the final drying operation, then dry sieve the sample as stipulated in 8.2 through 8.7 of this method.

7.7.2 For aggregates with a nominal maximum size greater than 12.5 mm (1/2 in.), a single test sample may be used as described in 7.7.1, or separate test samples may be used for Test Method C 117 and this method.

7.7.3 Where the specifications require determination of the total amount of material finer than the 75-μm sieve by washing and dry sieving, use the procedure described in 7.7.1.

8. Procedure

8.1 Dry the sample to constant mass at a temperature of 110 ± 5°C (230 ± 9°F).

NOTE 4—For control purposes, particularly where rapid results are desired, it is generally not necessary to dry coarse aggregate for the sieve analysis test. The results are little affected by the moisture content unless: (*1*) the nominal maximum size is smaller than about 12.5 mm (½ in.); (*2*) the coarse aggregate contains appreciable material finer than 4.75 mm (No. 4); or (*3*) the coarse aggregate is highly absorptive (a lightweight aggregate, for example). Also, samples may be dried at the higher temperatures associated with the use of hot plates without affecting results, provided steam escapes without generating pressures sufficient to fracture the particles, and temperatures are not so great as to cause chemical breakdown of the aggregate.

8.2 Select sieves with suitable openings to furnish the information required by the specifications covering the material to be tested. Use additional sieves as desired or necessary to provide other information, such as fineness modulus, or to regulate the amount of material on a sieve. Nest the sieves in order of decreasing size of opening from top to bottom and place the sample on the top sieve. Agitate the sieves by hand or by mechanical apparatus for a sufficient period, established by trial or checked by measurement on the actual test sample, to meet the criterion for adequacy or sieving described in 8.4.

8.3 Limit the quantity of material on a given sieve so that all particles have opportunity to reach sieve openings a number of times during the sieving operation. For sieves with openings smaller than 4.75-mm (No. 4), the quantity retained on any sieve at the completion of the sieving operation shall not exceed 7 kg/m² of sieving surface area (Note 5). For sieves with openings 4.75 mm (No. 4) and larger, the quantity retained in kg shall not exceed the product of 2.5 × (sieve opening, mm × (effective sieving area, m²)). This quantity is shown in Table 1 for five sieve-frame dimensions in common use. In no case shall the quantity retained be so great as to cause permanent deformation of the sieve cloth.

8.3.1 Prevent an overload of material on an individual sieve by one of the following methods:

8.3.1.1 Insert an additional sieve with opening size intermediate between the sieve that may be overloaded and the sieve immediately above that sieve in the original set of sieves.

8.3.1.2 Split the sample into two or more portions, sieving each portion individually. Combine the masses of the several portions retained on a specific sieve before calculating the percentage of the sample on the sieve.

⚒ C 136

TABLE 1   Maximum Allowable Quantity of Material Retained on a Sieve, kg

Sieve Opening Size, mm	Nominal Dimensions of Sieve[A]				
	203.2-mm dia[B]	254-mm dia[B]	304.8-mm dia[B]	350 by 350 mm	372 by 580 mm
	Sieving Area, m²				
	0.0285	0.0457	0.0670	0.1225	0.2158
125	c	c	c	c	67.4
100	c	c	c	30.6	53.9
90	c	c	15.1	27.6	48.5
75	c	8.6	12.6	23.0	40.5
63	c	7.2	10.6	19.3	34.0
50	3.6	5.7	8.4	15.3	27.0
37.5	2.7	4.3	6.3	11.5	20.2
25.0	1.8	2.9	4.2	7.7	13.5
19.0	1.4	2.2	3.2	5.8	10.2
12.5	0.89	1.4	2.1	3.8	6.7
9.5	0.67	1.1	1.6	2.9	5.1
4.75	0.33	0.54	0.80	1.5	2.6

[A] Sieve frame dimensions in inch units: 8.0-in. diameter; 10.0-in. diameter, 12.0-in. diameter; 13.8 by 13.8 in. (14 by 14 in. nominal); 14.6 by 22.8 in. (16 by 24 in. nominal).

[B] The sieve area for round sieve frames is based on an effective diameter 12.7 mm (½ in.) less than the nominal frame diameter, because Specification E 11 permits the sealer between the sieve cloth and the frame to extend 6.35 mm (¼ in.) over the sieve cloth. Thus the effective sieving diameter for a 203.2-mm (8.0-in.) diameter sieve frame is 190.5 mm (7.5 in.). Some manufacturers of sieves may not infringe on the sieve cloth by the full 6.35 mm (¼ in.).

c Sieves indicated have less than five full openings and should not be used for sieve testing except as provided in 8.6.

8.3.1.3 Use sieves having a larger frame size and providing greater sieving area.

NOTE 5—The 7 kg/m² amounts to 200 g for the usual 203.2-mm (8-in.) diameter sieve (with effective sieving surface diameter of 190.5 mm (7.5 in.)).

8.4 Continue sieving for a sufficient period and in such manner that, after completion, not more than 1 mass % of the residue on any individual sieve will pass that sieve during 1 min of continuous hand sieving performed as follows: Hold the individual sieve, provided with a snug-fitting pan and cover, in a slightly inclined position in one hand. Strike the side of the sieve sharply and with an upward motion against the heel of the other hand at the rate of about 150 times per minute, turn the sieve about one sixth of a revolution at intervals of about 25 strokes. In determining sufficiency of sieving for sizes larger than the 4.75-mm (No. 4) sieve, limit the material on the sieve to a single layer of particles. If the size of the mounted testing sieves makes the described sieving motion impractical, use 203-mm (8 in.) diameter sieves to verify the sufficiency of sieving.

8.5 In the case of coarse and fine aggregate mixtures, the portion of the sample finer than the 4.75-mm (No. 4) sieve may be distributed among two or more sets of sieves to prevent overloading of individual sieves.

8.5.1 Alternatively, the portion finer than the 4.75-mm (No. 4) sieve may be reduced in size using a mechanical splitter according to Practice C 702. If this procedure is followed, compute the mass of each size increment of the original sample as follows:

$$A = \frac{W_1}{W_2} \times B$$

where:

A = mass of size increment on total sample basis,

W_1 = mass of fraction finer than 4.75-mm (No. 4) sieve in total sample,

W_2 = mass of reduced portion of material finer than 4.75-mm (No. 4) sieve actually sieved, and

B = mass of size increment in reduced portion sieved.

8.6 Unless a mechanical sieve shaker is used, hand sieve particles larger than 75 mm (3 in.) by determining the smallest sieve opening through which each particle will pass. Start the test on the smallest sieve to be used. Rotate the particles, if necessary, in order to determine whether they will pass through a particular opening; however, do not force particles to pass through an opening.

8.7 Determine the mass of each size increment on a scale or balance conforming to the requirements specified in 5.1 to the nearest 0.1 % of the total original dry sample mass. The total mass of the material after sieving should check closely with original mass of sample placed on the sieves. If the amounts differ by more than 0.3 %, based on the original dry sample mass, the results should not be used for acceptance purposes.

8.8 If the sample has previously been tested by Test Method C 117, add the mass finer than the 75-μm (No. 200) sieve determined by that method to the mass passing the

TABLE 2   Precision

	Total Percentage of Material Passing		Standard Deviation (1s), %[A]	Acceptable Range of Two Results (d2s), %[A]
Coarse Aggregate:[B]				
Single-operator precision	<100	≥95	0.32	0.9
	<95	≥85	0.81	2.3
	<85	≥80	1.34	3.8
	<80	≥60	2.25	6.4
	<60	≥20	1.32	3.7
	<20	≥15	0.96	2.7
	<15	≥10	1.00	2.8
	<10	≥5	0.75	2.1
	<5	≥2	0.53	1.5
	<2	>0	0.27	0.8
Multilaboratory precision	<100	≥95	0.35	1.0
	<95	≥85	1.37	3.9
	<85	≥80	1.92	5.4
	<80	≥60	2.82	8.0
	<60	≥20	1.97	5.6
	<20	≥15	1.60	4.5
	<15	≥10	1.48	4.2
	<10	≥5	1.22	3.4
	<5	≥2	1.04	3.0
	<2	>0	0.45	1.3
Fine Aggregate:				
Single-operator precision	<100	≥95	0.26	0.7
	<95	≥60	0.55	1.6
	<60	≥20	0.83	2.4
	<20	≥15	0.54	1.5
	<15	≥10	0.36	1.0
	<10	≥2	0.37	1.1
	<2	>0	0.14	0.4
Multilaboratory precision	<100	≥95	0.23	0.6
	<95	≥60	0.77	2.2
	<60	≥20	1.41	4.0
	<20	≥15	1.10	3.1
	<15	≥10	0.73	2.1
	<10	≥2	0.65	1.8
	<2	>0	0.31	0.9

[A] These numbers represent, respectively, the (1s) and (d2s) limits described in Practice C 670.

[B] The precision estimates are based on aggregates with nominal maximum size of 19.0 mm (¾ in.).

 C 136

TABLE 3   Precision Data for 300-g and 500-g Test Samples

Fine Aggregate Proficiency Sample				Within Laboratory		Between Laboratory	
Test Result	Sample Size	Number Labs	Average	1s	d2s	1s	d2s
ASTM C136/AASHTO T27							
Total material passing the No. 4 sieve (%)	500 g	285	99.992	0.027	0.066	0.037	0.104
	300 g	276	99.990	0.021	0.060	0.042	0.117
Total material passing the No. 8 sieve (%)	500 g	281	84.10	0.43	1.21	0.63	1.76
	300 g	274	84.32	0.39	1.09	0.69	1.92
Total material passing the No. 16 sieve (%)	500 g	286	70.11	0.53	1.49	0.75	2.10
	300 g	272	70.00	0.62	1.74	0.76	2.12
Total material passing the No. 30 sieve (%)	500 g	287	48.54	0.75	2.10	1.33	3.73
	300 g	276	48.44	0.87	2.44	1.36	3.79
Total material passing the No. 50 sieve (%)	500 g	286	13.52	0.42	1.17	0.98	2.73
	300 g	275	13.51	0.45	1.25	0.99	2.76
Total material passing the No. 100 sieve (%)	500 g	287	2.55	0.15	0.42	0.37	1.03
	300 g	270	2.52	0.18	0.52	0.32	0.89
Total Material passing the No. 200 sieve (%)	500 g	278	1.32	0.11	0.32	0.31	0.85
	300 g	266	1.30	0.14	0.39	0.31	0.85

75-μm (No. 200) sieve by dry sieving of the same sample in this method.

9. Calculation

9.1 Calculate percentages passing, total percentages retained, or percentages in various size fractions to the nearest 0.1 % on the basis of the total mass of the initial dry sample. If the same test sample was first tested by Test Method C 117, include the mass of material finer than the 75-μm (No. 200) size by washing in the sieve analysis calculation; and use the total dry sample mass prior to washing in Test Method C 117 as the basis for calculating all the percentages.

9.1.1 When sample increments are tested as provided in 7.6, total the masses of the portion of the increments retained on each sieve, and use these masses to calculate the percentages as in 9.1.

9.2 Calculate the fineness modulus, when required, by adding the total percentages of material in the sample that is coarser than each of the following sieves (cumulative percentages retained), and dividing the sum by 100: 150-μm (No. 100), 300-μm (No. 50), 600-μm (No. 30), 1.18-mm (No. 16), 2.36-mm (No. 8), 4.75-mm (No. 4), 9.5-mm (3/8-in.), 19.0-mm (3/4-in.), 37.5-mm (1½-in.), and larger, increasing in the ratio of 2 to 1.

10. Report

10.1 Depending upon the form of the specifications for use of the material under test, the report shall include the following:

10.1.1 Total percentage of material passing each sieve, or

10.1.2 Total percentage of material retained on each sieve, or

10.1.3 Percentage of material retained between consecutive sieves.

10.2 Report percentages to the nearest whole number, except if the percentage passing the 75-μm (No. 200) sieve is

less than 10 %, it shall be reported to the nearest 0.1 %.

10.3 Report the fineness modulus, when required, to the nearest 0.01.

11. Precision and Bias

11.1 *Precision*—The estimates of precision for this test method are listed in Table 2. The estimates are based on the results from the AASHTO Materials Reference Laboratory Proficiency Sample Program, with testing conducted by Test Method C 136 and AASHTO Test Method T 27. The data are based on the analyses of the test results from 65 to 233 laboratories that tested 18 pairs of coarse aggregate proficiency test samples and test results from 74 to 222 laboratories that tested 17 pairs of fine aggregate proficiency test samples (Samples No. 21 through 90). The values in the table are given for different ranges of total percentage of aggregate passing a sieve.

11.1.1 The precision values for fine aggregate in Table 2 are based on nominal 500-g test samples. Revision of this test method in 1994 permits the fine aggregate test sample size to be 300 g minimum. Analysis of results of testing of 300-g and 500-g test samples from Aggregate Proficiency Test Samples 99 and 100 (Samples 99 and 100 were essentially identical) produced the precision values in Table 3, which indicate only minor differences due to test sample size.

NOTE 6—The values for fine aggregate in Table 2 will be revised to reflect the 300-g test sample size when a sufficient number of Aggregate Proficiency Tests have been conducted using that sample size to provide reliable data.

11.2 *Bias*—Since there is no accepted reference material suitable for determining the bias in this test method, no statement on bias is made.

12. Keywords

12.1 aggregate; coarse aggregate; fine aggregate; gradation; grading; sieve analysis; size analysis

ASTM C 136

 **Designation: C 618 – 97**

AMERICAN SOCIETY FOR TESTING AND MATERIALS
100 Barr Harbor Dr., West Conshohocken, PA 19428
Reprinted from the Annual Book of ASTM Standards. Copyright ASTM
If not listed in the current combined index, will appear in the next edition.

Standard Specification for
Coal Fly Ash and Raw or Calcined Natural Pozzolan for Use as a Mineral Admixture in Concrete[1]

This standard is issued under the fixed designation C 618; the number immediately following the designation indicates the year of original adoption or, in the case of revision, the year of last revision. A number in parentheses indicates the year of last reapproval. A superscript epsilon (ε) indicates an editorial change since the last revision or reapproval.

This specification has been approved for use by agencies of the Department of Defense. Consult the DoD Index of Specifications and Standards for the specific year of issue which has been adopted by the Department of Defense.

1. Scope

1.1 This specification covers coal fly ash and raw or calcined natural pozzolan for use as a mineral admixture in concrete where cementitious or pozzolanic action, or both, is desired, or where other properties normally attributed to finely divided mineral admixtures may be desired, or where both objectives are to be achieved.

NOTE 1—Finely divided materials may tend to reduce the entrained air content of concrete. Hence, if a mineral admixture is added to any concrete for which entrainment of air is specified, provision should be made to ensure that the specified air content is maintained by air content tests and by use of additional air-entraining admixture or use of an air-entraining admixture in combination with air-entraining hydraulic cement.

1.2 The values stated in SI units are to be regarded as the standard.

2. Referenced Documents

2.1 *ASTM Standards:*
C 311 Test Methods for Sampling and Testing Fly Ash or Natural Pozzolans for Use as a Mineral Admixture in Portland Cement Concrete[2]

3. Terminology

3.1 *Definitions:*
3.1.1 *fly ash*—finely divided residue that results from the combustion of ground or powdered coal.

NOTE 2—This definition of fly ash does not include, among other things, the residue resulting from: (*1*) the burning of municipal garbage or any other refuse with coal; (*2*) the injection of lime directly into the boiler for sulfur removal; or (*3*) the burning of industrial or municipal garbage in incinerators commonly known as "incinerator ash."

3.1.2 *pozzolans*—siliceous or siliceous and aluminous materials which in themselves possess little or no cementitious value but will, in finely divided form and in the presence of moisture, chemically react with calcium hydroxide at ordinary temperatures to form compounds possessing cementitious properties.

4. Classification

4.1 *Class N*—Raw or calcined natural pozzolans that comply with the applicable requirements for the class as given herein, such as some diatomaceous earths; opaline cherts and shales; tuffs and volcanic ashes or pumicites, calcined or uncalcined; and various materials requiring calcination to induce satisfactory properties, such as some clays and shales.

4.2 *Class F*—Fly ash normally produced from burning anthracite or bituminous coal that meets the applicable requirements for this class as given herein. This class fly ash has pozzolanic properties.

4.3 *Class C*—Fly ash normally produced from lignite or subbituminous coal that meets the applicable requirements for this class as given herein. This class of fly ash, in addition to having pozzolanic properties, also has some cementitious properties.

NOTE 3—Some Class C fly ashes may contain lime contents higher than 10 %.

5. Ordering Information

5.1 The purchaser shall specify any supplementary optional chemical or physical requirements.

5.2 The purchaser shall indicate which procedure, A or B, shall be used when specifying requirements for effectiveness in contribution to sulfate resistance under Table 2A.

6. Chemical Composition

6.1 Fly ash and natural pozzolans shall conform to the requirements as to chemical composition prescribed in Table 1. Supplementary optional chemical requirements are shown in Table 1A.

7. Physical Properties

7.1 Fly ash and natural pozzolans shall conform to the physical requirements prescribed in Table 2. Supplementary optional physical requirements are shown in Table 2A.

8. Methods of Sampling and Testing

8.1 Sample and test the mineral admixture in accordance with the requirements of Test Methods C 311.

8.2 Use cement of the type proposed for use in the work and, if available, from the mill proposed as the source of the cement, in all tests requiring the use of hydraulic cement.

9. Storage and Inspection

9.1 The mineral admixture shall be stored in such a manner as to permit easy access for proper inspection and

---

[1] This specification is under the jurisdiction of ASTM Committee C-9 on Concrete and Concrete Aggregates and is the direct responsibility of Subcommittee C09.24 on Ground Slag and Pozzolonic Admixtures.
Current edition approved Jan. 10, 1997. Published March 1997. Originally published as C 618 – 68 T to replace C 350 and C 402. Last previous edition C 618 – 96a.
[2] *Annual Book of ASTM Standards*, Vol 04.02.

ASTM C 618

TABLE 1   Chemical Requirements

	Mineral Admixture Class		
	N	F	C
Silicon dioxide (SiO_2) plus aluminum oxide (Al_2O_3) plus iron oxide (Fe_2O_3), min, %	70.0	70.0	50.0
Sulfur trioxide (SO_3), max, %	4.0	5.0	5.0
Moisture content, max, %	3.0	3.0	3.0
Loss on ignition, max, %	10.0	6.0[A]	6.0

[A] The use of Class F pozzolan containing up to 12.0 % loss on ignition may be approved by the user if either acceptable performance records or laboratory test results are made available.

TABLE 1A   Supplementary Optional Chemical Requirement

NOTE—This optional requirement applies only when specifically requested.

	Mineral Admixture Class		
	N	F	C
Available alkalies, as equivalent, as Na_2O, max, %[A]	1.5	1.5	1.5

[A] Applicable only when specifically required by the purchaser for mineral admixture to be used in concrete containing reactive aggregate and cement to meet a limitation on content of alkalies.

TABLE 2   Physical Requirements

	Mineral Admixture Class		
	N	F	C
Fineness:			
Amount retained when wet-sieved on 45 μm (No. 325) sieve, max, %[A]	34	34	34
Strength activity index: [B]			
With portland cement, at 7 days, min, percent of control	75[D]	75[D]	75[D]
With portland cement, at 28 days, min, percent of control	75[D]	75[D]	75[D]
Water requirement, max, percent of control	115	105	105
Soundness: [C]			
Autoclave expansion or contraction, max, %	0.8	0.8	0.8
Uniformity requirements:			
The density and fineness of individual samples shall not vary from the average established by the ten preceding tests, or by all preceding tests if the number is less than ten, by more than:			
Density, max variation from average, %	5	5	5
Percent retained on 45-μm (No. 325), max variation, percentage points from average	5	5	5

[A] Care should be taken to avoid the retaining of agglomerations of extremely fine material.

[B] The *strength* activity index with portland cement is not to be considered a measure of the compressive strength of concrete containing the mineral admixture. The mass of mineral admixture specified for the test to determine the *strength* activity index with portland cement is not considered to be the proportion recommended for the concrete to be used in the work. The optimum amount of mineral admixture for any specific project is determined by the required properties of the concrete and other constituents of the concrete and is to be established by testing. *Strength* activity index with portland cement is a measure of reactivity with a given cement and may vary as to the source of both the mineral admixture and the cement.

[C] If the mineral admixture will constitute more than 20 % by weight of the cementitious material in the project mix design, the test specimens for autoclave expansion shall contain that anticipated percentage. Excessive autoclave expansion is highly significant in cases where water to mineral admixture and cement ratios are low, for example, in block or shotcrete mixes.

[D] Meeting the 7 day or 28 day *strength* activity index will indicate specification compliance.

identification of each shipment.

9.2 Inspection of the material shall be made as agreed upon by the purchaser and the seller as part of the purchase contract.

10. Rejection

10.1 The purchaser has the right to reject material that fails to conform to the requirements of this specification. Rejection shall be reported to the producer or supplier promptly and in writing.

10.2 The purchaser has the right to reject packages varying more than 5 % from the stated weight. The purchaser also has the right to reject the entire shipment if the average weight of the packages in any shipment, as shown by weighing 50 packages taken at random, is less than that specified.

10.3 The purchaser has the right to require that mineral admixture in storage prior to shipment for a period longer than 6 months after testing be retested. The purchaser has the right to reject such material if it fails to meet the fineness requirements.

11. Packaging and Package Marking

11.1 When the mineral admixture is delivered in packages, the class, name, and brand of the producer, and the weight of the material contained therein, shall be plainly marked on each package. Similar information shall be provided in the shipping invoices accompanying the shipment of packaged or bulk mineral admixture.

12. Keywords

12.1 fly ash; mineral admixtures; natural pozzolan; pozzolans

C 618

TABLE 2A Supplementary Optional Physical Requirements

NOTE—These optional requirements apply only when specifically requested.

	Mineral Admixture Class		
	N	F	C
Multiple factor, calculated as the product of loss on ignition and fineness, amount retained when wet-sieved on 45-μm (No. 325) sieve, max, % [A]	....	255	....
Increase of drying shrinkage of mortar bars at 28 days, max, difference, in %, over control [B]	0.03	0.03	0.03
Uniformity Requirements:			
In addition, when air-entraining concrete is specified, the quantity of air-entraining agent required to produce an air content of 18.0 vol % of mortar shall not vary from the average established by the ten preceding tests or by all preceding tests if less than ten, by more than, %	20	20	20
Effectiveness in Controlling Alkali-Silica Reaction: [C]			
Expansion of test mixture as percentage of low-alkali cement control, at 14 days, max, %	100	100	100
Effectiveness in Contributing to Sulfate Resistance: [D]			
Procedure A:			
Expansion of test mixture:			
For moderate sulfate exposure after 6 months exposure, max, %	0.10	0.10	0.10
For high sulfate exposure after 6 months exposure, max, %	0.05	0.05	0.05
Procedure B:			
Expansion of test mixture as a percentage of sulfate resistance cement control after at least 6 months exposure, max, %	100	100	100

[A] Applicable only for Class F mineral admixtures since the loss on ignition limitations predominate for Class C.

[B] Determination of compliance or noncompliance with the requirement relating to increase in drying shrinkage will be made only at the request of the purchaser.

[C] Mineral admixtures meeting this requirement are considered as effective in controlling alkali aggregate reactions as the use of the low-alkali control cement used in the evaluation. However, the mineral admixture shall be considered effective only when the mineral admixture is used at percentages by mass of the total cementitious material equal to or exceeding that used in the tests and when the alkali content of the cement to be used with the mineral admixture does not exceed that used in the tests by more than 0.05 %. See Appendix XI, Test Methods C 311.

[D] Fly ash or natural pozzolan shall be considered effective only when the fly ash or natural pozzolan is used at percentages, by mass, of the total cementitious material within 2 % of those that are successful in the test mixtures or between two percentages that are successful, and when the C_3A content of the project cement is less than, or equal to, that which was used in the test mixtures. See Appendix X2 of Test Method C 311.

Designation: C 666 – 92

Standard Test Method for
Resistance of Concrete to Rapid Freezing and Thawing[1]

This standard is issued under the fixed designation C 666; the number immediately following the designation indicates the year of original adoption or, in the case of revision, the year of last revision. A number in parentheses indicates the year of last reapproval. A superscript epsilon (ε) indicates an editorial change since the last revision or reapproval.

This method has been approved for use by agencies of the Department of Defense. Consult the DoD Index of Specifications and Standards for the specific year of issue which has been adopted by the Department of Defense.

1. Scope

1.1 This test method covers the determination of the resistance of concrete specimens to rapidly repeated cycles of freezing and thawing in the laboratory by two different procedures: Procedure A, Rapid Freezing and Thawing in Water, and Procedure B, Rapid Freezing in Air and Thawing in Water. Both procedures are intended for use in determining the effects of variations in the properties of concrete on the resistance of the concrete to the freezing-and-thawing cycles specified in the particular procedure. Neither procedure is intended to provide a quantitative measure of the length of service that may be expected from a specific type of concrete.

1.2 The values stated in inch-pound units are to be regarded as the standard.

1.3 All material in this test method not specifically designated as belonging to Procedure A or Procedure B applies to either procedure.

1.4 *This standard does not purport to address all of the safety problems, if any, associated with its use. It is the responsibility of the user of this standard to establish appropriate safety and health practices and determine the applicability of regulatory limitations prior to use.*

2. Referenced Documents

2.1 *ASTM Standards:*
C 157 Test Method for Length Change of Hardened Hydraulic-Cement Mortar and Concrete[2]
C 192 Practice for Making and Curing Concrete Test Specimens in the Laboratory[2]
C 215 Test Method for Fundamental Transverse, Longitudinal, and Torsional Frequencies of Concrete Specimens[2]
C 233 Test Method for Testing Air-Entraining Admixtures for Concrete[2]
C 295 Guide for Petrographic Examination of Aggregates for Concrete[2]
C 341 Test Method for Length Change of Drilled or Sawed Specimens of Hydraulic–Cement Mortar and Concrete[2]
C 490 Practice for Use of Apparatus for Determination of Length Change of Hardened Cement Paste, Mortar, and Concrete[2]

C 494 Specification for Chemical Admixtures for Concrete[2]
C 670 Practice for Preparing Precision and Bias Statements for Test Methods for Construction Materials[2]
C 823 Practice for Examination and Sampling of Hardened Concrete in Constructions[2]

3. Significance and Use

3.1 As noted in the scope, the two procedures described in this test method are intended to determine the effects of variations in both properties and conditioning of concrete in the resistance to freezing and thawing cycles specified in the particular procedure. Specific applications include specified use in Specification C 494, Test Method C 233, and ranking of coarse aggregates as to their effect on concrete freeze-thaw durability, especially where soundness of the aggregate is questionable.

3.2 It is assumed that the procedures will have no significantly damaging effects on frost-resistant concrete which may be defined as (*1*) any concrete not critically saturated with water (that is, not sufficiently saturated to be damaged by freezing) and (*2*) concrete made with frost-resistant aggregates and having an adequate air-void system that has achieved appropriate maturity and thus will prevent critical saturation by water under common conditions.

3.3 If as a result of performance tests as described in this test method concrete is found to be relatively unaffected, it can be assumed that it was either not critically saturated, or was made with "sound" aggregates, a proper air-void system, and allowed to mature properly.

3.4 No relationship has been established between the resistance to cycles of freezing and thawing of specimens cut from hardened concrete and specimens prepared in the laboratory.

4. Apparatus

4.1 *Freezing-and-Thawing Apparatus:*

4.1.1 The freezing-and-thawing apparatus shall consist of a suitable chamber or chambers in which the specimens may be subjected to the specified freezing-and-thawing cycle, together with the necessary refrigerating and heating equipment and controls to produce continuously, and automatically, reproducible cycles within the specified temperature requirements. In the event that the equipment does not operate automatically, provision shall be made for either its continuous manual operation on a 24-h a day basis or for the storage of all specimens in a frozen condition when the equipment is not in operation.

4.1.2 The apparatus shall be so arranged that, except for

---

[1] This test method is under the jurisdiction of ASTM Committee C-9 on Concrete and Concrete Aggregates and is the direct responsibility of Subcommittee C09.67 on Resistance of Concrete to Its Environment.
Current edition approved Sept. 15, 1992. Published November 1992. Originally published as C 666 – 71. Last previous edition C 666 – 90.
[2] *Annual Book of ASTM Standards,* Vol 04.02.

necessary supports, each specimen is: (*1*) for Procedure A, completely surrounded by not less than ¹⁄₃₂ in. (1 mm) nor more than ⅛ in. (3 mm) of water at all times while it is being subjected to freezing-and-thawing cycles, or (*2*) for Procedure B, completely surrounded by air during the freezing phase of the cycle and by water during the thawing phase. Rigid containers, which have the potential to damage specimens, are not permitted. Length change specimens in vertical containers shall be supported in a manner to avoid damage to the gage studs.

NOTE 1—Experience has indicated that ice or water pressure, during freezing tests, particularly in equipment that uses air rather than a liquid as the heat transfer medium, can cause excessive damage to rigid metal containers, and possibly to the specimens therein. Results of tests during which bulging or other distortion of containers occurs should be interpreted with caution.

4.1.3 The temperature of the heat-exchanging medium shall be uniform within 6°F (3.3°C) throughout the specimen cabinet when measured at any given time, at any point on the surface of any specimen container for Procedure A or on the surface of any specimen for Procedure B, except during the transition between freezing and thawing and *vice versa*.

4.1.3.1 Support each specimen at the bottom of its container in such a way that the temperature of the heat-exchanging medium will not be transmitted directly through the bottom of the container to the full area of the bottom of the specimen, thereby subjecting it to conditions substantially different from the remainder of the specimen.

NOTE 2—A flat spiral of ⅛-in. (3-mm) wire placed in the bottom of the container has been found adequate for supporting specimens.

4.1.4 For Procedure B, it is not contemplated that the specimens will be kept in containers. The supports on which the specimens rest shall be such that they are not in contact with the full area of the supported side or end of the specimen, thereby subjecting this area to conditions substantially different from those imposed on the remainder of the specimen.

NOTE 3—The use of relatively open gratings, metal rods, or the edges of metal angles has been found adequate for supporting specimens, provided the heat-exchanging medium can circulate in the direction of the long axis of the rods or angles.

4.2 *Temperature-Measuring Equipment,* consisting of thermometers, resistance thermometers, or thermocouples, capable of measuring the temperature at various points within the specimen chamber and at the centers of control specimens to within 2°F (1.1°C).

4.3 *Dynamic Testing Apparatus,* conforming to the requirements of Test Method C 215.

4.4 *Optional Length Change Test Length Change Comparator,* conforming to the requirements of Specification C 490. When specimens longer than the nominal 11¼ in. (286 mm) length provided for in Specification C 490 are used for freeze-thaw tests, use an appropriate length reference bar, which otherwise meets the Specification C 490 requirements. Dial gage micrometers for use on these longer length change comparators shall meet the gradation interval and accuracy requirements for Specification C 490 for either the inch or millimetre calibration requirements. Prior to the start of measurements on any specimens, fix the comparator at an appropriate length to accommodate all of the speci-

mens to be monitored for length change.

4.5 *Scales,* with a capacity approximately 50 % greater than the weight of the specimens and accurate to at least 0.01 lb (4.5 g) within the range of ±10 % of the specimen weight will be satisfactory.

4.6 *Tempering Tank,* with suitable provisions for maintaining the temperature of the test specimens in water, such that when removed from the tank and tested for fundamental transverse frequency and length change, the specimens will be maintained within −2°F and +4°F (−1.1°C and +2.2°C) of the target thaw temperature for specimens in the actual freezing-and-thawing cycle and equipment being used. The use of the specimen chamber in the freezing-and-thawing apparatus by stopping the apparatus at the end of the thawing cycle and holding the specimens in it shall be considered as meeting this requirement, provided the specimens are tested for fundamental transverse frequency within the above temperature range. It is required that the same target specimen thaw temperature be used throughout testing of an individual specimen since a change in specimen temperature at the time of length measurement can affect the length of the specimen significantly.

5. Freezing-and-Thawing Cycle

5.1 Base conformity with the requirements for the freezing-and-thawing cycle on temperature measurements of control specimens of similar concrete to the specimens under test in which suitable temperature-measuring devices have been imbedded. Change the position of these control specimens frequently in such a way as to indicate the extremes of temperature variation at different locations in the specimen cabinet.

5.2 The nominal freezing-and-thawing cycle for both procedures of this test method shall consist of alternately lowering the temperature of the specimens from 40 to 0°F (4.4 to −17.8°C) and raising it from 0 to 40°F (−17.8 to 4.4°C) in not less than 2 nor more than 5 h. For Procedure A, not less than 25 % of the time shall be used for thawing, and for Procedure B, not less than 20 % of the time shall be used for thawing (Note 4). At the end of the cooling period the temperature at the centers of the specimens shall be 0 ± 3°F (−17.8 ± 1.7°C), and at the end of the heating period the temperature shall be 40 ± 3°F (4.4 ± 1.7°C), with no specimen at any time reaching a temperature lower than −3°F (−19.4°C) nor higher than 43°F (6.1°C). The time required for the temperature at the center of any single specimen to be reduced from 37 to 3°F (2.8 to −16.1°C) shall be not less than one half of the length of the cooling period, and the time required for the temperature at the center of any single specimen to be raised from 3 to 37°F (−16.1 to 2.8°C) shall be not less than one half of the length of the heating period. For specimens to be compared with each other, the time required to change the temperature at the centers of any specimens from 35 to 10°F (1.7 to −12.2°C) shall not differ by more than one sixth of the length of the cooling period from the time required for any specimen and the time required to change the temperature at the centers of any specimens from 10 to 35°F (−12.2 to 1.7°C) shall not differ by more than one third of the length of the heating period from the time required for any specimen.

ASTM C 666

NOTE 4—In most cases, uniform temperature and time conditions can be controlled most conveniently by maintaining a capacity load of specimens in the equipment at all times. In the event that a capacity load of test specimens is not available, dummy specimens can be used to fill empty spaces. This procedure also assists greatly in maintaining uniform fluid level conditions in the specimen and solution tanks.

The testing of concrete specimens composed of widely varying materials or with widely varying thermal properties, in the same equipment at the same time, may not permit adherence to the time-temperature requirements for all specimens. It is advisable that such specimens be tested at different times and that appropriate adjustments be made to the equipment.

5.3 The difference between the temperature at the center of a specimen and the temperature at its surface shall at no time exceed 50°F (27.8°C).

5.4 The period of transition between the freezing-and-thawing phases of the cycle shall not exceed 10 min, except when specimens are being tested in accordance with 8.2.

6. Sampling

6.1 Constituent materials for concrete specimens made in the laboratory shall be sampled using applicable standard methods.

6.2 Samples cut from hardened concrete are to be obtained in accordance with Practice C 823.

7. Test Specimens

7.1 The specimens for use in this test method shall be prisms or cylinders made and cured in accordance with the applicable requirements of Practice C 192 and Specification C 490.

7.2 Specimens used shall not be less than 3 in. (76 mm) nor more than 5 in. (127 mm) in width, depth, or diameter, and not less than 11 in. (279 mm) nor more than 16 in. (406 mm) in length.

7.3 Test specimens may also be cores or prisms cut from hardened concrete. If so, the specimens should not be allowed to dry to a moisture condition below that of the structure from which taken. This may be accomplished by wrapping in plastic or by other suitable means. The specimens so obtained shall be furnished with gage studs in accordance with Test Method C 341.

7.4 For this test the specimens shall be stored in saturated lime water from the time of their removal from the molds until the time freezing-and-thawing tests are started. All specimens to be compared with each other initially shall be of the same nominal dimensions.

8. Procedure

8.1 Immediately after the specified curing period (Note 5), bring the specimen to a temperature within −2°F and +4°F (−1.1°C and +2.2°C) of the target thaw temperature that will be used in the freeze-thaw cycle and test for fundamental transverse frequency, weigh, determine the average length and cross section dimensions of the concrete specimen within the tolerance required in Test Method C 215, and determine the initial length comparator reading (optional) for the specimen with the length change comparator. Protect the specimens against loss of moisture between the time of removal from curing and the start of the freezing-and-thawing cycles.

NOTE 5—Unless some other age is specified, the specimens should be removed from curing and freezing-and-thawing tests started when the specimens are 14 days old.

8.2 Start freezing-and-thawing tests by placing the specimens in the thawing water at the beginning of the thawing phase of the cycle. Remove the specimens from the apparatus, in a thawed condition, at intervals not exceeding 36 cycles of exposure to the freezing-and-thawing cycles, test for fundamental transverse frequency and measure length change (optional) with the specimens within the temperature range specified for the tempering tank in 4.6, weigh each specimen, and return them to the apparatus. To ensure that the specimens are completely thawed and at the specified temperature place them in the tempering tank or hold them at the end of the thaw cycle in the freezing-and-thawing apparatus for a sufficient time for this condition to be attained throughout each specimen to be tested. Protect the specimens against loss of moisture while out of the apparatus and turn them end-for-end when returned. For Procedure A, rinse out the container and add clean water. Return the specimens either to random positions in the apparatus or to positions according to some predetermined rotation scheme that will ensure that each specimen that continues under test for any length of time is subjected to conditions in all parts of the freezing apparatus. Continue each specimen in the test until it has been subjected to 300 cycles or until its relative dynamic modulus of elasticity reaches 60 % of the initial modulus, whichever occurs first, unless other limits are specified (Note 6). For the optional length change test, 0.10 % expansion may be used as the end of test. Whenever a specimen is removed because of failure, replace it for the remainder of the test by a dummy specimen. Each time the specimen is tested for fundamental frequency (Note 7) and length change, make a note of its visual appearance and make special comment on any defects that develop (Note 8). When it is anticipated that specimens may deteriorate rapidly, they should be tested for fundamental transverse frequency and length change (optional) at intervals not exceeding 10 cycles when initially subjected to freezing and thawing.

NOTE 6—It is not recommended that specimens be continued in the test after their relative dynamic modulus of elasticity has fallen below 50 %.

NOTE 7—It is recommended that the fundamental longitudinal frequency be determined initially and as a check whenever a question exists concerning the accuracy of determination of fundamental transverse frequency, and that the fundamental torsional frequency be determined initially and periodically as a check on the value of Poisson's ratio.

NOTE 8—In some applications, such as airfield pavements and other slabs, popouts may be defects that are a concern. A popout is characterized by the breaking away of a small portion of the concrete surface due to internal pressure, thereby leaving a shallow and typically conical spall in the surface of the concrete through the aggregate particle. Popouts may be observed as defects in the test specimens. Where popouts are a concern, the number and general description should be reported as a special comment. The aggregates causing the popout may be identified by petrographic examination as in Practice C 295.

8.3 When the sequence of freezing-and-thawing cycles must be interrupted store the specimens in a frozen condition.

NOTE 9—If, due to equipment breakdown or for other reasons, it becomes necessary to interrupt the cycles for a protracted period, store

C 666

the specimens in a frozen condition in such a way as to prevent loss of moisture. For Procedure A, maintain the specimens in the containers and surround them by ice, if possible. If it is not possible to store the specimens in their containers, wrap and seal them, in as wet a condition as possible, in moisture-proof material to prevent dehydration and store in a refrigerator or cold room maintained at $0 \pm 3°F$ ($-17.8 \pm 1.7°C$). Follow the latter procedure when Procedure B is being used. In general, for specimens to remain in a thawed condition for more than two cycles is undesirable, but a longer period may be permissible if this occurs only once or twice during a complete test.

9. Calculation

9.1 *Relative Dynamic Modulus of Elasticity*—Calculate the numerical values of relative dynamic modulus of elasticity as follows:

$$P_c = (n_1^2/n^2) \times 100$$

where:

P_c = relative dynamic modulus of elasticity, after c cycles of freezing and thawing, percent,

n = fundamental transverse frequency at 0 cycles of freezing and thawing, and

n_1 = fundamental transverse frequency after c cycles of freezing and thawing.

NOTE 10—This calculation of relative dynamic modulus of elasticity is based on the assumption that the weight and dimensions of the specimen remain constant throughout the test. This assumption is not true in many cases due to disintegration of the specimen. However, if the test is to be used to make comparisons between the relative dynamic moduli of different specimens or of different concrete formulations, P_c as defined is adequate for the purpose.

9.2 *Durability Factor*—Calculate the durability factor as follows:

$$DF = PN/M$$

where:

DF = durability factor of the test specimen,

P = relative dynamic modulus of elasticity at N cycles, %,

N = number of cycles at which P reaches the specified minimum value for discontinuing the test or the specified number of cycles at which the exposure is to be terminated, whichever is less, and

M = specified number of cycles at which the exposure is to be terminated.

9.3 *Length Change in Percent (optional)*—Calculate the length change as follows:

$$L_c = \frac{(l_2 - l_1)}{L_g} \times 100$$

where:

L_c = length change of the test specimen after C cycles of freezing and thawing, %,

l_1 = length comparator reading at 0 cycles,

l_2 = length comparator reading after C cycles, and

L_g = the effective gage length between the innermost ends of the gage studs as shown in the mold diagram in Specification C 490.

10. Report

10.1 Report the following data such as are pertinent to the variables or combination of variables studied in the tests:

10.2 *Properties of Concrete Mixture:*

10.2.1 Type and proportions of cement, fine aggregate, and coarse aggregate, including maximum size and grading (or designated grading indices), and ratio of net water content to cement,

10.2.2 Kind and proportion of any addition or admixture used,

10.2.3 Air content of fresh concrete,

10.2.4 Unit weight of fresh concrete,

10.2.5 Consistency of fresh concrete,

10.2.6 Air content of the hardened concrete, when available,

10.2.7 Indicate if the test specimens are cut from hardened concrete, and if so, state the size, shape, orientation of the specimens in the structure, and any other pertinent information available, and

10.2.8 Curing period.

10.3 *Mixing, Molding, and Curing Procedures*—Report any departures from the standard procedures for mixing, molding, and curing as prescribed in Section 7.

10.4 *Procedure*—Report which of the two procedures was used.

10.5 *Characteristics of Test Specimens:*

10.5.1 Dimensions of specimens at 0 cycles of freezing and thawing,

10.5.2 Weight of specimens at 0 cycles of freezing and thawing,

10.5.3 Nominal gage length between embedded ends of gage studs, and

10.5.4 Any defects in each specimen present at 0 cycles of freezing and thawing.

10.6 *Results:*

10.6.1 Values for the durability factor of each specimen, calculated to the nearest whole number, and for the average durability factor for each group of similar specimens, also calculated to the nearest whole number, and the specified values for minimum relative dynamic modulus and maximum number of cycles (Note 10),

10.6.2 Values for the percent length change of each specimen and for the average percent length change for each group of similar specimens (Note 11),

10.6.3 Values of weight loss or gain for each specimen and average values for each group of similar specimens, and

10.6.4 Any defects in each specimen which develop during testing, and the number of cycles at which such defects were noted.

NOTE 11—It is recommended that the results of the test on each specimen, and the average of the results on each group of similar specimens, be plotted as curves showing the value of relative modulus of elasticity or percent length change against time expressed as the number of cycles of freezing and thawing.

11. Precision

11.1 *Within-Laboratory Precision (Single Beams)*—Criteria for judging the acceptability of durability factor results obtained by the two procedures in the same laboratory on concrete specimens made from the same batch of concrete or from two batches made with the same materials are given in Table 1. Precision data for length change (optional) are not available at this time.

NOTE 12—The between-batch precision of durability factors has been found to be the same as the within-batch precision. Thus the limits

C 666

TABLE 1   Within-Laboratory Durability Factor Precision for Single Beams

NOTE—The values given in Columns 2 and 4 are the standard deviations that have been found to be appropriate for Procedures A and B, respectively, for tests for which the average durability factor is in the corresponding range given in Column 1. The values given in Columns 3 and 5 are the corresponding limits that should not be exceeded by the difference between the results of two single test beams.

Range of Average Durability Factor	Procedure A		Procedure B	
	Standard Deviation[A]	Acceptable Range of Two Results[A]	Standard Deviation[A]	Acceptable Range of Two Results[A]
0 to 5	0.8	2.2	1.1	3.0
5 to 10	1.5	4.4	4.0	11.4
10 to 20	5.9	16.7	8.1	22.9
20 to 30	8.4	23.6	10.5	29.8
30 to 50	12.7	35.9	15.4	43.5
50 to 70	15.3	43.2	20.1	56.9
70 to 80	11.6	32.7	17.1	48.3
80 to 90	5.7	16.0	8.8	24.9
90 to 95	2.1	6.0	3.9	11.0
Over 95	1.1	3.1	2.0	5.7

[A] These numbers represent the (1S) and (D2S) limits as described in Practice C 670.

given in this precision statement apply to specimens from different batches made with the same materials and mix design and having the same air content as well as to specimens from the same batch.

NOTE 13—The precision of this method for both procedures has been found to depend primarily on the average durability factor and not on the maximum N or minimum P specified for terminating the test nor

on the size of the beams within limits. The data on which these precision statements are based cover maximum N's from 100 to 300 cycles, and minimum P's from 50 to 70 percent of E_0. The indexes of precision are thus valid at least over these ranges.

11.1.1 The different specimen sizes represented by the data include the following: 3 by 3 by 16-in; 3 by 3 by 16¼-in.; 3 by 4 by 16-in.; 3½ by 4½ by 16-in.; 3 by 3 by 11-in.; and 3½ by 4 by 16-in.; and 4 by 3 by 16-in. The first dimension given represents the direction in which the specimens were vibrated in the test for fundamental transverse frequency. The most commonly used size was 3 by 4 by 16-in.

11.2 *Within-Laboratory Precision (Averages of Two or More Beams)*—Specifications sometimes call for comparisons between averages of two or more beams. Tables 2 and 3 give appropriate standard deviations and acceptable ranges for the two procedures for two averages of the number of test beams shown.

11.3 *Multilaboratory Precision*—No data are available for evaluation of multilaboratory precision. It is believed that a multilaboratory statement of precision is not appropriate because of the limited possibility that two or more laboratories will be performing freezing-and-thawing tests on the same concretes.

12. Keywords

12.1 accelerated testing; concrete-weathering tests; conditioning; freezing and thawing; resistance-frost

TABLE 2   Within-Laboratory Durability Factor Precision for Averages of Two or More Beams—Procedure A

Range of Average Durability Factor	Number of Beams Averaged									
	2		3		4		5		6	
	Standard Deviation[A]	Acceptable Range[A]	Standard Deviation[A]	Acceptable Range[A]	Standard Deviation[A]	Acceptable Range[A]	Standard Deviation[A]	Acceptable Range[A]	Standard Deviation[A]	Acceptable Range[A]
0 to 5	0.6	1.6	0.5	1.3	0.4	1.1	0.4	1.0	0.3	0.9
5 to 10	1.1	3.1	0.9	2.5	0.8	2.2	0.7	2.0	0.6	1.8
10 to 20	4.2	11.8	3.4	9.7	3.0	8.4	2.7	7.5	2.4	6.8
20 to 30	5.9	16.7	4.8	13.7	4.2	11.8	3.7	10.6	3.4	9.7
30 to 50	9.0	25.4	7.4	20.8	6.4	18.0	5.7	16.1	5.2	14.7
50 to 70	10.8	30.6	8.8	25.0	7.6	21.6	6.8	19.3	6.2	17.6
70 to 80	8.2	23.1	6.7	18.9	5.8	16.4	5.2	14.6	4.7	13.4
80 to 90	4.0	11.3	3.3	9.2	2.8	8.0	2.5	7.2	2.3	6.5
90 to 95	1.5	4.2	1.2	3.5	1.1	3.0	0.9	2.7	0.9	2.4
Above 95	0.8	2.2	0.6	1.8	0.5	1.5	0.5	1.4	0.4	1.3

[A] These numbers represent the (1S) and (D2S) limits as described in Practice C 670.

TABLE 3   Within-Laboratory Durability Factor Precision for Averages of Two or More Beams—Procedure B

Range of Average Durability Factor	Number of Beams Averaged									
	2		3		4		5		6	
	Standard Deviation[A]	Acceptable Range[A]	Standard Deviation[A]	Acceptable Range[A]	Standard Deviation[A]	Acceptable Range[A]	Standard Deviation[A]	Acceptable Range[A]	Standard Deviation[A]	Acceptable Range[A]
0 to 5	0.8	2.1	0.6	1.8	0.5	1.5	0.5	1.4	0.4	1.2
5 to 10	2.9	8.1	2.3	6.6	2.0	5.7	1.8	5.1	1.7	4.7
10 to 20	5.7	16.2	4.7	13.2	4.1	11.5	3.6	10.3	3.3	7.4
20 to 30	7.4	21.0	6.1	17.2	5.3	14.9	4.7	13.3	4.3	12.2
30 to 50	10.9	30.8	8.9	25.1	7.7	21.8	6.9	19.5	6.3	17.8
50 to 70	14.2	40.2	11.6	32.9	10.1	28.5	9.0	25.5	8.2	23.2
70 to 80	12.1	34.2	9.9	27.9	8.5	24.2	7.6	11.6	7.0	19.7
80 to 90	6.2	17.6	5.0	14.4	4.4	12.5	3.9	11.1	3.6	10.2
90 to 95	2.8	7.8	2.3	6.4	2.0	5.5	1.7	4.9	1.6	4.5
Above 95	1.4	4.1	1.2	3.3	1.0	2.9	0.9	2.6	0.8	2.3

[A] These numbers represent the (1S) and (D2S) limits as described in Practice C 670.

C 666

 **Designation: C 995 – 94**

Standard Test Method for
Time of Flow of Fiber-Reinforced Concrete Through Inverted Slump Cone[1]

This standard is issued under the fixed designation C 995; the number immediately following the designation indicates the year of original adoption or, in the case of revision, the year of last revision. A number in parentheses indicates the year of last reapproval. A superscript epsilon (ϵ) indicates an editorial change since the last revision or reapproval.

1. Scope

1.1 This test method covers the determination of the inverted slump-cone time of fiber-reinforced concrete, both in the laboratory and in the field.

1.2 This test method is considered applicable to freshly mixed concrete having coarse aggregate up to 1½ in. (38 mm) in size. It is not applicable to concrete that flows freely through the cone.

1.3 The values stated in inch-pound units are to be regarded as the standard. SI units are for information only.

1.4 *This standard does not purport to address all of the safety concerns, if any, associated with its use. It is the responsibility of the user of this standard to establish appropriate safety and health practices and determine the applicability of regulatory limitations prior to use.*

2. Referenced Documents

2.1 *ASTM Standards:*
C 29/ C 29M Test Method for Unit Weight and Voids in Aggregate[2]
C 31 Practice for Making and Curing Concrete Test Specimens in the Field[2]
C 143 Test Method for Slump of Hydraulic Cement Concrete[2]
C 172 Practice for Sampling Freshly Mixed Concrete[2]
C 192 Practice for Making and Curing Concrete Test Specimens in the Laboratory[2]
C 670 Practice for Preparing Precision and Bias Statements for Test Methods for Construction Materials[2]

3. Summary of Test Method

3.1 This test method determines the time required for fiber-reinforced concrete to flow through an inverted slump cone under internal vibration.

4. Significance and Use

4.1 This test method provides a measure of the consistency and workability of fiber-reinforced concrete.

4.2 The inverted slump-cone time is a better indicator than slump of the appropriate level of workability for fiber-reinforced concrete placed by vibration because such concrete can exhibit very low slump due to the presence of the fibers and still be easily consolidated.

4.3 The results may be used for mixture proportioning, quality control both in the laboratory and in the field, and in development and research.

4.4 The results obtained using this test method may be influenced by vibrator diameter, amplitude, and frequency.

4.5 This test method may not be applicable to some concretes reinforced with fibers flexible and long enough to wrap around the vibrating element and dampen vibration.

5. Apparatus

5.1 *Cone*, shall be the mold specified in Test Method C 143.

5.2 *Bucket*—The container to receive the concrete shall be the 1-ft³ (30-L) capacity bucket specified in Test Method C 29/ C 29M.

5.3 *Positioning Device*—A device of the type shown in Fig. 1 shall be provided to center the cone in the bucket, prevent it from tilting, and maintain the small end of the cone 4 ± ¼ in. (100 ± 5 mm) from the bottom of the bucket.

5.4 *Vibrator*, shall be of the internal type specified in Practices C 31 or C 192, except that the vibrating element shall be 1 ± ⅛ in. (25 ± 3 mm) in diameter.

5.5 *Stopwatch*—One that measures elapsed time to the nearest second or less.

5.6 *Screeding Rod*—The rod used for screeding shall be the tamping rod specified in Test Method C 143.

6. Sampling

6.1 The sample of concrete for the test shall be representative of the entire batch. It shall be obtained in accordance with Practice C 172, except that wet-sieving shall not be permitted.

7. Procedure

7.1 Dampen the bucket and place it on a level, rigid, horizontal surface free of vibration and other disturbances. Dampen the cone and place it in the positioning device, making sure it is level. From the sample obtained in accordance with Section 6, fill the cone in three layers, each approximately one third of the volume of the cone. Avoid compacting the concrete, but lightly level each layer with a scoop or trowel to minimize the entrapment of large voids. Strike off the surface of the top layer by means of a screeding and rolling motion of the tamping rod. Protruding fibers which inhibit screeding may be removed by hand. One third of the volume of the cone corresponds to a depth of 5.875 in. (149 mm); two thirds of the volume corresponds to a depth of 9.375 in. (237 mm).

---

[1] This test method is under the jurisdiction of ASTM Committee C-9 on Concrete and Concrete Aggregates and is the direct responsibility of Subcommittee C09.42 on Fiber-Reinforced Concrete.
Current edition approved March 15, 1994. Published May 1994. Originally published as C 995 – 83. Last previous edition C 995 – 91.
[2] *Annual Book of ASTM Standards*, Vol 04.02.

C 995

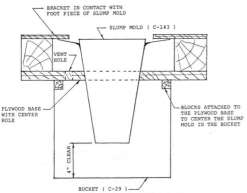

BRACKET IN CONTACT WITH
FOOT PIECE OF SLUMP MOLD

SLUMP MOLD (C-143)

VENT
HOLE

PLYWOOD BASE
WITH CENTER
HOLE

BLOCKS ATTACHED TO
THE PLYWOOD BASE
TO CENTER THE SLUMP
MOLD IN THE BUCKET

4" CLEAR

BUCKET (C-29)

FIG. 1   Method of Positioning the Inverted Slump Cone in the Bucket

NOTE—A small amount of the sample may fall through the cone during the filling process. This effect is minimized by ensuring the volume of the first layer of concrete placed is large enough to bridge the opening in the cone.

7.2 Start the vibrator. Simultaneously start the stopwatch and insert the vibrating element centrally and vertically into the top of the sample in the cone. Let it descend at a rate such that it touches the bottom of the bucket in 3 ±1 s. Maintain the vibrating element approximately vertical and in contact with the bottom of the bucket. Avoid touching the cone with the vibrator. Stop the stopwatch when the cone becomes empty which occurs when an opening becomes visible at the bottom of the cone. Complete the portion of the test from the start of filling the cone through allowing the vibrating element to descend through the sample without interruption and within an elapsed time of 2 min. If the cone becomes plugged during the test, or fails to empty because of an excess of material that has fallen through during filling, disregard the result and make a new test on another portion of the sample.

7.3 Record the time in seconds.

8. Report

8.1 Report the following information:

8.1.1 The time in seconds from initial immersion of the vibrating element to when the cone first becomes empty as the inverted slump-cone time.

8.1.2 The diameter, frequency, and amplitude of the vibrating element.

9. Precision and Bias

9.1 Based on limited data, the single-operator one-sigma limit has been found to be 1.0 s or less, and the between-operator one-sigma limit 1.5 s or less.[3]

9.2 A multilaboratory statement of precision is not appropriate for this test method.

10. Keywords

10.1 fiber; inverted slump cone; reinforced concrete; time of flow; workability

---

[3] These are the 1s limits described in Practice C 670.

AMERICAN SOCIETY FOR TESTING AND MATERIALS
1916 Race St. Philadelphia, Pa 19103
Reprinted from the Annual Book of ASTM Standards. Copyright ASTM
If not listed in the current combined index, will appear in the next edition.

Designation: D 198 – 94

Standard Methods of
Static Tests of Lumber in Structural Sizes[1]

This standard is issued under the fixed designation D 198; the number immediately following the designation indicates the year of original adoption or, in the case of revision, the year of last revision. A number in parentheses indicates the year of last reapproval. A superscript epsilon (ϵ) indicates an editorial change since the last revision or reapproval.

INTRODUCTION

Numerous evaluations of structural members of solid sawn lumber have been conducted in accordance with ASTM Methods D 198 – 27. While the importance of continued use of a satisfactory standard should not be underestimated, the original standard (1927) was designed primarily for sawn material such as solid wood bridge stringers and joists. With the advent of laminated timbers, wood-plywood composite members, and even reinforced and prestressed timbers, a procedure adaptable to a wider variety of wood structural members is required.

The present standard expands the original standard to permit its application to wood members of all types. It provides methods of evaluation under loadings other than flexure in recognition of the increasing need for improved knowledge of properties under such loadings as tension to reflect the increasing use of dimension lumber in the lower chords of trusses. The standard establishes practices that will permit correlation of results from different sources through the use of a uniform procedure. Provision is made for varying the procedure to take account of special problems.

1. Scope

1.1 These test methods cover the evaluation of lumber in structural size by various testing procedures.

1.2 The test methods appear in the following order:

	Sections
Flexure	4 to 11
Compression (Short Column)	12 to 19
Compression (Long Member)	20 to 27
Tension	28 to 35
Torsion	36 to 43
Shear Modulus	44 to 51

1.3 Notations and symbols relating to the various testing procedures are given in Table X1.1.

1.4 *This standard does not purport to address all of the safety concerns, if any, associated with its use. It is the responsibility of the user of this standard to establish appropriate safety and health practices and determine the applicability of regulatory limitations prior to use.*

2. Referenced Documents

2.1 *ASTM Standards:*

D 9 Terminology Relating to Wood[2]

D 1165 Nomenclature of Domestic Hardwoods and Softwoods[2]

D 2395 Test Methods for Specific Gravity of Wood and Wood-Base Materials[2]

D 4442 Test Methods for Direct Moisture Content Measurement of Wood and Wood-Base Materials[2]

D 4444 Test Methods for Use and Calibration of Hand-Held Moisture Meters[2]

E 4 Practices for Load Verification of Testing Machines[3]

E 6 Terminology Relating to Methods of Mechanical Testing[3]

E 83 Practice for Verification and Classification of Extensometers[3]

3. Terminology

3.1 *Definitions*—See Terminology E 6, Terminology D 9, and Nomenclature D 1165. A few related terms not covered in these standards are as follows:

3.1.1 *span*—the total distance between reactions on which a beam is supported to accommodate a transverse load (Fig. 1).

3.1.2 *shear span*—two times the distance between a reaction and the nearest load point for a symmetrically loaded beam (Fig. 1).

3.1.3 *depth of beam*—that dimension of the beam which is perpendicular to the span and parallel to the direction in which the load is applied (Fig. 1).

3.1.4 *span-depth ratio*—the numerical ratio of total span divided by beam depth.

3.1.5 *shear span-depth ratio*—the numerical ratio of shear span divided by beam depth.

3.1.6 *structural wood beam*—solid wood, laminated wood, or composite structural members for which strength or stiffness, or both, are primary criteria for the intended application and which usually are used in full length and in cross-sectional sizes greater than nominal 2 by 2 in. (38 by 38 mm).

3.1.7 *composite wood beam*—a laminar construction comprising a combination of wood and other simple or

---

[1] These methods are under the jurisdiction of ASTM Committee D-7 on Wood and are under the jurisdiction of Subcommittee D07.02 on Lumber and Engineered Wood Products.

Current edition approved May 15, 1994. Published July 1994. Originally published as D 198 – 24. Last previous edition D 198 – 84$^{\epsilon 1}$.

[2] *Annual Book of ASTM Standards,* Vol 04.10.

[3] *Annual Book of ASTM Standards,* Vol 03.01.

complex materials assembled and intimately fixed in relation to each other so as to use the properties of each to attain specific structural advantage for the whole assembly.

FLEXURE

4. Scope

4.1 This test method covers the determination of the flexural properties of structural beams made of solid or laminated wood, or of composite constructions. This test method is intended primarily for beams of rectangular cross section but is also applicable to beams of round and irregular shapes, such as round posts, I-beams, or other special sections.

5. Summary of Test Method

5.1 The structural member, usually a straight or a slightly cambered beam of rectangular cross section, is subjected to a bending moment by supporting it near its ends, at locations called reactions, and applying transverse loads symmetrically imposed between these reactions. The beam is deflected at a prescribed rate, and coordinate observations of loads and deflections are made until rupture occurs.

6. Significance and Use

6.1 The flexural properties established by this test method provide:

6.1.1 Data for use in development of grading rules and specifications.

6.1.2 Data for use in development of working stresses for structural members.

6.1.3 Data on the influence of imperfections on mechanical properties of structural members.

6.1.4 Data on strength properties of different species or grades in various structural sizes.

6.1.5 Data for use in checking existing equations or hypotheses relating to the structural behavior of beams.

6.1.6 Data on the effects of chemical or environmental conditions on mechanical properties.

6.1.7 Data on effects of fabrication variables such as depth, taper, notches, or type of end joint in laminations.

6.1.8 Data on relationships between mechanical and physical properties.

6.2 Procedures are described here in sufficient detail to permit duplication in different laboratories so that comparisons of results from different sources will be valid. Special circumstances may require deviation from some details of these procedures. Any variations shall be carefully described in the report (see Section 11).

7. Apparatus

7.1 *Testing Machine*—A device that provides (*1*) a rigid frame to support the specimen yet permit its deflection without restraint, (*2*) a loading head through which the force is applied without high-stress concentrations in the beam, and (*3*) a force-measuring device that is calibrated to ensure accuracy in accordance with Practices E 4.

7.2 *Support Apparatus:*

7.2.1 *Reaction Bearing Plates*—The beam shall be supported by metal bearing plates to prevent damage to the beam at the point of contact between beam and reaction support (Fig. 1). The size of the bearing plates may vary with the size and shape of the beam. For rectangular beams as large as 12 in. (305 mm) deep by 6 in. (152 mm) wide, the recommended size of bearing plate is ½ in. (13 mm) thick by 6 in. (152 mm) lengthwise and extending entirely across the width of the beam.

7.2.2 *Reaction Bearing Roller*—The bearing plates shall be supported by either rollers and a fixed knife edge reaction or a rocker type-knife edge reaction so that shortening and rotation of the beam about the reaction due to deflection will be unrestricted (Fig. 1).

7.2.3 *Reaction Bearing Alignment*—Provisions shall be made at the reaction to allow for initial twist in the length of the beam. If the bearing surfaces of the beam at its reactions are not parallel, the beam shall be shimmed or the individual bearing plates shall be rotated about an axis parallel to the span to provide full bearing across the width of the specimen (Fig. 2).

7.2.4 *Lateral Support*—Specimens that have a depth-to-width ratio of three or greater are subject to lateral instability during loading, thus requiring lateral support. Support shall be provided at least at points located about half-way between the reaction and the load point. Additional supports may be used as required. Each support shall allow vertical movement without frictional restraint but shall restrict lateral deflection (Fig. 3).

7.3 *Load Apparatus:*

7.3.1 *Load Bearing Blocks*—The load shall be applied through bearing blocks (Fig. 1) across the full beam width which are of sufficient thickness to eliminate high-stress

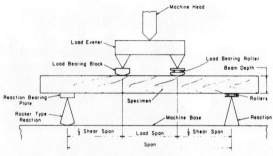

FIG. 1 Flexure Method

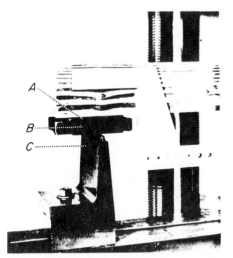

FIG. 2   Example of Bearing Plate, A, Rollers, B, and Reaction-
Alignment-Rocker, C, for Small Beams

concentrations at places of contact between beam and bearing blocks. The loading surface of the blocks shall have a radius of curvature equal to two to four times the beam depth for a chord length at least equal to the depth of the beam. Load shall be applied to the blocks in such a manner that the blocks may rotate about an axis perpendicular to the span (Fig. 4). Provisions such as rotatable bearings or shims shall be made to ensure full contact between the beam and both loading blocks. Metal bearing plates and rollers shall be used in conjunction with one load bearing block to permit beam deflection without restraint (Fig. 4). The size of these plates and rollers may vary with the size and shape of the beam, the same as for the reaction bearing plates. Beams having circular or irregular cross sections shall have bearing blocks which distribute the load uniformly to the bearing surface and permit, unrestrained deflections.

7.3.2 *Load Points*—The total load on the beam shall be applied equally at two points equidistant from the reactions. The two load points will normally be at a distance from their reaction equal to one third of the span, but for special purposes other distances may be specified.

NOTE 1—One of the objectives of two-point loading is to subject the portion of the beam between load points to a uniform bending moment, free of shear, and with comparatively small loads at the load points. For example, loads applied at one-third span length from reactions would be less than if applied at one-fourth span length from reaction to develop a moment of similar magnitude. When loads are applied at the one-third points the moment distribution of the beam simulates that for loads uniformly distributed across the span to develop a moment of similar magnitude. If loads are applied at the outer one-fourth points of the span, the maximum moment and shear are the same as the maximum moment and shear for the same total load uniformly distributed across the span.

7.4 *Deflection Apparatus:*

7.4.1 *General*—For either apparent or true modulus of elasticity calculations, devices shall be provided by which the deflection of the neutral axis of the beam at the center of the span is measured with respect to either the reaction or between cross sections free of shear deflections.

7.4.2 *Wire Deflectometer*—Deflection may be read directly by means of a wire stretched taut between two nails driven into the neutral axis of the beam directly above the reactions and extending across a scale attached at the neutral axis of the beam at midspan. Deflections may be read with a telescope or reading glass to magnify the area where the wire crosses the scale. When a reading glass is used, a reflective surface placed adjacent to the scale will help to avoid parallax.

7.4.3 *Yoke Deflectometer*—A satisfactory device commonly used for short, small beams or to measure deflection of the center of the beam with respect to any point along the neutral axis consists of a lightweight U-shaped yoke suspended between nails driven into the beam at its neutral axis and a dial micrometer attached to the center of the yoke with its stem attached to a nail driven into the beam at midspan at the neutral axis. Further modification of this device may be attained by replacing the dial micrometer with a deflection transducer for automatic recording (Fig. 4).

7.4.4 *Accuracy*—The devices shall be such as to permit

FIG. 3   Example of Lateral Support for Long, Deep Beams

FIG. 4   Example of Curved Loading Block, A, Load-Alignment Rocker, B, Roller-Curved Loading Block, C, Load Evener, D, and Deflection-Measuring Apparatus, E

measurements to the nearest 0.01 in. (0.25 mm) on spans greater than 3 ft (0.9 m) and 0.001 in. (0.03 mm) on spans less than 3 ft (0.9 m).

8. Test Specimen

8.1 *Material*—The test specimen shall consist of a structural member which may be solid wood, laminated wood, or a composite construction of wood or of wood combined with plastics or metals in sizes that are usually used in structural applications.

8.2 *Identification*—Material or materials of the test specimen shall be identified as fully as possible by including the origin or source of supply, species, and history of drying and conditioning, chemical treatment, fabrication, and other pertinent physical or mechanical details which may affect the strength. Details of this information shall depend on the material or materials in the beam. For example, the solid wooden beams would be identified by the character of the wood, that is, species, source, etc., whereas composite wooden beams would be identified by the characteristics of the dissimilar materials and their size and location in the beam.

8.3 *Specimen Measurements*—The weight and dimensions as well as moisture content of the specimen shall be accurately determined before test. Weights and dimensions (length and cross section) shall be measured to three significant figures. Sufficient measurements of the cross section shall be made along the length of the beam to describe the width and depth of rectangular specimen and to accurately describe the critical section or sections of nonuniform beams. The physical characteristics of the specimen as described by its density and moisture content may be determined in accordance with Test Methods D 2395 and Method A of Test Methods D 2016.

8.4 *Specimen Description*—The inherent imperfections or intentional modifications of the composition of the beam shall be fully described by recording the size and location of

such factors as knots, checks, and reinforcements. Size and location of intentional modifications such as placement of laminations, glued joints, and reinforcing steel shall be recorded during the fabrication process. The size and location of imperfections in the interior of any beam must be deduced from those on the surface, especially in the case of large sawn members. A sketch or photographic record shall be made of each face and the ends showing the size, location, and type of growth characteristics, including slope of grain, knots, distribution of sapwood and heartwood, location of pitch pockets, direction of annual rings, and such abstract factors as crook, bow, cup, or twist which might affect the strength of the beam.

8.5 *Rules for Determination of Specimen Length*—The cross-sectional dimensions of solid wood structural beams and composite wooden beams usually have established sizes, depending upon the manufacturing process and intended use, so that no modification of these dimensions is involved. The length, however, will be established by the type of data desired. The span length is determined from knowledge of beam depth, the distance between load points, as well as the type and orientation of material in the beam. The total beam length shall also include an overhang or extension beyond each reaction support so that the beam can accommodate the bearing plates and rollers and will not slip off the reactions during test.

NOTE 2—Some evaluations will require simulation of a specific design condition where nonnormal overhang is involved. In such instances the report shall include a complete description of test conditions, including overhang at each support.

8.5.1 The span length of beams intended primarily for evaluation of shear properties shall be such that the shear span is relatively short. Beams of wood of uniform rectangular cross section having the ratio of a/h less than five are in this category and provide a high percentage of shear failures.

NOTE 3—If approximate values of modulus of rupture S_R and shear strength τ_m are known, a/h values should be less than $S_R/4\tau_m$, assuming that when $a/h = S_R/4\tau_m$ the beam will fail at the same load in either shear or in extreme outer fibers.

8.5.2 The span length of beams intended primarily for evaluation of flexural properties shall be such that the shear span is relatively long. Beams of wood of uniform rectangular cross section having a/h ratios of from 5:1 to 12:1 are in this category.

NOTE 4—The a/h values should be somewhat greater than $S_R/4\tau_m$ so that the beams do not fail in shear but should not be so large that beam deflections cause sizable thrust of reactions and thrust values need to be taken into account. A suggested range of a/h values is between approximately 0.5 S_R/τ_m and 1.2 S_R/τ_m. In this category, shear distortions affect the total deflection, so that flexural properties may be corrected by formulae provided in the appendix.

8.5.3 The span length of beams intended primarily for evaluation of only the deflection of specimen due to bending moment shall be such that the shear span is long. Wood beams of uniform rectangular cross section in this category have a/h ratios greater than 12:1.

NOTE 5—The shear stresses and distortions are assumed to be small so that they can be neglected; hence the a/h ratio is suggested to be greater than S_R/τ_m.

9. Procedure

9.1 *Conditioning*—Unless otherwise indicated in the research program or material specification, condition the test specimen to constant weight so it is in moisture equilibrium under the desired environmental conditions. Approximate moisture contents with moisture meters or measure more accurately by weights of samples in accordance with Method A of Test Methods D 2016.

9.2 *Test Setup*—Determine the size of the specimen, the span, and the shear span in accordance with 7.3.2 and 8.5. Locate the beam symmetrically on its supports with load bearing and reaction bearing blocks as described in 7.2 to 7.4. The beams shall be adequately supported laterally in accordance with 7.2.4. Set apparatus for measuring deflections in place (see 7.4). Full contact shall be attained between support bearings, loading blocks, and the beam surface.

9.3 *Speed of Testing*—Conduct the test at a constant rate to achieve maximum load in about 10 min, but maximum load should be reached in not less than 6 min nor more than 20 min. A constant rate of outer strain, z, of 0.0010 in./in.·min (0.001 mm/mm·min) will usually permit the tests of wood members to be completed in the prescribed time. The rate of motion of the movable head of the test machine corresponding to this suggested rate of strain when two symmetrical concentrated loads are employed may be computed from the following equation:

$$N = Za(3L - 4a)/3h$$

9.4 *Load-Deflection Curves:*

9.4.1 Obtain load-deflection data with apparatus described in 7.4.1. Note the load and deflection at first failure, at the maximum load, and at points of sudden change. Continue loading until complete failure or an arbitrary terminal load has been reached.

9.4.2 If additional deflection apparatus is provided to measure deflection over a second distance, l, in accordance with 7.4.1, such load-deflection data shall be obtained only up to the proportional limit.

9.5 *Record of Failures*—Describe failures in detail as to type, manner and order of occurrence, and position in beam. Record descriptions of the failures and relate them to drawings or photographs of the beam referred to in 8.4. Also record notations as the order of their occurrence on such references. Hold the section of the beam containing the failure for examination and reference until analysis of the data has been completed.

10. Calculation

10.1 Compute physical and mechanical properties and their appropriate adjustments for the beam in accordance with the relationships in Appendix X2.

11. Report

11.1 Report the following information:

11.1.1 Complete identification of the solid wood or composite construction, including species, origin, shape and form, fabrication procedure, type and location of imperfections or reinforcements, and pertinent physical or chemical characteristics relating to the quality of the material,

11.1.2 History of seasoning and conditioning,

11.1.3 Loading conditions to portray the load, support

mechanics, lateral supports, if used, and type of equipment,

11.1.4 Deflection apparatus,

11.1.5 Depth and width of the specimen or pertinent cross-sectional dimensions,

11.1.6 Span length and shear span distance,

11.1.7 Rate of load application,

11.1.8 Computed physical and mechanical properties, including specific gravity and moisture content, flexural strength, stress at proportional limit, modulus of elasticity, and a statistical measure of variability of these values,

11.1.9 Data for composite beams include shear and bending moment values and deflections,

11.1.10 Description of failure, and

11.1.11 Details of any deviations from the prescribed or recommended methods as outlined in the standard.

COMPRESSION PARALLEL TO GRAIN (SHORT COLUMN, NO LATERAL SUPPORT, $l/r < 17$)

12. Scope

12.1 This test method covers the determination of the compressive properties of elements taken from structural members made of solid or laminated wood, or of composite constructions when such an element has a slenderness ratio (length to least radius of gyration) of less than 17. The method is intended primarily for members of rectangular cross section but is also applicable to irregularly shaped studs, braces, chords, round posts, or special sections.

13. Summary of Test Method

13.1 The structural member is subjected to a force uniformly distributed on the contact surface of the specimen in a direction generally parallel to the longitudinal axis of the wood fibers, and the force generally is uniformly distributed throughout the specimen during loading to failure without flexure along its length.

14. Significance and Use

14.1 The compressive properties obtained by axial compression will provide information similar to that stipulated for flexural properties in Section 6.

14.2 The compressive properties parallel to grain include modulus of elasticity, stress at proportional limit, compressive strength, and strain data beyond proportional limit.

15. Apparatus

15.1 *Testing Machine*—Any device having the following is suitable:

15.1.1 *Drive Mechanism*—A drive mechanism for imparting to a movable loading head a uniform, controlled velocity with respect to the stationary base.

15.1.2 *Load Indicator*—A load-indicating mechanism capable of showing the total compressive force on the specimen. This force-measuring system shall be calibrated to ensure accuracy in accordance with Methods E 4.

15.2 *Bearing Blocks*—Bearing blocks shall be used to apply the load uniformly over the two contact surfaces and to prevent eccentric loading on the specimen. At least one spherical bearing block shall be used to ensure uniform bearing. Spherical bearing blocks may be used on either or both ends of the specimen, depending on the degree of parallelism of bearing surfaces (Fig. 5). The radius of the

D 198

FIG. 5   Compression of a Wood Structural Element

sphere shall be as small as practicable, in order to facilitate adjustment of the bearing plate to the specimen, and yet large enough to provide adequate spherical bearing area. This radius is usually one to two times the greatest cross-section dimension. The center of the sphere shall be on the plane of the specimen contact surface. The size of the compression plate shall be larger than the contact surface. It has been found convenient to provide an adjustment for moving the specimen on its bearing plate with respect to the center of spherical rotation to ensure axial loading.

15.3 *Compressometer:*

15.3.1 *Gage Length*—For modulus of elasticity calculations, a device shall be provided by which the deformation of the specimen is measured with respect to specific paired gage points defining the gage length. To obtain test data representative of the test material as a whole, such paired gage points shall be located symmetrically on the lengthwise surface of the specimen as far apart as feasible, yet at least one times the larger cross-sectional dimension from each of the contact surfaces. At least two pairs of such gage points on diametrically opposite sides of the specimen shall be used to measure the average deformation.

15.3.2 *Accuracy*—The device shall be able to measure changes in deformation to three significant figures. Since gage lengths vary over a wide range, the measuring instruments should conform to their appropriate class in accordance with Practice E 83.

16. Test Specimen

16.1 *Material*—The test specimen shall consist of a structural member which may be solid wood, laminated wood, or a composite construction of wood or of wood combined with plastics or metals in sizes that are commercially used in structural applications, that is, in sizes greater than nominal 2 by 2-in. (38 by 38-mm) cross section (see 3.1.6).

16.2 *Identification*—Material or materials of the test specimen shall be as fully described as that for beams in 8.2.

16.3 *Specimen Dimensions*—The weight and dimensions, as well as moisture content of the specimen, shall be accurately measured before test. Weights and dimensions (length and cross section) shall be measured to three significant figures. Sufficient measurements of the cross section shall be made along the length of the specimen to describe shape characteristics and to determine the smallest section. The physical characteristics of the specimen, as described by its density and moisture content, may be determined in accordance with Test Methods D 2395 and Method A of Test Methods D 2016, respectively.

16.4 *Specimen Description*—The inherent imperfections and intentional modifications shall be described as for beams in 8.4.

16.5 *Specimen Length*—The length of the specimen shall be such that the compressive force continues to be uniformly distributed throughout the specimen during loading—hence no flexure occurs. To meet this requirement, the specimen shall be a short column having a maximum length, l, less than 17 times the least radius of gyration, r, of the cross section of the specimen (see compressive notations). The minimum length of the specimen for stress and strain measurements shall be greater than three times the larger cross section dimension or about ten times the radius of gyration.

17. Procedure

17.1 *Conditioning*—Unless otherwise indicated in the research program or material specification, condition the test specimen to constant weight so it is at moisture equilibrium, under the desired environment. Approximate moisture contents with moisture meters or measure more accurately by weights of samples in accordance with Method A of Test Methods D 2016.

17.2 *Test Setup:*

17.2.1 *Bearing Surfaces*—After the specimen length has been calculated in accordance with 17.5, cut the specimen to the proper length so that the contact surfaces are plane, parallel to each other, and normal to the long axis of the specimen. Furthermore, the axis of the specimen shall be generally parallel to the fibers of the wood.

NOTE 6—A sharp fine-toothed saw of either the crosscut or "novelty" crosscut type has been used satisfactorily for obtaining the proper end surfaces. Power equipment with accurate table guides is especially recommended for this work

NOTE 7—It is desirable to have failures occur in the body of the specimen and not adjacent to the contact surface. Therefore, the cross-sectional areas adjacent to the loaded surface may be reinforced.

17.2.2 *Centering*—First geometrically center the specimens on the bearing plates and then adjust the spherical seats so that the specimen is loaded uniformly and axially.

17.3 *Speed of Testing*—For measuring load-deformation data, apply the load at a constant rate of head motion so that the fiber strain is 0.001 in./in.·min ± 25 % (0.001 mm/mm·min). For measuring only compressive strength, the test may be conducted at a constant rate to achieve maximum load in about 10 min, but not less than 5 nor more than 20 min.

17.4 *Load-Deformation Curves*—If load-deformation data have been obtained, note the load and deflection at first failure, at changes in slope of curve, and at maximum load.

17.5 *Records*—Record the maximum load, as well as a description and sketch of the failure relating the latter to the location of imperfections in the specimen. Reexamine the section of the specimen containing the failure during analysis of the data.

18. Calculation

18.1 Compute physical and mechanical properties in accordance with Terminology E 6, and as follows (see compressive notations):

18.1.1 Stress at proportional limit = P'/A in pounds per square inch (MPa).

18.1.2 Compressive strength = P/A in pounds per square inch (MPa).

18.1.3 Modulus of elasticity = $P'/A\epsilon$ in pounds per square inch (MPa).

19. Report

19.1 Report the following information:

19.1.1 Complete identification,

19.1.2 History of seasoning and conditioning,

19.1.3 Load apparatus,

19.1.4 Deflection apparatus,

19.1.5 Length and cross-section dimensions,

19.1.6 Gage length,

19.1.7 Rate of load application,

19.1.8 Computed physical and mechanical properties, including specific gravity and moisture content, compressive strength, stress at proportional limit, modulus of elasticity, and a statistical measure of variability of these values,

19.1.9 Description of failure, and

19.1.10 Details of any deviations from the prescribed or recommended methods as outlined in the standard.

COMPRESSION PARALLEL TO GRAIN (CRUSHING STRENGTH OF LATERALLY SUPPORTED LONG MEMBER, EFFECTIVE $l'/r < 17$)

20. Scope

20.1 This test method covers the determination of the compressive properties of structural members made of solid or laminated wood, or of composite constructions when such a member has a slenderness ratio (length to least radius of gyration) of more than 17, and when such a member is to be evaluated in full size but with lateral supports which are spaced to produce an effective slenderness ratio, l'/r, of less than 17. This test method is intended primarily for members of rectangular cross section but is also applicable to irregularly shaped studs, braces, chords, round posts, or special sections.

21. Summary of Test Method

21.1 The structural member is subjected to a force uniformly distributed on the contact surface of the specimen in a direction generally parallel to the longitudinal axis of the wood fibers, and the force generally is uniformly distributed throughout the specimen during loading to failure without flexure along its length.

22. Significance and Use

22.1 The compressive properties obtained by axial compression will provide information similar to that stipulated for flexural properties in Section 6.

22.2 The compressive properties parallel to grain include modulus of elasticity, stress at proportional limit, compressive strength, and strain data beyond proportional limit.

23. Apparatus

23.1 *Testing Machine*—Any device having the following is suitable:

23.1.1 *Drive Mechanism*—A drive mechanism for imparting to a movable loading head a uniform, controlled velocity with respect to the stationary base.

23.1.2 *Load Indicator*—A load-indicating mechanism capable of showing the total compressive force on the specimen. This force-measuring system shall be calibrated to ensure accuracy in accordance with Methods E 4.

23.2 *Bearing Blocks*—Bearing blocks shall be used to apply the load uniformly over the two contact surfaces and to prevent eccentric loading on the specimen. One spherical bearing block shall be used to ensure uniform bearing, or a rocker-type bearing block shall be used on each end of the specimen with their axes of rotation at 0° to each other (Fig. 6). The radius of the sphere shall be as small as practicable, in order to facilitate adjustment of the bearing plate to the specimen, and yet large enough to provide adequate spherical bearing area. This radius is usually one to two times the greatest cross-section dimension. The center of the sphere shall be on the plane of the specimen contact surface. The size of the compression plate shall be larger than the contact surface.

23.3 *Lateral Support:*

23.3.1 *General*—Evaluation of the crushing strength of long structural members requires that they be supported laterally to prevent buckling during the test without undue

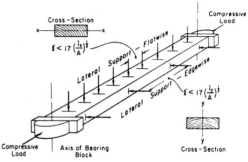

FIG. 6   Minimum Spacing of Lateral Supports of Long Columns

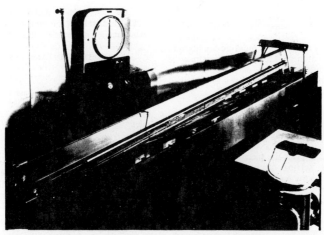

D 198

FIG. 7 Compression of Long Slender Structural Member

pressure against the sides of the specimen. Furthermore, the support shall not restrain either the longitudinal compressive deformation or load during test. The support shall be either continuous or intermittent. Intermittent supports shall be spaced so that the distance, l', between supports is less than 17 times the least radius of gyration of the cross section.

23.3.2 *Rectangular Members*—The general rules for structural members apply to rectangular structural members. However, the effective column length as controlled by intermittent support spacing on flatwise face need not equal that on edgewise face. The minimum spacing of the supports on the flatwise face shall be 17 times the least radius of gyration of the cross section which is about the centroidal axis parallel to flat face. And the minimum spacing of the supports on the edgewise face shall be 17 times the other radius of gyration (Fig. 6). A satisfactory method of providing lateral support for 2-in. (38-mm) dimension stock is shown in Fig. 7. A 27-in. (686-mm) I-beam provides the frame for the test machine. Small I-beams provide reactions for longitudinal pressure. A pivoted top I-beam provides lateral support on one flatwise face, while the web of the large I-beam provides the other. In between these steel members, metal guides on 3-in. (7.6-cm) spacing (hidden from view) attached to plywood fillers provide the flatwise support and contact surface. In between the flanges of the 27-in. I-beam, fingers and wedges provide edgewise lateral support.

23.4 *Compressometer:*

23.4.1 *Gage Length*—For modulus of elasticity calculations, a device shall be provided by which the deformation of the specimen is measured with respect to specific paired gage points defining the gage length. To obtain data representative of the test material as a whole, such paired gage points shall be located symmetrically on the lengthwise surface of the specimen as far apart as feasible, yet at least one times the larger cross-sectional dimension from each of the contact surfaces. At least two pairs of such gage points on diametrically opposite sides of the specimen shall be used to measure the average deformation.

23.4.2 *Accuracy*—The device shall be able to measure changes in deformation to three significant figures. Since gage lengths vary over a wide range, the measuring instruments should conform to their appropriate class in accordance with Practice E 83.

24. Test Specimen

24.1 *Material*—The test specimen shall consist of a structural member which may be solid wood, laminated wood, or it may be a composite construction of wood or of wood combined with plastics or metals in sizes that are commercially used in structural applications, that is, in sizes greater than nominal 2 by 2-in. (38 by 38-mm) cross section (see 3.1.6).

24.2 *Identification*—Material or materials of the test specimen shall be as fully described as that for beams in 8.2.

24.3 *Specimen Dimensions*—The weight and dimensions, as well as moisture content of the specimen, shall be accurately measured before test. Weights and dimensions (length and cross section) shall be measured to three significant figures. Sufficient measurements of the cross section shall be made along the length of the specimen to describe shape characteristics and to determine the smallest section. The physical characteristics of the specimen, as described by its density and moisture content, may be determined in accordance with Test Methods D 2395 and Method A of Test Methods D 2016, respectively.

24.4 *Specimen Description*—The inherent imperfections and intentional modifications shall be described as for beams in 8.4.

24.5 *Specimen Length*—The cross-sectional and length dimensions of structural members usually have established sizes, depending on the manufacturing process and intended use, so that no modification of these dimensions is involved. Since the length has been approximately established, the full length of the member shall be tested, except for trimming or squaring the bearing surface (see 25.2.1).

25. Procedure

25.1 *Preliminary*—Unless otherwise indicated in the research program or material specification, condition the test specimen to constant weight so it is at moisture equilibrium, under the desired environment. Moisture contents may be approximated with moisture meters or more accurately measured by weights of samples in accordance with Method A of Test Methods D 2016.

25.2 *Test Setup:*

25.2.1 *Bearing Surfaces*—Cut the bearing surfaces of the specimen so that the contact surfaces are plane, parallel to each other, and normal to the long axis of the specimen.

25.2.2 *Setup Method*—After physical measurements have been taken and recorded, place the specimen in the testing machine between the bearing blocks at each end and between the lateral supports on the four sides. Center the contact surfaces geometrically on the bearing plates and then adjust the spherical seats for full contact. Apply a slight longitudinal pressure to hold the specimen while the lateral supports are adjusted and fastened to conform to the warp, twist, or bend of the specimen.

25.3 *Speed of Testing*—For measuring load-deformation data, apply the load at a constant rate of head motion so that the fiber strain is 0.001 in./in. · min ± 25 % (0.001 mm/mm · min). For measuring only compressive strength, the test may be conducted at a constant rate to achieve maximum load in about 10 min, but not less than 5 nor more than 20 min.

25.4 *Load-Deformation Curves*—If load-deformation data have been obtained, note load and deflection at first failure, at changes in slope of curve, and at maximum load.

25.5 *Records*—Record the maximum load as well as a description and sketch of the failure relating the latter to the location of imperfections in the specimen. Reexamine the section of the specimen containing the failure during analysis of the data.

26. Calculation

26.1 Compute physical and mechanical properties in accordance with Terminology E 6 and as follows (see compressive notations):

26.1.1 Stress at proportional limit = P'/A in pounds per square inch (MPa).

26.1.2 Compressive strength = P/A in pounds per square inch (MPa).

26.1.3 Modulus of elasticity = $P'/A\epsilon$ in pounds per square inch (MPa).

27. Report

27.1 Report the following information:
27.1.1 Complete identification,
27.1.2 History of seasoning conditioning,
27.1.3 Load apparatus,
27.1.4 Deflection apparatus,
27.1.5 Length and cross-section dimensions,
27.1.6 Gage length,
27.1.7 Rate of load application,
27.1.8 Computed physical and mechanical properties, including specific gravity of moisture content, compressive strength, stress at proportional limit, modulus of elasticity, and a statistical measure of variability of these values,

27.1.9 Description of failure, and
27.1.10 Details of any deviations from the prescribed or recommended methods as outlined in the standard.

TENSION PARALLEL TO GRAIN

28. Scope

28.1 This test method covers the determination of the tensile properties of structural elements made primarily of lumber equal to and greater than nominal 1 in. (19 mm) thick.

29. Summary of Test Method

29.1 The structural member is clamped at the extremities of its length and subjected to a tensile load so that in sections between clamps the tensile forces shall be axial and generally uniformly distributed throughout the cross sections without flexure along its length.

30. Significance and Use

30.1 The tensile properties obtained by axial tension will provide information similar to that stipulated for flexural properties in Section 6.

30.2 The tensile properties obtained include modulus of elasticity, stress at proportional limit, tensile strength, and strain data beyond proportional limit.

31. Apparatus

31.1 *Testing Machine*—Any device having the following is suitable:

31.1.1 *Drive Mechanism*—A drive mechanism for imparting to a movable clamp a uniform, controlled velocity with respect to a stationary clamp.

31.1.2 *Load Indicator*—A load-indicating mechanism capable of showing the total tensile force on the test section of the tension specimen. This force-measuring system shall be calibrated to ensure accuracy in accordance with Practices E 4.

31.1.3 *Grips*—Suitable grips or fastening devices shall be provided which transmit the tensile load from the movable head of the drive mechanism to one end of the test section of the tension specimen, and similar devices shall be provided to transmit the load from the stationary mechanism to the other end of the test section of the specimen. Such devices shall not apply a bending moment to the test section, allow slippage under load, inflict damage, or inflict stress concentrations to the test section. Such devices may be either plates bonded to the specimen or unbonded plates clamped to the specimen by various pressure modes.

31.1.3.1 *Grip Alignment*—The fastening device shall apply the tensile loads to the test section of the specimen without applying a bending moment. For ideal test conditions, the grips should be self-aligning, that is, they should be attached to the force mechanism of the machine in such a manner that they will move freely into axial alignment as soon as the load is applied, and thus apply uniformly distributed forces along the test section and across the test cross section (Fig. 8(*a*)). For less ideal test conditions, each grip should be gimbaled about one axis which should be perpendicular to the wider surface of the rectangular cross

ⒶⓈⓉⓂ **D 198**

section of the test specimen, and the axis of rotation should be through the fastened area (Fig. 8(*b*)). When neither self-aligning grips nor single gimbaled grips are available, the specimen may be clamped in the heads of a universal-type testing machine with wedge-type jaws (Fig. 8(*c*)). A method of providing approximately full spherical alignment has three axes of rotation, not necessarily concurrent but, however, having a common axis longitudinal and through the centroid of the specimen (Figs. 8(*d*) and 9).

31.1.3.2 *Contact Surface*—The contact surface between grips and test specimen shall be such that slippage does not occur. A smooth texture on the grip surface should be avoided, as well as very rough and large projections which damage the contact surface of the wood. Grips that are surfaced with a coarse emery paper (60× aluminum oxide emery belt) have been found satisfactory for softwoods. However, for hardwoods, grips may have to be glued to the specimen to prevent slippage.

31.1.3.3 *Contact Pressure*—For unbonded grip devices, lateral pressure should be applied to the jaws of the grip so that slippage does not occur between grip and specimen. Such pressure may be applied by means of bolts or wedge-shaped jaws, or both. Wedge-shaped jaws, such as those shown on Fig. 10, which slip on the inclined plane to produce contact pressure have been found satisfactory. To eliminate stress concentration or compressive damage at the tip end of the jaw, the contact pressure should be reduced to zero. The variable thickness jaws (Fig. 10), which cause a variable contact surface and which produce a lateral pressure gradient, have been found satisfactory.

31.1.4 *Extensometer:*

FIG. 9   Horizontal Tensile Grips for 2 by 10-in. Structural Members

31.1.4.1 *Gage Length*—For modulus of elasticity determinations, a device shall be provided by which the elongation of the test section of the specimen is measured with respect to specific paired gage points defining the gage length. To obtain data representative of the test material as a whole, such gage points shall be symmetrically located on the lengthwise surface of the specimen as far apart as feasible, yet at least two times the larger cross-sectional dimension from each jaw edge. At least two pairs of such gage points on diametrically opposite sides of the specimen shall be used to measure the average deformation.

31.1.4.2 *Accuracy*—The device shall be able to measure changes in elongation to three significant figures. Since gage lengths vary over a wide range, the measuring instruments

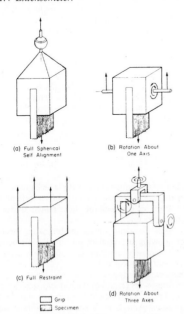

FIG. 8   Types of Tension Grips for Structural Members

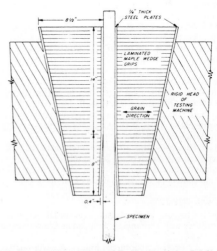

FIG. 10   Side View of Wedge Grips Used to Anchor Full-Size Structurally, Graded Tension Specimens

should conform to their appropriate class in accordance with Practice E 83.

32. Test Specimen

32.1 *Material*—The test specimen shall consist of a structural member which may be solid wood, laminated wood, or it may be a composite construction of wood or wood combined with plastics or metals in sizes that are commercially used in structural "tensile" applications, that is, in sizes equal to and greater than nominal 1-in. (32-mm) thick lumber.

32.2 *Identification*—Material or materials of the test specimen shall be fully described as beams in 8.2.

32.3 *Specimen Description*—The specimen shall be described in a manner similar to that outlined in 8.3 and 8.4.

32.4 *Specimen Length*—The tension specimen, which has its long axis parallel to grain in the wood, shall have a length between grips equal to at least eight times the larger cross-sectional dimension when tested in self-aligning grips (see 31.1.3.1). However, when tested without self-aligning grips, it is recommended that the length between grips be at least 20 times the greater cross-sectional dimension.

33. Procedure

33.1 *Conditioning*—Unless otherwise indicated, condition the specimen as outlined in 9.1.

33.2 *Test Setup*—After physical measurements have been taken and recorded, place the specimen in the grips of the load mechanism, taking care to have the long axis of the specimen and the grips coincide. The grips should securely clamp the specimen with either bolts or wedge-shaped jaws. If the latter are employed, apply a small preload to ensure that all jaws move an equal amount and maintain axial-alignment of specimen and grips. If either bolts or wedges are employed tighten the grips evenly and firmly to the degree necessary to prevent slippage. Under load, continue the tightening if necessary, even crushing the wood perpendicular to grain, so that no slipping occurs and a tensile failure occurs outside the jaw contact area.

33.3 *Speed of Testing*—For measuring load-elongation data, apply the load at a constant rate of head motion so that the fiber strain in the test section between jaws is 0.0006 in./in.·min ± 25 % (0.0006 mm/mm·min). For measuring only tensile strength, the load may be applied at a constant rate of grip motion so that maximum load is achieved in about 10 min but not less than 5 nor more than 20 min.

33.4 *Load-Elongation Curves*—If load-elongation data have been obtained throughout the test, correlate changes in specimen behavior, such as appearance of cracks or splinters, with elongation data.

33.5 *Records*—Record the maximum load, as well as a description and sketch of the failure relating the latter to the location of imperfections in the test section. Reexamine the section containing the failure during analysis of data.

34. Calculation

34.1 Compute physical and mechanical properties in accordance with Terminology E 6, and as follows (see tensile notations):

34.1.1 Stress at proportional limit = P'/A in pounds per square inch (MPa).

34.1.2 Tensile strength = P/A in pounds per square inch (MPa).

34.1.3 Modulus of elasticity = $P'/A\epsilon$ in pounds per square inch (MPa).

35. Report

35.1 Report the following information:

35.1.1 Complete identification,

35.1.2 History of seasoning,

35.1.3 Load apparatus, including type of end condition,

35.1.4 Deflection apparatus,

35.1.5 Length and cross-sectional dimensions,

35.1.6 Gage length,

35.1.7 Rate of load application,

35.1.8 Computed properties,

35.1.9 Description of failures, and

35.1.10 Details of any deviations from the prescribed or recommended methods as outlined in the standard.

TORSION

36. Scope

36.1 This test method covers the determination of the torsional properties of structural elements made of solid or laminated wood, or of composite constructions. This test method is intended primarily for structural element or rectangular cross section but is also applicable to beams of round or irregular shapes.

37. Summary of Test Method

37.1 The structural element is subjected to a torsional moment by clamping it near its ends and applying opposing couples to each clamping device. The element is deformed at a prescribed rate and coordinate observations of torque and twist are made for the duration of the test.

38. Significance and Use

38.1 The torsional properties obtained by twisting the structural element will provide information similar to that stipulated for flexural properties in Section 6.

38.2 The torsional properties of the element include an apparent modulus of rigidity of the element as a whole, stress at proportional limit, torsional strength, and twist beyond proportional limit.

39. Apparatus

39.1 *Testing Machine*—Any device having the following is suitable:

39.1.1 *Drive Mechanism*—A drive mechanism for imparting an angular displacement at a uniform rate between a movable clamp on one end of the element and another clamp at the other end.

39.1.2 *Torque Indicator*—A torque-indicating mechanism capable of showing the total couple on the element. This measuring system shall be calibrated to ensure accuracy in accordance with Practices E 4.

39.2 *Support Apparatus:*

39.2.1 *Clamps*—Each end of the element shall be securely held by metal plates of sufficient bearing area and strength to grip the element with a vise-like action without slippage, damage, or stress concentrations in the test section when the

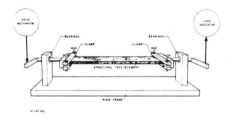

FIG. 11   Fundamentals of a Torsional Test Machine

torque is applied to the assembly. The plates of the clamps shall be symmetrical about the longitudinal axis of the cross section of the element.

39.2.2 *Clamp Supports*—Each of the clamps shall be supported by roller bearings or bearing blocks that allow the structural element to rotate about its natural longitudinal axis. Such supports may be ball bearings in a rigid frame of a torque-testing machine (Figs. 11 and 12) or they may be bearing blocks (Figs. 13 and 14) on the stationary and movable frames of a universal-type test machine. Either type of support shall allow the transmission of the couple without friction to the torque measuring device, and shall allow freedom for longitudinal movement of the element during the twisting. Apparatus of Fig. 13 is not suitable for large amounts of twist unless the angles are measured at each end to enable proper torque calculation.

39.2.3 *Frame*—The frame of the torque-testing machine shall be capable of providing the reaction for the drive mechanism, the torque indicator, and the bearings. The framework necessary to provide these reactions in a universal-type test machine shall be two rigid steel beams attached to the movable and stationary heads forming an X. The extremities of the X shall bear on the lever arms attached to the test element (Fig. 13).

39.3 *Troptometer:*

39.3.1 *Gage Length*—For modulus of rigidity calcula-

tions, a device shall be provided by which the angle of twist of the element is measured with respect to specific paired gage points defining the gage length. To obtain test data representative of the element as a whole, such paired gage points shall be located symmetrically on the lengthwise surface of the element as far apart as feasible, yet at least two times the larger cross-sectional dimension from each of the clamps. A yoke (Fig. 16) or other suitable device (Fig. 12) shall be firmly attached at each gage point to permit measurement of the angle of twist. The angle of twist is measured by observing the relative rotation of the two yokes or other devices at the gage points with the aid of any suitable apparatus including a light beam (Fig. 12), dials (Fig. 14), or string and scale (Figs. 15 and 16).

39.3.2 *Accuracy*—The device shall be able to measure changes in twist to three significant figures. Since gage lengths may vary over a wide range, the measuring instruments should conform to their appropriate class in accordance with Practice E 83.

40. Test Element

40.1 *Material*—The test element shall consist of a structural member, which may be solid wood, laminated wood, or a composite construction of wood or wood combined with plastics or metals in sizes that are commercially used in structural applications.

40.2 *Identification*—Material or materials of the test element shall be as fully described as for beams in 8.2.

40.3 *Element Measurements*—The weight and dimensions as well as the moisture content shall be accurately determined before test. Weights and dimensions (length and cross section) shall be measured to three significant figures. Sufficient measurements of the cross section shall be made along the length of the specimen to describe characteristics and to determine the smallest cross section. The physical characteristics of the element, as described by its density and moisture content, may be determined in accordance with Test Methods D 2395 and Test Methods D 2016, Method A, respectively.

40.4 *Element Description*—The inherent imperfections

FIG. 12   Example of Torque-Testing Machine (Torsion test in apparatus meeting specification requirements)

D 198

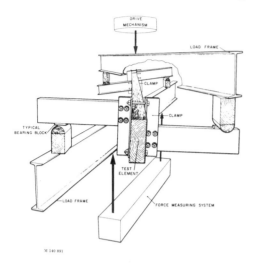

FIG. 13 Schematic Diagram of a Torsion Test Made in a Universal-Type Test Machine

and intentional modifications shall be described as for beams in 8.4.

40.5 *Element Length*—The cross-sectional dimensions of solid wood structural elements and composite elements usually are established, depending upon the manufacturing process and intended use so that normally no modification of these dimensions is involved. However, the length of the specimen shall be at least eight times the larger cross-

sectional dimension.

41. Procedure

41.1 *Conditioning*—Unless otherwise indicated in the research program or material specification, condition the test element to constant weight so it is at moisture equilibrium under the desired environment. Approximate moisture contents with moisture meters, or measure more accurately by weights of samples in accordance with Test Methods D 2016, Method A.

41.2 *Test Setups*—After physical measurements have been taken and recorded, place the element in the clamps of the load mechanism, taking care to have the axis of rotation of the clamps coincide with the longitudinal centroidal axis of the element. Tighten the clamps to securely hold the element in either type of testing machine. If the tests are made in a universal-type test machine, the bearing blocks shall be equal distances from the axis of rotation of the element.

41.3 *Speed of Testing*—For measuring torque-twist data, apply the load at a constant rate of head motion so that the angular detrusion of the outer fibers in the test section between gage points is about 0.004 radian per inch of length (0.16 radian per metre of length) per minute ±50 %. For measuring only shear strength, the torque may be applied at a constant rate of twist so that maximum torque is achieved in about 10 min but not less than 5 nor more than 20 min.

41.4 *Torque-Twist Curves*—If torque-twist data have been obtained, note torque and twist at first failure, at changes in slope of curve, and at maximum torque.

41.5 *Record of Failures*—Describe failures in detail as to type, manner and order of occurrence, angle with the grain, and position in the test element. Record descriptions relating to imperfections in the element. Reexamine the section of the element containing the failure during analysis of the data.

FIG. 14 Example of Torsion Test of Structural Beam in a Universal-Type Test Machine

(ASTM) D 198

FIG. 15   Torsion Test with Yoke-Type Troptometer

42. Calculation

42.1 Compute physical and mechanical properties in accordance with Terminology E 6 and relationships in Tables X3.1 and X3.2.

43. Report

43.1 Report the following information:
43.1.1 Complete identification,
43.1.2 History of seasoning and conditioning,
43.1.3 Apparatus for applying and measuring torque,
43.1.4 Apparatus for measuring angle of twist,
43.1.5 Length and cross-section dimensions,
43.1.6 Gage length,
43.1.7 Rate of twist applications,
43.1.8 Computed properties, and
43.1.9 Description of failures.

SHEAR MODULUS

44. Scope

44.1 This test method covers the determination of the modulus of rigidity (G) or shear modulus of structural beams made of solid or laminated wood. Application to composite constructions can only give a measure of the apparent or

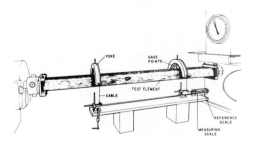

M 140 890

FIG. 16   Troptometer Measuring System

effective shear modulus. This test method is intended primarily for beams of rectangular cross section but is also applicable to other sections with appropriate modification of equation coefficients.

45. Summary of Test Method

45.1 The structural member, usually a straight or a slightly cambered beam of rectangular cross section, is subjected to a bending moment by supporting it at two locations called reactions, and applying a single transverse load midway between these reactions. The beam is deflected at a prescribed rate and a single observation of coordinate load and deflection is taken. This procedure is repeated on at least four different spans.

46. Significance and Use

46.1 The shear modulus established by this test method will provide information similar to that stipulated for flexural properties in Section 6.

47. Apparatus

47.1 The test machine and specimen configuration, supports, and loading are identical to Section 7 with the following exception:
47.1.1 The load shall be applied as a single, concentrated load midway between the reactions.

48. Test Specimen

48.1 See Section 8.

49. Procedure

49.1 *Conditioning*—See 9.1.
49.2 *Test Setup*—Position the specimen in the test machine as described in 9.2 and load in center point bending over at least four different spans with the same cross section at the center of each. Choose the spans so as to give approximately equal increments of $(h/L)^2$ between them, within the range from 0.035 to 0.0025. The applied load must be sufficient to provide a reliable estimate of the initial bending stiffness of the specimen, but in no instance shall

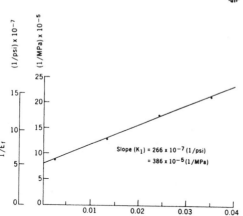

FIG. 17   Determination of Shear Modulus

NOTE 8—Span to depth ratios of 5.5, 6.5, 8.5, and 20 meet the $(h/L)^2$ requirements of this section.

49.3 *Load-Deflection Measurements*—Obtain load-deflection data with the apparatus described in 7.4.1. One data point is required on each span tested.

49.4 *Records*—Record span to depth ratios chosen and load levels achieved on each span.

49.5 *Speed of Testing*—See 9.3.

50. Calculation

50.1 Determine shear modulus by plotting $1/E_f$ (where E_f is the apparent modulus of elasticity calculated under center point loading) versus $(h/L)^2$ for each span tested. As indicated in Fig. 17 and in Appendix X4, shear modulus is proportional to the slope of the best-fit line between these points.

51. Report

51.1 See Section 11.

PRECISION AND BIAS

52. Precision and Bias

52.1 The precision and bias of these test methods are being established.

exceed the proportional limit or shear capacity of the specimen.

⑤ D 198

APPENDIXES

(Nonmandatory Information)

X1. PHYSICAL PROPERTIES

TABLE X1.1 Physical Properties

Specific gravity (at test), $G_g = CW_g/V$	Test Methods D 2395
Specific gravity (ovendry), $G_d = G_g/(100 + MC)$	
Moisture content (% of dry weight), $MC = 100 (W_g/W_d - 1)$	Method A of Test Methods D 2016

GENERAL NOTATIONS

A	Cross-sectional area, in.² (mm²).	L	Span of beam, in. (mm).
C	0.061, a constant for use when W_g is measured in grams in equation for specific gravity.	M	Maximum bending moment at maximum load, lbf·in. (N·m).
		M'	Maximum bending moment at proportional limit load, lbf·in. (N·m).
	22.7, a constant for use when W_g is measured in pounds in equation for specific gravity.	P	Maximum transverse load on beam, lbf (N).
ϵ	Strain at proportional limit, in./in. (mm/mm).	P'	Load on beam at proportional limit, lbf (N).
G_d	Specific gravity (ovendry).	S_f	Fiber stress at proportional limit, psi (MPa).
G_g	Specific gravity (at test).	S_R	Modulus of rupture.
I	Moment of inertia of the cross section about a designated axis, in.⁴ (mm⁴).	z	Rate of fiber strain, in./in. (mm/mm), of outer fiber length per min.
		Δ	Deflection of beam, in. (mm), at neutral axis between reaction and center of beam at the proportional limit, in. (mm).
N	Rate of motion of movable head, in./min (mm/min).	Δ_f	Deflection of the beam measured at midspan over distance l, in. (mm).
n	Number of specimens in sample.	τ_m	Maximum shear strength, psi (MPa).
S	Estimated standard deviation = $[(\Sigma X^2 - n\bar{X}^2)/(n - 1)]1/2$.		
V	Volume, in.³ (mm³).		**COMPRESSIVE NOTATIONS**
W_g	Weight of moisture specimen (at test), lb (g).	l	Length of compression column, in. (mm).
W_d	Weight of moisture specimen (ovendry), lb (g).	l'	Effective length of column between supports for lateral stability, in. (mm).
X	Individual values.		
$\bar{X}$	Average of n individual values.	P	Maximum compressive load, lbf (N).
		P'	Compressive load at proportional limit, lbf (N).
	FLEXURAL NOTATIONS	r	Radius of gyration = $[(I)/(A)]1/2$, in. (mm).
a	Distance from reaction to nearest load point, in. (mm) (½ shear span).		**TENSILE NOTATIONS**
A_m	Area of graph paper under the load-deflection curve from zero load to maximum load in.² (mm²) when deflection is measured between reaction and center of span.	P	Maximum tensile load, lbf (N).
		P'	Tensile load at proportional limit, lbf (N).
A_t	Area of graph paper under load-deflection curve from zero load to failing load or arbitrary terminal load, in.² (mm²), when deflection is measured between reaction and center of span.		**SHEAR NOTATIONS**
b	Width of beam, in. (mm).	E	Modulus of elasticity.
c	Distance from neutral axis of beam to extreme outer fiber, in. (mm).	E_f	Apparent E, center point loading.
G	Modulus of rigidity in shear, psi (MPa).	G	Modulus of rigidity (shear modulus).
h	Depth of beam, in. (mm).	I	Moment of inertia.
k	Graph paper scale constant for converting unit area of graph paper to load-deflection units.	P'	Load on beam at deflection, Δ', lbf (N) (below proportional limit).
		Δ'	Deflection of beam, in. (mm).
l	Span of the beam that is used to measure deflections caused only by the bending moment, that is, no shear distortions, in. (mm).	K	Shear coefficient. Defined in Table X4.1.
		K_1	Slope of line through multiple test data plotted on $(h/L)^2$ versus $(1/E_f)$.

D 198

X2. FLEXURE

TABLE X2.1 Flexure Formulas[A]

Mechanical Properties	General	Two-Point Loading Rectangular Beam	Third-Point Loading Rectangular Beam
Fiber stress at proportional limit, S_f	$\dfrac{M'c}{I}$	$\dfrac{3P'a}{bh^2}$	$\dfrac{P'L}{bh^2}$
Modulus of rupture, S_R	$\dfrac{Mc}{I}$	$\dfrac{3Pa}{bh^2}$	$\dfrac{PL}{bh^2}$
Modulus of elasticity, E_f (apparent E)	$\dfrac{P'a}{48I\Delta}(3L^2-4a^2)$	$\dfrac{P'a}{4bh^3\Delta}(3L^2-4a^2)$	$\dfrac{P'L^3}{4.7bh^3\Delta}$
Modulus of elasticity, E_G (true E, shear corrected)	$\dfrac{M'l^2}{8I\Delta_f}$	$\dfrac{P'a(3L^2-4a^2)}{4bhh^3\Delta}\left(1-\dfrac{3P'a}{5bhG\Delta}\right)$ $\dfrac{3P'al^2}{4bh^3\Delta_f}$	$\dfrac{P'L^3}{4.7bh^3\Delta\left(1-\dfrac{P'L}{5bhG\Delta}\right)}$ $\dfrac{P'Ll^2}{4bh^3\Delta_f}$
Work to proportional limit per unit of volume, W_k		$\dfrac{P'\Delta}{2Lbh}\left[\dfrac{4a(3L-4a)+\dfrac{24h^2E_G}{10G}}{3L^2-4a^2+\dfrac{24h^2E_G}{10G}}\right]$	$\dfrac{P'\Delta}{2Lbh}\left[\dfrac{\dfrac{20}{9}L^2+\dfrac{24h^2E_G}{10G}}{\dfrac{23}{9}L^2+\dfrac{24h^2E_G}{10G}}\right]$
Approximate work to maximum load per unit of volume, W_m		$\dfrac{KA_m}{Lbh}\left[\dfrac{4a(3L-4a)+\dfrac{24h^2E_G}{10G}}{3L^2-4a^2+\dfrac{24h^2E_G}{10G}}\right]$	$\dfrac{KA_m}{Lbh}\left[\dfrac{\dfrac{20}{9}L^2+\dfrac{24h^2E_G}{10G}}{\dfrac{23}{9}L^2+\dfrac{24h^2E_G}{10G}}\right]$
Approximate total work per unit of volume, W_t		$\dfrac{KA_t}{Lbh}\left[\dfrac{4a(3L-4a)+\dfrac{24h^2E_G}{10G}}{3L^2-4a^2+\dfrac{24h^2E_G}{10G}}\right]$	$\dfrac{KA_t}{Lbh}\left[\dfrac{\dfrac{20}{9}L^2+\dfrac{24h^2E_G}{10G}}{\dfrac{23}{9}L^2+\dfrac{24h^2E_G}{10G}}\right]$
Shear stress, τ_m		$\dfrac{3}{4}\dfrac{P}{bh}$	$\dfrac{0.75G}{bh}$

[A] For wooden beams having uniform cross section throughout their length.

X3. TORSION

TABLE X3.1 Torsion Formulas[A]

Mechanical Properties	Cross Section			
	Circle	Square	Rectangle	General[B]
Fiber shear stress of greatest intensity at middle of long side; at proportional limit, S_s'	$2T'/\pi r^3$ (1A)	$4.808\,T'/w^3$ (1B)	$8\gamma T'/\mu wt^2$ (1C)	T'/Q (1D)
Fiber shear strength of greatest intensity at middle of long side, S_s	$2T/\pi r^3$ (2A)	$4.808\,T/w^3$ (2B)	$8\gamma T/\mu wt^2$ (2C)	T/Q (2D)
Fiber shear strength at middle of short side, S_s''			$8\gamma_1 T/\mu t^3$ (3C)	
Apparent modulus of rigidity, G	$2L_gT'/\pi r^4\theta$ (4A)	$7.11\,L_gT'/w^4\theta$ (4B)	$16\,L_gT'/wt^3[(16/3)-\lambda(t/w)]\,\theta$ (4C)	$L_gT'/\theta K$ (4D)

[A] From NACA rep. 334.
[B] Values of "Q" and "K" may be found in Roark, R. J., *Formulas for Stress and Strain*, McGraw-Hill, 1965, p. 194.

D 198

TABLE X3.2 Factors for Calculating Torsional Rigidity and Stress of Rectangular Prisms[A]

Ratio of Sides Column 1	λ Column 2	μ Column 3	γ Column 4	γ₁ Column 5
1.00	3.08410	2.24923	1.35063	1.35063
1.05	3.12256	2.35908	1.39651	
1.10	3.15653	2.46374	1.43956	
1.15	3.18554	2.56330	1.47990	
1.20	3.21040	2.65788	1.51753	
1.25	3.23196	2.74772	1.55268	1.13782
1.30	3.25035	2.83306	1.58544	
1.35	3.26632	2.91379	1.61594	
1.40	3.28002	2.99046	1.64430	
1.45	3.29171	3.06319	1.67265	
1.50	3.30174	3.13217	1.69512	0.97075
1.60	3.31770	3.25977	1.73889	0.91489
1.70	3.32941	3.37486	1.77649	
1.75	3.33402	3.42843	1.79325	0.84098
1.80	3.33798	3.47890	1.80877	
1.90	3.34426	3.57320	1.83643	
2.00	3.34885	3.65891	1.86012	0.73945
2.25	3.35564	3.84194	1.90543	
2.50	3.35873	3.98984	1.93614	0.59347
2.75	3.36023	4.11143	1.95687	
3.00	3.36079	4.21307	1.97087	
3.33	...	...	...	0.44545
3.50	3.36121	4.37299	1.98672	
4.00	3.36132	4.49300	1.99395	0.37121
4.50	3.36133	4.58639	1.99724	
5.00	3.36133	4.66162	1.99874	0.29700
6.00	3.36133	4.77311	1.99974	
6.67	3.36133	...	...	0.22275
7.00	3.36133	4.85314	1.99995	
8.00	3.36133	4.91317	1.99999	0.18564
9.00	3.36133	4.95985	2.00000	
10.00	3.36133	4.99720	2.00000	0.14858
20.00	3.36133	5.16527	2.00000	0.07341
50.00	3.36133	5.26611	2.00000	
100.00	3.36133	5.29972	2.00000	
∞	3.36133	5.33333	2.00000	0.00000

[A] Table I, "Factors for Calculating Torsional Rigidity and Stress of Rectangular Prisms," from National Advisory Committee for Aeronautics Report No. 334, "The Torsion of Members Having Sections Common in Aircraft Construction," by G. W. Trayer and H. W. March about 1929.

Torsion Notations

G	Apparent modulus of rigidity, psi (MPa).
K	Stiffness-shape factor.
L_g	Gage length of torsional element, in. (mm).
Q	Stress-shape factor.
r	Radius, in. (mm).
S_s	Fiber shear stress of greatest intensity at middle of long side at proportional limit, psi (MPa).
S_s	Fiber shear strength of greatest intensity at middle of long side at maximum torque, psi (MPa).
S_s''	Fiber shear strength at middle of short side at maximum torque, psi (MPa).
T	Twisting moment or torque, lbf·in. (N·m).
T'	Torque at proportional limit, lbf·in. (N·m).
t	Thickness, in. (mm).
w	Width of element, in. (mm).
γ	St. Venant constant, Column 4, Table X4.1.
γ_1	St. Venant constant, Column 5, Table X4.1.
θ	Total angle of twist, radians (in./in. or mm/mm).
λ	St. Venant constant, Column 2, Table X4.1.
μ	St. Venant constant, Column 3, Table X4.1.

X4. SHEAR MODULUS

X4.1 The elastic deflection of a prismatic beam under a single center point load is:

$$\Delta = \frac{PL^3}{48EI} + \frac{PL}{4GA'} \qquad (X4.1)$$

where:

Δ = deflection at midspan,
P = applied load,
L = span,
E = modulus of elasticity,
I = moment of inertia,
G = modulus of rigidity (shear modulus), and
A' = modified shear area.

X4.2 All parameters are self-explanatory with the exception of the modified shear area. The modified shear area is the product of the cross-sectional area, A, and a shear coefficient, K.[5] The shear coefficient relates the effective transverse shear strain to the average shear stress on the section. "K" is defined as the ratio of average shear strain on a section to shear strain at the centroid. Shear coefficients have been calculated and tabulated for a variety of beam configurations.

[5] Cowper, G. R., "The Shear Coefficient in Timoshenko's Beam Theory," *Journal of Applied Mechanics*, ASME, 1966, pp. 335–340.

⬡ D 198

TABLE X4.1   Shear Modulus Formulas

Mechanical Property	Formula
Modulus of elasticity, E_f (apparent E, center point loading)	$\dfrac{P'L^3}{48/\Delta'}$
Shear modulus, G^A	
Rectangular section	$1.2/K_1{}^B$
Circular section	$1.55/K_1$

^{A} Based on solution of the equation $\Delta = (PL^3/48EI) + (PL/4KGA)$. K is tabulated for other cross sections by Cowper, G. R., "The Shear Coefficient in Timoshenko's Beam Theory," *Journal of Applied Mechanics*, ASME, 1966, pp. 335–340.
^{B} K_1 = Slope of the line plotted through the test values as shown in Fig. 17.

X4.2.1 Introducing K into Eq 1:

$$\Delta = \frac{PL^3}{48EI} + \frac{PL}{4GKA} \qquad (X4.2)$$

X4.3 Often the relationship between deflection and elastic constants is simplified by ignoring the shear contribution, or the second term in Eq 2. The remaining elastic constant is called the "apparent" modulus of elasticity, E_f:

$$\Delta = \frac{PL^3}{48E_f I} \qquad (X4.3)$$

X4.4 At the same deflection the apparent modulus of elasticity can be expressed in terms of the true elastic constants:

$$\frac{PL^3}{48E_f I} = \frac{PL^3}{48EI} + \frac{PL}{4GKA} \qquad (X4.4)$$

X4.5 For a rectangular section of width, b, and depth, h, Eq 4 reduces to:

$$\frac{L^2}{E_f h^2} = \frac{L^2}{Eh^2} + \frac{1}{KG} \qquad (X4.5)$$

X4.6 Multiplying both sides of Eq 5 by $(h/L)^2$ yields:

$$\frac{1}{E_f} = \frac{1}{E} + \frac{1}{KG}(h/L)^2 \qquad (X4.6)$$

X4.6.1 Equation 6 can be graphed by substituting $y = 1/E_f$ and $x = (h/L)^2$. In the resulting $y = mx + b$ graph, the slope of a line connecting multiple data points is equal to $1/KG$.

X4.7 For a circular section of diameter, h, Eq 4 reduces to:

$$\frac{1}{E_f} = \frac{1}{E} + \frac{3}{4KG}(h/L)^2 \qquad (X4.7)$$

X4.8 Using values for $K = (10(1 + v))/(12 + 11v)$ (rectangular) and $K = (6(1 + v))/(7 + 6v)$ (circular) and Poisson's ratios ranging from 0.05 to 0.5 yield:[4]

$$\begin{array}{l}\text{Rectangular: } K = 0.84 \text{ to } 0.86, \text{ and} \\ \text{Circular: } K = 0.86 \text{ to } 0.90.\end{array} \qquad (X4.8)$$

X4.9 On plots of $1/E_f$ versus $(h/L)^2$, shear modulus, G, can be expressed in terms of the slope of the line connecting multiple observations. If the slope is called K_1, then:

$$\begin{array}{l}G = 1.17/K1 \text{ to } 1.20/K_1 \text{ (rectangular), and} \\ G = 1.48/K1 \text{ to } 1.55/K_1 \text{ (circular)}.\end{array} \qquad (X4.9)$$

X4.10 As CIB/RILEM has already proposed $1.2/K_1$ for rectangular beams, the corresponding value for circular beams, $1.55/K_1$, should be used.

X4.11 Determination of shear modulus for other beam cross sections must start at Eq 4, substituting appropriate values for I, A, and K.

Designation: D 1559 – 89

Standard Test Method for
Resistance to Plastic Flow of Bituminous Mixtures Using Marshall Apparatus[1]

This standard is issued under the fixed designation D 1559; the number immediately following the designation indicates the year of original adoption or, in the case of revision, the year of last revision. A number in parentheses indicates the year of last reapproval. A superscript epsilon (ε) indicates an editorial change since the last revision or reapproval.

This standard has been approved for use by agencies of the Department of Defense. Consult the DoD Index of Specifications and Standards for the specific year of issue which has been adopted by the Department of Defense.

1. Scope

1.1 This test method covers the measurement of the resistance to plastic flow of cylindrical specimens of bituminous paving mixture loaded on the lateral surface by means of the Marshall apparatus. This test method is for use with mixtures containing asphalt cement, asphalt cut-back or tar, and aggregate up to 1-in. (25.4-mm) maximum size.

1.2 *This standard may involve hazardous materials, operations, and equipment. This standard does not purport to address all of the safety problems associated with its use. It is the responsibility of the user of this standard to establish appropriate safety and health practices and determine the applicability of regulatory limitations prior to use.*

2. Significance and Use

2.1 This test method is used in the laboratory mix design of bituminous mixtures. Specimens are prepared in accordance with the method and tested for maximum load and flow. Density and voids properties may also be determined on specimens prepared in accordance with the test method. The testing section of this method can also be used to obtain maximum load and flow for bituminous paving specimens cored from pavements or prepared by other methods. These results may differ from values obtained on specimens prepared by this test method.

3. Apparatus

3.1 *Specimen Mold Assembly*—Mold cylinders 4 in. (101.6 mm) in diameter by 3 in. (76.2 mm) in height, base plates, and extension collars shall conform to the details shown in Fig. 1. Three mold cylinders are recommended.

3.2 *Specimen Extractor*, steel, in the form of a disk with a diameter not less than 3.95 in. (100 mm) and ½ in. (13 mm) thick for extracting the compacted specimen from the specimen mold with the use of the mold collar. A suitable bar is required to transfer the load from the ring dynamometer adapter to the extension collar while extracting the specimen.

3.3 *Compaction Hammer*—The compaction hammer (Fig. 2) shall have a flat, circular tamping face and a 10-lb (4536-g) sliding weight with a free fall of 18 in. (457.2 mm).

[1] This test method is under the jurisdiction of ASTM Committee D-4 on Road and Paving Materials and is the direct responsibility of Subcommittee D04.20 on Mechanical Tests of Bituminous Mixes.
Current edition approved Oct. 18, 1989. Published December 1989. Originally published as D 1559 – 58. Last previous edition D 1559 – 82.

Two compaction hammers are recommended.

NOTE 1—The compaction hammer may be equipped with a finger safety guard as shown in Fig. 2.

3.4 *Compaction Pedestal*—The compaction pedestal shall consist of an 8 by 8 by 18-in. (203.2 by 203.2 by 457.2-mm) wooden post capped with a 12 by 12 by 1-in. (304.8 by 304.8 by 25.4-mm) steel plate. The wooden post shall be oak, pine, or other wood having an average dry weight of 42 to 48 lb/ft^3 (0.67 to 0.77 g/cm^3). The wooden post shall be secured by four angle brackets to a solid concrete slab. The steel cap shall be firmly fastened to the post. The pedestal assembly shall be installed so that the post is plumb and the cap is level.

3.5 *Specimen Mold Holder*, mounted on the compaction pedestal so as to center the compaction mold over the center of the post. It shall hold the compaction mold, collar, and base plate securely in position during compaction of the specimen.

3.6 *Breaking Head*—The breaking head (Fig. 3) shall consist of upper and lower cylindrical segments or test heads having an inside radius of curvature of 2 in. (50.8 mm) accurately machined. The lower segment shall be mounted on a base having two perpendicular guide rods or posts extending upward. Guide sleeves in the upper segment shall be in such a position as to direct the two segments together without appreciable binding or loose motion on the guide rods.

3.7 *Loading Jack*—The loading jack (Fig. 4) shall consist of a screw jack mounted in a testing frame and shall produce a uniform vertical movement of 2 in. (50.8 mm)/min. An electric motor may be attached to the jacking mechanism.

NOTE 2—Instead of the loading jack, a mechanical or hydraulic testing machine may be used provided the rate of movement can be maintained at 2 in. (50.8 mm)/min while the load is applied.

3.8 *Ring Dynamometer Assembly*—One ring dynamometer (Fig. 4) of 5000-lb (2267-kg) capacity and sensitivity of 10 lb (4.536 kg) up to 1000 lb (453.6 kg) and 25 lb (11.340 kg) between 1000 and 5000 lb (453.6 and 2267 kg) shall be equipped with a micrometer dial. The micrometer dial shall be graduated in 0.0001 in. (0.0025 mm). Upper and lower ring dynamometer attachments are required for fastening the ring dynamometer to the testing frame and transmitting the load to the breaking head.

NOTE 3—Instead of the ring dynamometer assembly, any suitable load-measuring device may be used provided the capacity and sensitivity meet the above requirements.

⊕ D 1559

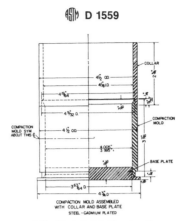

Table of Equivalents for Figs. 1 and 3

Inch-Pound Units, in.	Metric Equivalents, mm	Inch-Pound Units, in.	Metric Equivalents, mm	Inch-Pound Units, in.	Metric Equivalents, mm	Inch-Pound Units, in.	Metric Equivalents, mm
0.005	0.11	$11/16$	17.5	$2^5/16$	58.7	$4^1/8$	104.8
$1/32$	0.8	$3/8$	19.0	$2^1/2$	63.5	$4^9/32$	108.7
$1/16$	1.6	$7/8$	22.2	$2^3/4$	69.8	$4^{19}/64$	109.1
$1/8$	3.2	$15/16$	23.8	$2^7/8$	73.0	$4^1/2$	114.3
$3/16$	4.8	1	25.4	3	76.2	$4^5/8$	117.5
$1/4$	6.4	$1^7/8$	28.6	$3^1/4$	82.6	$4^3/4$	120.6
$9/32$	7.1	$1^1/4$	31.8	$3^7/16$	87.3	$5^1/16$	128.6
$3/8$	9.5	$1^3/8$	34.9	$3^7/8$	98.4	$5^1/8$	130.2
0.496	12.6	$1^1/2$	38.1	$3^{63}/64$	101.2	$5^3/4$	146.0
0.499	12.67	$1^5/8$	41.3	3.990	101.35	6	152.4
$1/2$	12.7	$1^3/4$	44.4	3.995	101.47	$6^1/4$	158.8
$9/16$	14.3	2	50.8	4	101.6	$7^5/8$	193.7
$5/8$	15.9	$2^1/4$	57.2	4.005	101.73	27	685.8

FIG. 1 Compaction Mold

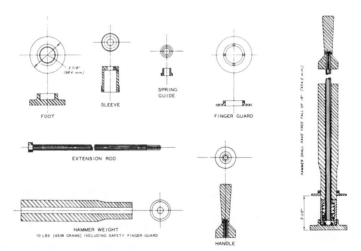

FIG. 2 Compaction Hammer

3.9 *Flowmeter*—The flowmeter shall consist of a guide sleeve and a gage. The activating pin of the gage shall slide inside the guide sleeve with a slight amount of frictional resistance. The guide sleeve shall slide freely over the guide rod of the breaking head. The flowmeter gage shall be adjusted to zero when placed in position on the breaking

ASTM **D 1559**

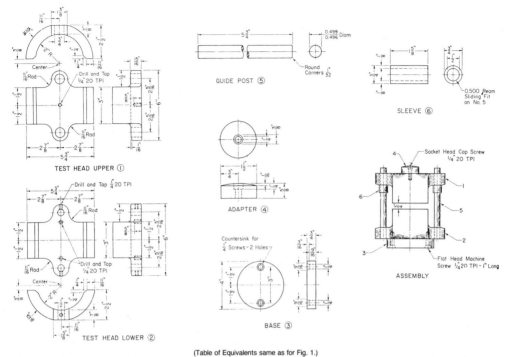

(Table of Equivalents same as for Fig. 1.)

FIG. 3   Breaking Head

head when each individual test specimen is inserted between the breaking head segments. Graduations of the flowmeter gage shall be in 0.01-in. (0.25-mm) divisions.

NOTE 4—Instead of the flowmeter, a micrometer dial or stress-strain recorder graduated in 0.001 in. (0.025 mm) may be used to measure flow.

3.10 *Ovens or Hot Plates*—Ovens or hot plates shall be provided for heating aggregates, bituminous material, specimen molds, compaction hammers, and other equipment to the required mixing and molding temperatures. It is recommended that the heating units be thermostatically controlled so as to maintain the required temperature within 5°F (2.8°C). Suitable shields, baffle plates or sand baths shall be used on the surfaces of the hot plates to minimize localized overheating.

3.11 *Mixing Apparatus*—Mechanical mixing is recommended. Any type of mechanical mixer may be used provided it can be maintained at the required mixing temperature and will provide a well-coated, homogeneous mixture of the required amount in the allowable time, and further provided that essentially all of the batch can be recovered. A metal pan or bowl of sufficient capacity and hand mixing may also be used.

3.12 *Water Bath*—The water bath shall be at least 6 in. (152.4 mm) deep and shall be thermostatically controlled so as to maintain the bath at 140 ± 1.8°F (60 ± 1.0°C) or 100 ±

1.8°F (37.8 ± 1°C). The tank shall have a perforated false bottom or be equipped with a shelf for supporting specimens 2 in. (50.8 mm) above the bottom of the bath.

3.13 *Air Bath*—The air bath for asphalt cut-back mixtures shall be thermostatically controlled and shall maintain the air temperature at 77°F ± 1.8°F (25 ± 1.0°C).

3.14 *Miscellaneous Equipment:*

3.14.1 *Containers* for heating aggregates, flat-bottom metal pans or other suitable containers.

3.14.2 *Containers* for heating bituminous material, either gill-type tins, beakers, pouring pots, or saucepans may be used.

3.14.3 *Mixing Tool*, either a steel trowel (garden type) or spatula, for spading and hand mixing.

3.14.4 *Thermometers* for determining temperatures of aggregates, bitumen, and bituminous mixtures. Armored-glass or dial-type thermometers with metal stems are recommended. A range from 50 to 400°F (9.9 to 204°C), with sensitivity of 5°F (2.8°C) is required.

3.14.5 *Thermometers* for water and air baths with a range from 68 to 158°F (20 to 70°C) sensitive to 0.4°F (0.2°C).

3.14.6 *Balance*, 2-kg capacity, sensitive to 0.1 g, for weighing molded specimens.

3.14.7 *Balance*, 5-kg capacity, sensitive to 1.0 g, for batching mixtures.

3.14.8 *Gloves* for handling hot equipment.

D 1559

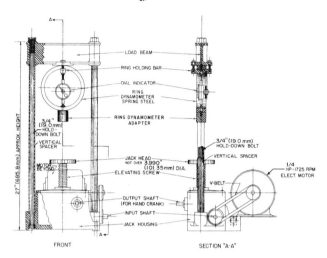

FIG. 4 Compression Testing Machine

3.14.9 *Rubber Gloves* for removing specimens from water bath.

3.14.10 *Marking Crayons* for identifying specimens.

3.14.11 *Scoop*, flat bottom, for batching aggregates.

3.14.12 *Spoon*, large, for placing the mixture in the specimen molds.

4. Test Specimens

4.1 *Number of Specimens*—Prepare at least three specimens for each combination of aggregates and bitumen content.

4.2 *Preparation of Aggregates*—Dry aggregates to constant weight at 221 to 230°F (105 to 110°C) and separate the aggregates to dry-sieving into the desired size fractions.[2] The following size fractions are recommended:

1 to ¾ in. (25.0 to 19.0 mm)
¾ to ⅜ in. (19.0 to 9.5 mm)
⅜ in. to No. 4 (9.5 mm to 4.75 mm)
No. 4 to No. 8 (4.75 mm to 2.36 mm)
Passing No. 8 (2.36 mm)

4.3 *Determination of Mixing and Compacting Temperatures:*

4.3.1 The temperatures to which the asphalt cement and asphalt cut-back must be heated to produce a viscosity of 170 ± 20 cSt shall be the mixing temperature.

4.3.2 The temperature to which asphalt cement must be heated to produce a viscosity of 280 ± 30 cSt shall be the compacting temperature.

4.3.3 From a composition chart for the asphalt cut-back used, determine from its viscosity at 140°F (60°C) the percentage of solvent by weight. Also determine from the chart the viscosity at 140°F (60°C) of the asphalt cut-back

after it has lost 50 % of its solvent. The temperature determined from the viscosity temperature chart to which the asphalt cut-back must be heated to produce a viscosity of 280 ± 30 cSt after a loss of 50 % of the original solvent content shall be the compacting temperature.

4.3.4 The temperature to which tar must be heated to produce Engler specific viscosities of 25 ± 3 and 40 ± 5 shall be respectively the mixing and compacting temperature.

4.4 *Preparation of Mixtures:*

4.4.1 Weigh into separate pans for each test specimen the amount of each size fraction required to produce a batch that will result in a compacted specimen 2.5 ± 0.05 in. (63.5 ± 1.27 mm) in height (about 1200 g). Place the pans on the hot plate or in the oven and heat to a temperature not exceeding the mixing temperature established in 4.3 by more than approximately 50°F (28°C) for asphalt cement and tar mixes and 25°F (14°C) for cut-back asphalt mixes. Charge the mixing bowl with the heated aggregate and dry mix thoroughly. Form a crater in the dry blended aggregate and weigh the preheated required amount of bituminous material into the mixture. For mixes prepared with cutback asphalt introduce the mixing blade in the mixing bowl and determine the total weight of the mix components plus bowl and blade before proceeding with mixing. Care must be exercised to prevent loss of the mix during mixing and subsequent handling. At this point, the temperature of the aggregate and bituminous material shall be within the limits of the mixing temperature established in 4.3. Mix the aggregate and bituminous material rapidly until thoroughly coated.

4.4.2 Following mixing, cure asphalt cutback mixtures in a ventilated oven maintained at approximately 20°F (11.1°C) above the compaction temperature. Curing is to be continued in the mixing bowl until the precalculated weight of 50 % solvent loss or more has been obtained. The mix may be stirred in a mixing bowl during curing to accelerate the solvent loss. However, care should be exercised to prevent

---

[2] Detailed requirements for these sieves are given in ASTM Specification E 11, for Wire-Cloth Sieves for Testing Purposes see *Annual Book of ASTM Standards*, Vol 14.02.

D 1559

loss of the mix. Weigh the mix during curing in successive intervals of 15 min initially and less than 10 min intervals as the weight of the mix at 50 % solvent loss is approached.

4.5 Compaction of Specimens:

4.5.1 Thoroughly clean the specimen mold assembly and the face of the compaction hammer and heat them either in boiling water or on the hot plate to a temperature between 200 and 300°F (93.3 and 148.9°C). Place a piece of filter paper or paper toweling cut to size in the bottom of the mold before the mixture is introduced. Place the entire batch in the mold, spade the mixture vigorously with a heated spatula or trowel 15 times around the perimeter and 10 times over the interior. Remove the collar and smooth the surface of the mix with a trowel to a slightly rounded shape. Temperatures of the mixtures immediately prior to compaction shall be within the limits of the compacting temperature established in 4.3.

4.5.2 Replace the collar, place the mold assembly on the compaction pedestal in the mold holder, and unless otherwise specified, apply 50 blows with the compaction hammer with a free fall in 18 in. (457.2 mm). During compaction, the operator shall hold the axis of the compaction hammer by hand as nearly perpendicular to the base of the mold assembly as possible. Remove the base plate and collar, and reverse and reassemble the mold. Apply the same number of compaction blows to the face of the reversed specimen. After compaction, remove the base plate and place the sample extractor on that end of the specimen. Place the assembly with the extension collar up in the testing machine, apply pressure to the collar by means of the load transfer bar, and force the specimen into the extension collar. Lift the collar from the specimen. Carefully transfer the specimen to a smooth, flat surface and allow it to stand overnight at room temperature. Weigh, measure, and test the specimen.

NOTE 5—The above procedure requires that the handle of the compaction hammer should be held freely by one hand in as nearly a vertical position as possible while hand compaction proceeds. Inevitably, this will result in some wobbling of the handle about the vertical as compaction takes place. This wobbling, however, provides a highly desirable kneading of the specimen during compaction, resulting in higher density. Consequently, for this standard test, no mechanical device of any kind should be used to restrict the handle of the hammer in the vertical position during compaction.

NOTE 6—In general, specimens shall be cooled as specified in 4.5.2. When more rapid cooling is desired, table fans may be used. Mixtures that lack sufficient cohesion to result in the required cylindrical shape on removal from the mold immediately after compaction may be cooled in the mold in air until sufficient cohesion has developed to result in the proper cylindrical shape.

5. Procedure

5.1 Bring the specimens prepared with asphalt cement or tar to the specified temperature by immersing in the water bath 30 to 40 min or placing in the oven for 2 h. Maintain the bath or oven temperature at 140 ± 1.8°F (60 ± 1.0°C) for the asphalt cement specimens and 100 ± 1.8°F (37.8 ± 1.0°C) for tar specimens. Bring the specimens prepared with asphalt cut-back to the specified temperature by placing them in the air bath for a minimum of 2 h. Maintain the air bath temperature at 77 ± 1.8°F (25 ± 1.0°C). Thoroughly clean the guide rods and the inside surfaces of the test heads prior to making the test, and lubricate the guide rods so that the upper test head slides freely over them. The testing-head

temperature shall be maintained between 70 to 100°F (21.1 to 37.8°C) using a water bath when required. Remove the specimen from the water bath, oven, or air bath, and place in the lower segment of the breaking head. Place the upper segment of the breaking head on the specimen, and place the complete assembly in position on the testing machine. Place the flowmeter, where used, in position over one of the guide rods and adjust the flowmeter to zero while holding the sleeve firmly against the upper segment of the breaking head. Hold the flowmeter sleeve firmly against the upper segment of the breaking head while the test load is being applied.

5.2 Apply the load to the specimen by means of the constant rate of movement of the load jack or testing-machine head of 2 in. (50.8 mm)/min until the maximum load is reached and the load decreases as indicated by the dial. Record the maximum load noted on the testing machine or converted from the maximum micrometer dial reading. Release the flowmeter sleeve or note the micrometer dial reading, where used, the instant the maximum load begins to decrease. Note and record the indicated flow value or equivalent units in hundredths of an inch (twenty-five hundredths of a millimetre) if a micrometer dial is used to measure the flow. The elapsed time for the test from removal of the test specimen from the water bath to the maximum load determination shall not exceed 30 s.

TABLE 1   Stability Correlation Ratios[A]

Volume of Specimen, cm³	Approximate Thickness of Specimen, in.[B]	mm	Correlation Ratio
200 to 213	1	25.4	5.56
214 to 225	1 1/16	27.0	5.00
226 to 237	1 1/8	28.6	4.55
238 to 250	1 3/16	30.2	4.17
251 to 264	1 1/4	31.8	3.85
265 to 276	1 5/16	33.3	3.57
277 to 289	1 3/8	34.9	3.33
290 to 301	1 7/16	36.5	3.03
302 to 316	1 1/2	38.1	2.78
317 to 328	1 9/16	39.7	2.50
329 to 340	1 5/8	41.3	2.27
341 to 353	1 11/16	42.9	2.08
354 to 367	1 3/4	44.4	1.92
368 to 379	1 13/16	46.0	1.79
380 to 392	1 7/8	47.6	1.67
393 to 405	1 15/16	49.2	1.56
406 to 420	2	50.8	1.47
421 to 431	2 1/16	52.4	1.39
432 to 443	2 1/8	54.0	1.32
444 to 456	2 3/16	55.6	1.25
457 to 470	2 1/4	57.2	1.19
471 to 482	2 5/16	58.7	1.14
483 to 495	2 3/8	60.3	1.09
496 to 508	2 7/16	61.9	1.04
509 to 522	2 1/2	63.5	1.00
523 to 535	2 9/16	65.1	0.96
536 to 546	2 5/8	66.7	0.93
547 to 559	2 11/16	68.3	0.89
560 to 573	2 3/4	69.8	0.86
574 to 585	2 13/16	71.4	0.83
586 to 598	2 7/8	73.0	0.81
599 to 610	2 15/16	74.6	0.78
611 to 625	3	76.2	0.76

[A] The measured stability of a specimen multiplied by the ratio for the thickness of the specimen equals the corrected stability for a 2 1/2-in. (63.5-mm) specimen.

[B] Volume-thickness relationship is based on a specimen diameter of 4 in. (101.6 mm).

D 1559

NOTE 7—For core specimens, correct the load when thickness is other than 2½ in. (63.5 mm) by using the proper multiplying factor from Table 1.

6. Report

6.1 The report shall include the following information:

6.1.1 Type of sample tested (laboratory sample or pavement core specimen).

NOTE 8—For core specimens, the height of each test specimen in inches (or millimetres) shall be reported.

6.1.2 Average maximum load in pounds-force (or newtons) of at least three specimens, corrected when required.

6.1.3 Average flow value, in hundredths of an inch, twenty-five hundredths of a millimetre, of three specimens, and

6.1.4 Test temperature.

Designation: E 8 – 96a

American Association State
Highway and Transportation Officials Standard
AASHTO No.: T 68

Standard Test Methods for
Tension Testing of Metallic Materials[1]

This standard is issued under the fixed designation E 8; the number immediately following the designation indicates the year of original adoption or, in the case of revision, the year of last revision. A number in parentheses indicates the year of last reapproval. A superscript epsilon (ϵ) indicates an editorial change since the last revision or reapproval.

This standard has been approved for use by agencies of the Department of Defense to replace method 211.1 of Federal Test Method Standard No. 151b. Consult the DoD Index of Specifications and Standards for the specific year of issue which has been adopted by the Department of Defense.

1. Scope*

1.1 These test methods cover the tension testing of metallic materials in any form at room temperature, specifically, the methods of determination of yield strength, yield point elongation, tensile strength, elongation, and reduction of area.

NOTE 1—A complete metric companion to Test Methods E 8 has been developed—E 8M; therefore, no metric equivalents are shown in these test methods.

NOTE 2—Gage lengths in these test methods are required to be 4D for most round specimens. Test specimens made from powder metallurgy (P/M) materials are exempt from this requirement by industrywide agreement to keep the pressing of the material to a specific projected area and density.

NOTE 3—Exceptions to the provisions of these test methods may need to be made in individual specifications or test methods for a particular material. For examples, see Test Methods and Definitions A 370 and Methods B 557.

NOTE 4—Room temperature shall be considered to be 50 to 100°F unless otherwise specified.

1.2 *This standard does not purport to address all of the safety concerns, if any, associated with its use. It is the responsibility of the user of this standard to establish appropriate safety and health practices and determine the applicability of regulatory limitations prior to use.*

2. Referenced Documents

2.1 *ASTM Standards:*

A 356/A 356M Specification for Heavy-Walled, Carbon, Low Alloy, and Stainless Steel Castings for Steam Turbines[2]

A 370 Test Methods and Definitions for Mechanical Testing of Steel Products[3]

B 557 Methods of Tension Testing Wrought and Cast Aluminum- and Magnesium-Alloy Products[4]

E 4 Practices for Force Verification of Testing Machines[5]

E 6 Terminology Relating to Methods of Mechanical Testing[5]

E 8M Test Methods for Tension Testing of Metallic Materials [Metric][5]

E 29 Practice for Using Significant Digits in Test Data to Determine Conformance with Specifications[6]

E 83 Practice for Verification and Classification of Extensometers[5]

E 345 Test Methods of Tension Testing of Metallic Foil[5]

E 691 Practice for Conducting an Interlaboratory Study to Determine the Precision of a Test Method[6]

E 1012 Practice for Verification of Specimen Alignment Under Tensile Loading[5]

3. Terminology

3.1 *Definitions*—The definitions of terms relating to tension testing appearing in Terminology E 6 shall be considered as applying to the terms used in these test methods of tension testing. Additional terms being defined are as follows:

3.1.1 *discontinuous yielding*—a hesitation or fluctuation of force observed at the onset of plastic deformation, due to localized yielding. (The stress-strain curve need not appear to be discontinuous.)

3.1.2 *lower yield strength, LYS [FL^{-2}]*—the minimum stress recorded during discontinuous yielding, ignoring transient effects.

3.1.3 *upper yield strength, UYS [FL^{-2}]*—the first stress maximum (stress at first zero slope) associated with discontinuous yielding.

3.1.4 *yield point elongation, YPE*—the strain (expressed in percent) separating the stress-strain curve's first point of zero slope from the point of transition from discontinuous yielding to uniform strain hardening. If the transition occurs over a range of strain, the YPE end point is the intersection between (*a*) a horizontal line drawn tangent to the curve at the last zero slope and (*b*) a line drawn tangent to the strain hardening portion of the stress-strain curve at the point of inflection. If there is no point at or near the onset of yielding at which the slope reaches zero, the material has 0 % YPE.

4. Significance and Use

4.1 Tension tests provide information on the strength and ductility of materials under uniaxial tensile stresses. This information may be useful in comparisons of materials, alloy development, quality control, and design under certain circumstances.

[1] These test methods are under the jurisdiction of ASTM Committee E-28 on Mechanical Testing and are the direct responsibility of Subcommittee E28.04 on Uniaxial Testing.

Current edition approved Dec. 10, 1996. Published February 1997. Originally published as E 8 – 24 T. Last previous edition E 8 – 96.

[2] *Annual Book of ASTM Standards*, Vol 01.02.
[3] *Annual Book of ASTM Standards*, Vol 01.03.
[4] *Annual Book of ASTM Standards*, Vol 02.02.
[5] *Annual Book of ASTM Standards*, Vol 03.01.

[6] *Annual Book of ASTM Standards*, Vol 14.02.

* A Summary of Changes section appears at the end of these test methods.

4.2 The results of tension tests of specimens machined to standardized dimensions from selected portions of a part or material may not totally represent the strength and ductility properties of the entire end product or its in-service behavior in different environments.

4.3 These test methods are considered satisfactory for acceptance testing of commercial shipments. The test methods have been used extensively in the trade for this purpose.

5. Apparatus

5.1 *Testing Machines*—Machines used for tension testing shall conform to the requirements of Practices E 4. The forces used in determining tensile strength and yield strength shall be within the verified force application range of the testing machine as defined in Practices E 4.

5.2 *Gripping Devices:*

5.2.1 *General*—Various types of gripping devices may be used to transmit the measured force applied by the testing machine to the test specimens. To ensure axial tensile stress within the gage length, the axis of the test specimen should coincide with the center line of the heads of the testing machine. Any departure from this requirement may introduce bending stresses that are not included in the usual stress computation (force divided by cross-sectional area).

NOTE 5—The effect of this eccentric force application may be illustrated by calculating the bending moment and stress thus added. For a standard $\frac{1}{2}$-in. diameter specimen, the stress increase is 1.5 percentage points for each 0.001 in. of eccentricity. This error increases to 2.24 percentage points/0.001 in. for a 0.350-in. diameter specimen and to 3.17 percentage points/0.001 in. for a 0.250-in. diameter specimen.

NOTE 6—Alignment methods are given in Practice E 1012.

5.2.2 *Wedge Grips*—Testing machines usually are equipped with wedge grips. These wedge grips generally furnish a satisfactory means of gripping long specimens of ductile metal and flat plate test specimens such as those shown in Fig. 1. If, however, for any reason, one grip of a pair advances farther than the other as the grips tighten, an undesirable bending stress may be introduced. When liners are used behind the wedges, they must be of the same thickness and their faces must be flat and parallel. For best results, the wedges should be supported over their entire lengths by the heads of the testing machine. This requires that liners of several thicknesses be available to cover the range of specimen thickness. For proper gripping, it is desirable that the entire length of the serrated face of each wedge be in contact with the specimen. Proper alignment of wedge grips and liners is illustrated in Fig. 2. For short specimens and for specimens of many materials it is generally necessary to use machined test specimens and to use a special means of gripping to ensure that the specimens, when under load, shall be as nearly as possible in uniformly distributed pure axial tension (see 5.2.3, 5.2.4, and 5.2.5).

5.2.3 *Grips for Threaded and Shouldered Specimens and Brittle Materials*—A schematic diagram of a gripping device for threaded-end specimens is shown in Fig. 3, while Fig. 4 shows a device for gripping specimens with shouldered ends. Both of these gripping devices should be attached to the heads of the testing machine through properly lubricated spherical-seated bearings. The distance between spherical bearings should be as great as feasible.

5.2.4 *Grips for Sheet Materials*—The self-adjusting grips

shown in Fig. 5 have proven satisfactory for testing sheet materials that cannot be tested satisfactorily in the usual type of wedge grips.

5.2.5 *Grips for Wire*—Grips of either the wedge or snubbing types as shown in Figs. 5 and 6 or flat wedge grips may be used.

5.3 *Dimension-Measuring Devices*—Micrometers and other devices used for measuring linear dimensions shall be accurate and precise to at least one half the smallest unit to which the individual dimension is required to be measured.

6. Test Specimens

6.1 *General:*

6.1.1 Test specimens shall be either substantially full size or machined, as prescribed in the product specifications for the material being tested.

6.1.2 Improperly prepared test specimens often are the reason for unsatisfactory and incorrect test results. It is important, therefore, that care be exercised in the preparation of specimens, particularly in the machining, to ensure the desired precision and bias in test results.

6.1.3 It is desirable to have the cross-sectional area of the specimen smallest at the center of the reduced section to ensure fracture within the gage length. For this reason, a small taper is permitted in the reduced section of each of the specimens described in the following sections.

6.1.4 For brittle materials it is desirable to have fillets of large radius at the ends of the gage length.

6.2 *Plate-Type Specimens*—The standard plate-type test specimen is shown in Fig. 1. This specimen is used for testing metallic materials in the form of plate, shapes, and flat material having a nominal thickness of $\frac{3}{16}$ in. or over. When product specifications so permit, other types of specimens may be used, as provided in 6.3, 6.4, and 6.5.

6.3 *Sheet-Type Specimens:*

6.3.1 The standard sheet-type test specimen is shown in Fig. 1. This specimen is used for testing metallic materials in the form of sheet, plate, flat wire, strip, band, hoop, rectangles, and shapes ranging in nominal thickness from 0.005 to $\frac{3}{4}$ in. When product specifications so permit, other types of specimens may be used, as provided in 6.2, 6.4, and 6.5.

NOTE 7—Test Methods E 345 may be used for tension testing of materials in thicknesses up to 0.0059 in.

6.3.2 Pin ends as shown in Fig. 7 may be used. In order to avoid buckling in tests of thin and high-strength materials, it may be neccessary to use stiffening plates at the grip ends.

6.4 *Round Specimens:*

6.4.1 The standard 0.500-in. diameter round test specimen shown in Fig. 8 is used quite generally for testing metallic materials, both cast and wrought.

6.4.2 Figure 8 also shows small-size specimens proportional to the standard specimen. These may be used when it is necessary to test material from which the standard specimen or specimens shown in Fig. 1 cannot be prepared. Other sizes of small round specimens may be used. In any such small-size specimen it is important that the gage length for measurement of elongation be four times the diameter of the specimen.

6.4.3 The shape of the ends of the specimen outside of the gage length shall be suitable to the material and of a shape to fit the holders or grips of the testing machine so that the forces may be applied axially. Figure 9 shows specimens with various types of ends that have given satisfactory results.

6.5 *Specimens for Sheet, Strip, Flat Wire, and Plate*—In testing sheet, plate, flat wire, and strip one of the following types of specimens shall be used:

6.5.1 For material ranging in nominal thickness from 0.005 to ¾ in., use the sheet-type specimen described in 6.3.

NOTE 8—Attention is called to the fact that either of the flat specimens described in 6.2 and 6.3 may be used for material from ³⁄₁₆ to ¾ in. in thickness, and one of the round specimens described in 6.4 may also be used for material ½ in. or more in thickness.

6.5.2 For material having a nominal thickness of ³⁄₁₆ in. or over (Note 8), use the plate-type specimen described in 6.2.

6.5.3 For material having a nominal thickness of ½ in. or over (Note 8), use the largest practical size of specimen described in 6.4.

6.6 *Specimens for Wire, Rod, and Bar:*

6.6.1 For round wire, rod, and bar, test specimens having the full cross-sectional area of the wire, rod, or bar shall be used wherever practicable. The gage length for the measurement of elongation of wire less than ⅛ in. in diameter shall be as prescribed in product specifications. In testing wire, rod, or bar that has a ⅛-in. or larger diameter, unless otherwise specified, a gage length equal to four times the diameter shall be used. The total length of the specimens shall be at least equal to the gage length plus the length of material required for the full use of the grips employed.

6.6.2 For wire of octagonal, hexagonal, or square cross section, for rod or bar of round cross section where the specimen required in 6.6.1 is not practicable, and for rod or bar of octagonal, hexagonal, or square cross section, one of the following types of specimens shall be used:

6.6.2.1 *Full Cross Section* (Note 9)—It is permissible to reduce the test section slightly with abrasive cloth or paper, or machine it sufficiently to ensure fracture within the gage marks. For material not exceeding 0.188 in. in diameter or distance between flats, the cross-sectional area may be reduced to not less than 90 % of the original area without changing the shape of the cross section. For material over 0.188 in. in diameter or distance between flats, the diameter or distance between flats may be reduced by not more than 0.010 in. without changing the shape of the cross section. Square, hexagonal, or octagonal wire or rod not exceeding 0.188 in. between flats may be turned to a round having a cross-sectional area not smaller than 90 % of the area of the maximum inscribed circle. Fillets, preferably with a radius of ⅜ in., but not less than ⅛ in., shall be used at the ends of the reduced sections. Square, hexagonal, or octagonal rod over 0.188 in. between flats may be turned to a round having a diameter no smaller than 0.010 in. less than the original distance between flats.

NOTE 9—The ends of copper or copper alloy specimens may be flattened 10 to 50 % from the original dimension in a jig similar to that shown in Fig. 10, to facilitate fracture within the gage marks. In flattening the opposite ends of the test specimen, care shall be taken to ensure that the four flattened surfaces are parallel and that the two parallel surfaces on the same side of the axis of the test specimen lie in the same plane.

6.6.2.2 For rod and bar, the largest practical size of round specimen as described in 6.4 may be used in place of a test specimen of full cross section. Unless otherwise specified in

the product specification, specimens shall be parallel to the direction of rolling or extrusion.

6.7 *Specimens for Rectangular Bar*—In testing rectangular bar one of the following types of specimens shall be used:

6.7.1 *Full Cross Section*—It is permissible to reduce the width of the specimen throughout the test section with abrasive cloth or paper, or by machining sufficiently to facilitate fracture within the gage marks, but in no case shall the reduced width be less than 90 % of the original. The edges of the midlength of the reduced section not less than ¾ in. in length shall be parallel to each other and to the longitudinal axis of the specimen within 0.002 in. Fillets, preferably with a radius of ⅜ in. but not less than ⅛ in. shall be used at the ends of the reduced sections.

6.7.2 Rectangular bar of thickness small enough to fit the grips of the testing machine but of too great width may be reduced in width by cutting to fit the grips, after which the cut surfaces shall be machined or cut and smoothed to ensure failure within the desired section. The reduced width shall be not less than the original bar thickness. Also, one of the types of specimens described in 6.2, 6.3, and 6.4 may be used.

6.8 *Shapes, Structural and Other*—In testing shapes other than those covered by the preceding sections, one of the types of specimens described in 6.2, 6.3, and 6.4 shall be used.

6.9 *Specimens for Pipe and Tube* (Note 10):

6.9.1 For all small tube (Note 10), particularly sizes 1 in. and under in nominal outside diameter, and frequently for larger sizes, except as limited by the testing equipment, it is standard practice to use tension test specimens of full-size tubular sections. Snug-fitting metal plugs shall be inserted far enough into the ends of such tubular specimens to permit the testing machine jaws to grip the specimens properly. The plugs shall not extend into that part of the specimen on which the elongation is measured. Elongation is measured over a length of 4D unless otherwise stated in the product specification. Figure 11 shows a suitable form of plug, the location of the plugs in the specimen, and the location of the specimen in the grips of the testing machine.

NOTE 10—The term "tube" is used to indicate tubular products in general, and includes pipe, tube, and tubing.

6.9.2 For large-diameter tube that cannot be tested in full section, longitudinal tension test specimens shall be cut as indicated in Fig. 12. Specimens from welded tube shall be located approximately 90° from the weld. If the tube-wall thickness is under ¾ in., either a specimen of the form and dimensions shown in Fig. 13 or one of the small-size specimens proportional to the standard ½-in. specimen, as mentioned in 6.4.2 and shown in Fig. 8, shall be used. Specimens of the type shown in Fig. 13 may be tested with grips having a surface contour corresponding to the curvature of the tube. When grips with curved faces are not available, the ends of the specimens may be flattened without heating. If the tube-wall thickness is ¾ in. or over, the standard specimen shown in Fig. 8 shall be used.

NOTE 11—In clamping of specimens from pipe and tube (as may be done during machining) or in flattening specimen ends (for gripping), care must be taken so as not to subject the reduced section to any deformation or cold work, as this would alter the mechanical properties.

⊕ E 8

TABLE 1 Details of Test Coupon Design for Castings (See Fig. 16)

NOTE 1—*Test Coupons for Large and Heavy Steel Castings:* The test coupons in Fig. 16 are to be used for large and heavy steel castings. However, at the option of the foundry the cross-sectional area and length of the standard coupon may be increased as desired. This provision does not apply to Specification A 356/A 356M.
NOTE 2—*Bend Bar:* If a bend bar is required, an alternate design (as shown by dotted lines in Fig. 16) is indicated.

Log Design (5 in.)		Riser Design	
1. *L* (length)	A 5-in. minimum length will be used. This length may be increased at the option of the foundry to accommodate additional test bars (see Note 1).	1. *L* (length)	The length of the riser at the base will be the same as the top length of the leg. The length of the riser at the top therefore depends on the amount of taper added to the riser.
2. End taper	Use of and size of end taper is at the option of the foundry.	2. Width	The width of the riser at the base of a multiple-leg coupon shall be *n* (2¼ in.) − ⅝ in. where *n* equals the number of legs attached to the coupon. The width of the riser at the top is therefore dependent on the amount of taper added to the riser.
3. Height	1¼ in.		
4. Width (at top)	1¼ in. (see Note 1)		
5. Radius (at bottom)	½ in. max		
6. Spacing between legs	A ½-in. radius will be used between the legs.		
7. Location of test bars	The tensile, bend, and impact bars will be taken from the lower portion of the leg (see Note 2).		
8. Number of legs	The number of legs attached to the coupon is at the option of the foundry providing they are equispaced according to Item 6.	3. *T* (riser taper) Height	Use of and size is at the option of the foundry. The minimum height of the riser shall be 2 in. The maximum height is at the option of the foundry for the following reasons: (*a*) many risers are cast open, (*b*) different compositions may require variation in risering for soundness, or (*c*) different pouring temperatures may require variation in risering for soundness.
9. *R$_s$*	Radius from 0 to approximately ¹⁄₁₆ in.		

6.9.3 Transverse tension test specimens for tube may be taken from rings cut from the ends of the tube as shown in Fig. 14. Flattening of the specimen may be either after separating as in *A*, or before separating as in *B*. Transverse tension test specimens for large tube under ¾ in. in wall thickness shall be either of the small-size specimens shown in Fig. 8 or of the form and dimensions shown for Specimen 2 in Fig. 13. When using the latter specimen, either or both surfaces of the specimen may be machined to secure a uniform thickness, provided not more than 15 % of the normal wall thickness is removed from each surface. For large tube ¾ in. and over in wall thickness, the standard specimen shown in Fig. 8 shall be used for transverse tension tests. Specimens for transverse tension tests on large welded tube to determine the strength of welds shall be located perpendicular to the welded seams, with the welds at about the middle of their lengths.

6.10 *Specimens for Forgings*—For testing forgings, the largest round specimen described in 6.4 shall be used. If round specimens are not feasible, then the largest specimen described in 6.5 shall be used.

6.11 *Specimens for Castings*—In testing castings either the standard specimen shown in Fig. 8 or the specimen shown in Fig. 15 shall be used unless otherwise provided in the product specifications.

6.11.1 Test coupons for castings shall be made as shown in Fig. 16 and Table 1.

6.12 *Specimen for Malleable Iron*—For testing malleable iron the test specimen shown in Fig. 17 shall be used, unless otherwise provided in the product specifications.

6.13 *Specimen for Die Castings*—For testing die castings the test specimen shown in Fig. 18 shall be used unless otherwise provided in the product specifications.

6.14 *Specimens for Powder Metallurgy (P/M) Materials*—For testing powder metallurgy (P/M) materials the test specimens shown in Figs. 19 and 20 shall be used, unless

otherwise provided in the product specifications. When making test specimens in accordance with Fig. 19, shallow transverse grooves, or ridges, may be pressed in the ends to allow gripping by jaws machined to fit the grooves or ridges. Because of shape and other factors, the flat unmachined tensile test specimen (Fig. 19) in the heat treated condition will have an ultimate tensile strength of 50 % to 85 % of that determined in a machined round tensile test specimen (Fig. 20) of like composition and processing.

6.15 *Gage Length of Test Specimens:*

6.15.1 The gage length for the determination of elongation shall be in accordance with the product specifications for the material being tested. Gage marks shall be stamped lightly with a punch, scribed lightly with dividers or drawn with ink as preferred. For material that is sensitive to the effect of slight notches and for small specimens, the use of layout ink will aid in locating the original gage marks after fracture.

6.15.2 Extensometers with gage lengths equal to or shorter than the nominal gage length (dimension shown as "*G*-Gage Length" in the accompanying figures) of the specimen may be used to determine the yield phenomenon.

6.16 *Location of Test Specimens:*

6.16.1 Unless otherwise specified, the axis of the test specimen shall be located as follows:

6.16.1.1 At the center for products 1½ in. or less in thickness, diameter, or distance between flats.

6.16.1.2 Midway from the center to the surface for products over 1½ in. in thickness, diameter, or distance between flats.

6.16.2 For forgings, specimens shall be taken as provided in the applicable product specifications, either from the predominant or thickest part of the forging from which a coupon can be obtained, or from a prolongation of the forging, or from separately forged coupons representative of the forging. When not otherwise specified, the axis of the specimen shall be parallel to the direction of grain flow.

6.17 *Surface Finish of Specimens*—When materials are tested with surface conditions other than as manufactured, the surface finish of the test specimens shall be as provided in the applicable product specifications.

NOTE 12—Particular attention should be given to the uniformity and quality of surface finishes of specimens for high strength and very low ductility materials since this has been shown to be a factor in the variability of test results.

7. Procedures

7.1 *Measurement of Dimensions of Test Specimens:*

7.1.1 To determine the cross-sectional area of a test specimen, measure the dimensions of the cross section at the center of the reduced section. For referee testing of specimens under $3/16$ in. in their least dimension, measure the dimensions where the least cross-sectional area is found. Measure and record the cross-sectional dimensions of tension test specimens 0.200 in. and over to the nearest 0.001 in.; the cross-sectional dimensions from 0.100 in. but less than 0.200 in., to the nearest 0.0005 in.; the cross-sectional dimensions from 0.020 in. but less than 0.100 in., to the nearest 0.0001 in.; and when practical, the cross-sectional dimensions less than 0.020 in., to at least the nearest 1 % but in all cases to at least the nearest 0.0001 in.

NOTE 13—Accurate and precise measurement of specimen dimensions can be one of the most critical aspects of tension testing, depending on specimen geometry. See Appendix X2 for additional information.

NOTE 14—Rough surfaces due to the manufacturing process such as hot rolling, metallic coating, etc., may lead to inaccuracy of the computed areas greater than the measured dimensions would indicate. Therefore, cross-sectional dimensions of test specimens with rough surfaces due to processing may be measured and recorded to the nearest 0.001 in.

7.1.2 Determine cross-sectional areas of full-size test specimens of nonsymmetrical cross sections by weighing a length not less than 20 times the largest cross-sectional dimension and using the value of density of the material. Determine the weight to the nearest 0.5 % or less.

7.1.3 When using specimens of the type shown in Fig. 13 taken from tubes, the cross-sectional area shall be determined as follows:

If $D/W \leq 6$:

$$A = [(W/4) \times (D^2 - W^2)^{1/2}] + [(D^2/4) \\ \times \arcsin(W/D)] - [(W/4) \times ((D - 2T)^2 - W^2)^{1/2}] \\ - [((D - 2T)/2)^2 \times \arcsin(W/(D - 2T))]$$

where:

A = exact cross-sectional area, in.^{2},
W = width of the specimen in the reduced section, in.,
D = measured outside diameter of the tube, in., and
T = measured wall thickness of the specimen, in.
arcsin values to be in radians
If $D/W > 6$, the exact equation or the following equation may be used:

$$A = W \times T$$

where:

A = approximate cross-sectional area, in.^{2},
W = width of the specimen in the reduced section, in., and
T = measured wall thickness of the specimen, in.

NOTE 15—See X2.8 for cautionary information on measurements and calculations for specimens taken from large-diameter tubing.

7.2 *Zeroing of the Testing Machine:*

7.2.1 The testing machine shall be set up in such a manner that zero force indication signifies a state of zero force on the specimen. Any force (or preload) imparted by the gripping of the specimen (see Note 16) must be indicated by the force measuring system unless the preload is physically removed prior to testing. Artificial methods of removing the preload on the specimen, such as taring it out by a zero adjust pot or removing it mathematically by software, are prohibited because these would affect the accuracy of the test results.

NOTE 16—Preloads generated by gripping of specimens may be either tensile or compressive in nature and may be the result of such things as:

— grip design
— malfunction of gripping apparatus (sticking, binding, etc.)
— excessive gripping force
— sensitivity of the control loop

NOTE 17—It is the operator's responsibility to verify that an observed preload is acceptable and to ensure that grips operate in a smooth manner. Unless otherwise specified, it is recommended that momentary (dynamic) forces due to gripping not exceed 20 % of the material's nominal yield strength and that static preloads not exceed 10 % of the material's nominal yield strength.

7.3 *Gripping of the Test Specimen:*

7.3.1 For specimens with reduced sections, gripping of the specimen shall be restricted to the grip section, because gripping in the reduced section or in the fillet can significantly affect test results.

7.4 *Speed of Testing:*

7.4.1 Speed of testing may be defined in terms of (*a*) rate of straining of the specimen, (*b*) rate of stressing of the specimen, (*c*) rate of separation of the two heads of the testing machine during a test, (*d*) the elapsed time for completing part or all of the test, or (*e*) free-running crosshead speed (rate of movement of the crosshead of the testing machine when not under load).

7.4.2 Specifying suitable numerical limits for speed and selection of the method are the responsibilities of the product committees. Suitable limits for speed of testing should be specified for materials for which the differences resulting from the use of different speeds are of such magnitude that the test results are unsatisfactory for determining the acceptability of the material. In such instances, depending upon the material and the use for which the test results are intended, one or more of the methods described in the following paragraphs is recommended for specifying speed of testing.

NOTE 18—Speed of testing can affect test values because of the rate sensitivity of materials and the temperature-time effects.

7.4.2.1 *Rate of Straining*—The allowable limits for rate of straining shall be specified in inches per inch per minute. Some testing machines are equipped with pacing or indicating devices for the measurement and control of rate of straining, but in the absence of such a device the average rate of straining can be determined with a timing device by observing the time required to effect a known increment of strain.

7.4.2.2 *Rate of Stressing*—The allowable limits for rate of stressing shall be specified in pounds per square inch per minute. Many testing machines are equipped with pacing or indicating devices for the measurement and control of the rate of stressing, but in the absence of such a device the average rate of stressing can be determined with a timing device by observing the time required to apply a known increment of stress.

7.4.2.3 *Rate of Separation of Heads During Tests*—The allowable limits for rate of separation of the heads of the testing machine, during a test, shall be specified in inches per inch of length of reduced section (or distance between grips for specimens not having reduced sections) per minute. The limits for the rate of separation may be further qualified by specifying different limits for various types and sizes of specimens. Many testing machines are equipped with pacing or indicating devices for the measurement and control of the rate of separation of the heads of the machine during a test, but in the absence of such a device the average rate of separation of the heads can be experimentally determined by using suitable length-measuring and timing devices.

7.4.2.4 *Elapsed Time*—The allowable limits for the elapsed time from the beginning of force application (or from some specified stress) to the instant of fracture, to the maximum force, or to some other stated stress, shall be specified in minutes or seconds. The elapsed time can be determined with a timing device.

7.4.2.5 *Free-Running Crosshead Speed*—The allowable limits for the rate of movement of the crosshead of the testing machine, with no force applied by the testing machine, shall be specified in inches per inch of length of reduced section (or distance between grips for specimens not having reduced sections) per minute. The limits for the crosshead speed may be further qualified by specifying different limits for various types and sizes of specimens. The average crosshead speed can be experimentally determined by using suitable length-measuring and timing devices.

NOTE 19—For machines not having crossheads or having stationary crossheads, the phrase "free-running crosshead speed" may be interpreted to mean the free-running rate of grip separation.

7.4.3 *Speed of Testing When Determining Yield Properties*—Unless otherwise specified, any convenient speed of testing may be used up to one half the specified yield strength or up to one quarter the specified tensile strength, whichever is smaller. The speed above this point shall be within the limits specified. If different speed limitations are required for use in determining yield strength, yield point elongation, tensile strength, elongation, and reduction of area, they should be stated in the product specifications. In the absence of any specified limitations on speed of testing, the following general rules shall apply:

NOTE 20—In the previous and following paragraphs, the yield properties referred to include yield strength and yield point elongation.

7.4.3.1 The speed of testing shall be such that the forces and strains used in obtaining the test results are accurately indicated.

7.4.3.2 When performing a test to determine yield properties, the rate of stress application shall be between 10 000 and 100 000 psi/min.

NOTE 21—When a specimen being tested begins to yield, the stressing rate decreases and may even become negative in the case of a specimen with discontinuous yielding. To maintain a constant stressing rate would require the testing machine to operate at extremely high

speeds and, in many cases, this is not practical. The speed of the testing machine shall not be increased in order to maintain a stressing rate when the specimen begins to yield. In practice, it is simpler to use either a strain rate, a rate of separation of the heads, or a free-running crosshead speed which approximates the desired stressing rate. As an example, use a strain rate that is less than 100 000 psi divided by the nominal Young's Modulus of the material being tested. As another example, find a rate of separation of the heads through experimentation which would approximate the desired stressing rate prior to the onset of yielding, and maintain that rate of separation of the heads through the region that yield properties are determined. While both of these methods will provide similar rates of stressing and straining prior to the onset of yielding, the rates of stressing and straining may be different in the region where yield properties are determined. This difference is due to the change in the rate of elastic deformation of the testing machine, before and after the onset of yielding. In addition, the use of any of the methods other than rate of straining may result in different stressing and straining rates when using different testing machines, due to differences in the stiffness of the testing machines used.

7.4.4 *Speed of Testing When Determining Tensile Strength*—In the absence of any specified limitations on speed of testing, the following general rules shall apply. When determining only the tensile strength, or after the yield properties have been determined, the speed of the testing machine may be increased to correspond to a strain rate between 0.05 and 0.5 in./in./min. An extensometer and strain rate indicator may be used to set the strain rate. If an extensometer and strain rate indicator are not used to set this strain rate, the speed of the testing machine shall be set between 0.05 and 0.5 in./in. of the length of the reduced section (or distance between the grips for specimens not having reduced sections) per minute.

7.5 *Determination of Yield Strength*—Determine yield strength by any of the methods described in 7.5.1 to 7.5.5.

7.5.1 *Offset Method*—To determine the yield strength by the offset method, it is necessary to secure data (autographic or numerical) from which a stress-strain diagram may be drawn. Then on the stress-strain diagram (Fig. 21) lay off *Om* equal to the specified value of the offset, draw *mn* parallel to *OA*, and thus locate *r*, the intersection of *mn* with the stress-strain diagram (Note 27). In reporting values of yield strength obtained by this method, the specified value of offset used should be stated in parentheses after the term yield strength. Thus:

$$\text{Yield strength (offset} = 0.2 \text{ \%)} = 52\ 000 \text{ psi}$$

In using this method, a Class B2 or better extensometer (see Practice E 83) shall be used.

NOTE 22—There are two general types of extensometers, averaging and non-averaging, the use of which is dependent on the product tested. For most machined specimens, there are minimal differences. However, for some forgings and tube sections, significant differences in measured yield strength can occur. For these cases, it is recommended that the averaging type be used.

NOTE 23—When there is a disagreement over yield properties, the offset method for determining yield strength is recommended as the referee method.

7.5.2 *Extension-Under-Load Method*—Yield strength by the extension-under-load method may be determined by: (*1*) using autographic or numerical devices to secure stress-strain data, and then analyzing this data (graphically or using automated methods) to determine the stress value at the specified value of extension, or (*2*) using devices that indicate when the specified extension occurs, so that the stress then occurring may be ascertained (Note 27). Any of these devices may be automatic. This method is illustrated in Fig. 22. The stress at the specified extension shall be reported as follows:

yield strength (EUL = 0.5 %) = 52 000 psi

Extensometers and other devices used in determination of the extension shall meet Class B2 requirements (see Practice E 83) at the strain of interest, except where use of low-magnification Class C devices is helpful, such as in facilitating measurement of YPE, if observed. If Class C devices are used, this must be reported along with the results.

NOTE 24—The appropriate value of the total extension must be specified. For steels with nominal yield strengths of less than 80 000 psi, an appropriate value is 0.005 in./in. (0.5 %) of the gage length. For higher strength steels, a greater extension or the offset method should be used.

NOTE 25—When no other means of measuring elongation are available, a pair of dividers or similar device can be used to determine a point of detectable elongation between two gage marks on the specimen. The gage length shall be 2 in. The stress corresponding to the load at the instant of detectable elongation may be recorded as the *approximate* extension-under-load yield strength.

7.5.3 *Autographic Diagram Method (for materials exhibiting discontinuous yielding)*—Obtain stress-strain (or force elongation) data or construct a stress-strain (or load-elongation) diagram using an autographic device. Determine the upper or lower yield strength as follows:

7.5.3.1 Record the stress corresponding to the maximum force at the onset of discontinuous yielding as the upper yield strength. This is illustrated in Figs. 23 and 24.

NOTE 26—If multiple peaks are observed at the onset of discontinuous yielding, the first is considered the upper yield strength. (See Fig. 24.)

7.5.3.2 Record the minimum stress observed during discontinuous yielding (ignoring transient effects) as the lower yield strength. This is illustrated in Fig. 24.

NOTE 27—Yield properties of materials exhibiting yield point elongation are often less repeatable and less reproducible than those of similar materials having no YPE. Offset and EUL yield strengths may be significantly affected by force fluctuations occurring in the region where the offset or extension intersects the stress-strain curve. Determination of upper or lower yield strengths (or both) may therefore be preferable for such materials, although these properties are dependent on variables such as test machine stiffness and alignment. Speed of testing may also have a significant effect, regardless of the method employed.

NOTE 28—Where low-magnification autographic recordings are needed to facilitate measurement of yield point elongation for materials which may exhibit discontinuous yielding, Class C extensometers may be employed. When this is done but the material exhibits no discontinuous yielding, the extension-under-load yield strength may be determined instead, using the autographic recording (see Extension-Under-Load Method).

7.5.4 *Halt-of-the-Force Method (for materials exhibiting discontinuous yielding)*—Apply an increasing force to the specimen at a uniform deformation rate. When the force hesitates, record the corresponding stress as the upper yield strength.

NOTE 29—The Halt-of-the-Force Method was formerly known as the Halt-of-the-Pointer Method, the Drop-of-the-Beam Method, and the Halt-of-the-Load Method.

7.5.5 *Strain Rate Method (for materials that do not exhibit well-defined discontinuous yielding)*—Attach a Class B2 or better extensometer to the specimen at the gage marks. Increase the force at a reasonably uniform rate and watch the

elongation of the specimen as indicated by the extensometer. Note the force at which the rate of elongation suddenly increases.

7.6 *Yield Point Elongation*—Calculate the yield point elongation from the stress-strain diagram or data by determining the difference in strain between the upper yield strength (first zero slope) and the onset of uniform strain hardening (see definition of YPE and Fig. 24).

NOTE 30—The stress-strain curve of a material exhibiting only a hint of the behavior causing YPE may have an inflection at the onset of yielding with no point where the slope reaches zero (Fig. 25). Such a material has no YPE, but may be characterized as exhibiting an *inflection*. Materials exhibiting inflections, like those with measurable YPE, may in certain applications acquire an unacceptable surface appearance during forming.

7.7 *Tensile Strength*—Calculate the tensile strength by dividing the maximum force carried by the specimen during the tension test by the original cross-sectional area of the specimen.

NOTE 31—If the upper yield strength is the maximum stress recorded, and if the stress-strain curve resembles that of Fig. 26, it is recommended that the maximum stress *after discontinuous yielding* be reported as the tensile strength. Where this may occur, determination of the tensile strength should be in accordance with the agreement between the parties involved.

7.8 *Elongation:*

7.8.1 In reporting values of elongation, give both the original gage length and the percentage increase. If any device other than an extensometer is placed in contact with the specimen's reduced section during the test, this also shall be noted.

Example: elongation = 30 % increase (2-in. gage length)

NOTE 32—Elongation results are very sensitive to variables such as: (*a*) speed of testing, (*b*) specimen geometry (gage length, diameter, width, and thickness), (*c*) heat dissipation (through grips, extensometers, or other devices in contact with the reduced section), (*d*) surface finish in reduced section (especially burrs or notches), (*e*) alignment, and (*f*) fillets and tapers. Parties involved in comparison or conformance testing should standardize the above items, and it is recommended that use of ancillary devices (such as extensometer supports) which may remove heat from specimens be avoided. See Appendix X1. for additional information on the effects of these variables.

7.8.2 When the specified elongation is greater than 3 %, fit ends of the fractured specimen together carefully and measure the distance between the gage marks to the nearest 0.01 in. for gage lengths of 2 in. and under, and to at least the nearest 0.5 % of the gage length for gage lengths over 2 in. A percentage scale reading to 0.5 % of the gage length may be used.

7.8.3 When the *specified* elongation is 3 % or less, determine the elongation of standard round specimens (see Fig. 8) using the following procedure, except that the procedure given in 7.8.2 may be used instead when the *measured* elongation is greater than 3 %.

7.8.3.1 Measure the original gage length of the specimen to the nearest 0.002 in.

7.8.3.2 Remove partly torn fragments that will interfere with fitting together the ends of the fractured specimen or with making the final measurement.

7.8.3.3 Fit the fractured ends together with matched surfaces and apply a force along the axis of the specimen sufficient to close the fractured ends together. If desired, this force may then be removed carefully, provided the specimen remains intact.

NOTE 33—The use of a force of approximately 2000 psi has been found to give satisfactory results on test specimens of aluminum alloy.

7.8.3.4 Measure the final gage length to the nearest 0.002 in. and report the elongation to the nearest 0.2 %.

7.8.4 Specimens other than the standard specimen described in Fig. 8 are exempt from the requirement of 7.8.3 except as required by the applicable product specification.

7.8.5 If any part of the fracture takes place outside of the middle half of the gage length or in a punched or scribed mark within the reduced section, the elongation value obtained may not be representative of the material. In acceptance testing, if the elongation so measured meets the minimum requirements specified, no further testing is required, but if the elongation is less than the minimum requirements, discard the test and retest.

7.8.6 Elongation at fracture is defined as the elongation measured just prior to the sudden decrease in force associated with fracture. For many ductile materials not exhibiting a sudden decrease in force, the elongation at fracture can be taken as the strain measured just prior to when the force falls below 10 % of the maximum force encountered during the test.

7.8.6.1 Elongation at fracture shall include elastic and plastic elongation and may be determined with autographic or automated methods using extensometers. Use a class B2 or better extensometer for materials having less than 5 % elongation, a class C or better extensometer for materials having elongation greater than or equal to 5 % but less than 50 %, and a class D or better extensometer for materials having 50 % or greater elongation. In all cases, the extensometer gage length shall be the nominal gage length required for the specimen being tested. Due to the lack of precision in fitting fractured ends together, the elongation after fracture using the manual methods of the preceding paragraphs may differ from the elongation at fracture determined with extensometers.

7.8.6.2 Percent elongation at fracture may be calculated directly from elongation at fracture data and be reported instead of percent elongation as calculated in paragraphs 7.8.2 to 7.8.3. However, these two parameters are not interchangeable. Use of the elongation at fracture method generally provides more repeatable results.

NOTE 34—When disagreements arise over the percent elongation results, agreement must be reached on which method to use to obtain the results.

7.9 *Reduction of Area:*

7.9.1 The reduced area used to calculate reduction of area (see 7.9.2 and 7.9.3) shall be the minimum cross section at the location of fracture.

7.9.2 *Specimens with Originally Circular Cross Sections*— Fit the ends of the fractured specimen together and measure the reduced diameter to the same accuracy as the original measurement.

NOTE 35—Because of anisotropy, circular cross sections often do not remain circular during straining in tension. The shape is usually elliptical, thus, the area may be calculated by $\pi \cdot d_1 \cdot d_2/4$, where d_1 and d_2 are the major and minor diameters, respectively.

7.9.3 *Specimens with Original Rectangular Cross Sec-*

tions—Fit the ends of the fractured specimen together and measure the thickness and width at the minimum cross section to the same accuracy as the original measurements.

NOTE 36—Because of the constraint to deformation that occurs at the corners of rectangular specimens, the dimensions at the center of the original flat surfaces are less than those at the corners. The shapes of these surfaces are often assumed to be parabolic. When this assumption is made, an effective thickness, t_e, may be calculated as follows: $(t_1 + 4t_2 + t_3)/6$, where t_1 and t_3 are the thicknesses at the corners, and t_2 is the thickness at mid-width. An effective width may be similarly calculated.

7.9.4 Calculate the reduced area based upon the dimensions determined in 7.9.2 or 7.9.3. The difference between the area thus found and the area of the original cross section expressed as a percentage of the original area is the reduction of area.

7.9.5 If any part of the fracture takes place outside the middle half of the reduced section or in a punched or scribed gage mark within the reduced section, the reduction of area value obtained may not be representative of the material. In acceptance testing, if the reduction of area so calculated meets the minimum requirements specified, no further testing is required, but if the reduction of area is less than the minimum requirements, discard the test results and retest.

7.9.6 Results of measurements of reduction of area shall be rounded using the procedures of Practice E 29 and any specific procedures in the product specifications. In the absence of a specified procedure, it is recommended that reduction of area test values in the range from 0 to 10 % be rounded to the nearest 0.5 % and test values of 10 % and greater to the nearest 1 %.

7.10 *Rounding Reported Test Data for Yield Strength and Tensile Strength*—Test data should be rounded using the procedures of Practice E 29 and the specific procedures in the product specifications. In the absence of a specified procedure for rounding the test data, one of the procedures described in the following paragraphs is recommended.

7.10.1 For test values up to 50 000 psi, round to the nearest 100 psi; for test values of 50 000 psi and up to 100 000 psi, round to the nearest 500 psi; for test values of 100 000 psi and greater, round to the nearest 1000 psi.

NOTE 37—For steel products, see Test Methods and Definitions A 370.

7.10.2 For all test values, round to the nearest 100 psi.

NOTE 38—For aluminum- and magnesium-alloy products, see Methods B 557.

7.10.3 For all test values, round to the nearest 500 psi.

7.11 *Replacement of Specimens*—A test specimen may be discarded and a replacement specimen selected from the same lot of material in the following cases:

7.11.1 The original specimen had a poorly machined surface,

7.11.2 The original specimen had the wrong dimensions,

7.11.3 The specimen's properties were changed because of poor machining practice,

7.11.4 The test procedure was incorrect,

7.11.5 The fracture was outside the gage length,

7.11.6 For elongation determinations, the fracture was outside the middle half of the gage length, or

7.11.7 There was a malfunction of the testing equipment.

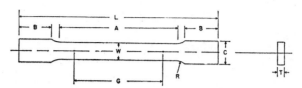

	Dimensions		
	Standard Specimens		Subsize Specimen
	Plate-Type, 1½-in. Wide	Sheet-Type, ½-in. Wide	¼-in. Wide
	in.	in.	in.
G—Gage length (Notes 1 and 2)	8.00 ± 0.01	2.000 ± 0.005	1.000 ± 0.003
W—Width (Notes 3 and 4)	1½ + ⅛, −¼	0.500 ± 0.010	0.250 ± 0.005
T—Thickness (Note 5)		thickness of material	
R—Radius of fillet, min (Note 6)	1	½	¼
L—Over-all length, min (Notes 2 and 7)	18	8	4
A—Length of reduced section, min	9	2¼	1¼
B—Length of grip section, min (Note 8)	3	2	1¼
C—Width of grip section, approximate (Notes 4 and 9)	2	¾	⅜

NOTE 1—For the 1½-in. wide specimen, punch marks for measuring elongation after fracture shall be made on the flat or on the edge of the specimen and within the reduced section. Either a set of nine or more punch marks 1 in. apart, or one or more pairs of punch marks 8 in. apart may be used.

NOTE 2—When elongation measurements of 1½-in. wide specimens are not required, a minimum length of reduced section (A) of 2¼ in. may be used with all other dimensions similar to those of the plate-type specimen.

NOTE 3—For the three sizes of specimens, the ends of the reduced section shall not differ in width by more than 0.004, 0.002 or 0.001 in., respectively. Also, there may be a gradual decrease in width from the ends to the center, but the width at each end shall not be more than 0.015, 0.005, or 0.003 in., respectively, larger than the width at the center.

NOTE 4—For each of the three sizes of specimens, narrower widths (W and C) may be used when necessary. In such cases the width of the reduced section should be as large as the width of the material being tested permits; however, unless stated specifically, the requirements for elongation in a product specification shall not apply when these narrower specimens are used.

NOTE 5—The dimension T is the thickness of the test specimen as provided for in the applicable material specifications. Minimum thickness of 1½-in. wide specimens shall be ³⁄₁₆ in. Maximum thickness of ½-in. and ¼-in. wide specimens shall be ¾ in. and ¼ in., respectively.

NOTE 6—For the 1½-in. wide specimen, a ½-in. minimum radius at the ends of the reduced section is permitted for steel specimens under 100 000 psi in tensile strength when a profile cutter is used to machine the reduced section.

NOTE 7—To aid in obtaining axial force application during testing of ¼-in. wide specimens, the over-all length should be as large as the material will permit, up to 8.00 in.

NOTE 8—It is desirable, if possible, to make the length of the grip section large enough to allow the specimen to extend into the grips a distance equal to two thirds or more of the length of the grips. If the thickness of ½-in. wide specimens is over ⅜ in., longer grips and correspondingly longer grip sections of the specimen may be necessary to prevent failure in the grip section.

NOTE 9—For the three sizes of specimens, the ends of the specimen shall be symmetrical in width with the center line of the reduced section within 0.10, 0.05 and 0.005 in., respectively. However, for referee testing and when required by product specifications, the ends of the ½-in. wide specimen shall be symmetrical within 0.01 in.

NOTE 10—For each specimen type, the radii of all fillets shall be equal to each other within a tolerance of 0.05 in., and the centers of curvature of the two fillets at a particular end shall be located across from each other (on a line perpendicular to the centerline) within a tolerance of 0.10 in.

NOTE 11—Specimens with sides parallel throughout their length are permitted, except for referee testing, provided: (a) the above tolerances are used; (b) an adequate number of marks are provided for determination of elongation; and (c) when yield strength is determined, a suitable extensometer is used. If the fracture occurs at a distance of less than 2W from the edge of the gripping device, the tensile properties determined may not be representative of the material. In acceptance testing, if the properties meet the minimum requirements specified, no further testing is required, but if they are less than the minimum requirements, discard the test and retest.

FIG. 1 Rectangular Tension Test Specimens

NOTE 39—The tension specimen is inappropriate for assessing some types of imperfections in a material. Other methods and specimens employing ultrasonics, dye penetrants, radiography, etc., may be considered when flaws such as cracks, flakes, porosity, etc., are revealed during a test and soundness is a condition of acceptance.

8. Report

8.1 Test information on materials not covered by a product specification should be reported in accordance with 8.2 or both 8.2 and 8.3.

8.2 Test information to be reported shall include the following when applicable:

8.2.1 Material and sample identification.

8.2.2 Specimen type (see Section 6).

8.2.3 Yield strength and the method used to determine yield strength (see 7.5).

8.2.4 Yield point elongation (see 7.6).

8.2.5 Tensile strength (see 7.7).

8.2.6 Elongation (report original gage length, percentage increase, and method used to determine elongation) (see 7.8).

8.2.7 Reduction of area (see 7.9).

8.3 Test information to be available on request shall include:

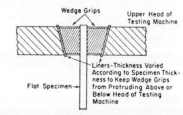

FIG. 2 Wedge Grips with Liners for Flat Specimens

🔺 **E 8**

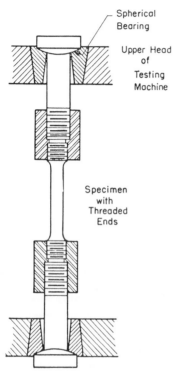

FIG. 3   Gripping Device for Threaded-End Specimens

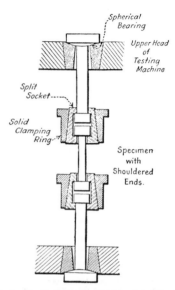

FIG. 4   Gripping Device for Shouldered-End Specimens

8.3.1 Specimen test section dimension(s).

8.3.2 Equation used to calculate cross-sectional area of rectangular specimens taken from large-diameter tubular products.

8.3.3 Speed and method used to determine speed of testing (see 7.4).

8.3.4 Method used for rounding of test results (see 7.10).

8.3.5 Reasons for replacement specimens (see 7.11).

9. Precision and Bias

9.1 *Precision*—An interlaboratory test program[7] gave the following values for coefficients of variation for the most commonly measured tensile properties:

Coefficient of Variation, %

	Tensile Strength	Yield Strength Offset= 0.02 %	Yield Strength Offset= 0.2 %	Elongation Gage Length= 4 Diameter	Reduction of Area
CV %$_r$	0.9	2.7	1.4	2.8	2.8
CV %$_R$	1.3	4.5	2.3	5.4	4.6

CV %$_r$ = repeatability coefficient of variation in percent within a laboratory
CV %$_R$ = repeatability coefficient of variation in percent between laboratories

9.1.1 The values shown are the averages from tests on six

---

[7] Supporting data can be found in Appendix I and additional data are available from ASTM Headquarters. Request RR: E28-1004.

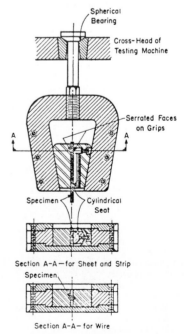

FIG. 5   Gripping Devices for Sheet and Wire Specimens

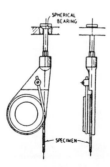

FIG. 6   Snubbing Device for Testing Wire

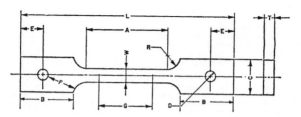

Dimensions	in.
G—Gage length	2.000 ± 0.005
W—Width (Note 1)	0.500 ± 0.010
T—Thickness, max (Note 2)	5/8
R—Radius of fillet, min (Note 3)	1/2
L—Over-all length, min	8
A—Length of reduced section, min	2¼
B—Length of grip section, min	2
C—Width of grip section, approximate	2
D—Diameter of hole for pin, min (Note 4)	1/2
E—Edge distance from pin, approximate	1½
F—Distance from hole to fillet, min	1/2

Note 1—The ends of the reduced section shall differ in width by not more than 0.002 in. There may be a gradual taper in width from the ends to the center, but the width at each end shall be not more than 0.005 in. greater than the width at the center.

Note 2—The dimension T is the thickness of the test specimen as stated in the applicable product specifications.

Note 3—For some materials, a fillet radius R larger than 1/2 in. may be needed.

Note 4—Holes must be on center line of reduced section, within ±0.002 in.

Note 5—Variations of dimensions C, D, E, F, and L may be used that will permit failure within the gage length.

FIG. 7   Pin-Loaded Tension Test Specimen with 2-in. Gage Length

frequently tested metals, selected to include most of the normal range for each property listed above. When these materials are compared, a large difference in coefficient of variation is found. Therefore, the values above should not be used to judge whether the difference between duplicate tests of a specific material is larger than expected. The values are provided to allow potential users of this test method to assess, in general terms, its usefulness for a proposed application.

9.2 *Bias*—The procedures in Test Methods E 8 for measuring tensile properties have no bias because these properties can be defined only in terms of a test method.

10. Keywords

10.1 accuracy; bending stress; discontinuous yielding; drop-of-the-beam; eccentric force application; elastic extension; elongation; extension-under-load; extensometer; force; free-running crosshead speed; gage length; halt-of-the force; percent elongation; plastic extension; preload; rate of stressing; rate of straining; reduced section; reduction of area; sensitivity; strain; stress; taring; tensile strength; tension testing; yield point elongation; yield strength

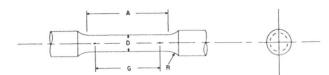

	Dimensions				
	Standard Specimen	Small-Size Specimens Proportional to Standard			
	in.	in.	in.	in.	in.
Nominal Diameter	0.500	0.350	0.250	0.160	0.113
G—Gage length	2.000 ± 0.005	1.400 ± 0.005	1.000 ± 0.005	0.640 ± 0.005	0.450 ± 0.005
D—Diameter (Note 1)	0.500 ± 0.010	0.350 ± 0.007	0.250 ± 0.005	0.160 ± 0.003	0.113 ± 0.002
R—Radius of fillet, min	3/8	1/4	3/16	5/32	3/32
A—Length of reduced section, min (Note 2)	2¼	1¾	1¼	3/4	5/8

Note 1—The reduced section may have a gradual taper from the ends toward the center, with the ends not more than 1 % larger in diameter than the center (controlling dimension).

Note 2—If desired, the length of the reduced section may be increased to accommodate an extensometer of any convenient gage length. Reference marks for the measurement of elongation should, nevertheless, be spaced at the indicated gage length.

Note 3—The gage length and fillets may be as shown, but the ends may be of any form to fit the holders of the testing machine in such a way that the load shall be axial (see Fig. 9). If the ends are to be held in wedge grips it is desirable, if possible, to make the length of the grip section great enough to allow the specimen to extend into the grips a distance equal to two thirds or more of the length of the grips.

Note 4—On the round specimens in Figs. 8 and 9, the gage lengths are equal to four times the nominal diameter. In some product specifications other specimens may be provided for, but unless the 4-to-1 ratio is maintained within dimensional tolerances, the elongation values may not be comparable with those obtained from the standard test specimen.

Note 5—The use of specimens smaller than 0.250-in. diameter shall be restricted to cases when the material to be tested is of insufficient size to obtain larger specimens or when all parties agree to their use for acceptance testing. Smaller specimens require suitable equipment and greater skill in both machining and testing.

Note 6—Five sizes of specimens often used have diameters of approximately 0.505, 0.357, 0.252, 0.160, and 0.113 in., the reason being to permit easy calculations of stress from loads, since the corresponding cross-sectional areas are equal or close to 0.200, 0.100, 0.0500, 0.0200, and 0.0100 in.^{2}, respectively. Thus, when the actual diameters agree with these values, the stresses (or strengths) may be computed using the simple multiplying factors 5, 10, 20, 50, and 100, respectively. (The metric equivalents of these five diameters do not result in correspondingly convenient cross-sectional areas and multiplying factors.)

FIG. 8   **Standard 0.500-in. Round Tension Test Specimen with 2-in. Gage Length and Examples of Small-Size Specimens Proportional to the Standard Specimen**

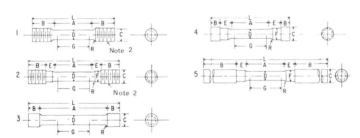

	Dimensions				
	Specimen 1	Specimen 2	Specimen 3	Specimen 4	Specimen 5
	in.	in.	in.	in.	in.
G—Gage length	2.000 ± 0.005	2.000 ± 0.005	2.000 ± 0.005	2.000 ± 0.005	2.000 ± 0.005
D—Diameter (Note 1)	0.500 ± 0.010	0.500 ± 0.010	0.500 ± 0.010	0.500 ± 0.010	0.500 ± 0.010
R—Radius of fillet, min	3/8	3/8	1/16	3/8	3/8
A—Length of reduced section	2¼, min	2¼, min	4, approximately	2¼, min	2¼, min
L—Over-all length, approximate	5	5½	5½	4¾	9½
B—Length of end section (Note 3)	1⅜, approximately	1, approximately	¾, approximately	½, approximately	3, min
C—Diameter of end section	¾	¾	23/32	⅞	¾
E—Length of shoulder and fillet section, approximate	. . .	5/8	. . .	¾	5/8
F—Diameter of shoulder	. . .	5/8	. . .	5/8	19/32

Note 1—The reduced section may have a gradual taper from the ends toward the center with the ends not more than 0.005 in. larger in diameter than the center.

Note 2—On Specimens 1 and 2, any standard thread is permissible that provides for proper alignment and aids in assuring that the specimen will break within the reduced section.

Note 3—On Specimen 5 it is desirable, if possible, to make the length of the grip section great enough to allow the specimen to extend into the grips a distance equal to two thirds or more of the length of the grips.

FIG. 9   **Various Types of Ends for Standard Round Tension Test Specimens**

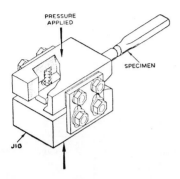

FIG. 10   Squeezing Jig for Flattening Ends of Full-Size Tension
Test Specimens

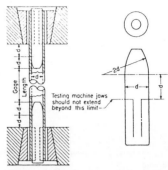

NOTE—The diameter of the plug shall have a slight taper from the line limiting
the test machine jaws to the curved section.

FIG. 11 Metal Plugs for Testing Tubular Specimens, Proper
Location of Plugs in Specimen and of Specimen in Heads of Testing
Machine

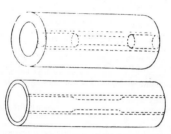

NOTE—The edges of the blank for the specimen shall be cut parallel to each
other.

FIG. 12   Location from Which Longitudinal Tension Test
Specimens Are to be Cut from Large-Diameter Tube

E 8

	Specimen 1	Specimen 2	Specimen 3	Specimen 4	Specimen 5	Specimen 6	Specimen 7
Dimensions							
	in.	in.	in.	in.	in.	in.	in.
G—Gage length	2.000 ± 0.005	2.000 ± 0.005	8.00 ± 0.01	2.000 ± 0.005	4.000 ± 0.005	2.000 ± 0.005	4.000 ± 0.005
W—Width (Note 1)	0.500 ± 0.010	1½ + ⅛, −¼	1½ + ⅛, −¼	0.750 ± 0.031	0.750 ± 0.031	1.000 ± 0.062	1.000 ± 0.062
T—Thickness	measured thickness of specimen						
R—Radius of fillet, min	½	1	1	1	1	1	1
A—Length of reduced section, min	2¼	2¼	9	2¼	4½	¼	4½
B—Length of grip section, min (Note 2)	3	3	3	3	3	3	3
C—Width of grip section, approximate (Note 3)	11/16	2	2	1	1	1½	1½

NOTE 1—The ends of the reduced section shall differ in width by not more than 0.002 in. for specimens 1 and 4, and not more than 0.005 in. for specimens 2, 3, 5, 6, and 7. There may be a gradual taper in width from the ends to the center, but the width at each end shall be not more than 0.005 in. greater than the width at the center for 2-in. gage length specimens, not more than 0.008 in. greater than the width at the center for 4-in. gage length specimens, and not more than 0.015 in. greater than the width at the center for 8-in. gage length specimens.

NOTE 2—It is desirable, if possible, to make the length of the grip section great enough to allow the specimen to extend into the grips a distance equal to two thirds or more of the length of the grips.

NOTE 3—The ends of the specimen shall be symmetrical with the center line of the reduced section within 0.05 in. for specimens 1, 4, and 5, and 0.10 in. for specimens 2, 3, 6, and 7.

NOTE 4—For each specimen type, the radii of all fillets shall be equal to each other within a tolerance of 0.05 in., and the centers of curvature of the two fillets at a particular end shall be located across from each other (on a line perpendicular to the centerline) within a tolerance of 0.10 in.

NOTE 5—For circular segments, the cross-sectional area may be calculated by multiplying W and T. If the ratio of the dimension W to the diameter of the tubular section is larger than about ⅙, the error in using this method to calculate the cross-sectional area may be appreciable. In this case, the exact equation (see 7.1.3) must be used to determine the area.

NOTE 6—Specimens with G/W less than 4 should not be used for determination of elongation.

NOTE 7—Specimens with sides parallel throughout their length are permitted, except for referee testing, provided: (a) the above tolerances are used; (b) an adequate number of marks are provided for determination of elongation; and (c) when yield strength is determined, a suitable extensometer is used. If the fracture occurs at a distance of less than 2W from the edge of the gripping device, the tensile properties determined may not be representative of the material. If the properties meet the minimum requirements, no further testing is required, but if they are less than the minimum requirements, discard the test and retest.

FIG. 13   Tension Test Specimens for Large-Diameter Tubular Products

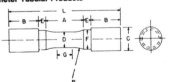

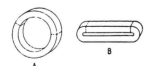

FIG. 14   Location of Transverse Tension Test Specimen in Ring Cut from Tubular Products

	Specimen 1	Specimen 2	Specimen 3
Dimensions			
	in.	in.	in.
G—Length of parallel section	Shall be equal to or greater than diameter D		
D—Diameter	0.500 ± 0.010	0.750 ± 0.015	1.25 ± 0.02
R—Radius of fillet, min	1	1	2
A—Length of reduced section, min	1¼	1½	2¼
L—Over-all length, min	3¾	4	6⅜
B—Length of end section, approximate	1	1	1¾
C—Diameter of end section, approximate	¾	1⅛	1⅞
E—Length of shoulder, min	¼	¼	5/16
F—Diameter of shoulder	⅝ ± 1/64	15/16 ± 1/64	1 7/16 ± 1/64

NOTE—The reduced section and shoulders (dimensions A, D, E, F, G, and R) shall be as shown, but the ends may be of any form to fit the holders of the testing machine in such a way that the force can be axial. Commonly the ends are threaded and have the dimensions B and C given above.

FIG. 15   Standard Tension Test Specimen for Cast Iron

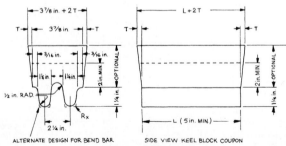

(a) Design for Double Keel Block Coupon

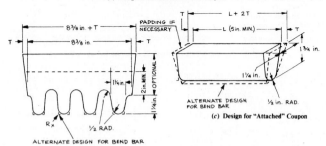

(c) Design for "Attached" Coupon

FIG. 16   Test Coupons for Castings (see Table 1 for Details of Design)

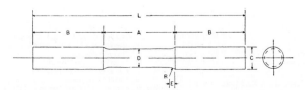

Dimensions	
	in.
D—Diameter	5/8
R—Radius of fillet	5/16
A—Length of reduced section	2½
L—Over-all length	7½
B—Length of end section	2½
C—Diameter of end section	¾
E—Length of fillet	3/16

FIG. 17   Standard Tension Test Specimen for Malleable Iron

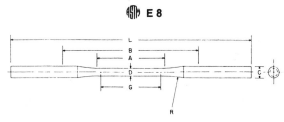

ASTM E 8

Dimensions	
	in.
G—Gage length	2.000 ± 0.005
D—Diameter (see Note)	0.250 ± 0.005
R—Radius of fillet, min	3
A—Length of reduced section, min	2¼
L—Over-all length, min	9
B—Distance between grips, min	4½
C—Diameter of end section, approximate	⅜

NOTE—The reduced section may have a gradual taper from the ends toward the center, with the ends not more than 0.005 in. larger in diameter than the center.

FIG. 18   Standard Tension Test Specimens for Die Castings

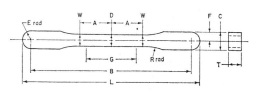

Pressing Area = 1.00 in.²

Dimensions	
	in.
G—Gage length	1.000 ± 0.003
D—Width at center	0.225 ± 0.001
W—Width at end of reduced section	0.235 ± 0.001
T—Compact to this thickness	0.140 to 0.250
R—Radius of fillet	1
A—Half-length of reduced section	⅝
B—Grip length	3.187 ± 0.001
L—Overall length	3.529 ± 0.001
C—Width of grip section	0.343 ± 0.001
F—Half-width of grip section	0.171 ± 0.001
E—End radius	0.171 ± 0.001

NOTE—Dimensions Specified, except G and T, are those of the die.

FIG. 19   Standard Flat Unmachined Tension Test Specimens for Powder Metallurgy (P/M) Products

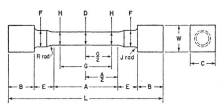

Approximate Pressing Area of Unmachined Compact = 1.166 in.²
Machining Recommendations

1. Rough machine reduced section to ¼-in. diameter
2. Finish turn 0.187/0.191-in. diameter with radii and taper
3. Polish with 00 emery cloth
4. Lap with crocus cloth

Dimensions	
	in.
G—Gage length	1.000 ± 0.003
D—Diameter at center of reduced section	0.187 ± 0.001
H—Diameter at ends of gage length	0.191 ± 0.001
R—Radius of fillet	0.250 ± 0.005
A—Length of reduced section	1.875 ± 0.003
L—Overall length (die cavity length)	3, nominal
B—Length of end section	0.310 ± 0.005
C—Compact to this end thickness	0.395 ± 0.005
W—Die cavity width	0.395 ± 0.003
E—Length of shoulder	0.250 ± 0.005
F—Diameter of shoulder	0.310 ± 0.001
J—End fillet radius	0.050 ± 0.005

NOTE 1—The gage length and fillets of the specimen shall be as shown. The ends as shown are designed to provide a practical minimum pressing area. Other end designs are acceptable, and in some cases are required for high-strength sintered materials.

NOTE 2—It is recommended that the test specimen be gripped with a split collet and supported under the shoulders. The radius of the collet support circular edge is to be not less than the end fillet radius of the test specimen.

Note 3—Diameters D and H are to be concentric within 0.001 in. total indicator runout (T.I.R.), and free of scratches and tool marks.

FIG. 20   Standard Round Machined Tension Test Specimen for Powder Metallurgy (P/M) Products

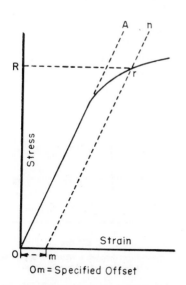

Om = Specified Offset

**FIG. 21   Stress-Strain Diagram for Determination of Yield
Strength by the Offset Method**

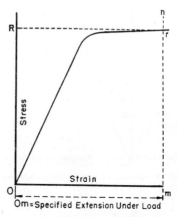

Om = Specified Extension Under Load

**FIG. 22   Stress-Strain Diagram for Determination of Yield
Strength by the Extension-Under-Load Method**

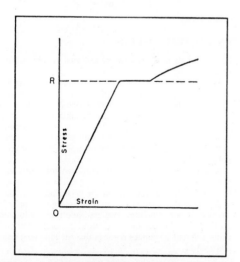

**FIG. 23   Stress-Strain Diagram Showing Upper Yield Strength
Corresponding with Top of Knee**

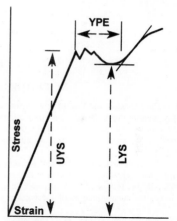

**FIG. 24   Stress-Strain Diagram Showing Yield Point Elongation
and Upper and Lower Yield Strengths**

🏛 E 8

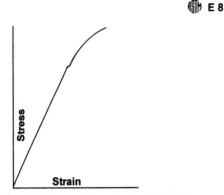

FIG. 25 Stress-Strain Diagram With an Inflection, But No YPE

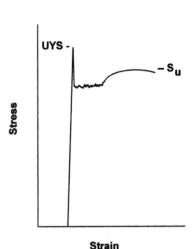

FIG. 26 Stress-Strain Diagram in Which the Upper Yield Strength is the Maximum Stress Recorded

APPENDIXES

(Nonmandatory Information)

X1. FACTORS AFFECTING TENSION TEST RESULTS

X1.1 The precision and bias of tension test strength and ductility measurements depend on strict adherence to the stated test procedure and are influenced by instrumental and material factors, specimen preparation, and measurement/testing errors.

X1.2 The consistency of agreement for repeated tests of the same material is dependent on the homogeneity of the material, and the repeatability of specimen preparation, test conditions, and measurements of the tension test parameters.

X1.3 Instrumental factors that can affect test results include: the stiffness, damping capacity, natural frequency, and mass of moving parts of the tensile test machine; accuracy of force indication and use of forces within the verified range of the machine; rate of force application, alignment of the test specimen with the applied force, parallelness of the grips, grip pressure, nature of the force control used, appropriateness and calibration of extensometers, heat dissipation (by grips, extensometers, or ancillary devices), and so forth.

X1.4 Material factors that can affect test results include: representativeness and homogeneity of the test material, sampling scheme, and specimen preparation (surface finish, dimensional accuracy, fillets at the ends of the gage length, taper in the gage length, bent specimens, thread quality, and so forth).

X1.4.1 Some materials are very sensitive to the quality of the surface finish of the test specimen (see Note 12) and must

be ground to a fine finish, or polished to obtain correct results.

X1.4.2 Test results for specimens with as-cast, as-rolled, as-forged, or other non-machined surface conditions can be affected by the nature of the surface (see Note 14).

X1.4.3 Test specimens taken from appendages to the part or component, such as prolongs or risers, or from separately produced castings (for example, keel blocks) may produce test results that are not representative of the part or component.

X1.4.4 Test specimen dimensions can influence test results. For cylindrical or rectangular specimens, changing the test specimen size generally has a negligible effect on the yield and tensile strength but may influence the upper yield strength, if one is present, and elongation and reduction of area values. Comparison of elongation values determined using different specimens requires that the following ratio be controlled:

$$L_o/(A_o)^{1/2}$$

where:
L_o = original gage length of specimen, and
A_o = original cross-sectional area of specimen.

X1.4.4.1 Specimens with smaller $L_o/(A_o)^{1/2}$ ratios generally give greater elongation and reduction in area values. This is the case for example, when the width or thickness of a rectangular tensile test specimen is increased.

X1.4.4.2 Holding the $L_o/(A_o)^{1/2}$ ratio constant minimizes, but does not necessarily eliminate, differences. Depending on material and test conditions, increasing the size of the proportional specimen of Fig. 8 may be found to increase or decrease elongation and reduction in area values somewhat.

X1.4.5 Use of a taper in the gage length, up to the allowed 1 % limit, can result in lower elongation values. Reductions of as much as 15 % have been reported for a 1 % taper.

X1.4.6 Some materials are highly strain-rate sensitive. Changes in the strain rate can affect the yield strength and elongation values, especially for strain-rate sensitive materials. In general, the yield strength and elongation will increase as the strain rate increases.

X1.4.7 Brittle materials require careful specimen preparation, high quality surface finishes, large fillets at the ends of the gage length, oversize threaded grip sections, and cannot

tolerate punch or scribe marks as gage length indicators.

X1.4.8 Flattening of tubular products to permit testing does alter the material properties, generally nonuniformly, in the flattened region which may affect test results.

X1.5 Measurement errors that can affect test results include: verification of the test force, extensometers, micrometers, dividers, and other measurement devices, alignment and zeroing of chart recording devices, and so forth.

X1.5.1 Measurement of the dimensions of as-cast, as-rolled, as-forged, and other test specimens with non-machined surfaces may be imprecise due to the irregularity of the surface flatness.

X1.5.2 Materials with anisotropic flow characteristics may exhibit non-circular cross sections after fracture and

TABLE X1.1 Precision Statistics—Tensile Strength, ksi

Material	X	s_r	s_r/X, %	s_R	s_R/X, %	r	R
EC-H19	25.66	0.63	2.45	0.63	2.45	1.76	1.76
2024-T351	71.26	0.88	1.24	0.96	1.34	2.47	2.68
ASTM A105	86.57	0.60	0.70	1.27	1.46	1.68	3.55
AISI 316	100.75	0.39	0.39	1.21	1.20	1.09	3.39
Inconel 600	99.48	0.42	0.43	0.72	0.72	1.19	2.02
SAE 51410	181.73	0.46	0.25	1.14	0.63	1.29	3.20
		Averages:	0.91		1.30		

NOTE: X is the average of the cell averages, that is, the grand mean for the test parameter,
s_r is the repeatability standard deviation (within-laboratory precision),
s_r/X is the coefficient of variation in %,
s_R is the reproducibility standard deviation (between-laboratory precision),
s_R/X is the coefficient of variation, %,
r is the 95 % repeatability limits,
R is the 95 % reproducibility limits.

TABLE X1.2 Precision Statistics—0.02 % Yield Strength, ksi

Material	X	s_r	s_r/X, %	s_R	s_R/X, %	r	R
EC-H19	16.17	0.65	3.99	1.19	7.36	1.81	3.33
2024-T351	51.38	0.84	1.64	0.89	1.73	2.36	2.49
ASTM A105	59.66	1.20	2.02	1.89	3.18	3.37	5.31
AISI 316	48.62	2.39	4.91	4.61	9.49	6.68	12.91
Inconel 600	38.74	0.46	1.18	0.76	1.96	1.28	2.13
SAE 15410	104.90	2.40	2.29	3.17	3.02	6.73	8.88
		Averages:	2.67		4.46		

TABLE X1.3 Precision Statistics—0.2 % Yield Strength, ksi

Material	X	s_r	s_r/X, %	s_R	s_R/X, %	r	R
EC-H19	22.98	0.47	2.06	0.48	2.07	1.33	1.33
2024-T351	52.64	0.74	1.41	0.79	1.49	2.08	2.20
ASTM A105	58.36	0.83	1.42	1.44	2.47	2.31	4.03
AISI 316	69.63	0.94	1.35	2.83	4.07	2.63	7.93
Inconel 600	38.91	0.36	0.93	0.85	2.18	1.01	2.37
SAE 51410	140.33	1.29	0.92	2.30	1.64	3.60	6.45
		Averages:	1.35		2.32		

TABLE X1.4 Precision Statistics—% Elongation in 4D

Material	X	s_r	s_r/X, %	s_R	s_R/X, %	r	R
EC-H19	17.45	0.64	3.69	0.92	5.30	1.80	2.59
2024-T351	19.75	0.59	2.99	1.58	8.00	1.65	4.43
ASTM A105	29.10	0.76	2.62	0.98	3.38	2.13	2.76
AISI 316	40.07	1.10	2.75	2.14	5.35	3.09	6.00
Inconel 600	44.27	0.66	1.50	1.54	3.48	1.86	4.31
SAE 51410	14.48	0.48	3.29	0.99	6.83	1.34	2.77
		Averages:	2.81		5.39		

NOTE A1—Length of reduced section = 6D.

⟨⁂⟩ E 8

TABLE X1.5   Precision Statistics—% Reduction in Area

Material	X	s_r	s_r/X, %	s_R	s_R/X, %	r	R
EC-H19	79.14	1.94	2.45	2.02	2.56	5.44	5.67
2024-T351	30.31	2.07	6.82	3.58	11.80	5.79	10.01
ASTM A105	65.59	0.84	1.28	1.26	1.92	2.35	3.53
AISI 316	71.49	0.99	1.39	1.61	2.25	2.78	4.50
Inconel 600	59.34	0.67	1.14	0.70	1.18	1.89	1.97
SAE 51410	50.49	1.86	<u>3.69</u>	3.95	<u>7.81</u>	5.21	11.05
		Averages:	2.80		4.59		

measurement precision may be affected, as a result (see Note 35).

X1.5.3 The corners of rectangular test specimens are subject to constraint during deformation and the originally flat surfaces may be parabolic in shape after testing which will affect the precision of final cross-sectional area measurements (see Note 36).

X1.5.4 If any portion of the fracture occurs outside of the middle of the gage length, or in a punch or scribe mark within the gage length, the elongation and reduction of area values may not be representative of the material. Wire specimens that break at or within the grips may not produce test results representative of the material.

X1.5.5 Use of specimens with shouldered ends ("button-head" tensiles) will produce lower 0.02 % offset yield strength values than threaded specimens.

X1.6 Because standard reference materials with certified tensile property values are not available, it is not possible to rigorously define the bias of tension tests. However, by the use of carefully designed and controlled interlaboratory studies, a reasonable definition of the precision of tension test results can be obtained.

X1.6.1 An interlaboratory test program[7] was conducted in which six specimens each, of six different materials were prepared and tested by each of six different laboratories. Tables X1.1 to X1.5 present the precision statistics, as defined in Practice E 691, for: tensile strength, 0.02 % yield strength, 0.2 % yield strength, % elongation in 4D, and % reduction in area. In each table, the first column lists the six materials tested, the second column lists the average of the

average results obtained by the laboratories, the third and fifth columns list the repeatability and reproducibility standard deviations, the fourth and sixth columns list the coefficients of variation for these standard deviations, and the seventh and eighth columns list the 95 % repeatability and reproducibility limits.

X1.6.2 The averages (below columns four and six in each table) of the coefficients of variation permit a relative comparison of the repeatability (within-laboratory precision) and reproducibility (between-laboratory precision) of the tension test parameters. This shows that the ductility measurements exhibit less repeatability and reproducibility than the strength measurements. The overall ranking from the least to the most repeatable and reproducible is: % elongation in 4D, % reduction in area, 0.02 % offset yield strength, 0.2 % offset yield strength, and tensile strength. Note that the rankings are in the same order for the repeatability and reproducibility average coefficients of variation and that the reproducibility (between-laboratory precision) is poorer than the repeatability (within-laboratory precision), as would be expected.

X1.6.3 No comments about bias can be made for the interlaboratory study due to the lack of certified test results for these specimens. However, examination of the test results showed that one laboratory consistently exhibited higher than average strength values and lower than average ductility values for most of the specimens. One other laboratory had consistently lower than average tensile strength results for all specimens.

X2. MEASUREMENT OF SPECIMEN DIMENSIONS

X2.1 Measurement of specimen dimensions is critical in tension testing, and it becomes more critical with decreasing specimen size, as a given absolute error becomes a larger relative (percent) error. Measuring devices and procedures should be selected carefully, so as to minimize measurement error and provide good repeatability and reproducibility.

X2.2 Relative measurement error should be kept at or below 1 %, where possible. Ideally, this 1 % error should include not only the resolution of the measuring device but also the variability commonly referred to as repeatability and reproducibility. (Repeatability is the ability of any operator to obtain similar measurements in repeated trials. Reproducibility is the ability of multiple operators to obtain similar measurements.)

X2.3 Formal evaluation of gage repeatability and reproducibility (GR and R) by way of a GR and R study is highly recommended. A GR and R study involves having multiple operators each take two or three measurements of a number

of parts—in this case, test specimens. Analysis, usually done by computer, involves comparing the observed measurement variations to a tolerance the procedure is to determine conformance to. High GR and R percentages (more than 20 %) indicate much variability relative to the tolerance whereas low percentages (10 % or lower) indicate the opposite. The analysis also estimates, independently, the repeatability and reproducibility.

X2.4 GR and R studies in which nontechnical personnel used different brands and models of hand-held micrometers have given results varying from about 10 % (excellent) to nearly 100 % (essentially useless), relative to a dimensional tolerance of 0.003 in. The user is therefore advised to be very careful in selecting devices, setting up measurement procedures, and training personnel.

X2.5 With a 0.003 in. tolerance, a 10 % GR and R result (exceptionally good, even for digital hand-held micrometers reading to 0.00005 in.) indicates that the total variation due

E 8

to repeatability and reproducibility is around 0.0003 in. This is less than or equal to 1 % only if all dimensions to be measured are greater than or equal to 0.03 in. The relative error in using this device to measure thickness of a 0.01 in. flat tensile specimen would be 3 %—which is considerably more than that allowed for load or strain measurement.

X2.6 Dimensional measurement errors can be identified as the cause of many *out-of-control* signals, as indicated by statistical process control (SPC) charts used to monitor tension testing procedures. This has been the experience of a production laboratory employing SPC methodology and the best hand-held micrometers available (from a GR and R standpoint) in testing of 0.018 in. to 0.25 in. flat rolled steel products.

X2.7 Factors which affect GR and R, sometimes dramatically, and which should be considered in the selection and evaluation of hardware and procedures include:

X2.7.1 Resolution,
X2.7.2 Verification,
X2.7.3 Zeroing,
X2.7.4 Type of anvil (flat, rounded, or pointed),
X2.7.5 Cleanliness of part and anvil surfaces,
X2.7.6 User-friendliness of measuring device,
X2.7.7 Stability/temperature variations,
X2.7.8 Coating removal,
X2.7.9 Operator techique, and
X2.7.10 Ratchets or other features used to regulate the clamping force.

X2.8 Flat anvils are generally preferred for measuring the dimensions of round or flat specimens which have relatively smooth surfaces. One exception is that rounded or pointed anvils must be used in measuring the thickness of curved specimens taken from large-diameter tubing (see Fig. 13), to prevent overstating the thickness. (Another concern for these curved specimens is the error that can be introduced through use of the equation $A = W \times T$; see 7.1.3.)

X2.9 Heavy coatings should generally be removed from at least one grip end of flat specimens taken from coated products to permit accurate measurement of base metal thickness, assuming (*a*) the base metal properties are what are desired, (*b*) the coating does not contribute significantly to the strength of the product, and (*c*) coating removal can be easily accomplished. Otherwise, it may be advisable to leave the coating intact. Where this issue may arise, all parties involved in comparison or conformance testing should agree as to whether or not coatings are to be removed before measurement.

X2.10 As an example of how the considerations identified above affect dimensional measurement procedures, consider the case of measuring the thickness of 0.015 in. painted, flat rolled steel specimens. The paint should be removed prior to measurement, if possible. The measurement device used should have flat anvils, must read to 0.0001 in. or better, and must have excellent repeatability and reproducibility. Since GR and R is a significant concern, it will be best to use a device which has a feature for regulating the clamping force used, and devices without digital displays should be avoided to prevent reading errors. Before use of the device, and periodically during use, the anvils should be cleaned, and the device should be verified or zeroed (if an electronic display is used) or both. Finally, personnel should be trained and audited periodically to ensure that the measuring device is being used correctly and consistently by all.

SUMMARY OF CHANGES

This section identifies the principal changes to this standard that have been incorporated since the last issue.

(1) Paragraph X2.10 was re-worded.

 **Designation: E 23 – 96**

Standard Test Methods for
Notched Bar Impact Testing of Metallic Materials[1]

This standard is issued under the fixed designation E 23; the number immediately following the designation indicates the year of original adoption or, in the case of revision, the year of last revision. A number in parentheses indicates the year of last reapproval. A superscript epsilon (ϵ) indicates an editorial change since the last revision or reapproval.

This standard has been approved for use by agencies of the Department of Defense to replace method 221.1 of Federal Test Method Standard No. 151b. Consult the DoD Index of Specifications and Standards for the specific year of issue which has been adopted by the Department of Defense.

1. Scope

1.1 These test methods describe notched-bar impact testing of metallic materials by the Charpy (simple-beam) apparatus and the Izod (cantilever-beam) apparatus. They give: (*a*) a description of apparatus, (*b*) requirements for inspection and calibration, (*c*) safety precautions, (*d*) sampling, (*e*) dimensions and preparation of specimens, (*f*) testing procedures, (*g*) precision and bias, and (*h*) appended notes on the significance of notched-bar impact testing. These test methods will in most cases also apply to tests on unnotched specimens.

1.2 The values stated in SI units are to be regarded as the standard. Inch-pound units are provided for information only.

1.3 This standard does not address the problems associated with impact testing at temperatures below 77°K.

1.4 *This standard does not purport to address all of the safety concerns, if any, associated with its use. It is the responsibility of the user of this standard to establish appropriate safety and health practices and determine the applicability of regulatory limitations prior to use.*

2. Referenced Documents

2.1 *ASTM Standards:*

E 177 Practice for Use of the Terms Precision and Bias in ASTM Test Methods[2]

E 399 Test Method for Plane-Strain Fracture Toughness of Metallic Materials[3]

E 604 Test Method for Dynamic Tear Testing of Metallic Materials[3]

E 691 Practice for Conducting an Interlaboratory Study to Determine the Precision of a Test Method[2]

E 1271 Practice for Qualifying Charpy Verification Specimens of Heat-Treated Steel[3]

3. Summary of Test Methods

3.1 The essential features of an impact test are: (*a*) a suitable specimen (specimens of several different types are recognized), (*b*) an anvil or support on which the test specimen is placed to receive the blow of the moving mass,

(*c*) a moving mass that has been released from a sufficient height to cause the mass to break the specimen placed in its path, and (*d*) a device for measuring the energy absorbed by the broken specimen.

4. Significance and Use

4.1 These test methods of impact testing relate specifically to the behavior of metal when subjected to a single application of a load resulting in multiaxial stresses associated with a notch, coupled with high rates of loading and in some cases with high or low temperatures. For some materials and temperatures, impact tests on notched specimens have been found to predict the likelihood of brittle fracture better than tension tests or other tests used in material specifications. Further information on significance appears in Appendix X1.

5. Apparatus

5.1 *General Requirements:*

5.1.1 The testing machine shall be a pendulum type of rigid construction and of capacity more than sufficient to break the specimen in one blow.

5.1.2 The machine frame shall be equipped with a bubble level or a machined surface suitable for establishing levelness of the axis of pendulum bearings or, alternatively, the levelness of the axis of rotation of the pendulum may be measured directly. The machine shall be level to within 3:1000 and securely bolted to a concrete floor not less than 150 mm (6 in.) thick or, when this is not practical, the machine shall be bolted to a foundation having a mass not less than 40 times that of the pendulum. The bolts shall be tightened as specified by the machine manufacturer.

5.1.3 The machine shall be furnished with scales graduated either in degrees or directly in energy on which readings can be estimated in increments of 0.25 % of the energy range or less. The scales may be compensated for windage and pendulum friction. The error in the scale reading at any point shall not exceed 0.2 % of the range or 0.4 % of the reading, whichever is larger. (See 6.2.6.2 and 6.2.7.)

5.1.4 The total friction and windage losses of the machine during the swing in the striking direction shall not exceed 0.75 % of the scale range capacity, and pendulum energy loss from friction in the indicating mechanism shall not exceed 0.25 % of scale range capacity.

5.1.5 When hanging free, the pendulum shall hang so that the striking edge is within 2.5 mm (0.10 in.) of the position where it would just touch the test specimen. When the indicator has been positioned to read zero energy in a free

---

[1] These test methods are under the jurisdiction of ASTM Committee E-28 on Mechanical Testing and are the direct responsibility of Subcommittee E28.07 on Impact Testing.

Current edition approved March 10, 1996. Published May 1996. Originally published as E 23 – 33 T. Last previous edition E 23 – 94b.

[2] *Annual Book of ASTM Standards*, Vol 14.02.

[3] *Annual Book of ASTM Standards*, Vol 03.01.

E 23

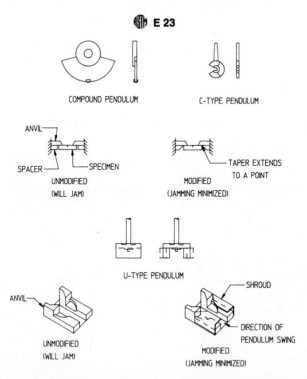

COMPOUND PENDULUM   C-TYPE PENDULUM

ANVIL

SPACER   SPECIMEN   TAPER EXTENDS
                                                   TO A POINT
UNMODIFIED   MODIFIED
(WILL JAM)   (JAMMING MINIMIZED)

U-TYPE PENDULUM

ANVIL   SHROUD

                                          DIRECTION OF
                                          PENDULUM SWING
UNMODIFIED   MODIFIED
(WILL JAM)   (JAMMING MINIMIZED)

FIG. 1   Typical Pendulums and Anvils for Charpy Machines, Shown with Modifications to Minimize Jamming

swing, it shall read within 0.2 % of scale range when the striking edge of the pendulum is held against the test specimen. The plane of swing of the pendulum shall be perpendicular to the transverse axis of the Charpy specimen anvils or Izod vise within 3:1000.

5.1.6 Transverse play of the pendulum at the striker shall not exceed 0.75 mm (0.030 in.) under a transverse force of 4 % of the effective weight of the pendulum applied at the center of strike. Radial play of the pendulum bearings shall not exceed 0.075 mm (0.003 in.). The tangential velocity (the impact velocity) of the pendulum at the center of the strike shall be not less than 3 nor more than 6 m/s (not less than 10 nor more than 20 ft/s).

5.1.7 Before release, the height of the center of strike above its free hanging position shall be within 0.4 % of the range capacity divided by the supporting force, measured as described in 6.2.3.3. If windage and friction are compensated for by increasing the height of drop, the height of drop may be increased by not more than 1 %.

5.1.8 The mechanism for releasing the pendulum from its initial position shall operate freely and permit release of the pendulum without initial impulse, retardation, or side vibration. If the same lever that is used to release the pendulum is also used to engage the brake, means shall be provided for preventing the brake from being accidentally engaged.

5.2 *Specimen Clearance*—To ensure satisfactory results when testing materials of different strengths and compositions, the test specimen shall be free to leave the machine

with a minimum of interference and shall not rebound into the pendulum before the pendulum completes its swing. Pendulums used on Charpy machines are of three basic designs, as shown in Fig. 1. When using a C-type pendulum or a compound pendulum, the broken specimen will not rebound into the pendulum and slow it down if the clearance at the end of the specimen is at least 13 mm (0.5 in.) or if the specimen is deflected out of the machine by some arrangement as is shown in Fig. 1. When using the U-type pendulum, means shall be provided to prevent the broken specimen from rebounding against the pendulum (Fig. 1). In most U-type pendulum machines, the shrouds should be designed and installed to the following requirements: (*a*) have a thickness of approximately 1.5 mm (0.06 in.), (*b*) have a minimum hardness of 45 HRC, (*c*) have a radius of less than 1.5 mm (0.06 in.) at the underside corners, and (*d*) be so positioned that the clearance between them and the pendulum overhang (both top and sides) does not exceed 1.5 mm (0.06 in.).

NOTE 1—In machines where the opening within the pendulum permits clearance between the ends of a specimen (resting on the anvil supports) and the shrouds, and this clearance is at least 13 mm (0.5 in.) requirements (*a*) and (*d*) need not apply.

5.3 *Charpy Apparatus:*

5.3.1 Means shall be provided (Fig. 2) to locate and support the test specimen against two anvil blocks in such a position that the center of the notch can be located within

E 23

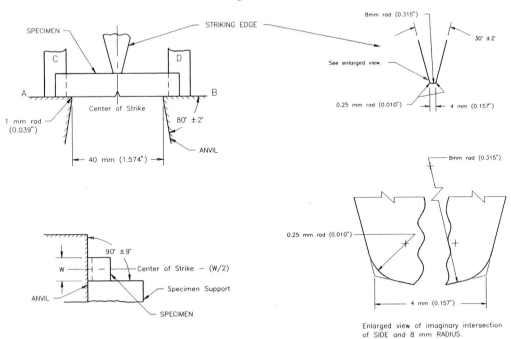

All dimensional tolerances shall be ±0.05 mm (0.002 in.) unless otherwise specified.

FIG. 2   Charpy Striking Tup

0.25 mm (0.010 in.) of the midpoint between the anvils (see 12.2.1.2).

5.3.2 The supports and striking edge shall be of the forms and dimensions shown in Fig. 2. Other dimensions of the pendulum and supports should be such as to minimize interference between the pendulum and broken specimens.

5.3.3 The center line of the striking edge shall advance in the plane that is within 0.40 mm (0.016 in.) of the midpoint between the supporting edges of the specimen anvils. The striking edge shall be perpendicular to the longitudinal axis of the specimen within 5:1000. The striking edge shall be parallel within 1:1000 to the face of a perfectly square test specimen held against the anvil.

5.4 *Izod Apparatus:*

5.4.1 Means shall be provided (Fig. 3) for clamping the specimen in such a position that the face of the specimen is parallel to the striking edge within 1:1000. The edges of the clamping surfaces shall be sharp angles of 90 ± 1° with radii less than 0.40 mm (0.016 in.). The clamping surfaces shall be smooth with a 2-μm (63-μin.) finish or better, and shall clamp the specimen firmly at the notch with the clamping force applied in the direction of impact. For rectangular specimens, the clamping surfaces shall be flat and parallel within 0.025 mm (0.001 in.). For cylindral specimens, the clamping surfaces shall be contoured to match the specimen and each surface shall contact a minimum of $\pi/2$ rad (90°) of the specimen circumference.

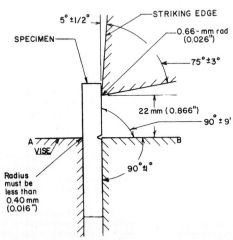

All dimensional tolerances shall be ±0.05 mm (0.002 in.) unless otherwise specified.

NOTE 1—The clamping surfaces of A and B shall be flat and parallel within 0.025 mm (0.001 in.).

NOTE 2—Finish on unmarked parts shall be 2 μm (63 μin.).

NOTE 3—Striker width must be greater than that of the specimen being tested.

FIG. 3   Izod (Cantilever-Beam) Impact Test

⚙ E 23

5.4.2 The dimensions of the striking edge and its position relative to the specimen clamps shall be as shown in Fig. 3.

5.5 *Energy Range*—Energy values above 80 % of the scale range are inaccurate and shall be reported as approximate. Ideally an impact test would be conducted at a constant impact velocity. In a pendulum-type test, the velocity decreases as the fracture progresses. For specimens that have impact energies approaching the capacity of the pendulum, the velocity of the pendulum decreases during fracture to the point that accurate impact energies are no longer obtained.

6. Inspection

6.1 *Critical Parts:*

6.1.1 *Specimen Anvils and Supports or Vise*—These shall conform to the dimensions shown in Fig. 2 or 3. To ensure a minimum of energy loss through absorption, bolts shall be tightened as specified by the machine manufacturer.

NOTE 2—The impact machine will be inaccurate to the extent that some energy is used in deformation or movement of its component parts or of the machine as a whole; this energy will be registered as used in fracturing the specimen.

6.1.2 *Pendulum Striking Edge*—The striking edge (tup) of the pendulum shall conform to the dimensions shown in Figs. 2 or 3. To ensure a minimum of energy loss through absorption, the striking edge bolts shall be tightened as

specified by the machine manufacturer. The pendulum striking edge (tup) shall comply with 5.3.3 (for Charpy tests) or 5.4.1 (for Izod tests) by bringing it into contact with a standard Charpy or Izod specimen.

6.2 *Pendulum Operation:*

6.2.1 *Pendulum Release Mechanism*—The mechanism for releasing the pendulum from its initial position shall comply with 5.1.8.

6.2.2 *Pendulum Alignment*—The pendulum shall comply with 5.1.5 and 5.1.6. If the side play in the pendulum or the radial plays in the bearings exceeds the specified limits, adjust or replace the bearings.

6.2.3 *Potential Energy*—Determine the initial potential energy using the following procedure when the center of strike of the pendulum is coincident with a radial line from the center line of the pendulum bearings (herein called the axis of rotation) to the center of gravity. (See Appendix X2.) If the center of strike is more than 1.0 mm (0.04 in.) from this line, suitable corrections in elevation of the center of strike must be made in 6.2.3.2, 6.2.3.3, 6.2.6.1, and 6.2.7, so that elevations set or measured correspond to what they would be if the center of strike were on this line.

6.2.3.1 For Charpy machines place a half-width specimen (see Fig. 4) 10 by 5 mm (0.394 by 0.197 in.) in test position. With the striking edge in contact with the specimen, a line

On subsize specimens the length, notch angle, and notch radius are constant (see Fig. 6); depth (*D*), notch depth (*N*), and width (*W*) vary as indicated below.

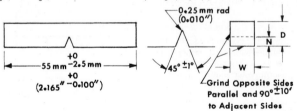

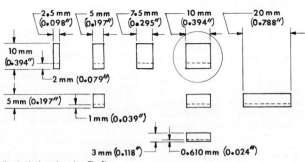

NOTE 1—Circled specimen is the standard specimen (see Fig. 6).
NOTE 2—Permissible variations shall be as follows:

Cross-section dimensions	±1 % or ±0.075 mm (0.003 in.), whichever is smaller
Radius of notch	±0.025 mm (0.001 in.)
Depth of notch	±0.025 mm (0.001 in.)
Finish requirements	2 μm (63 μin.) on notched surface and opposite face; 4 μm (125 μin.) on other two surfaces

FIG. 4   Charpy (Simple-Beam) Subsize (Type A) Impact Test Specimens

E 23

scribed from the top edge of the specimen to the striking edge
will indicate the center of strike on the striking edge.

6.2.3.2 For Izod machines, the center of strike may be
considered to be the contact line when the pendulum is
brought into contact with a specimen in the normal testing
position.

6.2.3.3 Support the pendulum horizontally to within
15:1000 with two supports, one at the bearings (or center of
rotation) and the other at the center of strike on the striking
edge (see Fig. 5). Arrange the support at the striking edge to
react upon some suitable weighing device such as a platform
scale or balance, and determine the weight to within 0.4 %.
Take care to minimize friction at either point of support.
Make contact with the striking edge through a round rod
crossing the edge at a 90° angle. The supporting force is the
scale reading minus the weights of the supporting rod and
any shims that may be used to maintain the pendulum in a
horizontal position.

6.2.3.4 Determine the height of pendulum drop for compliance with the requirement of 5.1.7. On Charpy machines
determine the height from the top edge of a half-width (or
center of a full-width) specimen to the elevated position of
the center of strike to 0.1 %. On Izod machines determine
the height from a distance 22.66 mm (0.892 in.) above the
vise to the release position of the center of strike to 0.1 %.
The height may be determined by direct measurement of the
elevation of the center of strike or by calculation from the
change in angle of the pendulum using the following
formulas: (See Fig. 5)

$$S = L (1 - \mathrm{Cos}\ \beta)\ \text{or}\ h_1 = L (1 - \mathrm{Cos}\ \alpha)"$$

6.2.3.5 The potential energy of the system is equal to the
height from which the pendulum falls, as determined in
6.2.3.4, times the supporting force, as determined in 6.2.3.3.

6.2.4 *Impact Velocity*—Determine the impact velocity, v,
of the machine, neglecting friction, by means of the following equation:

$$v = \sqrt{2\ gh}$$

where:
v = velocity, m/s (or ft/s),
g = acceleration of gravity, 9.81 m/s^2 (32.2 ft/s^2), and
h = initial elevation of the striking edge, m (or ft).

6.2.5 *Center of Percussion*—To ensure that minimum
force is transmitted to the point of rotation, the center of
percussion shall be at a point within 1 % of the distance from
the axis of rotation to the center of strike in the specimen.
Determine the location of the center of percussion as follows:

6.2.5.1 Using a stop watch or some other suitable time-
measuring device, capable of measuring time to within 0.2 s,
swing the pendulum through a total angle not greater than
15° and record the time for 100 complete cycles (to and fro).

6.2.5.2 Determine the center of percussion by means of
the following equation:

$$L = \frac{gp^2}{4\pi^2}$$

where:
L = distance from the axis to the center of percussion, m (ft),
g = local gravitational acceleration (accuracy of one part in
     one thousand), m/s^2 (ft/s^2),
π = 3.1416, and

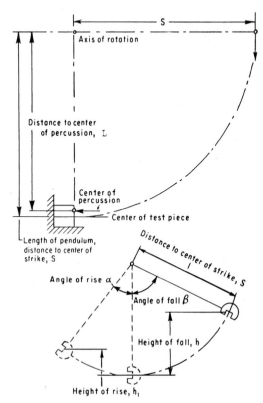

FIG. 5   Dimensions for Calculations

p = period of a complete swing (to and fro), s.

6.2.6 *Friction*—The energy loss from friction and windage
of the pendulum and friction in the recording mechanism, if
not corrected, will be included in the energy loss attributed to
breaking the specimen and can result in erroneously high
impact values. In machines recording in degrees, normal
frictional losses are usually not compensated for by the
machine manufacturer, whereas they are usually compensated for in machines recording directly in energy by
increasing the starting height of the pendulum. Determine
energy losses from friction as follows:

6.2.6.1 Without a specimen in the machine, and with the
indicator at the maximum energy reading, release the
pendulum from its starting position and record the energy
value indicated. This value should indicate zero energy if
frictional losses have been corrected by the manufacturer.
Raise the pendulum so it just contacts the pointer at the
value obtained in the free swing. Secure the pendulum at this
height and determine the vertical distance from the center of
strike to the top of a half-width specimen positioned on the
specimen rests (see 6.2.3.1). Determine the supporting force
as in 6.2.3.2 and multiply by this distance. The difference in
this value and the initial potential energy is the total energy

⚙ E 23

loss in the pendulum and indicator combined. Without resetting the pointer, repeatedly release the pendulum from its initial position until the pointer shows no further movement. The energy loss determined by the final position of the pointer is that due to the pendulum alone. The frictional loss in the indicator alone is then the difference between the combined indicator and pendulum losses and those due to the pendulum alone.

6.2.6.2 To ensure that friction and windage losses are within tolerances allowed (see 5.1.4), a simple weekly procedure may be adopted for direct-reading machines. The following steps are recommended: (*a*) release the pendulum from its upright position without a specimen in the machine, and the energy reading should be 0 J (0 ft·lbf); (*b*) without resetting the pointer, again release the pendulum and permit it to swing 11 half cycles; and after the pendulum starts its 11th cycle, move the pointer to between 5 and 10 % of scale range capacity and record the value obtained. This value, divided by 11, shall not exceed 0.4 % of scale range capacity. If this value does exceed 0.4 %, the bearings should be cleaned or replaced.

6.2.7 *Indicating Mechanism*—To ensure that the direct reading scale is recording accurately over the entire range, check it at graduation marks corresponding to approximately 0, 10, 20, 30, 50, and 70 % of each range. With the striking edge of the pendulum scribed to indicate the center of strike, lift the pendulum and set it in a position where the indicator reads, for example, 13 J (10 ft·lbf). Determine the height of center of strike to within 0.1 %. The height of center of strike multiplied by the supporting force, as determined in 6.2.3.3, is the residual energy. Increase this value by the total frictional and windage losses for a free swing (see 6.2.6.1) multiplied by the ratio of the angle of swing during a test to twice the angle of fall. Subtract the sum of the residual energy and proportional frictional and windage loss from the potential energy at the latched position. (See 6.2.3.) Make similar calculations at other points of the scale. The scale pointer shall not overshoot or drop back with the pendulum. Make test swings from various heights to check visually the operation of the pointer over several portions of the scale.

6.2.8 The impact value shall be taken as the energy absorbed in breaking the specimen and is equal to the difference between the energy in the striking member at the instant of impact with the specimen and the energy remaining after breaking the specimen.

7. Precaution in Operation of Machine

7.1 *Safety Precautions*—Precautions should be taken to protect personnel from the swinging pendulum, flying broken specimens, and hazards associated with specimen warming and cooling media.

8. Sampling

8.1 Specimens shall be taken from the material as specified by the applicable specification.

9. Test Specimens

9.1 *Material Dependence*—The choice of specimen depends to some extent upon the characteristics of the material to be tested. A given specimen may not be equally satisfactory for soft nonferrous metals and hardened steels; there-

fore, a number of types of specimens are recognized. In general, sharper and deeper notches are required to distinguish differences in the more ductile materials or with lower testing velocities.

9.1.1 The specimens shown in Figs. 6 and 7 are those most widely used and most generally satisfactory. They are particularly suitable for ferrous metals, excepting cast iron.[4]

9.1.2 The specimen commonly found suitable for die cast alloys is shown in Fig. 8.

9.1.3 The specimens commonly found suitable for powdered metals (P/M) are shown in Figs. 9 and 10. The specimen surface may be in the as-produced condition or smoothly machined, but polishing has proven generally unnecessary. Unnotched specimens are used with P/M materials. In P/M materials, the impact test results will be affected by specimen orientation. Therefore, unless otherwise specified, the position of the specimen in the machine shall be such that the pendulum will strike a surface that is parallel to the compacting direction.

9.2 *Sub-Size Specimen*—When the amount of material available does not permit making the standard impact test specimens shown in Figs. 6 and 7, smaller specimens may be used, but the results obtained on different sizes of specimens cannot be compared directly (X1.3). When Charpy specimens other than the standard are necessary or specified, it is recommended that they be selected from Fig. 4.

9.3 *Supplementary Specimens*—For economy in preparation of test specimens, special specimens of round or rectangular cross section are sometimes used for cantilever beam test. These are shown as Specimens X, Y, and Z in Figs. 11 and 12. Specimen Z is sometimes called the Philpot specimen after the name of the original designer. In the case of hard materials, the machining of the flat surface struck by the pendulum is sometimes omitted. Types Y and Z require a different vise from that shown in Fig. 3, each half of the vise having a semi-cylindrical recess that closely fits the clamped portion of the specimen. As previously stated, the results cannot be reliably compared to those obtained using specimens of other sizes or shapes.

9.4 *Specimen Machining:*

9.4.1 When heat-treated materials are being evaluated, the specimen shall be finish machined, including notching, after the final heat treatment, unless it can be demonstrated that there is no difference when machined prior to heat treatment.

9.4.2 Notches shall be smoothly machined but polishing has proven generally unnecessary. However, since variations in notch dimensions will seriously affect the results of the tests, it is necessary to adhere to the tolerances given in Fig. 6 (X1.2 illustrates the effects from varying notch dimensions on Type A specimens). In keyhole specimens, the round hole shall be carefully drilled with a slow feed. The slot may be cut by any feasible method. Care must be exercised in cutting the slot to see that the surface of the drilled hole opposite the slot is not marked.

9.4.3 Identification marks shall only be placed in the following locations on specimens: either of the 10-mm

---

[4] For testing cast iron, see 1933 Report of Subcommittee XV on Impact Testing of Committee A-3 on Cast Iron, *Proceedings,* Am. Soc. Testing Mats., Vol 33, Part 1, 1933.

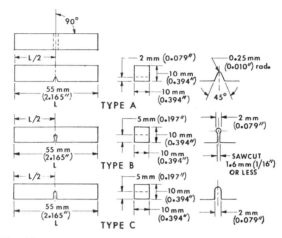

FIG. 6   Charpy (Simple-Beam) Impact Test Specimens, Types A, B, and C

NOTE—Permissible variations shall be as follows:

Notch length to edge	90 ±2°
Adjacent sides shall be at	90° ±10 min
Cross-section dimensions	±0.075 mm (±0.003 in.)
Length of specimen (*L*)	+0, −2.5 mm (+0, −0.100 in.)
Centering of notch (*L*/2)	±1 mm (±0.039 in.)
Angle of notch	±1°
Radius of notch	±0.025 mm (±0.001 in.)
Notch depth:	
Type A specimen	±0.025 mm (±0.001 in.)
Types B and C specimen	±0.075 mm (±0.003 in.)
Finish requirements	2 μm (63 μin.) on notched surface and opposite face; 4 μm (125 μin.) on other two surfaces

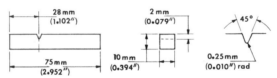

NOTE—Permissible variations shall be as follows:

Notch length to edge	90 ±2°
Cross-section dimensions	±0.025 mm (±0.001 in.)
Length of specimen	+0, −2.5 mm (±0, −0.100 in.)
Angle of notch	±1°
Radius of notch	±0.025 mm (±0.001 in.)
Notch depth	±0.025 mm (±0.001 in.)
Adjacent sides shall be at	90° ± 10 min
Finish requirements	2 μm (63 μin.) on notched surface and op posite face; 4 μm (125 μin.) on other two surfaces

FIG. 7   Izod (Cantilever-Beam) Impact Test Specimen, Type D

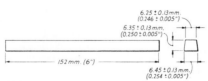

NOTE 1—Two test specimens may be cut from this bar.
NOTE 2—Blow shall be struck on narrowest face.

FIG. 8   Simple Beam Impact Test Bar for Die Castings Alloys

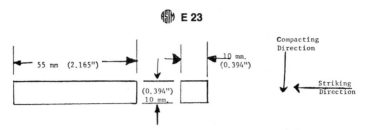

E 23

NOTE—Permissible variations shall be as follows:

Adjacent sides shall be at   90° ± 10 min
Cross section dimensions   ±0.125 mm (0.005 in.)
Length of specimen   +0, −2.5 mm (+0, −0.100 in.)

FIG. 9   Charpy (Simple Beam) Impact Test Specimens for Metal Powder Structural Parts

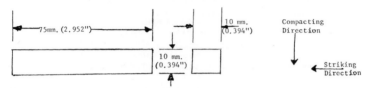

NOTE—Permissible variations shall be as follows:

Adjacent sides shall be at   90° ± 10 min.
Cross section dimensions   ±0.125 mm (0.005 in.)
Length of specimens   +0, −2.5 mm (+0, −0.100 in.)

FIG. 10   Izod (Cantilever-Beam) Impact Test Specimen for Metal Powder Structural Parts

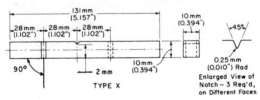

NOTE—Permissible variations for type X specimens shall be as follows:

Notch length to edge   90 ±2°
Adjacent sides shall be at   90° ±10 min
Notch depth of Type X specimen   ±0.025 mm (±0.001 in.)

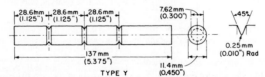

NOTE—Permissible variations for both specimens shall be as follows:

Cross-section dimensions   ±0.025 mm (±0.001 in.)
Lengthwise dimensions   +0, −2.5 mm (+0, −0.100 in.)
Angle of notch   ±1°
Radius of notch   ±0.025 mm (±0.001 in.)
Notch diameter of Type Y specimen   ±0.025 mm (±0.001 in.)

FIG. 11   Izod (Cantilever-Beam) Impact Test Specimens, Types X and Y

⚙ E 23

The flat shall be parallel to the longitudinal centerline of the specimen and shall be parallel to the bottom of the notch within 2:1000.

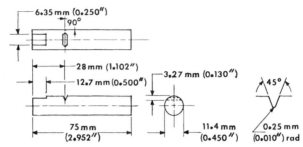

NOTE—Permissible variations shall be as follows:

Notch length to longitudinal centerline	90 ±2°
Cross-section dimensions	±0.025 mm (±0.001 in.)
Length of specimen	+0, −2.5 mm (+0 −0.100 in.)
Angle of notch	±1°
Radius of notch	±0.025 mm (±0.001 in.)
Notch depth	±0.025 mm (.130 ±0.001 in.)

FIG. 12 Izod (Cantilever-Beam) Impact Test Specimen (Philpot), Type Z

square ends; the side of the specimen which faces up when the specimen is positioned in the anvils (see Note 3); or the portion of the side opposite the notch which is at least 10 mm away from the center line of the notch. No marking shall be done on any portion of the specimen that is visibly deformed during fracture. An electrostatic pencil may be used for identification purposes, but caution must be taken to avoid excessive heat.

NOTE 3—Careful consideration must be given before placing identification marks on the side of the specimen to be placed up when positioned in the anvils. If the test operator is not careful, the specimen may be placed in the machine with the identification marking resting on the specimen supports. Under these circumstances, the energy value obtained will be unreliable.

10. Preparation of Apparatus

10.1 *Daily Checking Procedure*—After the testing machine has been ascertained to comply with Sections 5 and 6, the routine daily checking procedures shall be as follows:

10.1.1 Prior to testing a group of specimens and before a specimen is placed in position to be tested, check the machine by a free swing of the pendulum. With the indicator at the maximum energy position, a free swing of the pendulum shall indicate zero energy on machines reading directly in energy, which are compensated for frictional losses. On machines recording in degrees, the indicated values when converted to energy shall be compensated for frictional losses that are assumed to be proportional to the arc of swing.

11. Verification of Charpy Machines

11.1 Verification consists of inspecting those parts subjected to wear to ensure that the requirements of Sections 5 and 6 are met and the testing of standardized specimens (Notes 4 to 6). It is not intended that parts not subjected to wear (such as pendulum and scale linearity) need to be remeasured during verification unless a problem is evident. The average value at each energy level determined for the standardized specimens shall correspond to the nominal

values of the standardized specimens within 1.4 J (1.0 ft·lbf) or 5.0 %, whichever is greater.

11.2 Verification specimens are available directly from the National Institute of Standards and Technology (NIST) through the Standard Reference Materials program. Other sources of verification specimens may be used provided they conform to Practice E 1271 and their reference value has been established on the three reference machines owned, maintained, and operated by NIST in Boulder, CO.

NOTE 4—Standardized specimens are available for Charpy machines only.

NOTE 5—Information pertaining to the availability of standardized specimens may be obtained by addressing: National Institute of Standards and Technology, Office of Standard Reference Materials, B311 Chemistry Building, Gaithersburg, MD 20899. Telephone inquires may be made to the following number: (301) 975-6776.

NOTE 6—The National Institute for Standards and Technology conducts a Charpy machine qualification program originally developed by the Army.[5] Under this program standardized specimens are used to certify the machines of laboratories using the test as an inspection requirement on contracts. If the user desires, the results of tests with the standardized specimens will be evaluated. Participants desirous of the evaluation should complete the questionnaire provided with the standardized specimens. The questionnaire provides for information such as testing temperature, the dimensions of certain critical parts, the cooling and testing techniques, and the results of the test. The broken standardized specimens are to be returned along with the completed questionnaire for evaluation to Charpy Program Coordinator, NIST Code 853, 325 Broadway, Boulder, CO 80303. Upon completion of the evaluation, NIST will return a report. If a machine is producing values outside the standardized specimen tolerances, the report may suggest changes in machine design, repair or replacement of certain machine parts, a change in testing techniques, etc.

11.3 The verified-range of a Charpy impact machine shall be described by a lower value and a higher value. Both these values are determined from tests on sets of standardized specimens at two or more levels of absorbed energy.

---

[5] Driscoll, D. E., "Reproducibility of Charpy Impact Test," *Symposium on Impact Testing, ASTM STP 176*, ASTM, 1955, p. 170.

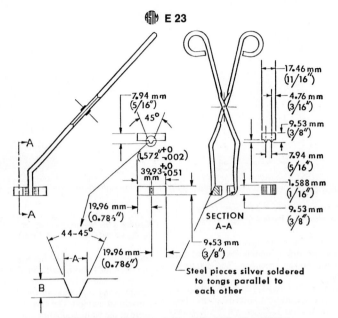

NOTE—Unless otherwise shown, permissible variation shall be ±1 mm (0.04 in.).

Specimen Depth, mm (in.)	Base Width (A), mm (in.)	Height (B), mm (in.)
10 (0.394)	1.60 to 1.70 (0.063 to 0.067)	1.52 to 1.65 (0.060 to 0.065)
5 (0.197)	0.74 to 0.80 (0.029 to 0.033)	0.69 to 0.81 (0.027 to 0.032)
3 (0.118)	0.45 to 0.51 (0.016 to 0.020)	0.36 to 0.48 (0.014 to 0.019)

FIG. 13 Centering Tongs for V-Notch Charpy Specimens

11.3.1 Values of impact energy outside the verified range shall be reported as approximate.

11.3.2 With the exception stated in 11.3.2.4, the lower limit of the verified range shall be the greater of:

11.3.2.1 One half the certified value of the lowest-value standardized specimen set tested, or,

11.3.2.2 On an analogue scale, five times the difference between the energy values represented by the graduation marks on either side of the pointer position when the lowest value from the standardized specimens is read, or,

11.3.2.3 On a digital scale, 25 times the least count of the digital display or printout when the lowest value from the standardized specimens is read.

11.3.2.4 If the lowest value of standardized specimens commercially available has been tested, then 11.3.2.1 does not apply.

11.3.3 With the exception stated in 11.3.3.2, the upper value of the verified range shall be the lesser of:

11.3.3.1 Eighty percent of the maximum value of the scale being used, or,

11.3.3.2 1.5 times the certified value of the highest level standardized specimens tested.

11.3.3.3 If the highest value of standardized specimens commercially available has been tested, then 11.2.3.2 does not apply.

11.3.4 Standardized specimens whose certified value is greater than 80 % of the maximum value on the scale being used shall not be tested.

11.3.4 If the ratio of the higher certified value to the lower certified value of the two levels of standardized specimens is greater than four, testing of a third set of intermediate energy level specimens, if commercially available, is required.

11.4 *Frequency of Verification*—Charpy machines shall be verified within one year prior to the time of testing. Charpy machines shall also be verified immediately after replacing parts that may affect the measured energy, after making repairs or adjustments, after they have been moved, or whenever there is reason to doubt the accuracy of the results, without regard to the time interval.

11.4.1 The accuracy of a Charpy machine shall be rechecked using standardization specimens (see 11.1) when parts that may affect the measured energy are removed from the machine and then reinstalled without modification (for example, when the tup or anvils are removed to permit use of a different tup or set of anvils and then are reinstalled).

12. Procedure

12.1 The Daily Checking Procedure (Section 10) shall be performed at the beginning of each day or each shift.

12.2 *Charpy Test Procedure*—The Charpy test procedure may be summarized as follows: the test specimen is removed from its cooling (or heating) medium, if used, and positioned on the specimen supports; the pendulum is released without

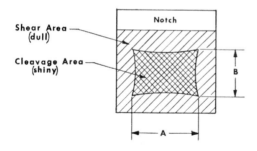

Shear Area
(dull)

Notch

Cleavage Area
(shiny)

B

A

NOTE 1—Measure average dimensions *A* and *B* to the nearest 0.5 mm or 0.02 in.

NOTE 2—Determine the percent shear fracture using Table 1 or Table 2.

FIG. 14 Determination of Percent Shear Fracture

FIG. 16 Halves of Broken Charpy V-Notch Impact Specimen Illustrating the Measurement of Lateral Expansion, Dimensions A_1, A_2, A_3, A_4 and Original Width, Dimension W

vibration, and the specimen is broken within 5 s after removal from the medium. Information is obtained from the machine and from the broken specimen. The details are described as follows:

12.2.1 *Temperature of Testing*—In most materials, impact values vary with temperature. Unless otherwise specified, tests shall be made at 15 to 32°C (60 to 90°F). Accuracy of results when testing at other temperatures requires the following procedure: For liquid cooling or heating fill a

suitable container, which has a grid raised at least 25 mm (1 in.) from the bottom, with liquid so that the specimen when immersed will be covered with at least 25 mm (1 in.) of the liquid. Bring the liquid to the desired temperature by any convenient method. The device used to measure the temperature of the bath should be placed in the center of a group of the specimens. Verify all temperature-measuring equipment at least twice annually. When using a liquid medium, hold

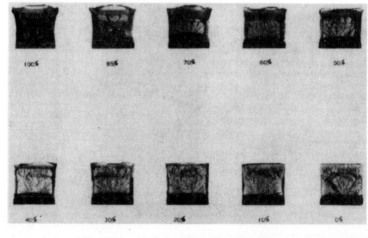

(a) Fracture Appearance Charts and Percent Shear Fracture Comparator

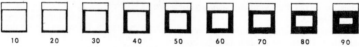

10   20   30   40   50   60   70   80   90

(b) Guide for Estimating Fracture Appearance

FIG. 15 Fracture Appearance

E 23

FIG. 17   Lateral Expansion Gage for Charpy Impact Specimens

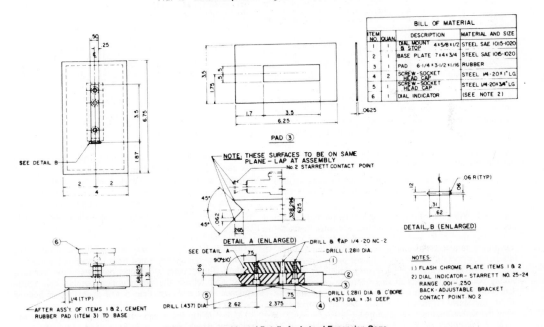

FIG. 18   Assembly and Details for Lateral Expansion Gage

⟨ASTM⟩ **E 23**

TABLE 1   Percent Shear for Measurements Made in Millimetres

NOTE—100 % shear is to be reported when either *A* or *B* is zero.

Dimension B, mm	Dimension A, mm																		
	1.0	1.5	2.0	2.5	3.0	3.5	4.0	4.5	5.0	5.5	6.0	6.5	7.0	7.5	8.0	8.5	9.0	9.5	10
1.0	99	98	98	97	96	96	95	94	94	93	92	92	91	91	90	89	89	88	88
1.5	98	97	96	95	94	93	92	92	91	90	89	88	87	86	85	84	83	82	81
2.0	98	96	95	94	92	91	90	89	88	86	85	84	82	81	80	79	77	76	75
2.5	97	95	94	92	91	89	88	86	84	83	81	80	78	77	75	73	72	70	69
3.0	96	94	92	91	89	87	85	83	81	79	77	76	74	72	70	68	66	64	62
3.5	96	93	91	89	87	85	82	80	78	76	74	72	69	67	65	63	61	58	56
4.0	95	92	90	88	85	82	80	77	75	72	70	67	65	62	60	57	55	52	50
4.5	94	92	89	86	83	80	77	75	72	69	66	63	61	58	55	52	49	46	44
5.0	94	91	88	85	81	78	75	72	69	66	62	59	56	53	50	47	44	41	37
5.5	93	90	86	83	79	76	72	69	66	62	59	55	52	48	45	42	38	35	31
6.0	92	89	85	81	77	74	70	66	62	59	55	51	47	44	40	36	33	29	25
6.5	92	88	84	80	76	72	67	63	59	55	51	47	43	39	35	31	27	23	19
7.0	91	87	82	78	74	69	65	61	56	52	47	43	39	34	30	26	21	17	12
7.5	91	86	81	77	72	67	62	58	53	48	44	39	34	30	25	20	16	11	6
8.0	90	85	80	75	70	65	60	55	50	45	40	35	30	25	20	15	10	5	0

TABLE 2   Percent Shear for Measurements Made in Inches

NOTE—100 % shear is to be reported when either *A* or *B* is zero.

Dimension B, in.	Dimension A, in.																
	0.05	0.10	0.12	0.14	0.16	0.18	0.20	0.22	0.24	0.26	0.28	0.30	0.32	0.34	0.36	0.38	0.40
0.05	98	96	95	94	94	93	92	91	90	90	89	88	87	86	85	85	84
0.10	96	92	90	89	87	85	84	82	81	79	77	76	74	73	71	69	68
0.12	95	90	88	86	85	83	81	79	77	75	73	71	69	67	65	63	61
0.14	94	89	86	84	82	80	77	75	73	71	68	66	64	62	59	57	55
0.16	94	87	85	82	79	77	74	72	69	67	64	61	59	56	53	51	48
0.18	93	85	83	80	77	74	72	68	65	62	59	56	54	51	48	45	42
0.20	92	84	81	77	74	72	68	65	61	58	55	52	48	45	42	39	36
0.22	91	82	79	75	72	68	65	61	57	54	50	47	43	40	36	33	29
0.24	90	81	77	73	69	65	61	57	54	50	46	42	38	34	30	27	23
0.26	90	79	75	71	67	62	58	54	50	46	41	37	33	29	25	20	16
0.28	89	77	73	68	64	59	55	50	46	41	37	32	28	23	18	14	10
0.30	88	76	71	66	61	56	52	47	42	37	32	27	23	18	13	9	3
0.31	88	75	70	65	60	55	50	45	40	35	30	25	20	18	10	5	0

the specimens in an agitated bath at the desired temperature within ±1°C (±2°F) for at least 5 min. When using a gas medium, position the specimens so that the gas circulates around them and hold the gas at the desired temperature within ±1°C (±2°F) for at least 30 min. Leave the mechanism used to remove the specimen from the medium in the medium except when handling the specimens.

NOTE 7—Temperatures up to +260°C (+500°F) may be obtained with certain oils, but "flash-point" temperatures must be carefully observed.

12.2.1.1 When the bath is near its boiling point, evaporative cooling can dramatically lower the specimen temperature during the interval between removal from the bath and fracture. A study has shown that a specimen heated to 100°C in water can cool 10°C in the 5 s allowed for transfer to the machine anvils.[6] When using a heating medium near its boiling point, use the data in this reference or calibration data with thermocouples to confirm that the specimen is within the stated tolerance from the desired temperature when the striker fractures the specimen.

12.2.2 *Placement of Test Specimen in Machine*—It is recommended that self-centering tongs similar to those

shown in Fig. 13 be used in placing the specimen in the machine (see 5.3.1). The tongs illustrated in Fig. 13 are for centering V-notch specimens. If keyhole specimens are used, modification of the tong design may be necessary. If an end-centering device is used, caution must be taken to ensure that low-energy high-strength specimens will not rebound off this device into the pendulum and cause erroneously high recorded values. Many such devices are permanent fixtures of machines, and if the clearance between the end of a specimen in test position and the centering device is not approximately 13 mm (0.5 in.), the broken specimens may rebound into the pendulum.

12.2.3 *Operation of the Machine:*

12.2.3.1 Set the energy indicator at the maximum scale reading; take the test specimen from its cooling (or heating) medium, if used; place it in proper position on the specimen anvils; and release the pendulum smoothly. This entire sequence shall take less than 5 s if a cooling or heating medium is used.

12.2.3.2 With the exception described as follows, any specimen, which when struck by a single blow does not separate into two pieces, shall be reported as unbroken. If the specimen can be separated by force applied by bare hands, the specimen may be considered as having been separated by the blow. Impact values from unbroken specimens with absorbed energy less than 80 % of the machine capacity may be averaged with values from broken specimens. If the

---

[6] Nanstad, R. K., Swain, R. L. and Berggren, R. G., 'Influence of Thermal Conditioning Media on Charpy Specimen Test Temperature,' *Charpy Impact Test: Factors and Variables, ASTM STP 1072*, Am. Soc. Testing Mats., 1990, p. 195.

E 23

individual values are not listed, the percent of unbroken specimens shall be reported with the average. If the absorbed energy exceeds 80 % of the machine capacity and the specimen passes completely between the anvils, the value shall be reported as approximate (see 5.5) and not averaged with others. If an unbroken specimen does not pass between the machine anvils, the result will be reported as exceeding the stated machine capacity. In no case shall the specimen be struck more than once.

12.2.3.3 If any specimen jams in the machine, disregard the results and check the machine thoroughly for damage or maladjustment, which would affect its calibration.

12.2.3.4 To prevent recording an erroneous value caused by jarring the indicator when locking the pendulum in its upright position, read the value from the indicator prior to locking the pendulum for the next test.

12.2.4 *Information Obtainable from the Test*:

12.2.4.1 *Absorbed Energy*—The amount of energy required to fracture the specimen is determined from the machine reading.

12.2.4.2 *Lateral Expansion*—The method for measuring lateral expansion must take into account the fact that the fracture path seldom bisects the point of maximum expansion on both sides of a specimen. One half of a broken specimen may include the maximum expansion for both sides, one side only, or neither. The technique used must therefore provide an expansion value equal to the sum of the higher of the two values obtained for each side by measuring the two halves separately. The amount of expansion on each side of each half must be measured relative to the plane defined by the undeformed portion of the side of the specimen, Fig. 16. Expansion may be measured by using a gage similar to that shown in Figs. 17 and 18. Measure the two broken halves individually. First, though, check the sides perpendicular to the notch to ensure that no burrs were formed on these sides during impact testing; if such burrs exist, they must be removed, for example, by rubbing on emery cloth, making sure that the protrusions to be measured are not rubbed during the removal of the burr. Next, place the halves together so that the compression sides are facing one another. Take one half (X in Fig.16) and press it firmly against the reference supports, with the protrusions against the gage anvil. Note the reading, then repeat this step with the other broken half (Y in Fig. 16), ensuring that the same side of the specimen is measured. The larger of the two values is the expansion of that side of the specimen. Next, repeat this procedure to measure the protrusions on the opposite side, then add the larger values obtained for each side. Measure each specimen. (See Note 8.)

NOTE 8—Examine each fracture surface to ascertain that the protrusions have not been damaged by contacting the anvil, machine mounting surface, etc. Such specimens should be discarded since this may cause erroneous readings. For example, if $A_1 > A_2$ and $A_3 = A_4$, then $LE = A_1 + (A_3$ or $A_4)$. If $A_1 > A_2$ and $A_3 > A_4$, then $LE = A_1 + A_3$.

12.2.4.3 *Fracture Appearance*—The percentage of shear fracture may be determined by any of the following methods: (*1*) measure the length and width of the cleavage portion of the fracture surface, as shown in Fig. 14, and determine the percent shear from either Table 1 or Table 2 depending on the units of measurement; (*2*) compare the appearance of the fracture of the specimen with a fracture appearance chart such as that shown in Fig. 15; (*3*) magnify the fracture surface and compare it to a precalibrated overlay chart or measure the percent shear fracture by means of a planimeter; or (*4*) photograph the fracture surface at a suitable magnification and measure the percent shear fracture by means of a planimeter.

NOTE 9—Because of the subjective nature of the evaluation of fracture appearance, it is not recommended that it be used in specifications.

12.3 *Izod Test Procedure*—The Izod test procedure may be summarized as follows: the test specimen is positioned in the specimen-holding fixture and the pendulum is released without vibration. Information is obtained from the machine and from the broken specimen. The details are described as follows:

12.3.1 *Temperature of Testing*—The specimen-holding fixture for Izod specimens is in most cases part of the base of the machine and cannot be readily cooled (or heated). For this reason, Izod testing is not recommended at other than room temperature.

12.3.2 Clamp the specimen firmly in the support vise so that the centerline of the notch is in the plane of the top of the vise within 0.125 mm (0.005 in.). Set the energy indicator at the maximum scale reading, and release the pendulum smoothly. Sections 12.2.3.2 to 12.2.3.4 inclusively, also apply when testing Izod specimens.

12.3.3 *Information Obtainable from the Test*—The impact energy, lateral expansion, and fracture appearance, may be determined as described in 12.2.4.

13. Report

13.1 For commercial acceptance testing, report the following information:

13.1.1 Specimen type (and size if not the full size specimen),

13.1.2 Test temperature of specimen, and

13.1.3 Energy absorbed.

13.2 For other than commercial acceptance testing, report the following information, when required, in addition to the information in 13.1:

13.2.1 Lateral expansion,

13.2.2 Fracture appearance, % shear (See Note 9),

13.2.3 Specimen orientation,

13.2.4 Specimen location,

13.2.5 Original specimen width, (*W* in Fig. 4), and

13.2.6 Original specimen depth, (*D* in Fig. 4).

14. Precision and Bias

14.1 *Interlaboratory Test Program*—An interlaboratory study used CVN specimens of low energy and of high energy to find sources of variation in the CVN absorbed energy. Data from 29 laboratories were included with each laboratory testing one set of five specimens of each energy level. Except for being limited to only two energy levels (by availability of reference specimens), Practice E 691 was followed for the design and analysis of the data, the details are given in ASTM Research Report No. RR:E28-1014.

14.2 *Precision*—The precision information given below (in units of J and ft·lb) is for the average CVN impact energy of five test determinations at each laboratory for each material.

E 23

Material	Low Energy		High Energy	
	J	ft·lb	J	ft·lb
Absorbed Energy	15.9	11.7	96.2	71.0
95 % Repeatability Limit	2.4	1.7	8.3	6.1
95 % Reproducibility Limits	2.7	2.0	9.2	6.8

The terms repeatability and reproducibility limit are used as defined in Practice E 177. The respective standard deviations among test results may be obtained by dividing the above limits by 2.8.

14.3 *Bias*—Bias cannot be defined for CVN absorbed energy. The physical simplicity of the pendulum design is complicated by complex energy loss mechanisms within the machine and the specimen. Therefore, there is no absolute standard to which the measured values can be compared.

15. Keywords

15.1 charpy test; fracture appearance; Izod test; impact test; notched specimens; pendulum machine

ANNEX

(Mandatory Information)

A1. PRECRACKING CHARPY V-NOTCH IMPACT SPECIMENS

A1.1 Scope

A1.1.1 This annex describes the procedure for the fatigue precracking of standard Charpy V-notch (CVN) impact specimens. The annex provides information on applications of precracked Charpy impact testing and fatigue-precracking procedures.

A1.2 Significance and Use

A1.2.1 Section 4 also applies to precracked Charpy V-notch impact specimens.

A1.2.2 It has been found that fatigue-precracked CVN specimens generally result in better correlations with other impact toughness tests such as Test Method E 604 and with fracture toughness tests such as Test Method E 399 than the standard V-notch specimens (**1–6**).[7] Also, the sharper notch yields more conservative estimations of the notched impact toughness and the transition temperature of the material (**7–8**).

A1.3 Apparatus

A1.3.1 The equipment for fatigue cracking shall be such that the stress distribution is uniform through the specimen thickness; otherwise the crack will not grow uniformly. The stress distribution shall also be symmetrical about the plane of the prospective crack; otherwise the crack will deviate unduly from that plane and the test result will be significantly affected.

A1.3.2 The recommended fixture to be used is shown in Fig. A1.1. The nominal span between support rollers shall be $4D \pm 0.2D$, where D is the depth of the specimen. The diameter of the rollers shall be between $D/2$ and D. The radius of the ram shall be between $D/8$ and D. This fixture is designed to minimize frictional effects by allowing the support rollers to rotate and move apart slightly as the specimen is loaded, thus permitting rolling contact. The rollers are initially positioned against stops that set the span length and are held in place by low-tension springs (such as rubber bands). Fixtures, rolls, and ram should be made of high hardness (greater than 40 HRC) steels.

A1.4 Test Specimens

A1.4.1 The dimensions of the precracked Charpy specimen are essentially those of type-A shown in Figs. 4 and A1.2. The notch depth plus the fatigue crack extension diameter shall be designated as N. When the amount of material available does not permit making the standard impact test specimen, smaller specimens may be made by reducing the width; but the results obtained on different sizes of specimens cannot be compared directly (see X1.3).

A1.4.2 The fatigue precracking is to be done with the material in the same heat-treated condition as that in which it will be impact tested. No intermediate treatments between fatigue precracking and testing are allowed.

A1.4.3 Because of the relatively blunt machined V-notch in the Charpy impact specimen, fatigue crack initiation can be difficult. Early crack initiation can be promoted by pressing or milling a sharper radius into the V-notch. Care must be taken to ensure that excessive deformation at the crack tip is avoided.

A1.4.4 It is advisable to mark two pencil lines on each side of the specimen normal to the anticipated paths of the surface traces of the fatigue crack. The first line should indicate the point at which approximately two-thirds of the crack extension has been accomplished. At this point, the stress intensity applied to the specimen should be reduced. The second line should indicate the point of maximum crack extension. At this point, fatigue precracking should be terminated.

A1.5 Fatigue Precracking Procedure

A1.5.1 Set up the test fixture so that the line of action of the applied load shall pass midway between the support roll center within 1 mm. Measure the span to within 1 % of the

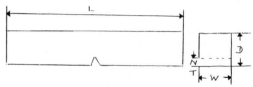

FIG. A1.1 Fatigue Precracking Fixture Design

---

[7] The boldface numbers in parentheses refer to the list of references at the end of this test method.

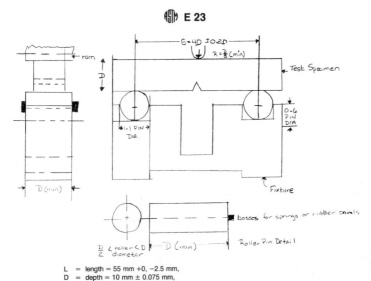

FIG. A1.2 Charpy (Simple-Beam, type A) Impact Test Specimen

L = length = 55 mm +0, −2.5 mm,
D = depth = 10 mm ± 0.075 mm,
W = variable width (see Fig. 4),
N = notch depth = 2 mm ± 0.025 mm, included angle of notch = 45° ± 1°, and
radius of notch = 0.25 mm ± 0.025 mm.

nominal length. Locate the specimen with the crack tip midway between the rolls within 1 mm of the span, and square to the roll axes within 2°.

A1.5.2 Select the initial loads used during precracking so that the remaining ligament remains undamaged by excessive plasticity. If the load cycle is maintained constant, the maximum K (stress intensity) and the K range will increase with crack length; care must be taken to ensure that the maximum K value is not exceeded in order to prevent excessive plastic deformation at the crack tip. This is done by continually shedding the load as the fatigue crack extends. The maximum load to be used at any instant can be calculated from Eqs A1.1 and A1.2 while the minimum load should be kept at 10 % of the maximum. Equation A1.1 relates the maximum load to a stress intensity (K) value for the material that will ensure an acceptable plastic-zone size at the crack tip. It is also advisable to check this maximum load to ensure that it is below the limit load for the material using Eq A1.2. When the most advanced crack trace has almost reached the first scribed line corresponding to approximately two-thirds of the final crack length, reduce the maximum load so that 0.6 K_{max} is not exceeded.

A1.5.3 Fatigue cycling is begun, usually with a sinusoidal waveform and near to the highest practical frequency. There is no known marked frequency effect on fatigue precrack formation up to at least 100 Hz in the absence of adverse environments; however, frequencies of 15 to 30 Hz are typically used. Carefully monitor the crack growth optically. A low-power magnifying glass is useful in this regard. If crack growth is not observed on one side when appreciable growth is observed on the first, stop fatigue cycling to determine the cause and remedy for the behavior. Simply turning the specimen around in relation to the fixture will often solve the

problem. When the most advanced crack trace has reached the half-way mark, turn the specimen around in relation to the fixture and complete the fatigue cycling. Continue fatigue cycling until the surface traces on both sides of the specimen indicate that the desired overall length of notch plus crack is

TABLE A1.1 Calculations of f(N/D)

N (mm)	D (mm)	N/D	f(N/D)
2.00	10.00	0.20	1.17
2.10	10.00	0.21	1.21
2.20	10.00	0.22	1.24
2.30	10.00	0.23	1.27
2.40	10.00	0.24	1.31
2.50	10.00	0.25	1.34
2.60	10.00	0.26	1.37
2.70	10.00	0.27	1.41
2.80	10.00	0.28	1.45
2.90	10.00	0.29	1.48
3.00	10.00	0.30	1.52
3.10	10.00	0.31	1.56
3.20	10.00	0.32	1.60
3.30	10.00	0.33	1.64
3.40	10.00	0.34	1.69
3.50	10.00	0.35	1.73
3.60	10.00	0.36	1.78
3.70	10.00	0.37	1.83
3.80	10.00	0.38	1.88
3.90	10.00	0.39	1.93
4.00	10.00	0.40	1.98
4.10	10.00	0.41	2.04
4.20	10.00	0.42	2.10
4.30	10.00	0.43	2.16
4.40	10.00	0.44	2.22
4.50	10.00	0.45	2.29
4.60	10.00	0.46	2.35
4.70	10.00	0.47	2.43
4.80	10.00	0.48	2.50
4.90	10.00	0.49	2.58
5.00	10.00	0.50	2.66

⚒️ E 23

reached. The fatigue crack should extend at least 1 mm beyond the tip of the V-notch but no more than 3 mm. A fatigue crack extension of approximately 2 mm is recommended.

A1.5.4 When fatigue cracking is conducted at a temperature T_1 and testing will be conducted at a different temperature T_2, and $T_1 > T_2$, the maximum stress intensity must not exceed 60 % of the K_{max} of the material at temperature T_1 multiplied by the ratio of the yield stresses of the material at the temperatures T_1 and T_2, respectively. Control of the plastic-zone size during fatigue cracking is important when the fatigue cracking is done at room temperature and the test is conducted at lower temperatures. In this case, the maximum stress intensity at room temperature must be kept to low values so that the plastic-zone size corresponding to the maximum stress intensity at low temperatures is smaller.

A1.6 Calculation

A1.6.1 Specimens shall be precracked in fatigue at load values that will not exceed a maximum stress intensity, K_{max}. For three-point bend specimens use:

$$P_{max} = [K_{max} * W * D^{3/2}]/[S * f(N/D)] \quad (A1.1)$$

where:
P_{max} = maximum load to be applied during precracking,
K_{max} = maximum stress intensity = $\sigma_{ys} * (2 * Pl * r_y)^{1/2}$, where r_y = is the radius of the induced plastic zone size which should be less than or equal to 0.5 mm,
D = specimen depth,
W = specimen width,
S = span, and
$f(N/D)$ = geometrical factor (see Table A1).

A1.6.2 See the appropriate section of Test Method E 399

for the $f(N/D)$ calculation. Table A1.1 contains calculated values for $f(N/D)$ for CVN precracking. Equation A1.2 should be used to ensure that the loads used in fatigue cracking are well below the calculated limit load for the material.

$$P_L = (4/3) * [D * (D - N)^2 * \sigma_{ys}]/S \quad (A1.2)$$

where P_L = limit load for the material.

A1.7 Crack Length Measurement

A1.7.1 After fracture, measure the initial notch plus fatigue crack length, N, to the nearest 1 % at the following three positions: at the center of the crack front and midway between the center and the intersection of the crack front with the specimen surfaces. Use the average of these three measurements as the crack length.

A1.7.2 If the difference between any two of the crack length measurements exceeds 10 % of the average, or if part of the crack front is closer to the machine notch root than 5 % of the average, the specimen should be discarded. Also, if the length of either surface trace of the crack is less than 80 % of the average crack length, the specimen should be discarded.

A1.8 Report

A1.8.1 Report the following information for each specimen tested: type of specimen used (and size if not the standard size), test temperatures, and energy absorption. Report the average precrack length in addition to these Test Method E 23 requirements.

A1.8.2 The following information may be provided as supplementary information: lateral expansion, fracture appearance, and also, it would probably be useful to report energy absorption normalized in some manner.

REFERENCES

(1) Wullaert, R. A., Ireland, D. R., and Tetelman, A. S., "Radiation Effects on the Metallurgical Fracture Parameters and Fracture Toughness of Pressure Vessel Steels," *Irradiation Effects on Structural Alloys for Nuclear Reactor Applications, ASTM STP 484,* ASTM, 1970, pp. 20–41.

(2) Sovak, J. F., "Correlation of Data from Standard and Precracked Charpy Specimens with Fracture Toughness Data for HY-130, A517-F, and HY-80 Steel," *Journal of Testing and Evaluation,* JTEVA, Vol 10, No. 3, May 1982, pp. 102–114.

(3) Succop, G. and Brown, W. F., Jr., "Estimation of K_{Ic} from Slow Bend Precracked Charpy Specimen Strength Ratios," *Developments in Fracture Mechanics Test Methods Standardization, ASTM STP 632,* W. F. Brown, Jr., and J. G. Kaufman, Eds., ASTM, 1977, pp. 179–192.

(4) Tauscher, S., "The Correlation of Fracture Toughness with Charpy V-notch Impact Test Data," Army Armament Research and

Development Command, Technical Report ARLCB-TR-81012, 1981.

(5) Wullaert, R. A., Ireland, D. R., and A. S. Tetelman, "Use of the Precracked Charpy Specimen in Fracture Toughness Testing," *Fracture Prevention and Control,* pp. 255–282.

(6) Barsom, J. M. and Rolfe, S. T., "Correlations Between K_{Ic} and Charpy V-notch Test Results in the Transition-Temperature Range," *Impact Testing of Metals, ASTM STP 466,* ASTM, 1970, pp. 281–302.

(7) Mikalac, S., Vassilaros, M. G., and H. C. Rogers, "Precracking and Strain Rate Effects on HSLA-100 Steel Charpy Specimens," *Charpy Impact Test: Factors and Variables, ASTM STP 1072,* J. M. Holt, Ed., ASTM, 1990.

(8) Sharkey, R. L. and Stone, D. H., "A Comparison of Charpy V-notch, Dynamic Tear, and Precracked Charpy Impact Transition-Temperature Curves for AAR Grades of Cast Steel."

E 23

APPENDIX

(Nonmandatory Information)

X1. NOTES ON SIGNIFICANCE OF NOTCHED-BAR IMPACT TESTING

X1.1 Notch Behavior

X1.1.1 The Charpy V-notch (CVN) impact test has been used extensively in mechanical testing of steel products, in research, and in procurement specifications for over three decades. Where correlations with fracture mechanics parameters are available, it is possible to specify CVN toughness values that would ensure elastic-plastic or plastic behavior for fracture of fatigue cracked specimens subjected to minimum operating temperatures and maximum in service rates of loading.

X1.1.2 The notch behavior of the face-centered cubic metals and alloys, a large group of nonferrous materials and the austenitic steels can be judged from their common tensile properties. If they are brittle in tension they will be brittle when notched, while if they are ductile in tension they will be ductile when notched, except for unusually sharp or deep notches (much more severe than the standard Charpy or Izod specimens). Even low temperatures do not alter this characteristic of these materials. In contrast, the behavior of the ferritic steels under notch conditions cannot be predicted from their properties as revealed by the tension test. For the study of these materials the Charpy and Izod type tests are accordingly very useful. Some metals that display normal ductility in the tension test may nevertheless break in brittle fashion when tested or when used in the notched condition. Notched conditions include restraints to deformation in directions perpendicular to the major stress, or multiaxial stresses, and stress concentrations. It is in this field that the Charpy and Izod tests prove useful for determining the susceptibility of a steel to notch-brittle behavior though they cannot be directly used to appraise the serviceability of a structure.

X1.2 Notch Effect

X1.2.1 The notch results in a combination of multiaxial stresses associated with restraints to deformation in directions perpendicular to the major stress, and a stress concentration at the base of the notch. A severely notched condition is generally not desirable, and it becomes of real concern in those cases in which it initiates a sudden and complete failure of the brittle type. Some metals can be deformed in a ductile manner even down to the low temperatures of liquid air, while others may crack. This difference in behavior can be best understood by considering the cohesive strength of a material (or the property that holds it together) and its relation to the yield point. In cases of brittle fracture, the cohesive strength is exceeded before significant plastic deformation occurs and the fracture appears crystalline. In cases of the ductile or shear type of failure, considerable deformation precedes the final fracture and the broken surface appears fibrous instead of crytalline. In intermediate cases, the fracture comes after a moderate amount of deformation and is part crystalline and part fibrous in appearance.

X1.2.2 When a notched bar is loaded, there is a normal stress across the base of the notch which tends to initiate fracture. The property that keeps it from cleaving, or holds it together, is the "cohesive strength." The bar fractures when the normal stress exceeds the cohesive strength. When this occurs without the bar deforming it is the condition for brittle fracture.

X1.2.3 In testing, though not in service because of side effects, it happens more commonly that plastic deformation precedes fracture. In addition to the normal stress, the applied load also sets up shear stresses which are about 45° to the normal stress. The elastic behavior terminates as soon as the shear stress exceeds the shear strength of the material and deformation or plastic yielding sets in. This is the condition for ductile failure.

X.1.2.4 This behavior, whether brittle or ductile, depends on whether the normal stress exceeds the cohesive strength before the shear stress exceeds the shear strength. Several important facts of notch behavior follow from this. If the notch is made sharper or more drastic, the normal stress at the root of the notch will be increased in relation to the shear stress and the bar will be more prone to brittle fracture (see Table X1.1). Also, as the speed of deformation increases, the shear strength increases and the likelihood of brittle fracture increases. On the other hand, by raising the temperature, leaving the notch and the speed of deformation the same, the shear strength is lowered and ductile behavior is promoted, leading to shear failure.

X1.2.5 Variations in notch dimensions will seriously affect the results of the tests. Tests on E 4340 steel

TABLE X1.1 Effect of Varying Notch Dimensions on Standard Specimens

	High-Energy Specimens, J (ft·lbf)	Medium-Energy Specimens, J (ft·lbf)	Low-Energy Specimens, J (ft·lbf)
Specimen with standard dimensions	103.0 ± 5.2 (76.0 ± 3.8)	60.3 ± 3.0 (44.5 ± 2.2)	16.9 ± 1.4 (12.5 ± 1.0)
Depth of notch, 2.13 mm (0.084 in.)[A]	97.9 (72.2)	56.0 (41.3)	15.5 (11.4)
Depth of notch, 2.04 mm (0.0805 in.)[A]	101.8 (75.1)	57.2 (42.2)	16.8 (12.4)
Depth of notch, 1.97 mm (0.0775 in.)[A]	104.1 (76.8)	61.4 (45.3)	17.2 (12.7)
Depth of notch, 1.88 mm (0.074 in.)[A]	107.9 (79.6)	62.4 (46.0)	17.4 (12.8)
Radius at base of notch 0.13 mm (0.005 in.)[B]	98.0 (72.3)	56.5 (41.7)	14.6 (10.8)
Radius at base of notch 0.38 mm (0.015 in.)[B]	108.5 (80.0)	64.3 (47.4)	21.4 (15.8)

[A] Standard 2.0 ± 0.025 mm (0.079 ± 0.001 in.).
[B] Standard 0.25 ± 0.025 mm (0.010 ± 0.001 in.).

specimens[8] have shown the effect of dimensional variations on Charpy results (see Table X1.1).

X1.3 Size Effect

X1.3.1 Increasing either the width or the depth of the specimen tends to increase the volume of metal subject to distortion, and by this factor tends to increase the energy absorption when breaking the specimen. However, any increase in size, particularly in width, also tends to increase the degree of restraint and by tending to induce brittle fracture, may decrease the amount of energy absorbed. Where a standard-size specimen is on the verge of brittle fracture, this is particularly true, and a doublewidth specimen may actually require less energy for rupture than one of standard width.

X1.3.2 In studies of such effects where the size of the material precludes the use of the standard specimen, as for example when the material is 6.35-mm (0.25-in.) plate, subsize specimens are necessarily used. Such specimens (Fig. 4) are based on the Type A specimen of Fig. 6.

X1.3.3 General correlation between the energy values obtained with specimens of different size or shape is not feasible, but limited correlations may be established for specification purposes on the basis of special studies of particular materials and particular specimens. On the other hand, in a study of the relative effect of process variations, evaluation by use of some arbitrarily selected specimen with some chosen notch will in most instances place the methods in their proper order.

X1.4 Temperature Effect

X1.4.1 The testing conditions also affect the notch behavior. So pronounced is the effect of temperature on the behavior of steel when notched that comparisons are frequently made by examining specimen fractures and by plotting energy value and fracture appearance versus temperature from tests of notched bars at a series of temperatures. When the test temperature has been carried low enough to start cleavage fracture, there may be an extremely sharp drop in impact value or there may be a relatively gradual falling off toward the lower temperatures. This drop in energy value starts when a specimen begins to exhibit some crystalline appearance in the fracture. The transition temperature at which this embrittling effect takes place varies considerably with the size of the part or test specimen and with the notch geometry.

X1.5 Testing Machine

X1.5.1 The testing machine itself must be sufficiently rigid or tests on high-strength low-energy materials will result in excessive elastic energy losses either upward through the pendulum shaft or downward through the base of the machine. If the anvil supports, the pendulum striking edge, or the machine foundation bolts are not securely fastened, tests on ductile materials in the range from 108 J (80 ft·lbf) may actually indicate values in excess of 122 to 136 J (90 to 100 ft·lbf)

X1.5.2 A problem peculiar to Charpy-type tests occurs when high-strength, low-energy specimens are tested at low temperatures. These specimens may not leave the machine in the direction of the pendulum swing but rather in a sidewise direction. To ensure that the broken halves of the specimens do not rebound off some component of the machine and contact the pendulum before it completes its swing, modifications may be necessary in older model machines. These modifications differ with machine design. Nevertheless the basic problem is the same in that provisions must be made to prevent rebounding of the fractured specimens into any part of the swinging pendulum. Where design permits, the broken specimens may be deflected out of the sides of the machine and yet in other designs it may be necessary to contain the broken specimens within a certain area until the pendulum passes through the anvils. Some low-energy high-strength steel specimens leave impact machines at speeds in excess of 15.2 m/s (50 ft/s) although they were struck by a pendulum traveling at speeds approximately 5.2 m/s (17 ft/s). If the force exerted on the pendulum by the broken specimens is sufficient, the pendulum will slow down and erroneously high energy values will be recorded. This problem accounts for many of the inconsistencies in Charpy results reported by various investigators within the 14 to 34-J (10 to 25-ft·lb) range. Figure 1 illustrates a modification found to be satisfactory in minimizing jamming.

X1.6 Velocity of Straining

X1.6.1 Velocity of straining is likewise a variable that affects the notch behavior of steel. The impact test shows somewhat higher energy absorption values than the static tests above the transition temperature and yet, in some instances, the reverse is true below the transition temperature.

X1.7 Correlation with Service

X1.7.1 While Charpy or Izod tests may not directly predict the ductile or brittle behavior of steel as commonly used in large masses or as components of large structures, these tests can be used as acceptance tests or tests of identity for different lots of the same steel or in choosing between different steels, when correlation with reliable service behavior has been established. It may be necessary to make the tests at properly chosen temperatures other than room temperature. In this, the service temperature or the transition temperature of full-scale specimens does not give the desired transition temperatures for Charpy or Izod tests since the size and notch geometry may be so different. Chemical analysis, tension, and hardness tests may not indicate the influence of some of the important processing factors that affect susceptibility to brittle fracture nor do they comprehend the effect of low temperatures in inducing brittle behavior.

---

[8] N. H. Fahey, "Effects of Variables in Charpy Impact Testing," *Materials Research & Standards*, Vol 1, No. 11, November 1961, p. 872.

E 23

X2. SUGGESTED METHODS OF MEASUREMENT

X2.1 *Position of the Center of Strike Relative to the Center of Gravity:*

X2.1.1 Since the center of strike can only be marked on an assembled machine, only the methods applicable to an assembled machine are described as follows:

X2.1.1.1 The fundamental fact on which all the methods are based is that when the friction forces are negligible, the center of gravity is vertically below the axis of rotation of a pendulum supported by the bearings only, (herein referred to as a free hanging pendulum). Paragraph 5.1.4 limits the friction forces in impact machines to a negligible value. The required measurements may be made using specialized instruments such as transits, clinometers, or cathometers. However, simple instruments have been used as described in the following to make measurements of sufficient accuracy.

X2.1.1.2 Suspend a plumb bob from the frame. The plumb line should appear visually to be in the plane of swing of the striking edge.

X2.1.1.3 Place a massive object on the base close to the latch side of the tup. Adjust the position of this object so that when back lighted, a minimal gap is visable between it and the tup when the pendulum is free hanging.

X2.1.1.4 With a scale or depth gage pressed lightly against the striking edge at the center of strike, measure the horizontal distance between the plumb line and striking edge. (The dimension *B* in Fig. X2.1.)

X2.1.1.5 Similarly, measure the distance in a horizontal plane through the axis of rotation from the plumb line to the clamp block or enlarged end of the pendulum stem. (Dimension *A* in Fig. X2.1.)

X2.1.1.6 Use a depth gage to measure the radial distance from the surface contacted in measuring *A* to a machined surface of the shaft which connects the pendulum to the bearings in the machine frame. (Dimension *C* in Fig. X2.1.)

X2.1.1.7 Use an outside caliper or micrometer to measure

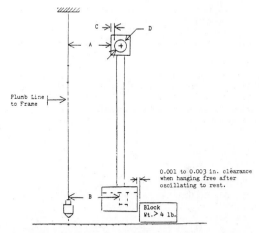

FIG. X2.1 Measurement of Deviation of Center of Strike from Vertical Plane through Axis of Rotation when Pendulum is Hanging Free

the diameter of the shaft at the same location contacted in measuring *C*. (Dimension *D* in Fig. X2.1.)

X2.1.1.8 Substitute the measured dimensions in the equation

$$X = A + C + D/2 - B$$

where:

X = deviation of the center of strike from a line from the center of rotation through the center of gravity.

INDEX

A

AASHTO Classification System, 20

Abrams' water–cement ratio law, 131

ACI method of concrete mixture proportioning, 135–41

Admixtures, 162–66, 470
  accelerating, 164, 165
  air–entraining, 162–63
  defined, 162
  mineral, 165–66
  retarding, 164
  and waste impoundments, 313–14
  water–reducing, 163–64
    and set–accelerating, 165
    and set–retarding, 165

Adsorption, 12

Advanced Fiber Reinforced Plastic (FRP) Composites, 284–85

Aeolian deposits, 2

Aggregate benefication, 62–63
  elastic fractionation, 63
  heavy–media separation, 63

Aggregate benefication, *continued*
  jigging, 63
  washing, 62–63

Aggregates:
  artificial, 55
  ASTM test specifications for, 63–75
  as a base–course material, 55–56
  benefication, 62–63
  for bituminous mixtures, 61–62
  coarse, 52
  and concrete, 128
  defined, 51
  fine, 51–52
  gradation, 53–56
  manufactured, 51, 55
  particles, 52–53
  particle strength/shape/texture, 56
  for portland cement concrete, 56–61
  processing, 52
  types of, 51–55

A horizon, 3

Air–cooled blast–furnace slag, 55

Air–entraining admixtures, 162–63

Allotropes, 222–23

Alloy steels, 266–68
  chromium, 267
  copper, 268
  manganese, 267
  molybdenum, 267
  nickel–chromium, 267
  silicon, 267
  tungsten, 268
  vanadium, 268

Alumina cements, 92–93

American Association of State Highway Transport Officials (AASHTO) Classification System, 20–21

Annealed glass, 287

Antibacterial cement, 94

Asphalt and cement admixture liners, and waste impoundments, 313

Asphalt cements, 202–19

Asphalt cements, *continued*
  asphaltenes, 203
  bitumen, 202–3
  ituminous materials
    defined, 202–3
    ductility, 206–7
    float test, 207
    penetration, 205–6
    sampling, 207
  mixtures:
    Marshall mix design,
      216–17
    mix design, 213–16
    properties, 212–13
    proportioning, 212–17
  modifications of, 208–10
    asphalt emulsions, 209–10
    liquid asphalts (cutbacks),
      208–9
    road oils, 209
  production/distillation, 203–4
  resins, 203
  road tars, 210–11
    manufacture of, 210–11
    test metyhods, 211
  specific gravity, 206
  testing methods, 205–8
  viscosity, 207–8
Asphalt emultsions, 209–10
  testing, 210
Asphaltenes, 203
Asphaltic binders, 273
ASTM test specifications for
    aggregates, 63–75
  ASTM C29, 75
  ASTM C33, 69
  ASTM C40, 71
  ASTM C88, 64–65
  ASTM C117, 70
  ASTM C123, 71
  ASTM C127, 73–74
  ASTM C128, 74–75
  ASTM C131, 64
  ASTM C136, 73
  ASTM C142, 69–70
  ASTM C215, 66
  ASTM C227, 71–72
  ASTM C289, 72
  ASTM C295, 68
  ASTM C597, 66

ASTM test specifications for
    aggregates, *continued*
  ASTM C666, 65
  ASTM C671, 66–67
  ASTM C672, 67
  ASTM C682, 67
  ASTM D75, 72–73
  ASTM D1075, 68–69
Atterberg Limits, 13–14
Austenite, 254

B

Barium and strontium cements,
  94
Bark pockets, lumber, 188
Basic oxygen furnace (BOF),
  252–53
Basic oxygen steelmaking
  process, 251–53
Bentonite amended soil liners,
  and waste impoundments,
  313
Bessemer process, 248–49
B horizon, 3
Bioremediation, 316–17
Bituminous mixtures:
  aggregates for, 61–62
  gradation/size, 61
  strength/toughness/shape/
    porosity, 62
Blast furnaces, 241, 242–45
Blast–furnace slags, 55, 80
Bond patterns, brick, 290
Borosilicates, 288
Brick, 289–90
Brick checkerwork, 242
Brinell hardness mehtod, 236
Brucite sheets, 11
Building brick, 290

C

Capillary fringe, 28
Capillary rise in soils, 27–30
Carbon steel, 254
  tensile strength of, 263
Carpenter ants, and wood, 196
Cast iron, defined, 242
Cellulose–filled compositions,
  278

Cement–based S/S, 315
Cementite, 254–55
Cements, 77–94
  alumina cements, 92–93
  expansive cements, 93
  lime, 77–80
  natural cements, 80–81
  portland blast–furnace–slag
    cements, 92
  portland cement, 81–90
  portland–pozzolan cements,
    90–92
  special portland cements,
    93–94
Checker chambers, 246
Checks, 187
Chemical damage, to wood, 199
Chemical weathering, 2
C horizon, 3
Chromium, 267
Clay mica, 11
Clay minerals, 10–13
  illite, 11
  kaolinite, 11
  montmorillonite, 11–12
Coarse aggregates, 52
Coarse–grained soils, 17–18
Cohesion limit, 13
Cohesive/cohesionless soils, 10
Colored cements, 93
Combustion chamber, 242
Compacted clay, and waste
  impoundments, 312–13
Compacted fill, 42
Compaction, 42–46
  deep, 46–47
Compressibility, 32–35
Compression wood, 188
Concrete:
  and abrasion, 116–17
  acid attack, 120
  aggregate chemical reactions,
    122–24
  alkali–carbonate reaction, 124
  alkali–silica reaction, 123
  biaxial behavior, 105–6
  bonding, 142
  bond strength, 111–14
  carbonation, 120–21
  and cavitation, 116–17

Concrete, *continued*
  compresson strength, 95–102
  creep, 106
  curing, 143
  depositing, 142
  deterioration:
    chemical causes of, 119–24
    physical causes of, 116–18
  durability, 114–24
  economy of, 129
  and erosion, 116–17
  fatigue, 110–11
  finishing, 143
  and freezing/thawing, 117–18
  high–strength, 166–70
  materials, 127–28
    aggregates, 128, 130
    cement, 127, 130–31
    influence on, 129–31
    portioning, 141
    water, 128, 130
  mixes, proportioning, 131–41
  mixing, 141
  and pozzolan/fly ash, 146–61
    background, 146–48
    lightweight concrete,
      applicability to, 157–58
    procedure, 148–57
    sample computations,
      158–61
  principal requirements for,
      128–29
  quality of, 129
  reinforced, corrosion of,
      121–22
  shearing/torsional strengths,
      114
  shrinkage, 106–10
  strength/behavior, 95–126
  sulfate attack, 119–20
  tension strength, 102–5
  transporting, 141–42
  uniaxial behavior, 95–105
  vibrated, 142
  waste/by–products used in,
      178–80
    coal ashes, 179
    rice husk ash, 179
    silica fume, 179–80
  workability of, 129

Concrete, *continued*
  See also High–strength
      concrete
Concrete block, 288–89
Concrete Masonry Units
    (CMUs), 289
Concrete mixture proportioning,
    131–41
  job–curve method, 134–35
  method of Goldbeck and
      Gray, 135–41
  mortar–voids methods, 135
  trial–batch method, 131–34
Consolidated–drained triaxial
    compression test (CD test),
    41
Consolidated–undrained triaxial
    compression test (CU test),
    41
Consolidation, 32
Consolidation settlement, 32
Consolidometers, 32–33
Contaminated soils, 314–17
  bioremediation, 316–17
  cement–based S/S, 315
  incineration/thermal
      destruction, 317
  lime–based S/S, 315
  pozzolanic S/S, 315
  soil vapor extraction (SVE),
      316
  soil washing/soil flushing
      technologies, 316
  thermoplastic techniques, 315
  vitrification, 316
Copper, 268
Cored brick, 290
Corrosion, metals, 229–30
Creep, metals, 229

D

Darcy's Law, 23–27
Darcy velocity, 25
Deep compaction, 46–47
Degree of saturation, 8
Design for the Environment
    (DfE), 295–308
  asbestos, 304
  building material compounds,
      304–5

Design for the Environment
    (DfE), *continued*
  defined, 295
  environmental quality
      standards, 307–8
  lead/lead compounds, 304
  Life Cycle Assessment (LCA),
      297–301
  material minimization, 304–7
  material selection, 301–3
  materials extraction impacts,
      303
  mercury, 304
  recycling, 305–6
  resource depletion, 301–3
  sustainable development,
      principle of, 296–97
  toxic materials, reduction of,
      304–5
Deviator stress, 40
Diffuse double layer, 12
Direct shear test, 38–39
Discharge velocity, 25
Dispersed structures, 15
Distortion, 32
D–line cracking, 59
Double–slag method, 251
Dry rot, 195–96

E

Earth reinforcement, 48–49
Effective stress concept, 20–23
Elastic deformation, 32
Elastic fractionation, 63
Electric furnaces, 249–51
Electromotive series, 225
Electron cloud, 220
Endogenous trees, 183
Environment, 309–17
  bulk chemical analysis, 311
  contaminated soils/remedial
      options, 314–17
    bioremediation, 316–17
    cement–based S/S, 315
    incineration/thermal
        destruction, 317
    lime–based S/S, 315
    pozzolanic S/S, 315
    soil vapor extraction (SVE),
        316

Environment, *continued*
    soil washing/soil flushing
      technologies, 316
    thermoplastic techniques,
      315
    vitrification, 316
    hazardous materials/hazardous
      waste, 310–12
    leaching test, 311–12
    regulatory issues, 309–12
    waste impoundments,
      materials for construction
      of, 312–14
    waste sampling, 311
Equilibrium diagram, 226–27
Eutectics, 227, 254
Eutectoids, 227, 254
Exogenous trees, 182–83
Expansive cements, 93

F

Face brick, 290
Fatigue, metals, 228
Ferrite, properties of, 254–55
Ferrous metals, 241–69
    alloy steels, 266–68
    heat treatments, 256–62
      annealing, 261
      hardening/quenching,
        256–61
      tempering, 261
    iron:
      classification of, 241–42
      structure of, 254–55
    steel:
      classification of, 241–42
      physical properties of,
        263–66
      structure of, 254–55
Fiber–reinforced cementitious
    composites, 170–78
    asbestos fibers, 177
    defined, 170–71
    glass fibers, 176–77
    natural fibers, 178
    steel fibers, 175–76
    synthetic fibers, 177–78
Fine aggregates, 52
Fine–grained soils, 17–18
Fineness, portland cement, 89

Fineness modulus, 57
Fire brick, 290
Flocculated structures, 15
Flocs, 15
Floor brick, 290
Flow, metals, 229
Flowing water in soil, 20, 23–27
    factors influencing, 23
    field tests, 27
    laboratory tests, 26–27
    permeability testing, 25–26
Fly ash:
    and concrete, 146–61
    defined, 147
Free moisture, 186
Friction angle, 36
Frost action in soils, 30–32
Fuller's maximum density curve,
    54

G

Geostatic pressure, 20
Gibbsite sheets, 11
Glacial deposits, 2
Glass, 287–88
Glazed brick, 290
Gradation, 4, 17
    aggregates, 53–56
Grain packing, 15–16
Grains, metal, 223
Grain size distribution curve,
    4–5
Graphite, 254
Gravity–transported soils, 3
Green brick, 289

H

Hardening, 256–61
Hardwood trees, 183
Head of hydration, portland
    cement, 89
Heartwood, 184
Heat treatments:
    ferrous metals, 256–62
      annealing, 261
      hardening/quenching,
        256–61
      tempering, 261
Heavy–media separation, 63
Highly organic soils, 20

High–strength concrete, 166–70
    applications of, 169–70
    material properties, 167–69
    materials, 166–67
Hydraulic conductivity, 24
Hydraulic head, 23
Hydraulic limes, 78–79
Hydrometer analysis, 4, 5
Hydrophobic cements, 93
Hydrostatic stress, 22

I

Ice lenses, 28
Igneous rocks, 1–3
Illite, 13
Impact damage, to wood, 199
Impact–resistant materials, 278
Impact testing, metals, 236–37
Inelastic action, metals, 227–28
Ingot iron, defined, 242
Insect damage, to wood, 196
Intergranular stress, 22
Interstitial allowing, 225
Iron:
    classification of, 241–42
    structure of, 254–55

J

Jigging, 63
Job–curve method, of concrete
    mixture proportioning,
    134–35

K

Kaolinite, 11
Knots, lumber, 187–88

L

Laminated glass, 287
Laminated plastic, 278
Leaching test, 311–12
Lime, 77–80, 287–88
    hydraulic limes, 78–79
    pozzolan cements, 79
    quicklime, 78
    slag cements, 79–80
Lime–based S/S, 315

Lime boil, 247
Liquid limit (LL), 13
Loose knot, lumber, 187
Loss of ignition, portland
    cement, 89

M

Malleable cat iron, defined, 242
Manganese, 267
Manufactured aggregates, 51, 55
Map cracking, 59
Marine borer, and wood, 195–99
Mechanical analysis, 4–5
Mechanical weathering, 2
Meniscus, 28
Merchant pig iron, 244–45
Metalloid elements, 225
Metals:
    allotropic behavior of, 222–23
    alloying, 224–26
    corrosion/wear, 229–30
    creep, 229
    deformation, 233
    ductility, 234
    elastic limit/inelastic limit,
        235
    fatigue, 228
    ferrous, 241–69
    flow, 229
    grain formation/growth,
        223–24
    impact, 228–29
    mechanical behavior of,
        227–30
    metallic bond, 220–21
    metallic state, 220–38
    modulus of elasticity, 234
    modulus of resilience, 236
    modulus of rigidity, 235
    modulus of toughness, 235
    offset yield strength, 234–35
    phase diagram, 226–27
    production of, 240–41
    proportional limit, 235
    slip, 227–28
    sources of, 239–40
    stress rupture, 229
    tensile strength, 233
    tensile test, 231–33
    testing/evaluating, 231–37

Metals, *continued*
    ASTM E8, 231–36
    ASTM E10, 236
    ASTM E18, 236
    ASTM E23, 236–37
    unit cells, 221–22
    yield point, 234
    See also Ferrous metals
Metamorphic rocks, 2
Metamorphism, 2
Method of Goldbeck and Gray
    (concrete mixture
    proportioning), 135–41
Mineral admixtures, 165–66
Mineral aggregates, 51–76
    aggregate types, 51–55
    See also Aggregates
Mineral–filled compositions, 278
Minerals, 1–3
    clay, 10–13
Modified Proctor Test, 43
Mohr–Coulomb failure
    envelope, 36–40
Moisture content, 8
Molding sheets, 278
Mollusk borers, and wood,
    195–99
Molybdenum, 267
Montmorillonite, 11–12
Mortar, 291

N

Natural cements, 80–81
Natural pozzolan, 147
Neutral stress, 22
Nickel–chromium, 267
Noncellulose plastic, 273
Nonferrous metals, 268
Nonhydraulic cements, 77
Normally consolidated soils, 34
Nuclear Density Method, 46

O

Oedometers, 32–33
Oil boil, 247
Oil–well cements, 93
Open–hearth furnaces, 245–48
Optimum moisture content, 42
Organic plastics:
    manufacture of, 278–79

Organic plastics, *continued*
    properties of, 279–84
Overburden pressure, 2, 20
Overconsolidated soils, 34

P

Packing, 15–16
Particle size, 3
Particle size distribution, 3–4
    grain size distribution curve,
        4–5
Paving brick, 290
Pearlite, 254
    properties of, 254–55
Permeability, 23, 25–26
Permeameters, 26
Phase diagrams, iron–carbon,
    254–55
Pig iron, 241
    merchant, 244–45
Pigs, 241, 244
Pitch pockets, lumber, 188
Pitting, 59
Plasticity index (PI), 13–14
Plastic limit (PL), 13
Plastics, 270–86
    Advanced Fiber Reinforced
        Plastic (FRP)
        Composites, 284–85
    cellulose plastic, 273
    chemically setting plastics,
        278
    classification of, 270–78
    condensation polymerides,
        273
    forming/fabricating methods,
        278–79
    laminated, 278
    natural thermoplastics, 273
    organic, manufacture of,
        278–79
    organic, properties of, 279–84
    phenol–formaldehyde resins,
        278
    polymerization, 270–72
    synthetic thermoplastics, 273
    thermoplastics, 273
    thermosetting plastics, 273–78
Pneumatically Applied Concrete,
    292

Pneumatic rubber tire rollers, 44
Polymorphism, 222
Popouts, 59
Pore water pressure, 22
Porosity, 7
Portland blast–furnace–slag
        cements, 92
Portland cement, 48, 53, 81–90
    composition of, 85–86
    defined, 81
    history of, 82
    manufacture of, 83–84
    properties of main
        compounds, 86–87
    properties of, 88–90
    raw materials, 82–83
    types of, 87–898
Portland cement concrete:
        aggregates for, 56–61
        freezing/thawing, 58–61
Portland–pozzolan cements,
        90–92
Potash, 287–88
Pozzolan:
    and concrete, 146–61
    defined, 146–47
    types of, 147
Pozzolan cements, 79
Pozzolanic S/S, 315
Preservative, for timber, 199.
Proctor Compaction Test, 42

Q

Quenching, 256–61
Quicklime, 78

R

Regulated cements, 93
Reinforce concrete, corrosion of,
    121–22
Reinforcing bars, kinds/grades
    of, 262
Reinforcing steel, 256
Relative density, 16
Residual soils, 2
Resin binders, 273
Resins, 203
Rift cutting, 185
Rigid plastic, 287
Rocks, 1–3

Rockwell hardness test, 238
Rollers, 44–46
Rubber Balloon Method, 46

S

Sand Cone Method, 46
Sapwood, 184
Searching process, 251
Sedimentary rocks, 2
Sewer brick, 290
Shakes, 187
Sheep's–foot rollers, 44–45
Shotcrete, 291–94
Shrinkage limit (SL), 13
Sieve analysis, 4–5
Silica, 287–88
Silica sheets, 10–11
Silicon, 267
Single–slag method, 251
Slag, 244
Slag cements, 79–80
Slaking, 78
Slash cutting, 185
Slip, metals, 227–228
Softwood trees, 183
Soil:
    classification, 1-3, 17–20
    clay and clay minerals, 10–13
    cohesive, 10
        structure of, 14–15
    cohesive/cohesionless soils,
        10
    composition/properties, 3–17
    compressibility, 32–35
    consistency, 13–14
    geologic properties, 1–3
    granular, structure of, 15–16
    improvement, 41–48
    origin, 1–3
    particle size, 3
    particle size distribution, 3–4
    soil–water interaction, 20–32
    strength, 35–41
    weight–volume relationships,
        5–10
Soil classification, 17–20
AASHTO Classification
        System, 20–21
Unified Soils Classification
        System, 17–20

Soil compressibility, 32–35
    normally consolidated soils, 34
    overconsolidated soils, 34
Soil improvement, 41–48
    compaction, 42–46
    deep compaction, 46–47
    stabilization, 47–48
Soil profile, 3
Soils, 1–3
Soil strength, 35–41
    cohesionless soils, shearing
        strength of, 35–37
    cohesive soils, shearing
        strength of, 37–38
    laboratory testing, 38–41
        direct shear test, 38–39
        triaxial shear test, 40–41
        unconfined compression
            test, 39–40
Soil vapor extraction (SVE), 316
Soil washing/soil flushing
        technologies, 316
Soil–water interaction, 20–32
    capillary rise, 27–30
    Darcy's Law, 23–27
    effective stress concept, 20–23
    frost action, 30–32
Soundness, portland cement, 89
Special portland cements, 93–94
Specific gravity, 8
    portland cement, 89
Specific surface, 11
Spike knot, lumber, 187
Stabilization, 47–48
Stationary water in soil, 20
Steel:
    carbon, 254
    carbon content of, 263
    classification of, 241–42
    defined, 242
    impurities of, 255–56
    and manganese, 263
    manufacture of, 242–53
        basic oxygen process,
            251–53
        Bessemer process, 248–49
        blast furnaces, 242–45
        electric furnaces, 249–51
        open–hearth furnaces,
            245–48

Steel, *continued*
 and phosphorus, 263
 physical properties of, 263–66
  effect of heat treatment on,
   263
 reinforcing, 256
 and silicon, 263
 structural, 256, 266
 structure of, 254–56
 and sulfur, 263
 tensile strength of, 263
 torsional, 266
Stick limit, 13
Stoves, 242–43
Stress–deformation curve, 33
Stress rupture, metals, 229
Structural fill, 42
Structural steel, 256, 266
Substitutional alloys, 225
Surface tension, 27–28
Synthetic membranes, and waste
  impoundments, 313–14
Synthetic pozzolan, 147

T

Taphole, 246
Tempered glass, 287
Tempering, 261
Termites, 196
Timber, 182–201
 checking, 187
 compression strength, 189
 decay, mechanism of, 194–99
 defects, 187–88
 elastic properties of, 189–90
 flexural strength/stiffness, 189

Timber, *continued*
 grain, 190–91
 hardwoods, 183
 mechanical properties of,
  189–94
  tabulation of, 181–94
 moisture content, 185–87
 rift cutting, 185
 seasoning of, 185–87
 shrinkage, 187
 slash cutting, 185
 softwoods, 183
 strength, factors affecting,
  190–91
 tensile strength, 189
 trees, classes of, 182–83
 warping, 187
 wood, physical characteristics,
  185–88
 wood structure, 183–84
Time–deformation curve, 33
Time of setting, portland cement,
  89
Torsional steel, 266
Transported soils, 2
Trial–batch method, of concrete
  mixture proportioning,
  131–34
Triaxial shear test, 40–41
Tungsten, 268
Turboblowers, 242
Tuyeres, 244

U

Unconfined compression test,
 39–40

Unconsolidated–undrained
  triaxial compression test
  (UU test), 41
Undrained unconsolidated
  condition, 39
Unified Soils Classification
  System, 17–20
Unit weight, 8

V

Vanadium, 268
Vermiculite, 11
Vertical stress, 20
Vibrated concrete, 142
Vibratory rollers, 45–46
Void ratio, 7

W–Z

Washing, 62–63
Waste impoundments, materials
  for construction of, 312–14
Waste sampling, 311
Water content, 8
Waterproofed cement, 93
Water–reducing admixtures,
  163–64, 165
Waynes, lumber, 188
Wear, metals, 229–30
Weathering profile, 3
Weight–volume relationships,
  5–10
White portland cement, 93
Wood, *See* Timber
Wood louse, 198–99
Wrought iron, defined, 242
Wythes, 289